CHEVROLET CORVETTE 1968-82 RESTORATION MANUAL

by LINDSAY PORTER

with TOM FALCONER, JOHN PFANSTIEHL & DAVE POLLARD

First published in the UK 1996 by Haynes and in USA by Motorbooks Intl; Jan 1, 1997). as Chevrolet Corvette Restoration Guide
Republished by Porter Publishing Ltd, 2024

Library of Congress Catalogue Card Number: 95-79126

A catalogue record for this book is available from the British Library

ISBN 978-1-89-923865-1

While every effort is taken to ensure the accuracy of the information given in this book, no liability can be accepted by the author or publishers for any loss, damage or injury caused by errors in, or omissions from, the information given.

Front cover illustrations:
Main picture: John Fletcher's 1969 Chevrolet Corvette Stingray - the project vehicle featured in this book. (Jeremy Gale) Below, left: Corvette Stingray Coupe of 1964. (Phil Talbot) Below, right: Corvette Targa of 1987. (Phil Talbot)

IN MEMORIAM
On 21 April 1996 Zora Arkus-Duntov died at the age of 86. Duntov was a hero to Corvette fans and was the man who masterminded the developments that made the Corvette America's favourite sportscar.
As an enthusiast and a Chevrolet employee, he worked on the Corvette from its earliest days and was responsible for almost all that made the Corvette great.
He left his wife, Elfriede "Elfi" Wolff and over a million Corvette owners.

Zora Arkus-Duntov (image, Wikepedia)

More from Lindsay Porter and Porter Publishing

...and for latest titles see LP's author page @ Amazon: http://tiny.cc/LPbooks

CLASSIC CAR BODYWORK MANUAL by Lindsay Porter ***MORE THAN 75,000 COPIES SOLD!*** **1,450 illustrations. More than 300 pages** B&W photographs.

Forty years of photos and facts from the '80s to the present day from **Lindsay Porter**, the UK's premier car technical author.

- Traditional skills to latest techniques.
- Panel beating and lead loading to plasma cutting and modern spray paints.
- Mig and Gas Welding and Brazing
- Plasma and Gas Cutting
- Masking and Spraying Techniques
- Rust Treatment and Proofing
- Windscreens, Window Winders and Locks
- Mixing and Applying Body Fillers • Repairing GRP Bodywork • Polishing and Cleaning

CLASSIC CARS AUTO-ELECTRICS MANUAL by Rodney Jaques & Lindsay Porter

Highly detailed, easy to follow instructions on how to repair all of your classic car's electrical systems.

Around 400 clear illustrations.

FAULT FINDING -track down and fix those elusive faults with ease! Step-by-Step instructions, all the way through.

- STARTING PROBLEMS
- CHARGING AND IGNITION SYSTEMS
- LIGHTING AND INSTRUMENT FAULTS
- HOW TO FIT classic I.C.E.
- ***DAB retrofit.***
- ELECTRICAL ACCESSORIES

LAND ROVER SERIES I, II, IIA & III Service Guide & Owner's Manual

Around 280 clear illustrations.

• Covers each service interval over your Land Rover's lifetime.

* Plus, regular body maintenance and preservation,

Body Repairs, Buying Spares, Fault Finding,• All the settings and data you'll need.

Make your Land Rover more reliable and maintain its value with the help of Inside Information - trade secrets and tips that workshop manuals never tell you about!

Keep a full service history of every job, adding to the value of your vehicle.

Land Rover Auto-Biography: a personal log of information about your vehicle,.

Covers all DIESEL and PETROL engines, inc. V8. • All wheelbases and bodies of all non-military models. Series I 1948 to 1958, Series II 1958 to 1961, Series HA 1961 - 1971, Series III 1971 to 1985. *

Now with latest info on how to pass the modern MoT

CLASSIC MINI SERVICE GUIDE & Owner's Manual 1959-2000* inc. Mini Cooper by Lindsay Porter

Highly detailed, easy ta follow service* instructions. *(* 'Tune-up' in USA)*

Over 280 clear photographs and line drawings.

Covers each service interval over the lifetime of your car.

Plus regular body maintenance and preservation, HELPFUL CHAPTERS on Body Repairs, Buying Secondhand, Buying Spares, Fault Finding, Safety and Getting Through the MoT (updated for the modern 'Test').

All the settings and data you'll need.

SERVICE RECORD: This unique feature enables you to keep a fully documentea service history of every job carried out, adding to the value of your vehicle.

With Mini and Mini Cooper Auto-Biography, you can compile a personal log of information about your car, providing invaluable assistance when ordering spare parts.

CLASSIC CARS RESTORATION MANUAL by Lindsay Porter

*** MORE THAN 900 ILLUSTRATIONS ***

261 pages

This manual distils the wisdom of the author's many years of restoring classic cars

MORE THAN 900 black & white illustrations.

Includes:- * Interiors & trim * Bodywork repairs *Buyers' Guide - highly detailed *Classic-car mechanical components *Setting up the workshop- and every other aspect of what's involved in restoring your favourite classic car.

- The definitive manual from the UK's top-selling practical car author.

GOLF/ RABBIT MK 1, MK 2 and JETTA RESTORATION MANUAL by Lindsay Porter.

- Over 1,000 illustrations and 250 pages
- Heritage and Buying
- Bodywork
- Engine and Transmission
- Suspension and Steering
- Braking System
- Electrical, Ignition
- Fuel and Exhaust
- Interior, Glass and Soft-top
- Modifications and Improvement,

CHEVROLET CORVETTE 1968-82 (C3) RESTORATION MANUAL by Lindsay Porter

OVER 1,000 ILLUSTRATIONS

Detailed, how-to manual for Corvette C3 restorers and repairers plus full model history.

★ Panels: Remove and refit the body, repair rusted frames and fiberglass and how to re-paint.

★ Mechanical components - from engine; gear shifter; steering box overhaul to suspension.

★ Uprates; including repairing brakes and clutch.

★ Overhaul electrics, vacuum systems and fiber optics and interior trim, tops and seats.

Contents

Acknowledgements

Back in the days when a stick shift was something you used to encourage your horse to pull your chariot with greater urgency, a guy called Sisyphus was given the somewhat bizarre job of pushing a giant rock up a hill. Trouble was, every time he made an inch of uphill progress, the boulder rolled right back down again and over his big toe. Never did discover how Sisyphus made out – could still be going strong for all I know, but writing this book gave me a deeper and better understanding of just what Mr S had to cope with – but with one major difference. At the end of it all, Sisyphus ended up with the same old rock. I, on the other hand, observed the transformation of a so-so example of four-wheeled fiberglass into a perfect, snarling, rumbling, gleaming, tire-smoking sports car – a wonderful piece of automotive sculpture – a breathtaking sight and a stunning performer – a Corvette!

It has been a great privilege to have been able to create this guide to the restoration of the greatest sports car the world has ever seen. As the author of this book, I have been no more than a conduit for other people's skills and abilities. It's a good old cliché, but nonetheless true for all that, to say that without them none of this would have been possible. But with them, this book is a compendium of knowledge from some of the finest experts in the field of Corvette restoration. They've all been incredibly patient and incredibly good at what they do: the crème de la Corvette crème, in fact. Going through the order in which I made contact with them – Tom Falconer of Claremont Corvette in Snodland, England, appears first in this book's Roll of Honour. Tom is Europe's Mr Corvette: he drives them, rebuilds them; sells parts and complete cars and has written a number of definitive books on the history of them. Quite apart from the invaluable time, advice and support he gave to the project, he was instrumental in locating the car in the first place and making the restoration part of this project 'work' by knowing all the right levers to pull. Thanks, Tom!

The second point of contact I made was with classic car enthusiast John Fletcher. I have known John for a good many years as a collector of every kind of classic vehicle imaginable, from a road-going steam-driven traction engine, through ex-WWII Jeep to Mini Cooper S. Lucky for this book, John leapt at the opportunity to have yet another of his projects immortalized in print. Lucky, too, that John has as much energy as any three other people you might know! This came in handy when he set about sourcing parts from 3,000 and more miles away, plus overseeing the restoration itself. All this, and the minor matter of restoring and maintaining umpteen other great cars and running his rapidly growing industrial heating company, Ambi-Rad.

While John's energy leaves you tired just to think of it, his ace mechanic, Em Fryer, is deceptively laid back. Deceptive because at the suction end of his old briar pipe is a man with the most amazing inventiveness and persistence. John's Corvette is well-nigh perfect, thanks to Em's amazing and unique skills, and it has been a very great pleasure to have spent so many hours on this project with both John and Em. Incidentally, I finally figured why Em has a unique ability to make a 'Vette's air-operated gear work properly. It's the insight gained from all those years sucking on the ol' briar pipe.

Once the project car was selected and purchased by John, it was then a matter of sourcing the parts. The great majority of the parts used in the project car's restoration came from two sources, both of which we are happy to commend. Auto Accessories of America's catalogue is an amazing cornucopia of Corvette components, and it was there that many of John Fletcher's car's parts were sourced. Vette Brakes and Products also produce a wonderful catalogue, majoring mainly on running gear components – and what a comprehensive selection of stock and custom parts! It was great to make friends with Judi Quinn-Routzong and Angelo Gonzalez: true 'Vette enthusiasts both! Much more of the sourcing and all of the shipping was arranged by Ken Jestis of KLJ Auto Parts in Hanover, Maryland, and all parts arrived via (and are available in Europe through) Tom Falconer's Claremont Corvette. All addresses are shown in the Appendix to this book.

Equally invaluable for their assistance were Avonvale Autobodies in Evesham, England, who re-painted John Fletcher's car to show standards, and *Corvette Fever* Magazine who allowed us to use some of their wonderful technical stuff in this book.

And that, finally, brings me to two of the most important 'star turns' of these Acknowledgements. One is my old friend Dave Pollard, who found time in his busy writing schedule to contribute a huge amount of photo research and caption writing in his usually hard-working and undaunted way. Without Dave the project would surely never have been finished! And the other is a much newer friend (one of the great pleasures of a project such as this!), the well-known Corvette author John Pfanstiehl, who contributed insightful and informative information – thousands of words and hundreds of pictures – to round off the story of Corvette restoration and make this the most complete of Corvette rebuild guides.

Lindsay Porter
Bromyard, Herefordshire, England

Post Script from Tom Falconer of Claremont Corvette

ABOUT FINDING THE CORVETTE:

John always insisted, quite rightly, that he wanted a 1969 427 Convertible 4-Speed. Unfortunately this was the summer of 1989 when classic car prices had never been higher, and this particular combination was most people's idea of the ultimate third generation Corvette.

I had been looking everywhere for this car, and having been at the summer Carlisle Swap meet since the Wednesday, had already missed two examples. Waking early Saturday morning from my sleeping bag in the back of a friend's Chevrolet pick-up truck, I was wandering towards the rest-room block with soap, towel and shaver, when in the distance I saw a shabby Big Block convertible with its characteristic bulging hood driving towards the car sales area. I ran all the way and caught the owner as he arrived at his allotted spot on the Hill.

What he must have thought as an unshaven breathless limey, carrying shaving gear, asked that he open the hood, I know not, but he complied with the request. I checked the cast and stamped engine numbers, asked his price, offered a little less, shook hands in agreement and left $20 and my Gillette as deposit.

We raised the money to pay the balance and spent the rest of that hot Pennsylvania weekend refusing higher offers on the car!

PUBLISHER'S NOTE

The variation in spelling and terminology in the UK and USA means that it is not possible to produce a book entirely suited to both markets. This volume includes a mixture of the two as it is felt this will be best accepted by the book's trans-Atlantic readers. Attention is drawn to the glossary of terms at the end of the book which provides a cross-reference of English and American technical terms.

Chapter 1

Heritage

by Tom Falconer of Claremont Corvette

1953–1955

Early in 1953 it could not be said that Chevrolet were making a particularly inspired range of motor cars. Their line-up of sedans comprised the 150, 210 and the Bel Air luxury range. On the engine front they were losing out to both Chrysler, with the hemi-headed Fire Power V8, and Ford, who also had a V8 engine on offer, even though it was the pre-war side valve unit. By comparison, Chevrolet could only offer potential performance car buffs a 6-cylinder engine. Indeed, they had not at this point built a V8 engine for over 30 years. In 1953, however, a chain of events began that was to change the face of American motoring history.

At that time, it was General Motors' commercial policy to run a series of travelling showcases called Motoramas, the aim of which was to present new ideas to the public and gauge reaction. In January 1953, the Motorama held at the Waldorf hotel in New York showed startled and enthusiastic onlookers a 'dream' of a sports car – the Corvette. It was based on GM's EX-122 show car, but the company had taken more than a passing interest in the British Jaguar XK120; in fact, they bought several and dismantled them to see just what made them tick!

There's no doubt that the first Corvette looked the part of the Great American Sports Car. The drophead fiberglass-reinforced plastic body was a purposeful yet stylish amalgam of practicality and the designer's art, and at the front a huge chrome grille shouted to the world to 'let me through'!

1954

Though it was built from fiberglass, it was designed so that it could have been made out of steel were it necessary. This can be seen to best effect in the rolled mouldings of the wheel arches on those cars produced between 1953 and 1955, which are similar to most mass-produced, steel-bodied sedans.

As the Motoramas moved around the major cities of America the initial enthusiasm grew apace, and by June of that year a limited run of 300 cars was being hand-built at a pilot plant in Flint, Michigan. The gestation time of the Corvette was remarkably short; just 15 months from clay model to limited production.

The prime movers behind the Corvette at this time were Harley Earl, the General Motors stylist supremo, and the newly appointed Chief of Engineering at Chevrolet, Ed Cole (who would eventually become president of General Motors). One of the key factors in achieving such a short lead time was that the time and vast amounts of money required to tool-up for steel body production were not necessary. Because of the nature of its bodywork, production of the Corvette was always going to be limited, and an all-new plant was never a realistic possibility.

By January 1954 the first proper assembly line for the car was introduced at Chevrolet's St Louis plant, a facility originally designed for building wooden car bodies in the 1920s. Most of this factory was used for manufacturing the more common Chevvy sedans and light trucks. But, compared with the 'Vettes produced at the Flint plant, very little was changed; although the body panels at St Louis were now made in matched moulds, resulting in better quality fiberglass with smooth inside surfaces.

However, though the St Louis plant easily had the capability to produce tens of thousands of Corvettes, in 1954 only 3,640 were made. This was simply

because the 1953 models, produced at Flint, were selling poorly, with the result that the dispatch area at St Louis was awash with 1954 models. There were a number of reasons for the poor sales. The soft-top was not best known for its resistance to the ingress of water, and while detachable Plexiglas side-screens fitted in with a sports car image, the 2-speed auto-transmission and Cadillac price-tag pointed to potential owners demanding something more of a luxury nature. Even the Corvette's standard-setting folding roof which folded into its own compartment under a rear body panel failed to balance out the minus points.

When the car was launched, the marketing men thought that they had manna from heaven with, apparently, people queuing up to be owners. With the initial run of 300 cars they limited sales to six cars per state, with VIPs and other celebrities being at the top of the preference list. However, the demand was not maintained, and with Chevrolet failing to address those weather protection problems, the Corvette picked up a nail in the tire of its desirability; it acquired a bad reputation.

The Corvette had image problems under the hood, too. The ageing six-cylinder Blue Flame 'Stovebolt' engine was a solid, reliable engine, but was not seen in a good light when both major competitors were offering muscular V8 engines. Even though the six-pot unit was 'breathed on' a little in 1955 to conjure up another five bhp, the axe of obsolescence loomed large over the car. But for one of those tricks of fate, the Corvette could have faded, Edsel-like, into motoring folklore.

Ironically, it was Ford who (inadvertently) played a large part in Chevrolet's decision to persevere with the Corvette. By launching the V8-powered Thunderbird they laid down the sports car gauntlet and, with the revised Corvette, Chevrolet took it up in some style. The company introduced their 'small block' Chevvy V8 engine, which went on to become one of the most popular production engines of all time, and was fitted to more than 500,000 Corvettes.

The lightweight, overhead-valve engine was first fitted in 1955 and the chrome 'Chevrolet' script was blessed with a large 'V' to mark the fact. Although in 1955 just 700 Corvettes were built.

1956

The year of 1956 was truly one of change – not that you could detect that there was much difference when sitting in the car. The main interior change featured the plastic two-spoke steering wheel, which was replaced by a drilled aluminium version. Outside, however, the stylist's pen had been used to great effect, and with just a few sweeping strokes the Corvette's lines had been radically altered. At the rear the tail lamp fins disappeared, the lamps finding new homes in flush chrome housings. Along each side of the car the flares around the wheel openings had been removed, and a 'scoop' had been taken out of the heavy flank all the way from the front wheel-arch almost to the back of the door. The characteristic chrome strip, which had extended from end to end down the flanks of the earlier models, disappeared and the result was that the car looked altogether more of-a-piece.

By this time the doors had wind-down windows, and external door handles and locks. At each side of the windshield there was an air inlet – though for display only. At the rear of the car the license plate was to be found in the more conventional place in the rear bumper rather than the glass box in the trunk lid on earlier models.

The '56 Corvette got a power up-rate to 210 horsepower with an increase in compression ratio to 9.25:1 and featuring a revised camshaft. Carburation came courtesy of 'dual-quad' four-barrel Carters, and the three-speed manual transmission, found on a few '55 Corvettes, became a standard fitment. (Though effective in some ways, the Carter carburettors were always difficult to tune and a little unruly in the mid-range. They were deleted from the options list in 1962.) The engine featured attractive aluminium-finned rocker covers, a feature more for cosmetic purposes than anything else; despite claims that they were fitted for weight-saving, they actually weighed about 1lb more than steel versions! It was in 1956 that the car gained a manual transmission, a vital part of the 'sports car' equation to the extent that, without it, the public and motoring press alike had little trouble in restraining their enthusiasm for the car.

The motoring press soon started to take notice of the dramatic effect the three-speed manual shift transmission had on the car, with *Road and Track* magazine noting that second gear was good for over 100 mph, with the car showing a top speed of 130 mph. And, as the year progressed, the Corvette was making inroads into Chevrolet's benign family sedan image, giving it a sporting connotation the company had been searching for. Sales of around 4,000 vehicles were still down on its overall potential, though massively up from previous years.

1956 was the year when the Corvette started to turn the corner. In the 12-hour Sebring marathon John Fitch and Walt Hansgen came in ninth in a Corvette, and the year is also notable for the arrival on the Corvette scene of Zora Arkus Dunstov, an innovative Belgian-born engineer who was to be responsible for the development of the car's chassis, transmission and engine for 20 years. His first idea was to modify the camshaft design, enabling the engine to produce 240 bhp, and with just such an engine fitted he drove a Corvette to a two-way average of 150.583 mph.

1957

By 1957 one of the original motive forces behind the Corvette, Ed Cole, was in charge of Chevrolet. Like many motoring entrepreneurs, he realized the benefit of motor sport in developing the Corvette into a true thoroughbred, and of giving the whole model range the spin-off benefits of image. Naturally enough, his main target was Ford.

One of the chief factors in making the '57 Corvette a true classic among classics was the introduction of fuel injection. The GM engineering mechanical system was fitted to the small block unit, bored out to 283 cubic inches, which quite handily gave 283 horsepower. Of the 2,570 fuel-injected passenger cars Chevrolet built in 1957, 1,040 were Corvettes.

Compared to today's high-tech micro-chipped wonders, the mechanical fuel injection was a simple affair, though it was robust and a snap to repair. Since its main weak link was the drive shaft from the main distributor, and as the spares were cheap and easy to fit, owners had few qualms about the shift from carburated engines; the fuel-injected cars were more powerful and frugal, and on the track they solved the problem of carburettors cutting out because of surge during heavy cornering. In addition to this, the Borg Warner T10 four-speed manual transmission became available (as an option to the three-speed unit which remained as a base). This combination, allied to the skinny bias-ply tires (6.70-15), was enough to see the car spinning its wheels in the first three gears during harsh acceleration! Most modern owners of 1957 cars make life easier on themselves by fitting radial tires which, of course, also offer better grip in the wet.

In 1957, just one in five Corvette owners ordered their new cars to come equipped with automatic transmission – another pointer that the car was being accepted as a *real* sports car.

The down side of the Corvette at this time, and indeed until the early sixties, was its handling. Its basic tendency was towards understeer, though the combination of inadequate rubber and excess power made power oversteer simple to attain.

1958

As the 1950s styling boom continued, 1958 saw various changes to the Corvette. The car received the four-headlamp treatment (aping the sedan of that year), and the hood and trunk a set of lateral 'washboard' ribs. The fake air vents of the 1957 car were deleted, though outlet vents just behind the front wheel-arches were a new addition. At the rear of the car, plastic covers smoothed over the tail lamps. The front bumpers were heftily up-rated, offering much needed protection from parking 'dings', and the headlamps offered much more in the way of night-time illumination. Inside the car the dashboard design and placing of instrumentation was totally revised, with the apparently sporadic and random positioning of instruments giving way to a much more ergonomic layout.

1959–1960

In 1959 the ribbed hood and trunk lids disappeared along with the total revolution counter, previously one of the pointers to the car's eccentric personality. The top performing engine remained the 290 bhp, fuel injected RPO579D variant.

A new locating bar for the rear axle improved suspension (a feature upgraded yet again the following year) with twin traction bars to prevent axle wind-up during hard acceleration. Externally, there is nothing to choose between 1959 and 1960 models,

1959

1960

1960

1960

though the road-going, fuel injected engines of 1960 were quoted as producing 315 bhp.

In that year European motoring enthusiasts had a taste of the Corvette sporting potential when racing driver Briggs Cunningham entered a team of three Corvettes in the Le Mans 24-hour race. The Chevrolet factory involvement was considerable and the cars featured, among other things, huge fuel tanks, Halibrand knock-off wheels and specially developed oil coolers to deal with the strains that only long distance racing can bring. John Fitch and Bob Grossman finished eighth, clocking up an average speed of almost 98 mph; an impressive result by anyone's standard, even more so in such exotic and exalted company.

1961

In 1961 styling revisions centred around the car's rear section, giving some hints as to what the 1963 model Sting Ray would look like. All in all, the 1961 car offered the Corvette buyer a host of minor cosmetic alterations; headlamp bezels, for example, lost their chrome plate and were painted instead. The trunk could handle a little more luggage and the car was protected at either end by simple 'quarter' bumpers, rather than the heavy, bulky items seen previously. Backup lock-out was introduced to the automatic transmission shift for the first time.

At the rear, the earlier system of passing the exhaust through the centre of the bumper was abandoned (not least because of rot affecting the bumper from the acidic exhaust gases), and inside, the Corvette gained as standard equipment, windshield washers and sun visors, while passengers had a little more legroom, with the transmission tunnel being reduced in width by around 20 per cent. 1961 was the fifth and final year the Corvette used the 283 ci engine.

1962

1962 was the first year that narrow whitewall tires were fitted (between ⅞ in. and 1 in.) and the last in which exposed headlights, a solid rear axle and a powered top option were available. 1962 models featured detailed styling changes, such as the omission of the chrome strip around the side body cove and the addition of a new ribbed rocker panel covering the tuckunder of the body. The elegant new shape was complemented beneath the hood by a bored and stroked 327 ci engine producing 250 bhp in base form (an increase of 20 bhp). This was fuelled by a single Carter AFB carburettor, whereas the sporty engine option used fuel injection to produce 295 bhp, with variants producing up to 360 bhp.

Where the Corvette was ordered with

just the hard-top (i.e. not the optional folding soft-top) the area on top of the fuel tank (the soft-top stowage area) was used for extra luggage space. Alternatively, for those who needed it, the space could be occupied by a 24-gallon fiberglass fuel tank for racing purposes.

1963 and 1964

1963 was a year of radical change. It was the year in which the name 'Sting Ray' first saw the light of day on production cars, having first been used on the Corvette racer as driven by Dick Thompson. With the Corvette production total topping 90,000 cars, a completely new frame was introduced, one which was to remain in service, regardless of body style for over 20 years!

This frame was essentially a steel box construction, strengthened by five crossmembers. The transmission crossmember featured holes through which the twin exhaust pipes passed, and the exhaust was directed under the differential to avoid contact with the fully extended transverse spring. On manual transmission cars, that crossmember was welded *in situ*, whereas on vehicles with automatic transmission it was extractable in order that the Turbo-Hydramatic could be removed without the necessity to displace the engine.

At the rear of the car, the perimeter rail angled up vertically and then continued back to support the rear bumper and fuel tank. At the front of the car it went ahead of the doors and picked up attachments for upper/lower arms.

The differential can be found on its own crossmember which was itself mounted on a higher part of the frame. It has two bushes which are extremely effective in absorbing vibration. Front suspension comes courtesy of unequal length wishbones (or A-arms), similar to the Chevrolet sedan cars of the era.

Whilst the chassis is the Achilles heel on many fiberglass-bodied cars, it has to be said that the Corvette frame was particularly rust resistant, even though it was never particularly well rustproofed from the factory. The steel was of high quality and, more important, there was plenty of ventilation, and mud traps, where moisture could rust away the metal in its own time, were few and far between. It's ironic, then, that few cars – if any – make frame replacement as easy as that of the Corvette. The fiberglass body attaches to the frame at just eight points and, with the securing bolts being well out of the elements, there is little danger of them being rusted solid. Once the body has been hoisted off the car, there's the rolling chassis in all its glory!

Because it's so easy, there's really no excuse for not removing the body for replacement of fuel/brake lines etc., and there's certainly no excuse for not doing so in a full restoration.

The other reason for wanting to get at the frame is to repair accident damage which, if it is at the front, is likely to involve the lower A-arms and inner bushing points, and, if the impact is hard enough, the curve on the frame rail. If damage here is relatively mild, it may be possible to use a jig to pull the frame back into shape. However, if the steel box section has been creased sufficiently, then a new frame will be necessary. Though accident damage at the front is often serious, side impacts are usually much less so, mostly involving the easily replaced trailing arms.

This model was clothed in one of two body styles – fastback coupé or convertible and though, throughout the year, coupé production was fully half of the total Corvette production, a year later it was only a third. This ratio remained stable for the remainder of the production run of this body shape.

Engine specifications remained as 1962, with a choice of manual transmission (three or four speed) or the powerglide automatic two-speed. Though the similarity with the 1962 model is obvious (particularly at the rear in the tail styling) none of the body panels was interchangeable. In addition, the luggage lid had disappeared and luggage had to be stored in the space behind the seats. Also, inside the car, the driver and passenger got more room to themselves, even though the wheelbase was reduced from 102 in. to 98 in. – something achieved by a concerted design effort.

Unlike the earlier models, the fuel tank sat at the very rear of the car and was marked by the huge (4 in. diameter), racing-style, hinged fuel flap. Gone, too, was the wraparound windshield. Whilst this was an undoubted improvement, the car also gained crank operated venti-panes which, though popular with drivers, were also highly popular with car thieves who used them as the easiest point of access. They were deleted from the car in 1968.

The new chassis enabled the Corvette to do away with the kingpins used on previous models, to be replaced by upper and lower ball joints on its modern front suspension. GM plundered their parts bin to find the engine-driven

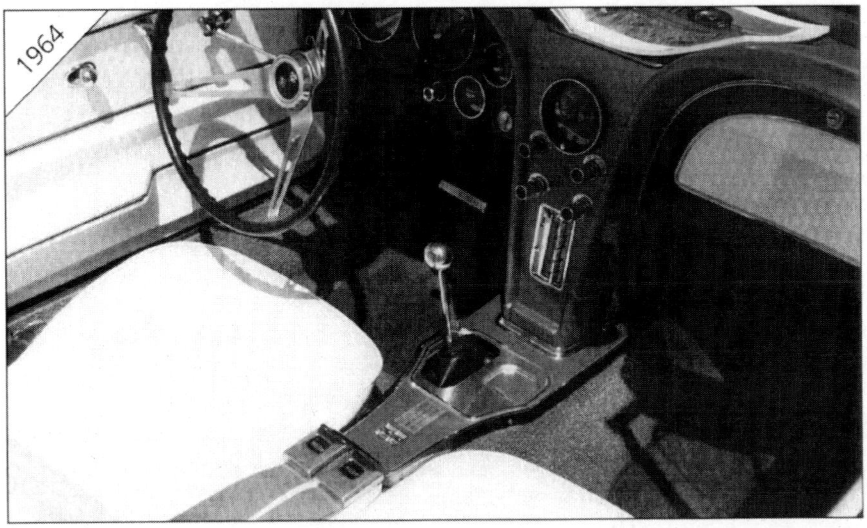

hydraulic power steering pump, the fitment of which, especially when allied to ever more powerful engines, made the cars much more usable particularly in urban environments.

At this time the car was still only running on 6.70 x 15 in. crossply tires. The power steering system was notable in that it offered the driver a remarkable amount of feedback, essential in any 'driver's' car but, incredibly, sadly lacking in many, even today. In fact, the system was so good that it stayed as a feature with no major changes for over 20 years.

The rear suspension system, being underneath the chassis, didn't use any spring mountings on the chassis itself, and the rear spring (hung transversely) helped polarize the weight low in the car. With a coupé being offered for the first time, it tended to emphasize how many squeaks and rattles are standard with the convertible versions; as ever, it is impossible for a convertible to be anywhere near as stiff as one with a full roof.

The flip-up headlamps (powered by electric motors) were unique at that time, and apart from keeping the aerodynamic lines of the car intact, being in the 'down' position during daylight hours meant that the lenses stayed cleaner, ready for night-time use. It's a curious fact that, probably because of increased resistance in the longer cable, the right-hand headlamp is always slower to lift than the left. Because of the new body design there was room to incorporate factory-fitted air conditioning, something which enabled GM to open up markets in warmer southern States. The GM options listing included leather seats for the first time. Ford-like, they offered only one color – saddle – though the following year saw a much-expanded color range, together with improvements in terms of backrest design and rake adjustment.

Early models of this year featured a storage well under the seats, though this was deleted later on. 1963 was the first year that aluminium knock-off wheels were offered and the only year to feature the idiosyncratic split rear window design, because in 1964 it was replaced by a single piece item. As many owners had a penchant for exploiting their cars' performance, it was politic to maintain a weather-eye on the rear-view mirror at all times for the tell-tale flashing blue lights; apparently the division in the rear screen neatly hid the outline of a motorcycle cop until it was too late! Inside the car, unique-to-the-year seats were fitted. Another criticism of 1963 cars was the lack of ventilation. This was subsequently solved by the fitment of a three-speed fan into the rear area of all coupé cars.

The engine power for the base model remained at 250 bhp, though there were versions of the 327 ci engine producing 300, 365 and 375 bhp. The latter was fitted with fuel injection and the 365 bhp unit featured a Holley carburettor, rather than the more usual Carter variant, for the first time.

1965

A major addition for 1965 was the introduction of four-wheel disc brakes, something which most owners felt was long overdue. Certainly, the availability of decent stopping power was essential, given the impressive line-up of engines awaiting the lucky new-car purchaser; the 327 ci V8 engines were the same as the previous year, but there was a newcomer – a 396 ci big block engine fitted with a single Holley carburettor, producing some 425 bhp.

Weighing in at around 680 lb, the engine created handling differences when compared with cars equipped with the standard lighter engine, though it came as standard with beefed-up front and rear suspension and rear sway bar. The reward for the extra weight was power and powerful performance; the 396 ci could usher the Corvette along from 0–60 mph in under six seconds – hardly hanging around, whichever way you look at it!

However, the increased power brought penalties, notably an increase in fuel consumption, in the order of 25 per cent, and, more seriously, a large measure of unreliability. To be fair, many

of the engine's problems could be laid firmly at the leaden feet of mechanically unsympathetic owners, who, among other bad habits, insisted on using high revs when the engine was cold. As such, tear-downs and major rebuilds were a common part of life for many big block owners. Nevertheless, in 1965, the big block engine was ordered by over three times more people than the fuel injected 327 ci unit, and by 1967, nearly half of all the Corvettes sold were ordered with big block engines.

The wide variety of engine and transmission variants (automatic or manual), trim and accessory options (air conditioning, power assistance etc.) and choice of convertible (with optional detachable hard top) or coupé available at this time, it was possible to 'factory customise' a Corvette to suit just about any taste; it was equally feasible to tailor a perfectly practical and relatively mild shopper or a tire-lighting, street-legal racer.

Visually, the 1965 cars reflected their designers' admiration for a certain Italian sports car manufacturer – the triple louvres behind the front wheels took their styling cues in no small amount from the Ferrari 250GTO. In addition, the steel road wheels featured a new style of triple-eared wheel trim.

The interior featured new, flat-faced instruments, and the trim was upgraded to the more modern vinyl/foam type. Unfortunately, much exposure to the sun caused the latter to crack unduly.

1966

The '66 model 'Vette was similar in style to that of the previous year, with the main differences being trim; a new vertical style Corvette script was added to the hood, the exhaust bezels were chrome plated alloy (rather than the previous stainless steel) and a new mesh style, plated grille was fitted. Under the hood, the big-block engine became 427 ci and there were versions producing 390 and 425 bhp. The 327 ci engine continued, with the base model producing 300 bhp and an option producing 350 bhp. Cars fitted with the bigger engines were instantly recognizable from the special hood that was fitted, with the engine size clearly marked. Inside the car, the seats were the same as those of the previous year (although additional pleats were fitted) and coupés got vinyl covered foam head liners rather than the fibreboard of previous years. For the first time, seat headrests became optional extras.

1967

For 1967 many of the differences were again trim, though this time much of the exterior brightwork went – notably emblems disappeared from hood and fenders. The number of functional fender vents was increased to five (rather than three) and safety regulations meant that the attractive knock-off wheels could no longer be offered as an option. Instead, cast alloy bolt-on wheels were offered, with special centre caps which masked the lug nuts. The standard wheel size increased from 5.5 in. to 6 in.

Inside the car, new seats were fitted and, for the first time, the parking brake was moved from the dash to live between them. As previously, the 327 ci engine was available in two (300 and 350 bhp) versions, and the big block 427 ci engine was available in four power outputs: 390, 400, 430 and 435 bhp.

1968

The new Corvette body style which appeared in 1968 was stunning, bowling over enthusiasts and non-believers alike! However, for those present at the 1965 London Motor Show held at the Earls Court Exhibition Centre in London, the design was a little déjà vu. That's because it was based on GM stylist, David Holls's, Mako Shark II dream car, shown there by Chevrolet. In fact, the company was repeating the promotional tactic that had paid off so well some 15 years earlier when the first Corvette was launched – by showing a massively 'over-the-top' stylist's creation at a show and then launching a moderately toned down version as a production model, the buying public is already half-way to accepting a new and radical design. In many ways it's a good job the new shape was so well received, as it was to be another 15 years before a completely new design appeared. The new model was available in coupé or convertible configuration. For the latter, a rather stylish hard top was available, secured in no fewer than eight positions to ensure a snug, leak-proof fit. With a raked rear panel and window, it created the illusion that the car was more capacious than the vertical-windowed coupé. When not in use, the folding top could be rolled right back out of sight.

The coupé incorporated a large, steel roll cage and the roof was of the 'T-bar' design, with twin, detachable panels. These could be stored in the luggage area behind the rear seats in protective vinyl carriers. A kit was introduced in 1977 which enabled them to be fastened to the exterior luggage carrier, but proved so inefficient, it was withdrawn, a workable replacement not appearing until 1980. As well as the roof panels, the vertical rear window of the coupé was also removable, offering almost convertible wind-in-the-hair motoring with most of the benefits of a coupé. Owners usually differ wildly in their opinions as to which is best, and so we'll leave you to make your choice!

The long, low look of the car was created to some extent by the particularly low roofline. In order to keep the occupants heads from denting it, the seats were raked at an unprecedented 33 degrees, from the previous 25 degrees. The frame for the '68 model was exactly the same as that for the '63–'67 series cars, and with the same wheelbase (98 in.), the extra 7 in. in length mostly being accounted for by extra front overhang. There was extra weight, too: around 150 lb over the previous model, and despite these increases there was actually less interior and luggage space.

The venti-pane front windows, which offered so much to the car thief, disappeared and were replaced by single piece units. Unlike the huge front bumpers on some earlier models, the new ones were positively skimpy and led the eye up and along, over a power bulge and on to the windshield. It was at this point that wistful onlookers, busy casting appreciative eyes along its flowing, curvaceous lines, would pause; no wipers! To create cleaner lines Chevrolet had hidden them in a covered compartment at the back of the hood.

When driving in the rain, operating the cockpit control caused the flap to lift and the two-speed wipers to emerge from their 'hibernation'. Though the mechanism was criticized, it proved to be reasonably efficient as long as it was not disturbed. This said, many owners complained of problems with the device in extremely cold weather, and there are numerous tales of the flap developing a mind of its own and raising without warning and closing whilst the wipers were operating! In the cabin, the passenger glovebox disappeared and was not to return until 1978.

Despite its many and numerous plus points, there were still complaints about build quality, with rather too many squeaks and rattles present. As a result, the St Louis factory instigated a rigorous revamp of production procedures.

Whilst the Corvette was badged the Sting Ray from 1963 to 1967, the name did not appear on the car during 1968. The engine options for 1968 were more or less those available in the previous year. All small block engines as fitted to Corvette from 1968–82 featured hydraulic valve adjustment, which prevents over-revving. (The sole exception to this was the LT-1, solid lifter engine, produced between 1970 and 1972.) The L-88 engine (introduced in 1967) featured a Holley 850 carb, alloy heads and a heavy-duty clutch, the latter being required to handle some 560 bhp!

Its prodigious thirst could only be quenched by 103 octane fuel, thanks to the compression ratio of 12.5:1.

1969

The basic body style remained in 1969, though the detail changes were legion: new door handles, passenger-side map pocket, column-mounted ignition/steering lock, headlamp washing system and a smaller diameter steering wheel (down by an inch to 15 in.) were fitted. The road wheels increased in width to 8 in. but retained the trims of the '68 car. No further changes were made until 1981.

Perhaps the most obvious external cosmetic change for 1969 was the replacement of the four rear tail lights and separate reversing lights. Inside the car, the newly introduced door panels gave increased shoulder room for the occupants but, unfortunately, production methods meant that they were quick to deteriorate. The door-opening mechanisms, too, changed slightly. In 1968 the doors were opened by using a button and finger depression plate, whereas in 1969 the depression plate itself operated the door lock mechanism. Missing from '68 models, the nomenclature 'Stingray' once more appeared on the Corvette's flanks. (Note that earlier models carried the name as two words – 'Sting Ray'.) The wide 8 in. wheels extended outside of the rear wheel arches and the result was that road dirt and grit tended to get thrown up onto the bodywork. This was changed the following year when the relevant areas of the arches were flared slightly.

A popular factory option for the '69 car was a side-mount exhaust system complete with protective grille to prevent driver and passenger burning their legs on entry and exit. Many owners of '68 model cars also fitted this exhaust option. Though the new-for-1968 model had attracted no small amount of criticism from the motoring press at the time, *Car Life* still reckoned that the Corvette driver '... will be tired of smiling long before he's tired of the car.'

The engine range, basically unchanged since '67, came in for attention during 1969. The small block engine was stroked from 3.25 to 3.5 in. to become the 350 ci. In standard trim, with 10.25:1 compression ratio, it produced 300 bhp, though an optional 11.00:1 compression ratio version was available and produced 350 bhp.

The big block (427 ci) engines were available in six versions, ranging from the L-36 390 bhp model through the cast-iron L-88 to the ZL-1. The latter was a

package comprising uprated engine and suspension which amounted to what was basically a race car for the road. The alloy engine was 100 lb lighter than it should have been but, though it churned out around 500 bhp, fears of mayhem on the streets were unfounded, as only two retail sales are recorded. All automatic transmissions were fitted with a safety feature, making it necessary to engage 'park' or backup gear before the ignition key could be removed.

1970

In 1970 a much improved front grille was introduced, replacing the plastic version of the previous two years. At the rear, new square exhaust outlets on the silencers were cosmetically effective but, unfortunately, they were also subject to corrosion problems. Inside the car, the revised seats were positioned lower, thus improving headroom, and they featured slots through which the seat belts passed.

The speakers were moved from their original position in the front foot wells to positions on top of the dash at either end – ideal for reproduction of higher frequencies and stereo imaging, though at the loss of bass response.

The 427 ci big block engine had its stroke increased by 0.25 in., which increased the cubic capacity to 454 ci (7442 cc). Their designation became 'LS'. The LS-5, with a compression ratio of 10.25:1, produced 390 bhp (and an incredible 500 lbs ft of torque!) whilst an increase in compression ratio to 11.25:1, meant that the LS-7 could turn out an amazing 460 bhp.

The LT-1 small block engine of that year, featuring mechanical valve operation, Holley carburettor and 11.0:1 compression ratio, produced 370 bhp. The use of a high-rise aluminium manifold, with the Holley carburettor, meant that the bulging hood from the 454 ci-engined vehicles had to be used. These were graced with a set of pinstripes and LT-1 logos and are now the dream 'find' of many a Corvette enthusiast.

1971

The 1971 cars were scarcely changed from 1970 cars, at least as far as appearance is concerned. In fact, because of a labour dispute, 1970 production overran well into the '71 model year and GM decided to treat all 1971 production as an extension of 1970. In addition GM decreed that the engines should be able to run on lead-free fuel, despite the fact that catalytic converters would not be fitted until the 1975 model year. As a result, both the 350 ci and 454 ci engines produced less power across the board – a first in Corvette year-to-year terms. The 350 ci was available with either 270 or 330 bhp, and the big block engine with either 365 or 425 bhp. Aluminium cylinder heads were as fitted to the LS-6 engine and the factory offered a racing option, the ZR1. It included the LT-1 engine, four-speed manual shift, power brakes and a host of suspension uprates. It was stripped of 'extras' such as power windows, air conditioning, radio and even wheel covers.

1972

In 1972, engine options were as in 1971 although the power outputs were quoted in SAE terms rather than gross as previously. Using the gross power method, the readings were taken from the engine minus certain power-sapping ancillaries, such as alternator, air cleaner and the regular exhaust system. In addition, the fuel mixture and timing were adjusted to produce a higher figure than the norm. But from 1972 the power quoted was SAE (Society of Automotive Engineers), which meant that the engine had to be tested in exactly the same specification it would be when installed in a road-going, production model and under specified conditions. Comparing an engine, apparently internally unchanged, between these years can reveal 'discrepancies' in power outputs of

between 60 and 120 bhp. In truth, the change is academic, as long as you are comparing like with like!

The Corvette's desirability among those who steal, rather than buy, their cars prompted the standard fitment of the previously optional alarm system on all models. The cost to the factory was somewhat offset by the deletion of the fiber optic light bulb monitor system. This 'first' fitment was balanced by a number of 'lasts' for 1972 – it was the last year of the 'egg-crate' grille, the last year that the coupé featured a removable rear window, and the last year you could order the LT-1 solid lifter small block engine. Moreover, it marked the end of a Corvette chrome era, for it was the end of the line for chrome bumpers front and rear, a feature since its inception.

Sales of the 'new' Corvette varied. A record was set in 1968 when 28,566 cars were sold, but this was immediately beaten in 1969 when 38,762 new owners were found. In 1970, a dramatic fall to just 17,316 units was countered in the following two years when sales rallied to 21,801 and 26,994 cars, respectively.

1973

In 1973 the Corvette began an evolutionary process to take it from a charging aggressive sports car to a grand tourer par excellence. The latter demands that the miles should be eaten up somewhat more quietly than the former, and the body was mounted in a different way with cushioned mounts, and thicker insulation was fitted along with sprayed-on sound deadening. For the first time, Corvettes were fitted with radial tires, offering better grip in the wet and increased longevity.

The hide-away windshield wipers disappeared after a run of five years. The stylists took the opportunity of the changes necessary in the hood to fit a hood bulge and an air duct system to pull cold air from the base of the windshield. Under normal driving (i.e. at low throttle openings) air is drawn into the carburettor from the engine bay. When the throttle opened sufficiently to activate the second chokes, a solenoid, operated by a switch under the gas pedal, opened a flap and admitted fresh, cold air. The induction roar this causes is a spine-tingling experience for any engine enthusiast! The car was also made quieter by the deletion of the removable rear window and the attendant sealing problems.

Externally, the 1973 car differed little from 1972 other than the new plastic-covered Federal front bumper and plain side vents. In order to meet the Federal impact regulations, energy dissipating draw bolts were used. This had the practical effect of lengthening the car's nose slightly over a design which was already a little nose-heavy.

Power, as expected, was reduced considerably, though fuel consumption improved. The high performance engine was designated the L-82, replacing the now-deleted solid lifter LT-1.

1974

In 1974 more Federal legislation affected the Corvette. Inside, the seat belts had to be integrated with the engine starter circuit in such a way that the engine could not be started unless the seat belt buckles were engaged. This was a year-only fitment, as the law was amended in 1975 and many '74 owners electronically bypassed this fitment. In addition, the Federal rear bumper was fitted. The 'softer' panels were now color-coded to the body color of the cars and were produced as a two-piece mould with a central split. The very clean and smooth look of the '74 rear end was enhanced by the exhaust outlets being positioned out of sight.

There were more 'lasts', too: this was the last time you could look down into the gasoline tank to see the fuel level and the last time that the big block engine was installed. Producing 380 lb ft at 2800 rpm it was available, in deference to the Federal regulations, only in low compression form. With the fuel crisis that affected the whole world in the early 1970s, the 55 mph speed limit was enforced most rigorously across the country, notably by the use of radar-based technology. It was at this point that Corvette owners realized a 'hidden' advantage of that fiberglass body; together with a 'laid back' radiator and concealed headlamps, the Corvette's radar 'silhouette' was far less noticeable than any other car at the time!

In 1975 the front and rear bumpers were revised once more as the design of the front bumper was amended slightly and small simulated 'pads' added. At the rear, the two-piece moulding was replaced by an altogether more pleasing single-piece item. The rear bumper now featured hydraulic rams as fitted to most other GM cars. Federal regulations also affected the engine in terms of emissions. In order to produce a 'cleaner' engine, the Corvette was fitted with a high-energy ignition system and a single 'cat' converter. In order to Siamese the twin exhaust outlets into the catalyst, a corrosion-prone 'Y' piece was fitted, though the original twin holes in the chassis remained and many owners took the route of replacing the 'cat' system with a standard exhaust system.

1975

In 1975, cars fitted with the L-82 performance option engine had that nomenclature on the hood. Until 1980 this was the only optional performance engine. It had shot-peened rods, forged crankshaft, rejetted Quadrajet and cast aluminium valve covers. On average, the L-82 offered more bhp (at higher revs) than the L-48, though torque suffered a little. All models produced since 1975 have electronic ignition.

It was thought by many that production of the Corvette would cease in 1975 because of the effect of the proposed barrier impact tests. If they had wanted to, GM could have used this as a perfect excuse to stop production. As it turned out, of course, that didn't happen, but government regulations *were* to affect the Corvette's power output and performance via the anti-pollution laws. However, the demise of the convertible, expected by many, did arrive. Though most thought that it would be killed off by the safety laws which were becoming more and more stringent, it was ironic that a simple lack of sales eventually convinced GM that it

was not viable to continue producing drop-top cars. In 1975 less than 5,000 were produced, compared with almost 34,000 coupés.

1976

In 1976 the car continued much as it had been in 1975. Inside, connoisseurs of other GM products will recognize the four-spoke steering wheel which more usually saw service in the Vega compact sedan. It was slightly smaller than the previous three-spoke wheel which had been fitted since 1969. The designers took the opportunity of fitting a conventional rear screen with in-built heating elements, and thus the blower fan was discontinued.

The complex air induction system for the engine was changed and air was drawn directly in from the front of the car via a plastic duct routed across the top of the radiator. Naturally, as a result of this, the hood became much lighter, and consequently easier to open. The use of a catalytic converter meant that much more heat was generated and a steel floor was fitted for the first time to help dissipate this. Previously a fibérglass floor had been fitted.

Many owners took up an option available for the '76 car and rapidly improved the looks of their Corvette by fitting GM's eight-spoke aluminium slotted wheels, which also had the advantage of saving some 35 lb in unsprung weight. (Although originally listed as options in 1973, they were not actually available at the time.) They were of the same width as the standard steel wheels.

1977

For 1977 the engine line-up was the same as in the previous year, with the base model 350 ci engine producing 180 bhp, and an optional 210 bhp version being available. Inside, a new console was fitted and the adoption of a standard size aperture allowed a choice from a wide range of radios/tape decks.

The Stingray name was dropped, and for the first part of the production year their logo positions on the side of the fenders were left bare, though latterly

they were filled by a pair of crossed flags, mimicking the new badge on the nose of the car. Also moved from its position on the left front fender to the driver's door lock was the burglar alarm switch. Streamlined sports mirrors were offered for the first time.

Most of the more important changes for this year were inside the car. The four-spoke Vega wheel was replaced by a three-spoke item in leather colour-matched to the interior trim. Leather seating was now standard, with cloth seats being available only as an option. By moving the steering column some 2 in. nearer the dashboard, more interior room was created and a more comfortable driving position was available. For the first time since 1969 the necessity to engage backup or park in order to remove the ignition key was deleted. The wiper control and headlamp dip switch were moved from the dash and floor respectively onto more conventional column stalks. Also on the column was the cruise control switch (automatic transmission only) and a taller seat divider console was fitted, incorporating the parking brake lever. Power brakes and power steering were standard.

In general, the three years 1975, 1976 and 1977 were ones of marking time, with the general feeling being that something big was soon to happen, though quite what no one was sure.

1978

The 1978 model year was special for two reasons: not only was it the Corvette's 25th Anniversary year but the Corvette was also given the honour of being chosen as the official Indy pace car for that year. Chevrolet responded by introducing the most extensive redesign since 1968. The new 'fast back' rear end styling and large rear window improved appearance and increased interior space, some of which was promptly taken up again by the fitment of two standard-fit rear stereo speakers. Owners were now able to store luggage inside the car rather than on an external luggage rack. To conceal the luggage from nefarious eyes, a roller blind was fitted. Appearance and performance gains also came from the wider wheels and sexy, low-profile tires which meant that the fenders were adapted to improve clearance.

The Anniversary was marked by Chevrolet by making the first two-tone paint option available for 16 years. Needless to say, the Silver Anniversary was marked by silver paint: dark at the bottom and lighter at the top, divided by dark silver striping. This was known as the 'Silver Anniversary' paint option.

All '78 Corvettes were fitted with '25th Anniversary' badges on the rear, but this just denotes the model year and not any other special options.

To commemorate the use of a Corvette as pace car for the 1978 Indy 500 race, Chevrolet built the Limited Edition Corvette with their own vehicle identification number sequence. All 6,502 Pace Car Replicas were finished in black over silver with contrasting decals, the 'Official Pace Car' decal for the doors was supplied for the owner to fit if he wished. 8-slot alloy wheels with a special red stripe were standard equipment. Mirrored glass roofs, lightweight plastic based seats and front and rear spoilers were fitted as standard and not available for other 1978's. Engine and transmission options were unaffected.

Those cars fitted with manual transmissions were also fitted with a harder suspension set-up, the reasoning being, quite logically, that the buyer opting for a manual gear shift would be a more sporting driver than his auto box compatriot. However, this was not an option for the following year.

1979

Little was changed in 1979, though many of the optional features of the previous year became standard, notably the front and rear spoilers. On the exterior, the badges on the front and rear of the car reverted once more to the crossed flags of 1977 and for the first time, Corvette production reached the 50,000 unit mark.

Inside the car an AM/FM radio came as standard, as did the high-back seats from the Limited Edition pace car replica – the use of plastics in their construction led to a weight-saving of some 12 lb. These seats were better designed, with more thought being given to passenger comfort and safety, and the new seat tracks offered another inch of forward travel. For the first time, an illuminated visor/mirror combination was available, though for the passenger side only. The passenger sun visor gained an illuminated vanity mirror. Corvette drivers could see better in dark thanks to the phased introduction of tungsten-halogen for the high-beam headlights.

1979

The base (L-48) 350 ci engine gained around 10 bhp; five from the fitment of a new, open flow muffler design (which also bestowed a similar power increase in the optional L-82 engine) and five from the fitment of a dual-snorkle air intake.

1980

The year of 1980 saw more radical changes aimed at maintaining the car's popularity prior to the introduction of the totally new model in 1984. The front three-piece spoiler of the previous year became an integral part of the new 'soft' bumper arrangement, which extended around the front fenders.

The front-end design changes resulted in an increase in radiator airflow of around 50 per cent, and a reduction in drag coefficient from 0.503 to 0.443. Beneath the spoiler, a rubber extension section was fitted together with a steel skid to protect the spoiler when the springs were fully compressed.

New side-marker and cornering lights were fitted, the latter having been used by Cadillac for some years. Another 'soft' spoiler was integrated into the new rear bumper design which also featured new recessed tail lights and no overriders. The driver and passenger were being ever-more pampered with the addition of several luxury items fitted as standard, which included twin remote-control door mirrors, electric windows, a tilting steering wheel and air conditioning.

The three storage compartments became two – the one for the battery remained as previously, but the two compartments behind the passenger seat were combined with one access door. Air conditioning became a standard feature of all 1980 cars. Federal regulations meant that the car was fitted with an 85 mph speedometer – a strange sight, given the actual speed capability of the car.

The effects of the emission regulations were evident from the fact that, for the first time since 1974, the Corvette was available with a choice of two engine displacements. The L-82 (230 bhp) engine could only be specified in conjunction with an automatic transmission and a 'special' 305 ci engine was introduced solely for California, where the emission laws were the strictest of all. This was fitted with stainless steel, tubular exhaust manifolds (headers) and an oxygen sensor in a closed loop system.

The 350 ci L-48 engine received the aluminium intake manifold, previously only fitted to the L-82 engine. Bucking the trend started in 1968, the 1980 car was actually lighter than the previous year's model – by some 250 lb! Weight savings were achieved in numerous ways. For example, thinner glass was fitted and the hood and roof panels and the doors were reduced in thickness, but the biggest weight reduction stemmed from the fact that the new nose was constructed from fiberglass and so did not need the substantial support structure required by the previous steel version. In addition, the car featured the use of aluminium for the support cross-member and the differential housing. As is usual with any weight saving measures, there were measurable benefits to the owners, in that fuel consumption, handling and overall performance were all improved.

A new optional extra was a rear deck mounted carrier for the removable roof panels. It was not the most successful of options and was available only from 1980 until 1982.

1980

At the rear of the car there was a new badge (lower case letters spelling 'Corvette') whereas there was no badging at the front other than the crossed badges on the hood unless the L-82 engine was fitted, when there would be a fender-mounted logo to that effect.

1981

In 1981, externally, the cars were changed very little. The crossed flag badges differed, front and rear, and the between-the-wheels mouldings were finished in mono-color black rather than the earlier two-tone black/silver. After the considerable weight-saving of the previous year, the 17 lb reduction in this year's model seemed a bit of an anti-climax. However, it was actually much higher than this but was offset by the addition of extra emission control equipment.

The most important aspect of the '81 cars was the introduction of the fiberglass rear spring, which offered a massive weight saving. It weighed 8 lb, compared with the previous multi-leaf steel spring, in use since 1963, which tipped the scales at some 44 lb! To lose 36 lb at a single stroke is some going! With no inter-leaf friction, it gave a better ride and, of course, the reduced weight at the rear of the car contributed to better handling characteristics. During 1981 it was only used for cars fitted with the automatic transmission, those opting for manual transmission and/or the 'gymkhana' suspension package were stuck with the original steel version.

Under the hood, the L-82 engine was dropped and the L-81 introduced. This featured a number of items which made it lighter than its predecessor and more environment-friendly, so much so it was certified for use in California, thus eliminating the need for a special one-state engine. An aluminium intake manifold and tubular exhaust manifold saved weight, and an advanced ECM (Electronic Control Module) controlled the fuel and air mixture together with the ignition timing, using information collected from sensors in the engine and catalytic exhaust system. By maximizing the fuel/air mixture, not only did the car run more economically, but its emissions performance was improved. Also contributing to fuel economy was a locking clutch torque converter which enabled the auto transmission to be held in second or third gears.

Inside the car the options list showed a six-position electric driver's seat, electric remote control exterior mirrors and an FM radio with LED time display. The alarm system was uprated to include an ignition interrupt, preventing the engine from being started by a thief.

In June of 1981, production moved to the Bowling Green Plant in Kentucky. Production overlapped between there and St Louis for two months during a transitory period until the end of July. The new plant featured an advanced painting facility, which Chevrolet utilized to the full by offering a new range of two-color options: light colours above the waistline; darker ones below it.

1982

In 1982, there were positive improvements under the hood and fuel injection was reintroduced, this time electronic not mechanical, as had previously been available on the small block V8 from 1957 through 1965. The new system, rather than use one injector per cylinder, used one injector per bank of four cylinders. Allied to a 9:1 compression ratio, it facilitated a 10 per cent increase in power to 200 bhp and there can be no doubt which engine you are looking at as the valve covers carry the legend 'Cross-Fire Injection' and the same words can be found on the side body panels of all 1982 cars. The increased efficiency of the engines also meant that some gains in fuel consumption were available, especially on long runs where figures exceeding 25 mpg were often the norm.

The manual transmission was not available during 1982, all cars leaving the factory with a four-speed overdrive unit with converter lock-up on third, fourth and fifth.

The 1982 Corvette was intended to be the last of this particular series, since the basic body shape dated to 1968 and the chassis to 1963. Before the introduction of a brand new model, a special edition car, the 'Collectors Edition', was produced, featuring silver leather seats, silver/beige metallic paint, 'Collectors Edition' badges and pin stripes, alloy wheels with radial fins and, perhaps most important, the large rear window was hinged to become a hatchback – a feature much appreciated by most owners.

1983

Although some pilot 1984 models were produced with 1983 vin numbers, and one has been retained by GM as a display vehicle, none was sold. Instead the 1984 model was released in March 1983 and continued in production until September 1984.

1984

Whilst the 1984 model was recognisably a Corvette, it was an almost completely redesigned car. It was leaner, more aerodynamic and faster. Compared with its predecessor, it had some 2 in. trimmed from the wheelbase and the overall length was 8.8 in. shorter. It had lost around 250 lb in curb weight and about 1.1 in. in height. The 'flattened' windshield (to an angle of 64 degrees) contributed to a much smoother shape and a CD factor of 0.34, an improvement of almost 25 per cent.

The new car was styled by Jerry Palmer and it was his first attempt at designing a vehicle from the ground up. He succeeded in creating an all-new shape, but one which still retained the inherent qualities of the original Corvettes. The design work started in 1978 and took a lot of its styling cues from the rotary engined concept cars of 1972. Such advanced features as gull-wings – and, of course, the rotary engine itself – were discarded as being impractical for use on a production road car.

Inside, there was more headroom and leg room and 6.5 in. more shoulder room. However, interior footroom suffered a little because of the need to increase headroom but decrease the overall height of the car – this was achieved largely by rerouting the exhaust system through the centre tunnel.

The driver and passenger were almost spoilt for choice with a list of standard fitments which never seems to end: 4-speaker FM radio with electric antenna, stylish new bucket seats, leather rimmed steering wheel and gear shift boot, luggage security cover, remote operated exterior mirrors, air-conditioning, halogen head/fog lights and built-in alarm with starter motor immobilizer.

The digital information display was, and still is, a source of some debate: some love it, some hate it, but no-one ignores it. Whatever opinion was held, there was no option to have the old analogue system. All models also featured a one-piece removable roof panel and a lifting rear hatch window. At this time, the rear window glass was the largest ever compound glass installed in an American car.

At the front, a major change was the introduction of the 'clamshell' hood, enabling far better access to the Cross-Fire injected engine, its ancillaries and the new aluminium suspension work. The styling team also had a hand in the designing of the engine bay, to the point where Delco were asked to produce a black/grey coloured battery in order to blend in. The 1984 model was equipped with a 5.7 litre V8 Cross-Fire engine with fuel injection and electronic spark control. With a 9:1 compression ratio, the power output was 205 bhp and 290 lb/ft of torque was produced at 2,800 rpm. The kerb weight was 3117 lb. This, and the sophisticated engine management systems and improved aerodynamics, helped to achieve excellent EPA mileage figures: city 20 mpg, highway 35 mpg and overall 25 mpg (all figures with automatic transmission). Hauling the car down from speed was performed by a new disc brake set-up at each corner, which included aluminium calipers.

The four-speed transmission returned part way through the model year, complete with computer-managed overdrive in the three gears in order to circumvent the government's so-called 'gas-guzzler' tax. Under rapid acceleration, the overdrive was automatically overridden.

The frame was galvanized and painted and totally new. The perimeter frame birdcage structure used high strength steel as a weight-saving measure, and various sections of the bodywork were bonded to the birdcage including the roof, dash and some of the underbody. The rear bumper was mounted on a bolt-on alloy frame. This model is noticeable for its extensive use of aluminium, with the independent front suspension utilising aluminium arms and knuckles, and the five-link independent rear suspension using aluminium for knuckles, struts and control arms. Front and rear leaf springs were single plastic transverse units. If the performance handling package or the power seat were specified, the drive/propshafts were also aluminium.

The fit and finish of the new model was, without doubt, the best of any Corvette to that point.

1985

From the outside, there was little to tell the '84 and '85 models apart, unless you looked for the straight tailpipes and the words 'Tuned-Port injection' along the fenders rather than the previous model's 'Cross-Fire Injection'. This was the first Corvette engine in over 20 years to feature fuel injection, and the new system translated into a big improvement in engine power, torque, economy and top speed. At 230 bhp it was 25 bhp up on the '84 model, torque was improved from 290 to 330 lb/ft and Chevrolet claimed an improvement in fuel economy of around 11 per cent. Better yet, this extra urge meant that the car could break the 150 mph barrier.

Not surprisingly, the '85 model had an uprated cooling system comprising a special radiator with higher boost fan and an uprated (18 psi) radiator pressure cap.

The Z51 handling package was retained for 1985, but with the option of 9.5 in. alloy wheels at the front as well as the rear, the diameter of the sway bars was increased and it also included Delco-Bilstein shock absorbers. Some drivers had criticized the '84 car for having too harsh a ride – the downside

of having a car described by some of the Motoring Press as 'The Best Handling Car in the World'. As a result, suspension rates were lowered on the standard set-up by 25 per cent at the front and 26 per cent at the rear.

The already impressive braking ability was improved by using a larger bore master cylinder (up from $\frac{13}{16}$ in. to $\frac{7}{8}$ in.) and increasing the capacity of the power booster. The latter created another 'first' for American motoring, being made of plastic.

The option list was expanded by the addition of leather-trimmed sports seats and automatic temperature controlled air-conditioning. The controversial instrument graphics were made clearer and more readable and the removable

glass top received a stronger sun screen, after comments from well-cooked owners of the '84 model!

1986

The year of 1986 is significant in Corvette terms as it saw the return of the Convertible after an absence of 11 years, a move welcomed by all enthusiasts. However, a removable hard top was not on offer. Early 1986 saw life for the new Corvette accelerating at a furious pace. The Bowling Green, Kentucky assembly plant started production in January, models hit the dealers' showrooms in February and in May a bright yellow Convertible was used as the pace car in the Indy 500 – and all '86 Corvette convertibles were designated as Pace Car replicas whether or not fitted with the optional decal packages supplied with, but not fitted to, all '86 cars. High level stop lights were featured on all models, however. The engine for the topless car was the same unit used in the coupés the previous year, i.e. the Tuned-Port V8.

The new lightweight alloy cylinder heads increased the compression ratio to 9.5:1 and resulted in a weight-saving of 40 lb, though initial problems with cracking meant late-in-the-day design revisions and they were not introduced until convertible production started, halfway through the model year. It drove through a standard four-speed automatic transmission (with overdrive) with a four-on-the-floor manual (overdrive on the top three gears) as a no-cost option.

All this 'go' meant that there had to be plenty of 'stop' available. A Corvette 'first' on '86 cars was the fitment of ABS anti-lock braking, adapted from the Bosch system, and working on all four discs. The braking ability was transmitted to the tarmac courtesy 255/50 VR Goodyear Eagle tires, wrapped around 16 x 8.5 in. alloy wheels. As with any convertible, torsional rigidity was more of a problem than with the coupé models and, therefore, reinforcement was required to cut down 'scuttle-shake' to a minimum. There was a strong brace placed under the mid-section and a crossbeam behind the seats, while the door latch mechanism was uprated and the seat back riser was a double steel panel. At the front, attention was paid to the steering column, its mounts and the front torque box. Incredibly, it was found that the convertible was actually more rigid than its coupé sibling! The Z51 handling package was only available on coupés.

Having made the new model more desirable, a new anti-theft system was introduced, going under the acronym VATS – Vehicle Anti-Theft System. It used a special ignition key with contacts in the lock cylinder to check to see if the right key was being used.

If it decided it wasn't, the engine would not start and three minutes would have to elapse before a further try was permitted.

At the 1986 Detroit automobile show, the Corvette show car was the 'Indy', a stunning and futuristic vision which incorporated a number of interesting features, such as all-wheel steering, traction control and active suspension. The engine was a direct derivative of the twin overhead cam Chevrolet Indianapolis V8 developed for the 1986 Indy CART racing season. Though it had a relatively small 2.65 litre capacity, it featured twin turbochargers, twin air-air intercoolers and four valves per cylinder. The driver and passenger sat in a tub made of Kevlar, an incredibly strong but light material used in the construction of Formula One racing cars. The Formula One link continued with the involvement of Lotus, the English sports and race car constructors, famous for their innovative technical achievements, particularly in the areas of handling and suspension. Also involved were the Italian Cecomp company. Further racing development ideas shown in this car were the so called fly-by-wire electronic actuations where, for example, the throttle is not operated by a cable, but rather by an electronic sensor.

1987

There were few changes to the '87 model. The RPO Z51 performance handling kit was still available, though once again only for manual transmission coupé models.

A new 'sports' handling package was offered with the designation of RPO Z52 and was available for manual or automatic, coupé or convertible cars. It comprised the standard suspension units but with Bilstein shock absorbers, 9.5 in. wide alloy wheels and slightly different steering gear and axle ratios. It also included a faster 13:1 steering ratio, larger front stabilizer bar and heavy-duty radiator.

Under the hood there was an increase in bhp but a decrease in the amount of fuel used and in engine noise. This was achieved by fitting the V8 engine with anti-friction roller hydraulic lifters, together with 18 needle roller bearings to support the main roller of each lifter. The power output was 240 bhp, up 5 bhp on the previous model, while for a more significant increase owners could opt for the Callaway Twin-Turbo engine. With 345 bhp and 465 lb ft of torque, it endowed the Corvette with the ability to touch the hem of 180 mph, and took it well into the 'supercar' league. The Callaway engine was not available as a factory option, but could be ordered through Chevrolet dealers. The very first engines used the short blocks from trucks (LF5) but, after that, reworked Corvette engines were the basis for the uprate. Some 184 cars were so fitted during the year.

An interesting option for this year was an electronic low tire-pressure warning system. It was far from reliable, though, and this, combined with the high price, tended to limit sales – less than 50 recorded for the whole year.

More mainstream options were a light for the driver's vanity mirror and an electrically powered passenger seat. In an effort to counter the 'Vette's title as America's most stolen car, and in addition to the standard alarm, '87 model key components carried individual identification.

1988

In 1988 the Corvette development pattern was again one of steady refinement rather than changes for their own sake. The horsepower was increased by 5 bhp to 245 bhp, though this was only for coupés with the higher of two axle ratios, as the power increase was obtained by fitting less restrictive mufflers which were deemed too loud for convertibles and low ratio coupés.

The handling was improved, largely by the fitment of new 'ZR'-rated P255/50ZR tires on the new six-slot standard wheels. New on the option list were 9.5 in. x 17 wheels, supplied with 270/40 ZR Goodyear Eagles. Slowing the action down were new dual-piston front brakes. In addition, the parking brake was modified so that it operated the rear brake pads rather than the old system of working on small separate parking drum brakes.

RPO Z51 & Z52 performance suspension packages were offered, but with slight changes. Both featured the new 12-slot, 17 x 9.5 in. wheels and P275/40ZR17 tires, whereas the Z51 now featured bigger front brake rotors and calipers in addition to the previous specification, and was available only on manual transmission coupés.

To celebrate the 35th anniversary of the Corvette, a limited production run of coupés was produced featuring two-tone paint finish comprising white with black roof along with white leather seats, console mounted plaque and various other 'anniversary only' additions.

Inside the car, improvements included better coupé ventilation, relocation of the parking brake (further back and lower) and carpeted door sills. 1988 also saw the Corvette Challenge racing series, for which GM built 56 street-legal engines at the Flint plant before being shipped to Bowling Green for production.

The Callaway Twin-Turbo cars were again available, the engines of which, reworked at the Old Lyme plant in Connecticut, produced 362 bhp and 562 lb ft of torque. The link between Chevrolet and Callaway was close inasmuch as the car retained its official GM 12 month (12,000 mile) warranty on all parts except the power train.

1989

An interesting no-cost option for '89 was the six-speed manual transmission which came about as a result of collaboration between Chevrolet and the German transmission manufacturers, ZF. They were built in the latter's facility in Germany and included the annoying Computer Aided Gear Selection (CAGS) system, which prevented selection of 2nd gear, following a standing start from 1st, under part throttle opening. This was to improve fuel economy during the Corporate Average Fuel Economy (CAFE) government test. Fortunately this system was easily unplugged to restore proper shifting operation.

The six-slot road wheel, introduced the previous year, was discontinued and replaced by the 17 in. twelve-slot version, previously available only as part of a suspension uprate package. Convertible owners were treated to a simplified manual top mechanism and a couple of new options – a removable hardtop and a rear luggage rack.

The RPO Z51 handling package was available only for manual transmission

coupés. An interesting option was that of a three-way variable suspension control, operated from the console. The RPO Z52 suspension option was deleted for '89, though the Z51 packages included Z52 springs and stabilizer bars.

Promised for 1989 was the King-of-the-Hill ZR-1 model but, although 100 cars were prepared for dealer and Press evaluation, none was available for sale to the public until 1990. According to Chevrolet, they were late because of a lack of engines caused by additional development.

1990

A camshaft revision, an increase in compression ratio and a speed density engine control system meant that the 1990 base model engine produced an extra 5 bhp – 245 or 250 bhp, depending on the axle ratios. Stopping the '90 cars was easier thanks to improved ABS brakes and the handling benefited from better yaw control. Inside the car, the instruments became a combination of digital and analogue – the speedometer was digital but the rest of the gauges, including the tachometer, were analogue. The driver's side received an air bag and passengers now had a box in which to store their gloves.

The biggest Corvette event of 1990 was the arrival of the much-trumpeted ZR-1, boasting 375 bhp from its engine, designated LT5. Though the 350 ci capacity, V8 configuration and bore size were the same as the standard engine, there was plenty that was new. Manufactured by Mercury Marine (in Stillwater, Oklahoma) the engines were shipped to the Bowling Green plant for fitting into the cars. The units used aluminium in their construction but that didn't prevent them from tipping the scales at some 40 lb more than the standard L98 engines, complete with iron block.

Computer technology was used to improve engine longevity; the life of the oil could be calculated by comparing engine temperatures, revolutions and stop/start data. When the computer judged an oil change was due, a light would flash to alert the driver.

1991

The effect of the ZR-1 can be seen in the exterior appearance of all '91 models, which featured four rectangular tail lights in a convex rear fascia, while the centre stop lamp was integrated into those tail lights for all variants except the ZR-1, which retained the roof mounted fitting from the previous year.

All models (ZR-1 included) had revised styling with wrap around light units, wider body mouldings along the side and a new style of panel louvres. However, even though its 'lesser' offspring now bore a distinct resemblance, the ZR-1 still used unique doors and rear body panels, changed to facilitate the 11 in. wide wheels. Base road wheels stayed the same size as 1990 (17 in. x 9.5 in.), though the styling changed to a 'turbo-vane' pattern.

The RPO Z07 package was introduced, which combined the essence of the Z51 performance handling package with the selective ride capabilities of the FX3 package. Intended for 'serious' road driving or even competition use, all suspension parts were heavy duty and thus limited to coupé only fitment.

Larger section mufflers with lower back pressure were fitted, though no power output increase was claimed. Finned power steering coolers were standard on all '91 models, and a sensor was fitted into the oil pan which operated a warning light with the words 'LOW OIL' when the level dropped below the requisite amount.

The 1991 model year saw the last of the Callaway Twin-Turbo conversions. The 500th was produced in September '91 and subsequent cars were badged and optioned as 'Callaway 500' models. On the track, the World Challenge race series continued but, unlike previous years, Bowling Green had no involvement in terms of car preparation – all modifications were the responsibility of the driver. Convenience features included a 12v power wire (for a cellular phone or other device) and a delay was added so that items such as the stereo or

electric windows would still work after the key had been turned. It would all cut out after 15 minutes or prior to that if the driver's door was opened.

1992

From the outside, little changed in 1992. At the rear, square exhaust outlets were standard on all models, with the ZR-1 getting 'ZR-1' badges just above the fender vents.

In July 1992 a momentous occasion occurred – the one millionth Corvette, a white convertible, rolled off the production line.

1992 saw the installation of a new base small block engine, the LT1, a unit which put out 300 bhp and 330 lb ft of torque. Contributory factors to the 20 per cent increase in power included computer controlled ignition timing, multiport fuel injection (MFI), a new camshaft profile and a new exhaust system featuring two catalysts and oxygen sensors. The unit could rev higher, with the red line being moved up by 700 rpm to 5,700 rpm, though a fuel cut-off was fitted to call a halt to any over-exuberance at 5,850 rpm.

At 452 lb, it was 21 lb heavier than the previous base engine (the LT98), something attributed largely to the replacement of stainless steel exhaust manifolds with cast iron versions. A feature copied from the 4-cam Lotus designed LT-5/ZR-1 engine was the reverse flow cooling, whereby the coolant was directed firstly to the cylinder heads then to the block, rather than the other way round as previously. The results were better cooling around the valve seats and spark plug bosses, reduced piston ring temperatures and higher bore temperatures. The company recommended the use of synthetic oil in the engines and reckoned that, because it ran cooler than 'normal' oils, it could do without the oil cooler.

The job of effecting more control over the 'Vette's rear end was taken up by the newly introduced switchable ASR system. Standing for Acceleration Slip Regulation, it was a Bosch product developed by Corvette engineers to limit wheel spin under rapid acceleration by using spark retard, brakes and throttle control.

Inside the car, the digital speedo was moved to be above the fuel gauge and the graphics were generally tidied up to make them easier to read. The black/grey face plates and buttons of 1991 became all black for '92.

The Corvette's cockpit was a quieter place to be, thanks to the fitment of improved weatherstrip seals and improved and additional insulation in the doors and over the transmission tunnel. The power delay feature was retained, but modified slightly so that it was also cut when the passenger door was opened.

1993

1993 brought with it the 40th anniversary of the Corvette's arrival into the motoring world and, not surprisingly, there was a special 40th Anniversary Package available. It included Ruby Red metallic paint finish, Ruby Red leather sport seats (electric driver's side) and special wheel centre trim and emblems.

The base LT1 engine was made quieter by changes to the heat shield design, the introduction of thermoset polyester valve covers and modifications to the camshaft exhaust lobe. The latter resulted in a reduction of the exhaust valve closing velocity and closing the inlet valve sooner gave an increase in torque – up 10 lb ft to 340 lb ft at 3,600 rpm.

The LT5 engine in the ZR-1 now produced 405 bhp, an increase of 30 bhp over the previous version, largely through revisions in the cylinder head design. Other changes included a requirement for Mobil 1 synthetic oil, platinum-tipped spark plugs and four-bolt main bearings.

The wheels appeared the same as in 1992, though the appearance was changed slightly, thanks to a change in the finish machining. At the front there was a 1 in. reduction in wheel width to 8.5 in. and a similarly reduced tire size was required, in this case a P255/45ZR17. The rear tire size was increased to P285/40ZR17.

A recurring problem throughout the Corvette's history has been security, since for obvious reasons it has always been at the top of the car thief's list! In 1993 the Corvette became the first GM car to feature an innovative new security system. The PKE (Passive Keyless Entry) system relied on the proximity of a battery powered fob to lock and unlock the car and simultaneously operate the alarm system. Two antennae in the car (in the coupé they were in the door and rear deck, while in the convertible they were in both doors) picked up the signal as the driver approached the car, which automatically (passively) unlocked the doors and switched off the alarm. When the driver left the car, the doors would lock and the alarm arm itself as he got more than a few metres from the car. Naturally, it could be turned off altogether, and the transmitters could be programmed to operate either both or just one door.

1994

A range of improvements included wider and more comfortable seats, omitting the stylish but restrictive side bolsters, and new door panel design. The convertible was improved by fitting a heated glass rear window in place of the flexible plastic used previously. The passenger's glove box was replaced by an airbag.

Goodyear Extended Mobility tires were offered as an option, capable of being driven at least 50 miles deflated. Other driving improvements included lower recommended tire pressures of 30psi instead of 35, adopting a practice learned by owners from experience. The 35psi recommendation remained for ZR-1 owners and should always be observed when sustained speeds over 75 mph are expected.

The engine efficiency was improved by sequential fuel injection and a more powerful ignition system, while more electronic control was incorporated into the automatic transmission.

1995

With an echo of 1962, the gills in the panels behind the front wheels were restyled for the first time since 1991, for a more dramatic exterior change it was necessary to procure one of the only 527 Indy Pace Car replicas produced. This was only available as a convertible in a stunning combination of dark purple and white with special accents.

For the driver, windshield wiper operation was revised with higher contact pressure and angles to reduce the high speed lift that had always been a problem with this model since 1984. If the Extended Mobility tires were specified, then a weight saving and a $100 credit were available by omitting the spare wheel.

Sadly, 1995 was the last year for production of the ZR-1. In its six years of production it achieved legendary status, with just 6,939 cars built.

With this model now in its 12th year of production, and still fundamentally unchanged, great pains were taken to reduce rattles and interior noise levels to keep pace with competing cars in the Corvette's market sector, and successfully improved the car's rating in owner surveys on perceived quality.

1996

Chevrolet announced that production of the 1996 model would finish in early Summer 1996 to allow retooling and retraining for the early 1997 introduction of a radically new Corvette.

The 1984–96 model having been the best performing and best built thus far, the opportunity was taken to celebrate this success with two special editions, the Grand Sport and Collector Edition.

Also new for 1996 was the LT4 engine option, only available with the 6-speed manual transmission, and rated at 330bhp, a 30bhp improvement on the LT1 which remained the standard motor if an automatic was specified.

The Grand Sport was available only in Admiral blue with white centre stripe, a red flash above the front left wheel, and featured a black-painted 1995 ZR1 wheel and tire combination, with flares to cover the wide wheels on the rear fenders, if the car was a coupé. The convertible used four black-finished ZR1 front wheels and dispensed with rear flares. All Grand Sports had either red or black interior and a Grand Sport logo on the throttle body. 6-speed manual combined with the LT4 motor was the only powertrain available.

The Collector Edition was finished only in Sebring silver, could be auto or 6-speed, and had silver-finished ZR1 style wheels. It too was available as a coupé or convertible, and like the Grand Sport, featured Corvette lettering on the brake calipers.

A new Real Time selective ride option was introduced, adjusting the damping of each shock absorber every 10 to 15 milliseconds, and the codes stored in the On Board Diagnostic system were increased from 60 to 140.

With the new model eagerly anticipated, it is interesting to consider what it will be like to restore a 1984–96 model. Already, many improved or replica parts are available from aftermarket sources but it is also obvious that when the time comes to restore these later models, competence with electronics will be essential!

Chapter 2

Buying a Used Corvette

PART 1 THE RULES

by Tom Falconer of Claremont Corvette

Buy the right Corvette and you'll love it! Buy badly and you could end up with a car you'll hate – which would be a tragedy since we'd much rather you shared our love for the finest sports car of all time! We strongly recommend that you read this chapter with enormous care before buying your car, and make sure you buy a Corvette that you can enjoy. After all, that is what this wonderful car is all about!

In 25 years of buying and selling Corvettes I have owned hundreds of lovely examples, some for years and others for just a few hours. I have bought cars in the United States, from all corners of the United Kingdom, from France, Germany and Switzerland and from the Gulf States. When the exchange rate has been favourable, I have even shipped them back across the Atlantic to their country of origin and to continental Europe.

As a parts specialist, and Classic Corvette Club UK member, I have advised in the purchase of thousands of Corvettes. So am I an expert? Probably not. Do I buy successfully? No, not always. Do friends and customers take my advice? Sometimes. Is it worth reading what follows and remembering the rules? Yes, most certainly!

THE RULES

Over 25 years of Corvette dealing I have learned by my mistakes and other people's, and I've developed certain rules which should lead to contented and fruitful ownership.

They are important truths for sellers as well as buyers, because all buyers must one day be sellers unless their car is to be crashed, bequeathed, repossessed or, worst of all, passed to the spouse in a divorce settlement. So, for better or worse, I'll start with:

Rule One – 'A buyer is most likely to buy the Corvette closest to where he or she lives, and it will probably be the first one seen for sale.'

This rule is the most dominant factor in choosing a Corvette, and the most frequently observed. First-time buyers see the Corvette, fall in love and hand over the money. Readers who already have bought a Corvette will know the truth of this rule: if not, why are you reading a book about restoration?

In countries where Corvettes are rare, this is understandable. Many buyers would not have thought of owning a Corvette until they saw one for sale. It should be obvious that the market is worth researching, and that it is worth seeking other cars for sale further away, but the local car is always more convenient. Fortunately, Rule One is only observed in purchasing the first Corvette and is the only Rule which should not be obeyed!

Rule Two – 'The Corvette interior reveals more about the car than anything else.'

A careless owner does more harm to the cockpit than to any other part of the car. If the ashtray hasn't been emptied for years, or the rear compartment frames have been cracked, then it is more likely he has never serviced the car properly, has let the power steering fluid level drop until the pump groans, or let the engine oil level drop until the valve lifters rattle.

Stained carpets reveal expensive water leaks from areas such as rusty windshield pillar frames; worn carpets indicate high mileages; and faded carpets suggest cars that have been kept outside. Similarly, seats and 1977-and-later leather-covered steering wheels can tell their own stories.

Over the decades and the years, Corvettes have been produced with more and more luxurious interiors, and the cost of replacing them reflects that fact. A glance at the Auto Accessories of America catalogue shows that a 1984 interior is more than twice the cost of a 1964. Therefore the more recent the car, the more important the Interior Rule becomes.

And never forget that any full interior replacement costs more than a full engine rebuild. So, faced with a choice between two cars, one with a wrecked interior and one with a wrecked engine, buy the latter!

Rule Three – 'Bad Corvettes smell bad.'

I once jokingly told a customer that the secret of finding a good Corvette was to open the door and to sniff. With experience I realize that I was speaking more truth than I then knew! In spite of 'New Car Smell' aerosols, decay will declare itself to the well tuned nostril.

Rotting seat foams, damp carpets, mice in the heater (surprisingly common), badly grommeted firewalls, leaking oil pressure gauge hoses, leaking heater matrices and even overheated wiring looms can be identified with a good sniff, even if subconsciously.

The engine compartment reveals more, such as fuel smells from worn carbs or leaking injectors, coolant smells from leaking hoses, core plugs or even cracked cylinder blocks, burnt oil smells from worn bores or crankshaft seals spraying the exhaust system. At the back of the car, don't ignore the sulphurous smell of a leaking differential casing.

A good Corvette smells delicious, even under the hood. Believe what your nose tells you!

Rule Four – 'Corvette weatherstrips cost a fortune.'

Unless you knew better, you could well assume that a full set of weatherstrips could be bought off the reel for less than £70 ($100). Don't be caught out. Corvettes use 'latex' weatherstrips which have a foam rubber core with a harder rubber skin. They are very effective at sealing out wind noise and weather, beautifully made with moulded-in mounting brackets and very expensive. They look terrible when worn, while cheaper replicas are fine until you try to close the windows or drive in rain!

Rule Five – 'Crashed Corvettes are always rebuilt.'

All Corvettes are valuable, and it pays to repair them after a crash. Buy one that it is badly repaired and you will regret your error not only while you own but, if you are honest, even more when you sell.

First take your tape measure with you and remember that '53–'62 has a 102 in. wheelbase, '63–'82 is 98 in., and '84 and newer is 96 in. On earlier cars with dished wheels it is easier to measure front of rim to front of rim. Be suspicious of variations of more than ¼ in. side to side. I have seen 2 in. variation!

Early models don't get crashed much these days, and they used to say that if a '58–'62 hit a house, then you had to call the builder not the bodyshop!

'63–'79s love to bend at the dogleg in the chassis frame behind the front wheel, and these are the cars that most need the tape measure test. On '68–'82 coupés it is easy to check visually that the point of the nose is in line with the central split of the targa roof. '80–'82s have thinner section chassis frame material which is even more prone to bend at the dogleg. If you buy a car on which the body has been repaired to fit a bent chassis frame, then you have a big problem.

Cheap aftermarket replacement panels, fenders in particular, can easily be identified by their rough or hairy underside. GM panels are smooth.

When examining '84 and newer Corvettes, first check the door shuts and then, if suspicious, ask the vendor to remove and then refit the targa roof. If the car is out of shape, he may struggle to replace it.

On Tuned Port Injection, 1985 to 1991 models, check the clearance between the rubber seal at the base of the windshield and the finned aluminium cover over the distributor. On cars which have been badly front-ended, these two components can squash together, instead of just touching. Generally, the build quality on the '84-and-newer cars is so good that bad repairs are easy to spot. Remember, too, that there is nothing inferior about a car that has been accident damaged provided that the repair work is first class.

Rule Six – 'Corvettes can be cooked.'

Corvettes have always been popular in the Sunbelt States, not least because they have a full size air conditioning system which can easily cope with cooling a small two passenger interior. But that doesn't help if the car is parked outside all day.

Hot state Corvettes that are not garaged all day suffer from faded paintwork, dried up and sawtoothed weatherstrips, sweat damaged seats and seat foams, and faded and dried up carpets, particularly in the '78 and later coupés.

When they travel north to the Rustbelt or to northern Europe, exhaust systems, brake pipes and other underbody components quickly rust over, since the dry heat and dust have long since dispelled any protective oil or grease. This particularly affects door lock mechanisms which can freeze solid after a few days of rain.

The sun states may be a wonderful source of unrusted steel bodied classics, but Corvettes are a different kind of car, made of fiberglass. Many of the nicest come from the northern enthusiasts who garage their cars for the length of the snowy winters, and therefore put their cars away for half the year. This means that their cars do only half the mileage, too!

Hot sunshine also dries up body sealer, makes the fiberglass more brittle, and damages the polyurethane nose and tail sections fitted to Corvettes starting with the 1973 soft-nose model.

1984-and-later sun cooked cars can be identified at a glance by their dash top speaker grilles which are always faded to bright blue from their original black. Earlier model windshields often have their blue top tint section faded to brown.

Rule Seven – 'fiberglass cars can't rust.'

More than one million Corvettes have been built, which makes it the best-selling two seat sports car of all time. It is also the largest production fibreglass-bodied vehicle ever.

Since production began in 1953, Chevrolet have always been the leaders in GRP (glass reinforced plastic) techniques. In 1954, the second year of production, matched pressure moulds were introduced which result in a smooth finish to both sides of the panel. Other manufacturers of fiberglass cars still lay up panels into a female mould using matting and resin, which is less strong and leaves a rough finish one side. Most aftermarket Corvette panels are made this way.

So, how do Corvettes rust if they are made of fiberglass? The answer is that there are many ways which are important to the potential buyer, but far fewer than there might be for a steel bodied car. Indeed, the Corvette is undoubtedly one of the most endurable vehicles ever built.

On all models, brake lines, fuel pipes and exhaust pipes will rust. Ideally these are replaced in stainless steel. Though more expensive, stainless steel exhausts and mufflers will never rust through and must be the logical choice to fit to a non-rusting car.

The 1963–1982 chassis frame, from which the body can be detached by undoing eight principal mounting bolts (plus a whole lot of other stuff!), is sometimes prone to rust where the trailing arm enters the rear kickup just ahead of the rear wheels. While this can be repaired *in situ*, it is a much easier job with the body removed.

On all models from 1953 to 1974 fuel tanks rust, usually from the inside. From 1975 safety bladders were fitted inside tanks to improve crash-worthiness, and this fixed the rust problem, too.

The radiator support frame is prone to rust on 1963–1982 models. These are hard to repair and expensive to replace. However, this is a key component to the front end structure and cannot be ignored, because it attaches the front end of the body to the chassis frame.

Two of the rust areas that are hardest – or even impossible – to identify are the windshield pillars. Water flows behind the outer stainless steel trims and corrodes the pillars behind the windshield adhesive. The symptom of this problem is that water sprayed onto the windshield will leak in and drip from behind the dash. If the windshield is to be replaced, then this is the ideal time to weld up the pillars. Also, the base of the pillar support, where it disappears from view behind the fender, can rust out. See the relevant section of this book for how to fix it.

1976 and later models have a half-steel floor which will rust through if the screen pillars have been leaking. Doors, too, are liable to rust from 1968 to 1982 if the drain holes are not clear. The only GRP on the doors is the outside body panel itself. Doors are not cheap and repair is not easy.

Rule Eight – 'Rusty bores mean endless chores.'

The most notorious rust on Corvettes is inside the cast-iron brake calipers used on both front and rear wheels from 1965 to 1982. These are great brakes, powerful and with more than adequate pad area to stop one-and-a-half tons of Corvette from very high speeds. They would probably last for ever if the brake fluid were changed annually, as all brake fluid should be. Unfortunately, most owners didn't, and reconditioning these calipers has become a multi-million dollar industry.

Unlike the aluminium calipers of the 1984 and newer Corvettes, '65–'82 cars have calipers made of cast iron. This specification of brake fluid means that it absorbs water from the air (Why? Don't ask me, ask the Big Three!) and it is the water absorbed into the fluid which rusts the four smooth bores of each caliper, and prevents an air-tight and fluid-tight seal.

Be warned that calipers that appear to be free of leaking brake fluid and are dry externally can still have badly rusted bores. A soft brake pedal followed progressively by complete brake failure can happen because the corroded caliper will breathe in air through the piston seal every time the brakes are released, but will not leak fluid when the brake pedal is pressed. New owners of these models are sometimes mystified by a softening brake pedal and a rising fluid level in the brake master cylinder. The explanation is that air is being drawn in through the calipers, displacing the fluid.

Similarly, a vendor's assurance that the brakes are all right because all the calipers have been resealed may be true, but the soft pedal will probably return in a hundred miles or so.

The solution to all these problems came in the mid-1970s with the invention of the stainless steel caliper. These are not actually solid stainless, but old calipers which are remanufactured with stainless steel sleeves in the bores. Always fit a good brand of exchange stainless calipers such as those from Vette Brakes & Products Inc. (used on one of the project cars in this book). Stainless steel calipers, combined with silicon brake fluid is the only permanent fix for Corvette brake problems.

Rule Nine – 'Rather a good L-48 than a bad L-82.'

Don't get hung up on specifications. It is so easy to read all the books, visualize your dream year and the options you want, and then embark on a hopeless search for it. You will miss out on the best cars. An amazing proportion of buyers of 1975 to 1979 cars who come to Claremont Corvette say that they must have the L-82 high performance engine option. Over that period this engine was fitted to only 10 to 30 per cent of production depending on the year, so these customers are going to miss out on some great cars. While the L-82 engine has a higher compression ratio, some great internals and finned alloy valve covers, rather than stamped steel, the pleasure derived from this option could not really compensate for worse condition in other areas.

I have always wanted a 1973 big block convertible in Elkhart Green with Saddle leather and a four-speed, with power steering but no air conditioning. Not only have I never seen one but it is possible that, although all these were normal options, no such car was ever built. And if I finally found this '73 with a high mileage and some rust in the frame and for the same money there was a mint '74 350 auto Convertible, then that would be the one to buy.

Rule Ten – 'Corvettes are a passion so ignore the other nine rules.'

All right, I admit it. I would buy the big block 1973 in Rule 9 above, even though the restoration would cost a lot. Big blocks are expensive engines to rebuild properly, and manual transmissions are a lot more trouble in Corvettes than autos.

You can be confident that any Corvette is so much more satisfying than any other sports car for the same money, once you've experience a good one, you'll never regret it.

Go get your 'Vette!

PART 2 WHICH CORVETTE? – AND HOW?

All Corvettes are great cars, it is just that for some people, their greatness lies in the car's antique appeal, but for others there is just no substitute for neck-wrenching performance when compared with the sheer luxury of any other car. Before you travel a million miles to find the Corvette that is right for you, check out the following to give you an idea of which generation has most appeal.

FROM 1953 TO 1962

You will pay hot money for an early model, although '59–'61 models are great value. If you like driving an antique on wheels – albeit a hell of an exciting antique on wheels! – these could be the 'Vettes for you. There is a big steering wheel, plenty of turn from lock to lock and handling that makes life interesting. But if this isn't the most fun you can have surrounded by fiberglass and chrome, we don't know what is.

FROM 1963 TO 1967

Just like their older relatives, fewer and fewer of these cars are being employed on the road because owners seem to be too worried about their investment potential – which is considerable! But these cars have got something special going for them, with the addition of horse power and handling to match.

FROM 1968 TO 1975

With these cars you can front-end a wall safe in the knowledge that you have a collapsible steering column and side guard door beams in case you get T-boned in the process. Seriously, though no one wants to crash their car, many will be concerned about the presence of such elementary safety fixtures and fittings. Others will take more interest in the Mako Shark design of the '68–'73 cars which sets them apart from everything else on the road, and the three-speed automatic transmissions and tilt steering columns that make these cars yet more civilized.

FROM 1976 TO 1982

Like the Coke bottle and Mickey Mouse's ears, this car is one of the world's classic shapes. Big-block, high-optioned roadsters are going to make it in the investment stakes and, in the meantime, you can enjoy excellent performance from the earlier cars in this series.

FROM 1984 TO 1989

These are the best handling, best driving sports cars on the face of the earth. The styling is understated, but with the total package that this car provides, who cares? You don't get old-timer appeal; you do get American muscle with style and comfort attached.

WHERE TO BUY A CORVETTE

On the face of it, you will have to pay more to buy a Corvette through a specialty shop or dealership, but there can be great advantages attached, particularly if the dealership is reasonably close to where you live. The first good reason for visiting a dealer is that you can try out the cars there and make up your mind which model you prefer. Just don't let yourself get drawn into the sales spiel and DON'T let yourself buy a car on your first visit because you are sure to get it wrong. Also, be aware that not all dealers are as pure as the driven snow, so it helps to take someone with you who really does know the cars inside out and can identify good from bad. The second good reason for going to a dealer is that he may accept your car as a trade-in, although if you have the patience and resource to do your own selling, you can always negotiate a better deal with the dealership. And the third is that you can save tax on the full amount if you trade in.

The next best way of buying a Corvette is through advertisements, either in your local classifieds or in national publications. One of the great things about the Corvette is that there are so many of them, you can always find one without too much difficulty. Bear in mind that the best deals sell immediately, so either get your local Sunday paper as early as you can or search through *Hemmings Motor News*, *Old Cars Weekly* or *'Vette Vues* to find the car you want. If you're not one of those lucky people who can get up at 3 am, collect the Sunday paper from a local newsagent and drive to wherever to beat the other early birds to the worm (you call that lucky, I hear you ask?), you could always try the opposite approach. Look in the papers a few weeks or months after the ads have first appeared and call again to see if the cars are still for sale. Overpriced cars will always tend to stick, and owners may be willing to accept more sensible offers.

If you must have a certain rare model of vehicle, try joining Corvette clubs in your area or nationally and place adverts to say that you are looking for such and such a car. You never know what might turn up. Some collector car events also feature collector car auctions and, provided you know what you are looking for, you may be able to pull off a great deal if you come across an owner who needs to sell his car right now and so is willing to bargain.

It is a sad fact of life that a large proportion of Corvettes are abused some time or another and there is more chaff than wheat out there. Accept the fact that you will have to look at a good number of cars before you find the one that is in the right condition and the right price. Also make a big point of not buying one of the first few cars you see. We have all fallen into that trap at some time or other – the gleam of sunshine on the chrome work, the brilliance of the paint and the throb of the engine are there just to seduce the unwary – and it surely never works out. Ignore the shine, ignore the allure – in short, don't let the mermaid pull you onto the rocks! Look at what lies beneath, and the majority of cars will be eliminated in the first few minutes of your inspection. That leaves you to look more deeply at the best cars that are left. Here are the main areas to check:

1 NUMBERS MATCH

No matter what common sense may tell you, a non-original engine will hit resale value. If the pad on which the engine ID number is stamped has been filed or ground off, assume the worst – and you will be right! If you can, check the block number and date code cast into the rear of the block.

Even more important, check the ID number on the body, write it down and 'phone the local police to see if this car has been reported as stolen. If it has, and the car is recovered, the original owner gets the car back AND YOU LOSE ALL YOUR MONEY! Also, if the car is a recovered theft it might have a non-original state-issued ID number and title. Also look out for a state ID tag or 'salvage' stamped on the title to see if the car has been written off (totalled) and then patched together – maybe from two or more cars.

You must also be aware that some American states require that at least two numbers out of engine, frame and body numbers must match or you won't be able to register the car.

So, as you can see, checking the numbers is not just vanity; it's the first essential when looking over any Corvette for sale. If fact, you could say that it is more important on a 'Vette than almost any other car on the road since Corvettes are America's most stolen car – and one of the most commonly crashed. It goes without saying that both types of Corvette are the ones to avoid like the plague.

2 RUST CHECK

Just because a Corvette's bodywork is fiberglass does not mean that the car is immune to rust. The basis of the car's strength is its steel frame, and the frame and its ancillary components are rust-prone. They are also the most expensive parts of the car to renew. A quick check is to reach down and feel the frame on Corvettes from 1963 to 1982 – see if you can feel jagged edges, rough patch repairs, scale or undercoating. The only way to be sure, however, is to have the car put up on a lift. Which leads us onto another Good Idea! If you are having to travel some distance to see a car and you would like a second opinion, why not call a local repair shop near to where the car is situated – and get the owner to allow you to have the car inspected there. The cost of an hour or so of workshop time could the best investment you will make.

3 BODYWORK CHECKS

You are checking here for poor quality repairs and collision damage. Two cardinal rules: never inspect a car at night and never when it's wet from rain or having just been washed. (In fact, if the car is wet all over from just having been washed, your Suspicion Sensor should light up at maximum intensity.) Insist on seeing the car leathered off before you carry out the following checks.

- Look out for seams, sunken areas or lumps and look out, too, for wavy panels – cheap quality replacement panels won't have GM quality behind them. All of these indicate second-class repair work, or worse, and always require that the car is clean and dry and in bright conditions for you to check it out.
- Look out for the width of panel gaps and levelness of adjacent panels – but don't expect them to be Porsche-like because they never were.
- Look for over-spray on weather seals and even on tires – which proves that the car has recently been blown-over just for your benefit.
- Look for fresh bonding materials behind a panel – another certain indicator of recent repair work.
- Take a look at the car from a few yards back, from the front and from the back. Is the car level and square? Measure the height of the wheel wells if the car is parked on level ground. A crooked nose is a sure sign of a serious crash which was not properly repaired and is quite expensive to fix properly. On the other hand, a weak spring or even a flat tire can make things look worse than they are, but from your point of view, it is safe to assume the worst.

4 CHECK THE INTERIOR

The first thing to check is the trim tag. Check out the interior and paint codes and if they are not original, suspect the worst. At the very least, the value of the car will be hit by trim or color changes and if the trim has been changed, you really want to know why. Has the car been cobbled together after a severe accident or even after theft?

- Lift up the mats to find out if the carpet is damp or rotted. There could be a leak in the cowl or windshield frame, and the latter could involve extensive repairs, not just a tube of sealer.
- Look hard at all of the major trim items and in particular door panels. Check windows to ensure that they operate smoothly and check door latches and hinges. Ignore odd odometer mileage and use the evidence of your senses to tell you whether the car is worn or not. Also, take a look at trim replacement costs – and you will then look even harder at the condition of the trim on the car you're inspecting!
- Check seat belts and mechanisms, and bear in mind that replacements are expensive.
- Don't bother checking about wear on the pedals because replacements come cheap and give no indication as to the amount of wear that has taken place. Take an insider tip instead and check for wear on the steering wheel. A worn-through steering wheel surface is a sure indication of high mileage, and steering wheels are rarely replaced as part of a con-artist's repair kit.

5 MECHANICAL CHECKS

A full range of mechanical checks requires an experienced mechanic to carry out properly, but here are some quick checks which will give you a good indication of whether all is well or not.

- Take off the oil filler cap and see if there is a creamy deposit on the inside of the cap. If so, you could be looking at a car with badly worn cylinder bores.
- Lift off the coolant filler cap (with engine and coolant cold) and check for the possible presence of oil droplets in the water – a sure sign of gasket/warped cylinder head problems resulting probably from past overheating. Check also for rusty water – poor maintenance; and for the presence of anti-freeze – a sign of good maintenance no matter where the car has lived in since anti-freeze cuts out internal corrosion.
- With the engine fully warmed-up, try having a helper blip the throttle hard several times while you look at the outpourings from the exhaust tailpipe. If you see big black clouds of smoke, you are probably looking at the symptoms of a worn engine.
- On an automatic transmission, delayed or jerky shifts could indicate expensive-to-repair problems. Or they could indicate an absence of oil or blocked filters which might not be expensive to repair now but suggest poor maintenance.
- Manual gearbox transmissions should not jump out of gear when you accelerate and lift off (in each gear) and each ratio selected should run without undue noise.
- Look for oil stains on the ground where the car is usually parked – if you are not sure where, ask. Oil seals are cheap but the cost of a major tear down to get at them is not! Valve covers on Chevy V8 engines always seep a little, so don't worry about traces of oil around the edges of the valve cover. Look for oil underneath the bell housing which suggests a weak rear engine seal – expensive to replace! Look for oil around the bottom of the radiator hose where the front seal throws oil when it gets tired.
- Check hard for cooling system leaks because heater coils and radiators are expensive to replace, especially on air conditioned cars.
- Check out the price of tires and then you will spend extra time checking the condition of the tires themselves. Look out for lumps or bulges, especially on the insides of the tires, and make sure that you rotate the wheel so that you can check the part of the wheel that has been in contact with the ground. Tires worn more on one side than the other tell you that, at best, there is some suspension adjustment needed and, at worst, there has been bad crash damage which has left the car out of alignment.
- While you are looking at the inboard side of the tires, look out for wet streaks extending radially outwards on the insides of the tires. This will be brake fluid waving a little flag at you telling you that the calipers are in need of overhaul or replacement. Expensive!

PART 3 BUYING THE RIGHT CAR FOR REBUILD

If there's one thing more difficult than choosing a car with which to share your life, it's choosing a car to *restore* and share your life! By definition, it's going to take up a huge number of spare hours and money, so it's worth spending some considerable time pondering your choice before you take the plunge.

Selecting a very early (pre-1960) car is not for the faint of heart – they aren't exactly common on today's roads! Moreover, what cars there are to be found will usually have been restored or be in good, original fettle. Even a basket case is likely to be big bucks.

So this leaves us with the '60s, '70s and '80s as, without doubt, the most popular decades (at present) for restorers. How much you pay for your base car depends first, of course, on its general condition but also on factors such as rarity, availability of parts and desirability. This latter point is one which can vary from year to year, the trick being to restore a car which is not highly regarded now but which gains in credibility over the years. It's not an easy trick to master.

At the end of the day, you've got to buy the car that you want, and that is going to make you feel good on a grand scale. That's how it was for John Fletcher, the proud owner of the Corvette you see in this section and which was used at various times throughout the book to illustrate various restoration techniques. His first task was to isolate which particular model Corvette he wanted and this was fairly simple: he wanted a convertible, a big block engine and a manual gearbox transmission. Which year to choose was not quite so easy. Early models are, of course, particularly desirable. Trouble is, they can cost much more to restore and, having finished the work, their restored value (and accompanying car-park paranoia) makes it difficult to drive and really *enjoy* the fruits of your labours. The more recent models should not really need extensive work and will cost more to buy in any case.

Though 1968 was the year that saw the introduction of the 'Coke bottle' shape cars, there were plenty of teething problems and John saw that buying a year younger car would probably result in much less hassle, not least since GM made a whole host of improvements as a result of pressure from the motoring press and buying public alike. So, a '69 was John's first choice.

He also knew that once the '70s got under way, the engines became ever more cluttered with power-sapping, anti-pollution gear along with the 'soft' bumpers and other details that he, personally, does not prefer.

Initial adverts for his dream car brought little response, but eventually he found just the car through Udo Horn of KLJ Corvettes and Tom Falconer of Claremont Corvette in England.

In theory, it should then have been straight to work, pulling down the car ready for its ground-up rebuild. In fact, John couldn't resist driving the car around for a few weeks just for the sheer hell of it! (In fact, you are strongly recommended to do the same, before you pull down a car for rebuild. That way you figure out anything that makes the car a bad driver, before you start the rebuild.)

Many of the restoration techniques used on this car are shown in this book, and at the end you will see the finished result – the transformation from a rather tired Corvette into a stunning yet practical classic car.

The later illustrations show the sorts of problems you can have when rust takes a hold – and what to avoid. See the relevant sections of this book for what to do in order to repair rust and, once you've bought yourself a car, carry out preventive maintenance to stop your car from becoming like those shown here!

▲ *CH1. The flowing 'Coke bottle' lines still pull the eye even today, confirming just how advanced the Mako Shark show car really was. At a distance of a few yards, the car looked (and sounded!) quite presentable, but deeper investigation showed otherwise.*

▲ *CH2. The wheel size had increased from 6 in. in 1967, to 7 in. in 1968 to 8 in. for our car. Note the particularly wide extensions to the front and rear fenders ...*

▲ *CH3. ... and the side exiting exhausts – an option only on 1969 cars and one which was to be deleted during the course of restoration.*

▲ *CH4. This was the year that the 'Stingray' name returned, after a year's sabbatical, albeit as one word, rather than the previous 'Sting Ray'.*

▲ CH5. The folding top was somewhat the worse for wear, though it was fully operational and didn't actually leak.

▼ CH6. The interior was in reasonable condition and the standard seats were OK, if well used.

▲ CH7. The relative simplicity of earlier Corvettes was long gone by 1969 and 'our' model featured a plethora of dials, gauges and warning lights, all anxious to feed the driver totally up-to-date information.

▲ CH9. ... reminds those who need to know of the car's vital statistics: 390 horsepower, 427 ci, 480 lb ft of torque and 10.25:1 compression ratio. It saves answering the same old questions from wide-eyed passengers! Of course, being a pre-1972 car, the figures quoted are gross rather than the later (and lower) SAE.

▲ CH10. The pneumatically opening headlamp doors are legendary for this kind of thing and we were far from astonished to find that we had a 'lazy eye'.

◀ CH8. Owner John Fletcher particularly wanted manual transmission. The meaty shifter has four forward gears at its command, and the legend below it ...

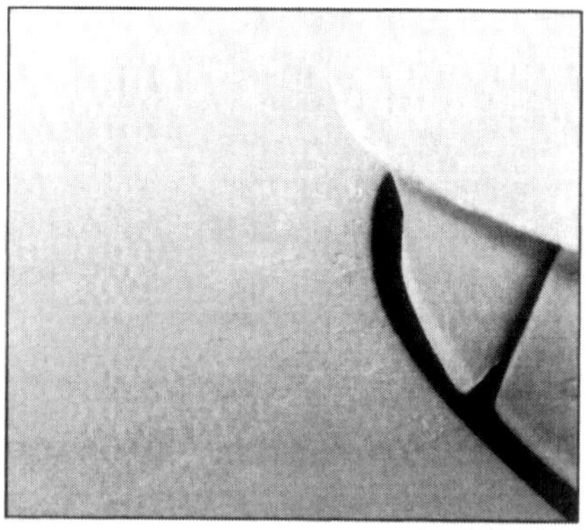

▲ CH11. The fiberglass bodywork can't rust, of course, but it can suffer in other ways, most notably 'crazing', as seen here around the headlamp doors ...

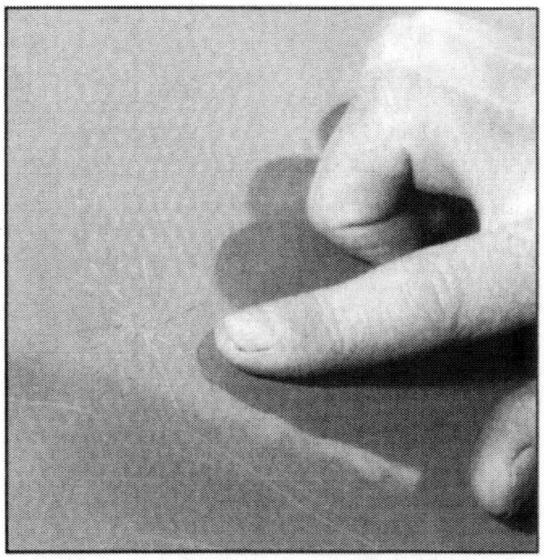

▲ *CH12. … and here on the front fender, where the characteristic 'star' effect can be seen. Look closely and you'll also see a couple of 'envy' scratches made by owners of lesser vehicles!*

▲ *CH13. The stresses and strains of 23 years of opening and closing the fuel tank filler cap show in the fiberglass.*

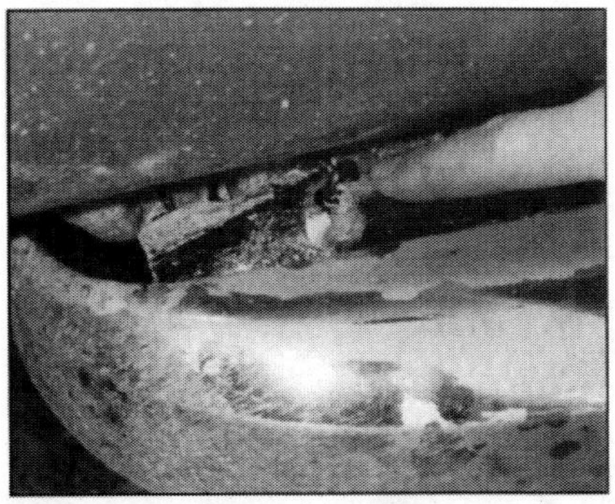

▲ *CH14. fiberglass is also prone to splitting if it is not treated right. Here, by the bumper, it can be seen that the screw fixing hole has split, allowing the screw to work free.*

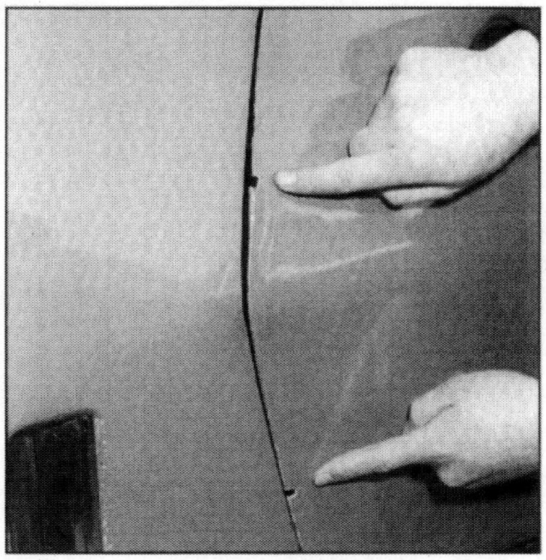

▲ *CH15. The Corvette has long, heavy (steel reinforced) doors which need to be shimmed just right to avoid this kind of nastiness. Adjusted incorrectly, the hinges have bitten into the door in a big way.*

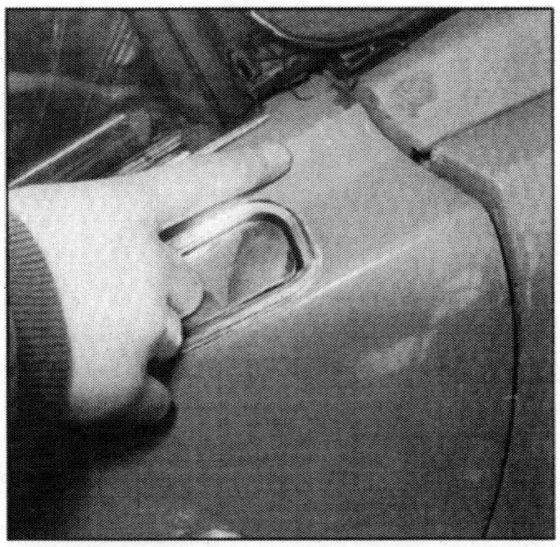

▲ *CH16. Apart from the obvious misalignment, the chips and knocks around the top of the door are numerous and a poor attempt at patching and painting has been made. Note the exterior door opener – it changed from the previous year by deleting the thumb button, opening simply by pressing the depression plate.*

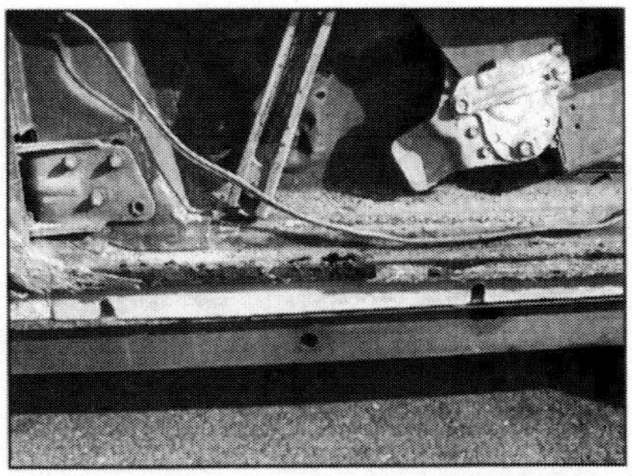

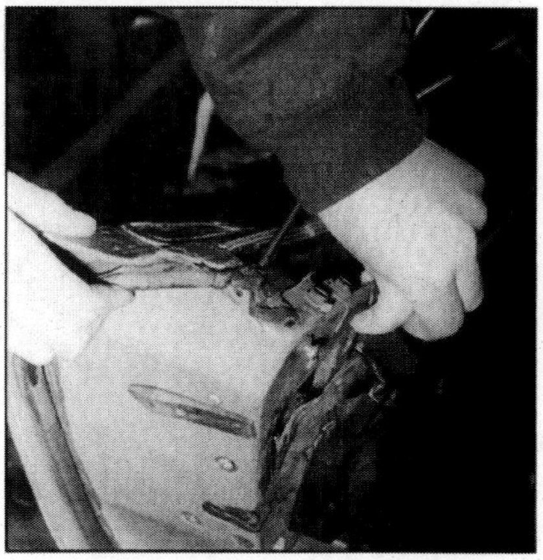

▲ *CH17. Though the inner door trim for '69 cars was revised, over two decades of hard use had led to it working loose on the driver's side.*

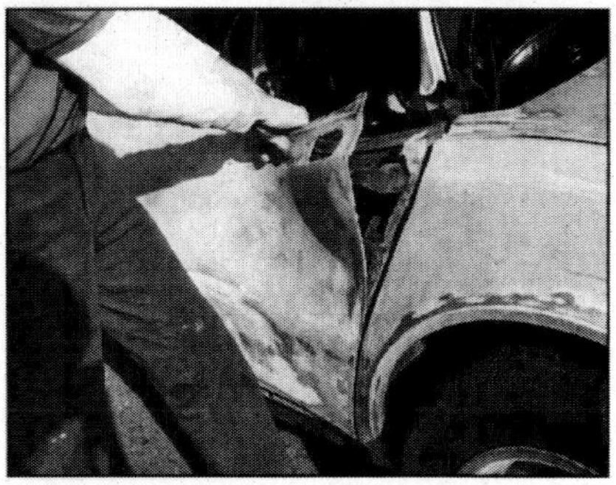

▲ *CH18. Here's the extent you need to go to expose the forbidden subject of Corvette rust. Beneath the ▲ door skin is a steel door made of sheet metal pieces spot-welded together. Although the fiberglass parts can be bought new, the steel parts can't. Open your door and look at the areas around the latch mechanism, especially on 1968–82 models. Even a little blistering of paint warns that the door could already be badly rusted.*

◀ *CH19. With the door skin removed, it is easy to see how far the rust has progressed. The lower lip is completely rotted through in many places. Look on the inside of the door at its lower edge for the lumpy evidence of body filler, which is often used to cover up rust damage. By the way, this was the original door on a car which had an absolutely rust-free original frame. Therefore, even if the frame looks great, check closely for rust elsewhere.*

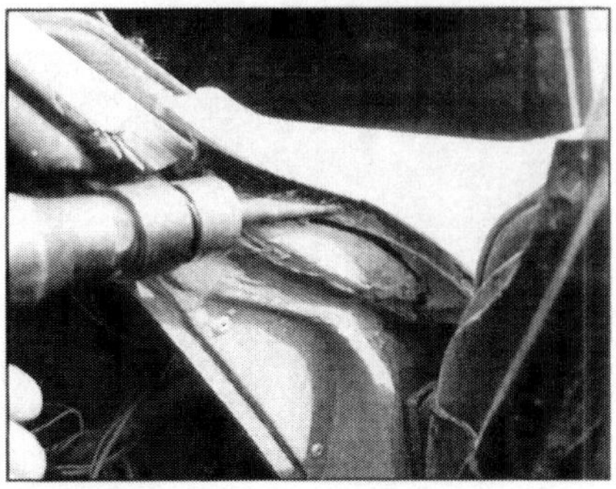

▲ CH20. Another important area to inspect is around the windshield post. Look for blistering paint or rusted-through spots below the weather-stripping (just behind the air-chisel in this shot). Here, we are removing the top panel to expose another rust-prone area.

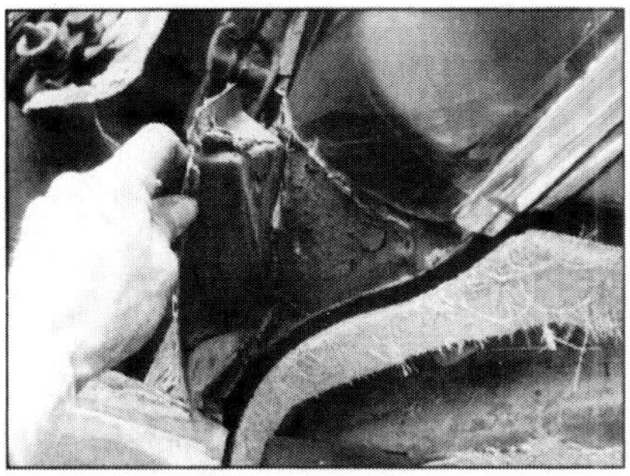

▲ CH21. After the top panel is removed, its fiberglass support is lifted up to show the rust beneath it. Originally, this fiberglass was glued and riveted to the steel portion of the cowl or lower windshield support. After this sealing deteriorated, water seeped in and caused rusting. Once the steel has rusted through, water can drip down so, if you feel water dripping on your foot, this is a likely culprit.

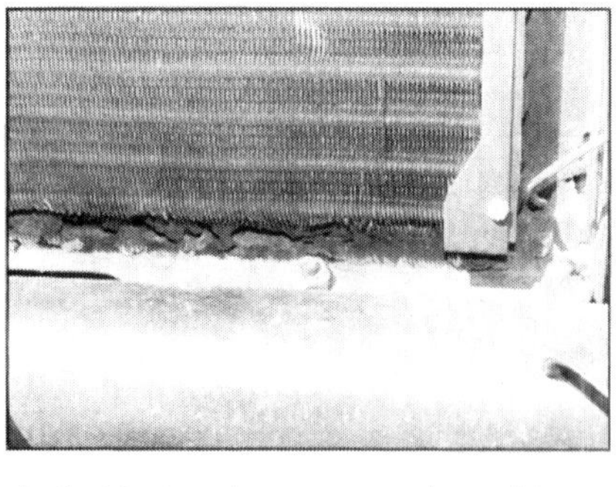

▲ CH22. Another common but seldom-seen-until-it's-too-late rust retreat is the radiator support. The frequent heating and cooling of this part, along with its susceptibility to getting splashed with road spray, accelerates its demise. The boxed construction allows moisture to be drawn inside to create a fertile garden for rust. Again, such parts can be found on a Corvette which otherwise displays a very rust-free frame.

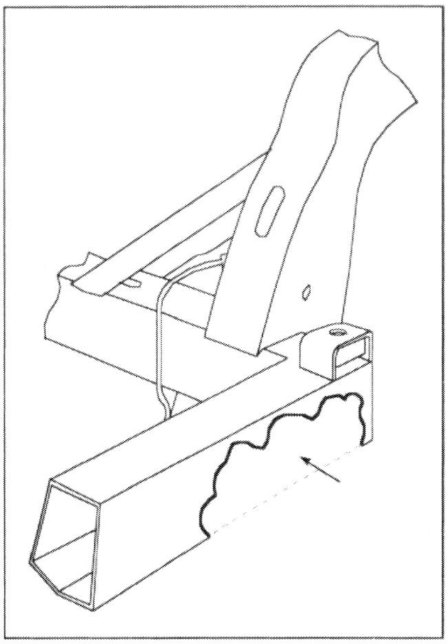

▲ CH23. The most severe rust problem occurs on 1963–82 Corvettes. The area to rust through first is usually the frame side rail just in front of the rear wheel. Make certain to thoroughly inspect this spot. We've seen very high-price, low-mileage, mid-year cars whose frames were dangerously eaten away. A buyer who doesn't look below the body is just asking for trouble!

▲ CH24. Replacement side rails are available in different lengths from aftermarket suppliers. A few other frame components, such as the corner brace and frame mount brackets, are also available.

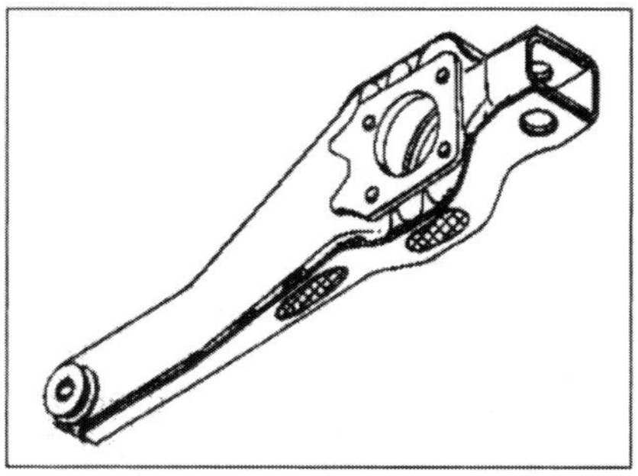

▲ CH25. The trailing arms on 1963–82 Corvettes are also a common and expensive site for rust. The overlapping steel layers trap moisture. If the flat areas look puffy or lumpy, or the steel is beginning to bulge between the skip-welds, beware – it means the rust is pushing the layers apart.

▼ CH26. Some restorers have developed techniques to replace sections of rusted-out side rails without the expense of removing the body.

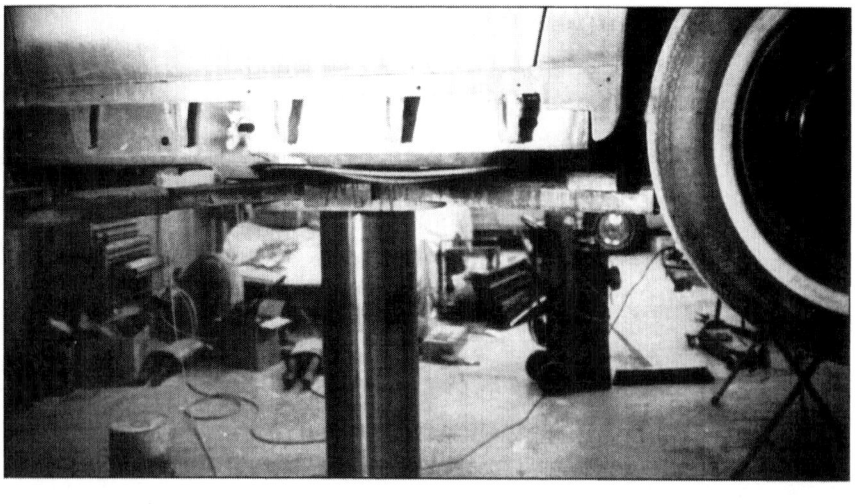

PART 4 CORVETTE BUYER'S CHECK POINTS

Place a tick alongside each 'check point' so you can be sure nothing gets missed!

CHECK NUMBERS

Do they look untampered and original? Do they match.

Body ID:

Engine No:

Trim Tag:

- [ ] Owner Title?

Name, address and phone number of past owner (on title):

BODYWORK

- [ ] Looks square?
- [ ] Nose alignment

Wheel well height:

- [ ] Front left
- [ ] Front right
- [ ] Rear left
- [ ] Rear right

Body surface:

- [ ] No ripples?
- [ ] Visible seams?
- [ ] Sinking beneath paint?

Paintwork:

- [ ] Original color?
- [ ] Overspray?
- [ ] Paint surface problems?
- [ ] color match?

Check gaps and panel levels, around:

- [ ] Front bumper
- [ ] Hood
- [ ] Doors
- [ ] T-top/deck lid
- [ ] Rear bumper

STEELWORK

Check frame for:

- [ ] Undercoat
- [ ] Surface rust
- [ ] Flaking rust
- [ ] Patch repairs
- [ ] Holes
- [ ] Check windshield post

UNDER HOOD AND ENGINE

- [ ] Condition of painted surfaces
- [ ] Condition of chrome

Check for originality:

- [ ] Exhaust
- [ ] Carburettor
- [ ] Intake manifold
- [ ] Air cleaner
- [ ] Check general condition of wire ends and harness
- [ ] Heater hose conditions
- [ ] Drive belts

Check for maintenance:

- [ ] Coolant level
- [ ] Engine oil level
- [ ] Hydraulic fluid
- [ ] Transmission fluid
- [ ] Power steering fluid level

Check engine for condition:

- [ ] Engine smoke at exhaust when revved?
- [ ] Examine inside of oil filler cap
- [ ] Examine coolant for oil in water
- [ ] Check radiator for leaks

Check engine for leaks:

- [ ] Bottom hose area
- [ ] Distributor
- [ ] Valve covers
- [ ] Rear of intake
- [ ] Transmission bell housing

INTERIOR AND TRIM

- [ ] Carpets – check for condition and dryness (damp indicates leaks)
- [ ] Condition of console
- [ ] Left-hand door panel
- [ ] Right-hand door panel
- [ ] Dash panel
- [ ] Left-hand seat
- [ ] Right-hand seat
- [ ] Seals
- [ ] Steering wheel
- [ ] Driver side seat belt
- [ ] Passenger side seat belt
- [ ] Left-hand door locks and latches
- [ ] Right-hand door locks and latches
- [ ] Soft top
- [ ] Soft top mechanism
- [ ] Steering tilt and telescopic
- [ ] Compartment doors, hinges and latches bracket
- [ ] Windshield
- [ ] Radio originality
- [ ] Radio function (all features)
- [ ] Tape player
- [ ] Speakers
- [ ] Power antenna

ROAD TEST

Before starting away check the following:

- ☐ Windshield wipers (check blades, function of different speeds and for hitting)
- ☐ Windshield washers
- ☐ Headlight mechanisms
- ☐ Override switch
- ☐ Light Units
- ☐ All other exterior lights
- ☐ Turn signals
- ☐ Dash lights/dimmer
- ☐ Interior light

Check gauges:

- ☐ Tachometer
- ☐ Water temperature
- ☐ Oil pressure
- ☐ Volt or amp meter
- ☐ Clock
- ☐ Fuel

NB: It is most important that all of the tests described here are carried out in good weather, in broad daylight and well away from people, animals and other traffic. Even though you are paying attention to the car on test, don't have an accident by letting your concentration waver away from other road users. Ensure that you are insured for driving that particular vehicle, or let the owner do all the driving.

Check with the car ready to move and then on the move:

- ☐ Does park brake hold car on incline?
- ☐ Manual transmission jumps out of gear?
- ☐ Shifts smoothly
- ☐ No clutch slip?
- ☐ Auto-transmission shifts smoothly?
- ☐ Down shifts quickly
- ☐ Thumps or clunks when shifting?
- ☐ Air conditioning
- ☐ Speedometer
- ☐ Odometer

Braking:

- ☐ Pulls to one side?
- ☐ Squealing?
- ☐ Pedal feels too hard?
- ☐ With car stationary again pedal sinks slowly to floor when pushed

At normal driving speeds:

- ☐ Wind noise?
- ☐ Vibration?
- ☐ Does steering self-centre itself after a turn? (it should!)
- ☐ Wheel shimmies

Do wheels rub, turning tight circles?:

- ☐ Turning left
- ☐ Turning right
- ☐ Excessive power steering noise, turn from side to side
- ☐ After the road test and after the car has stood for several minutes, will engine restart?

EXTRA CHECKS TO CARRY OUT – USE A LIFT FOR SOME, IF AVAILABLE

- ☐ Spare tire inflated
- ☐ Good tread
- ☐ Side walls good, both sides
- ☐ Jack present and effective
- ☐ Exhaust good all through?

Check tires:

- ☐ Uneven wear?
- ☐ Bulges on both side walls
- ☐ Splits surface cracks or any other damage

Mechanical checks:

- ☐ Check driveshaft universal joints
- ☐ Brake rotors
- ☐ Brake fluid on calipers?
- ☐ Front suspension bushings
- ☐ Idler arm
- ☐ Condition of tie rods
- ☐ Wheel bearings
- ☐ Bent, rusty or damaged suspension components

Chapter 3
Bodywork and Frame

FRONT END

BUMPERS AND BUMPER SUB-ASSEMBLY DISMANTLING

▲ *FE1. Over the years, less and less exterior chromework was to be found on Corvettes, so it's a point of honour for owners of most older cars to ensure that their chrome sparkles reflectively wherever possible. But there will probably come a time when no amount of polish and elbow grease can restore that once new shine.*

As you might expect on a car some 24 years old, the (original) chromework on John Fletcher's '69 model left something to be desired. This meant that, having removed the relevant trim, it had to be rechromed or replaced altogether. In the latter case, it would mean purchasing new trim from a company such as Auto Accessories of America. Which you choose depends on how important originality is to you and how you choose to define it.

▲ *FE2. We removed the license plate and bracket and then the two screws that secure the parking light on a bracket.*

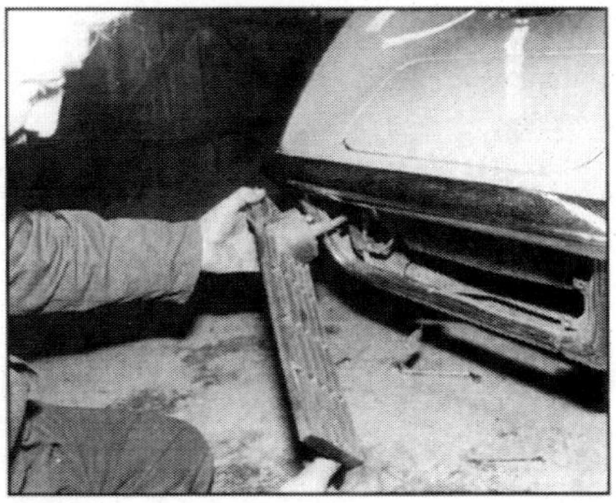

▲ *FE3. Three screws hold on the two outer grilles. Having unplugged the parking light, the grille/light assembly can then be pulled away.*

▲ *FE4. Here, with the grille removed, you can see the parking light bracket. Note the parking light connector hanging down. Wherever you have wires like this it's a good idea to label the wires and plug to make reassembly easier.*

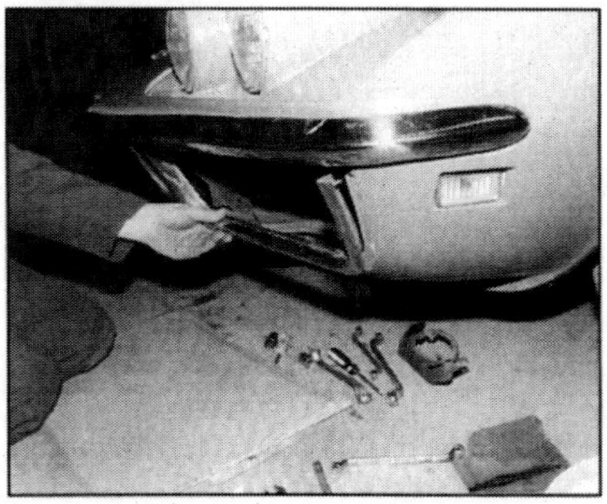

▲ *FE5. We removed the L-shaped chrome moulding from around the outer grilles and the chrome guards at either end.*

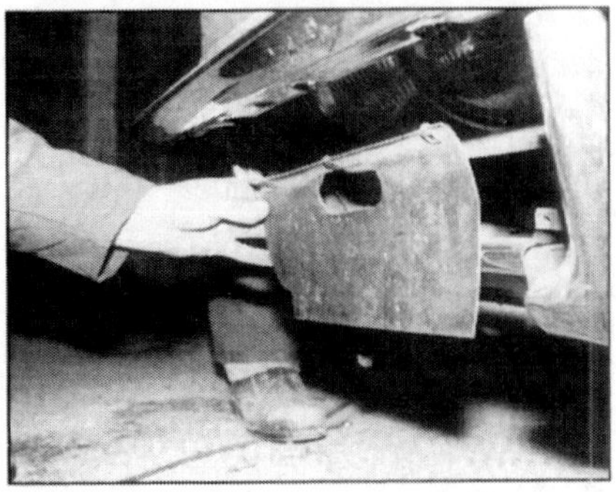

▲ FE6. This rubber flap is fitted to protect the headlamps (when in the closed position) from stone chip damage.

▲ FE7. The bumper securing bolts were a little reluctant to leave home (not surprising after more than two decades!) and so plenty of penetrating oil was called for.

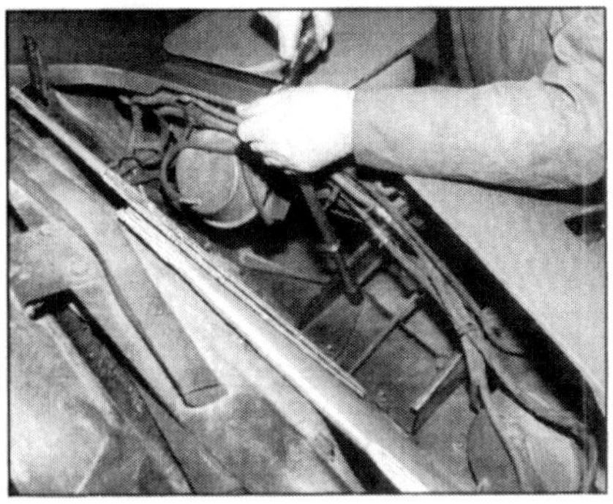

▲ FE8. The centre front brace is secured to the bumper cross-member by four bolts. It's not a necessity, but access in our case was made considerably simpler by having already removed the hood.

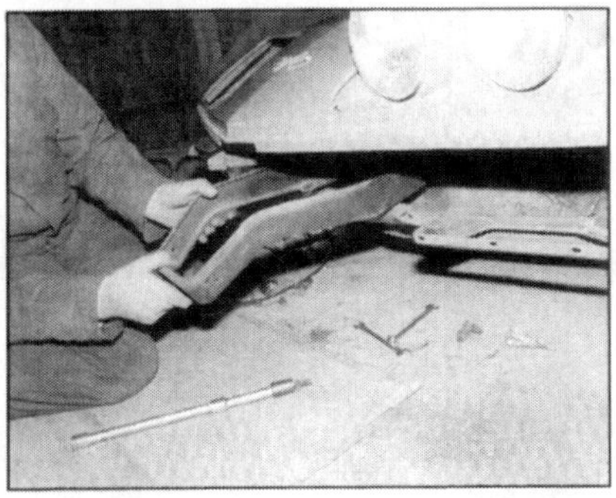

▲ FE9. It was then pulled out and away.

▲ FE10. This bracket links the bumper …

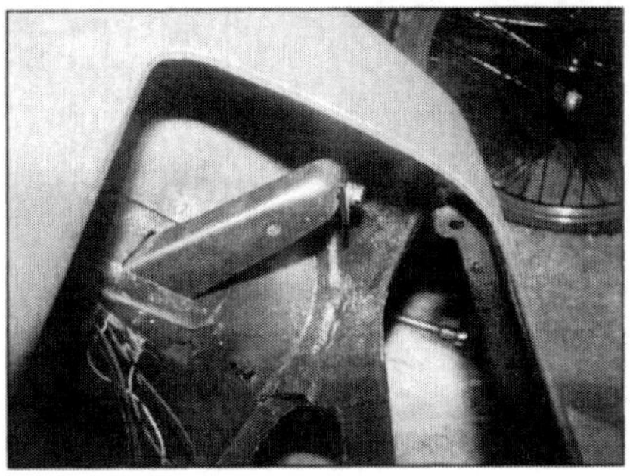

▲ FE11. … to the frame-to-bumper brace. This in turn is bolted …

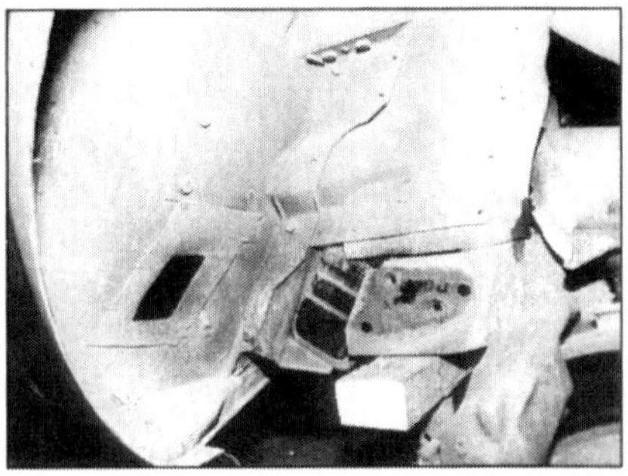

▲ FE12. …to the bumper-to-frame extension and it is this last item which is attached …

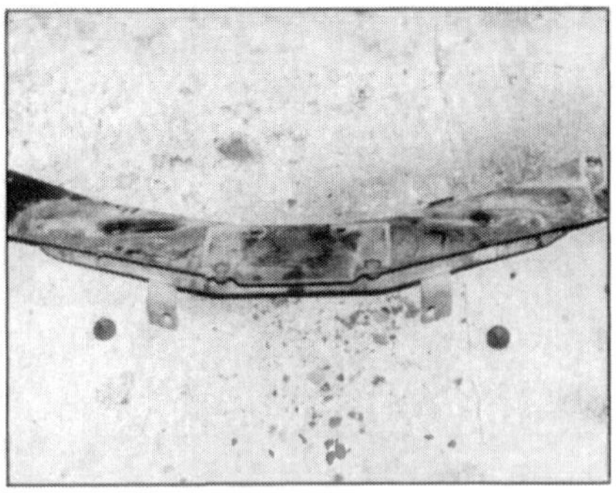

▲ FE13. … to both the bumper cross-member, seen here removed from the car …

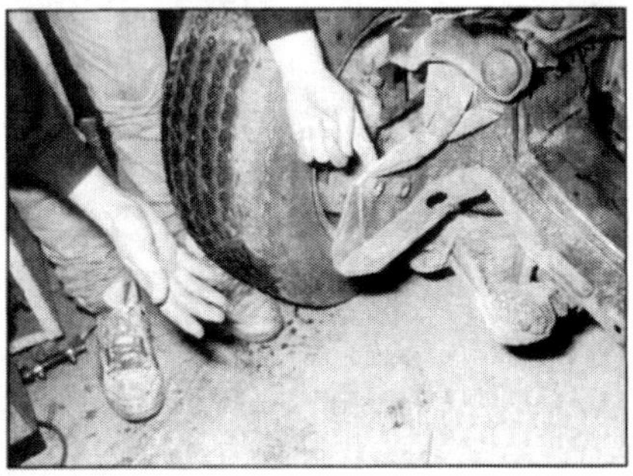

▲ FE14. … and the 'horns' of the frame itself. As you can see, there was plenty of dirt and grime, even though it had been hosed down prior to tear-down.

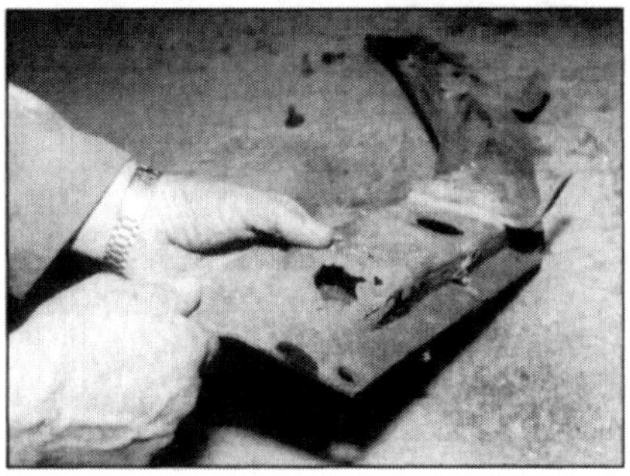

▲ FE15. The long pointed nose of the '68–'82 Corvettes (echoing the lines of the Mako Shark) gave the car plenty of overhang at the front. This resulted in the need for a bumper sub assembly which bolts onto the front frame 'horns'. As with all metal components, this is prone to corrosion over a period of time, as will be seen in this section. Exposure to the elements causes rust, and any caked-on mud just forms a poultice enabling the rust bug to bite even quicker and harder than usual. We found that the bumper cross-member had suffered in a big way.

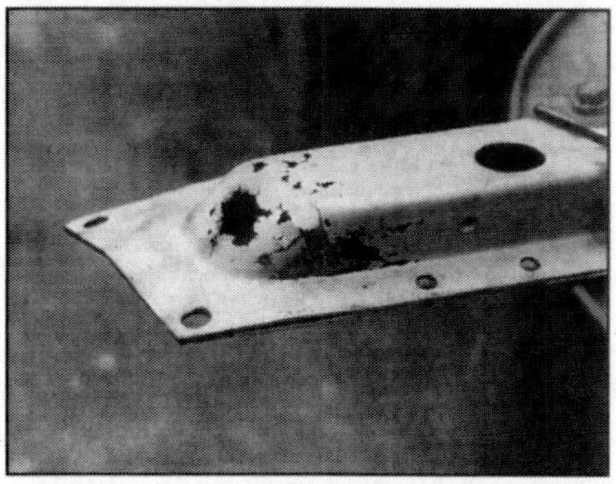

▲ FE16. The severe rot on the underside was even worse than it looked after Em had sand-blasted the end. (Note that it had been sprayed with primer straight away after sand-blasting – which is how soon rust starts to form again on a freshly blasted metal surface.) The original hole was surrounded by a latticework of smaller ones.

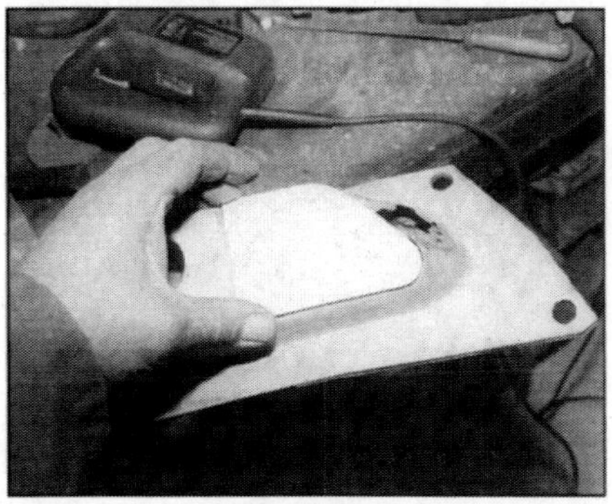

▲ FE17. Em worked in three sections – here, cutting away the rotten metal and then cutting a patch to suit the new 'hole'.

▲ FE18. He welded the patch in, as shown here ...

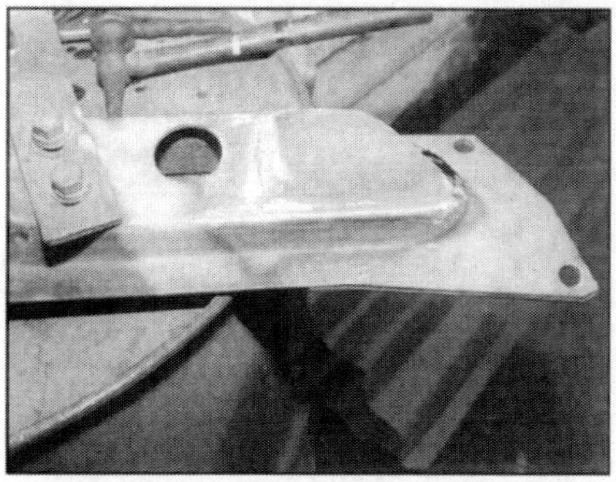

▲ FE19. ... and then power sanded off the rough welds to make a smooth surface.

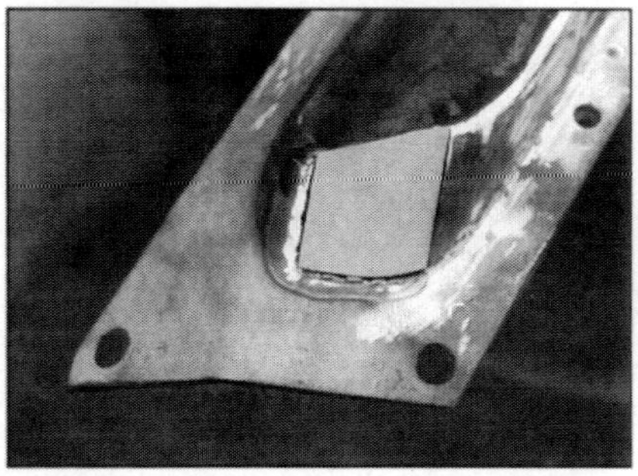

▲ FE20. This process was repeated at the side of the problem area and at the front, where another patch was required ...

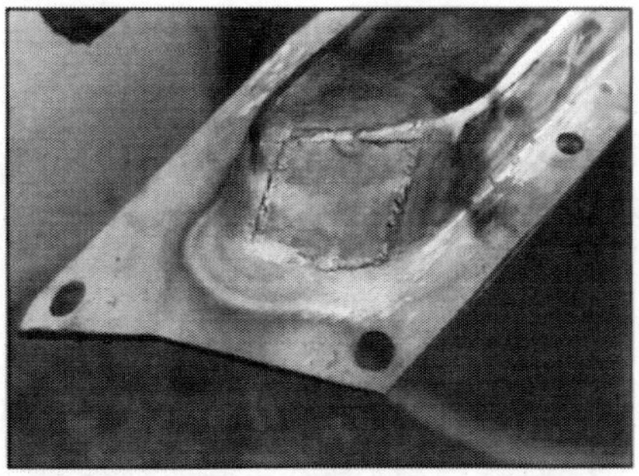

▲ FE21. ... followed by more welding ...

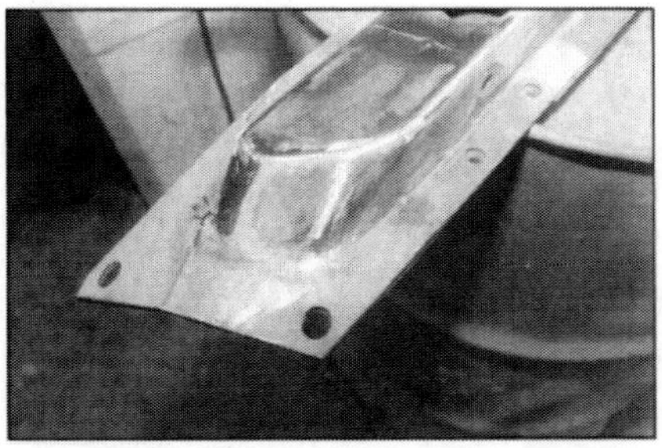

▲ FE22. ... and more sanding. That wasn't the end of the problems on this particular hollow section.

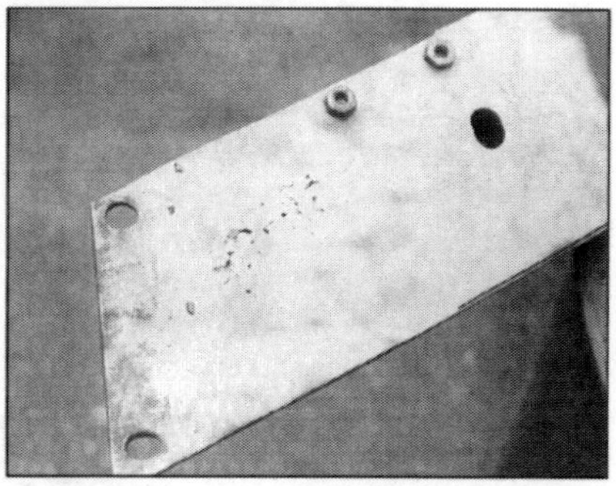

▲ FE23. On the top side of the cross-member these few holes indicated that all was not well.

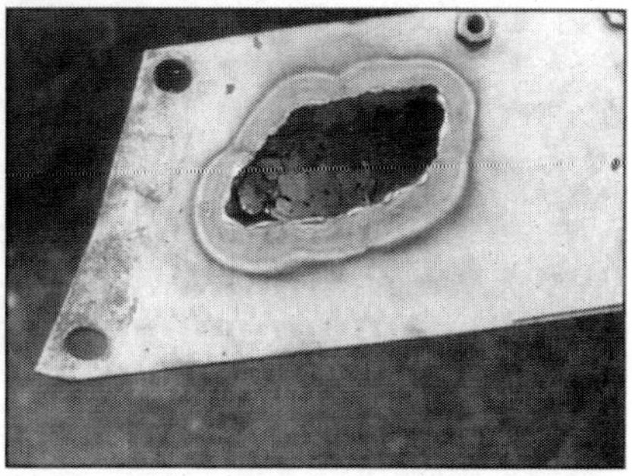

▲ FE24. Investigation showed that there was rot aplenty, and a sizable hole had to be cut away in order to weld to good metal all round.

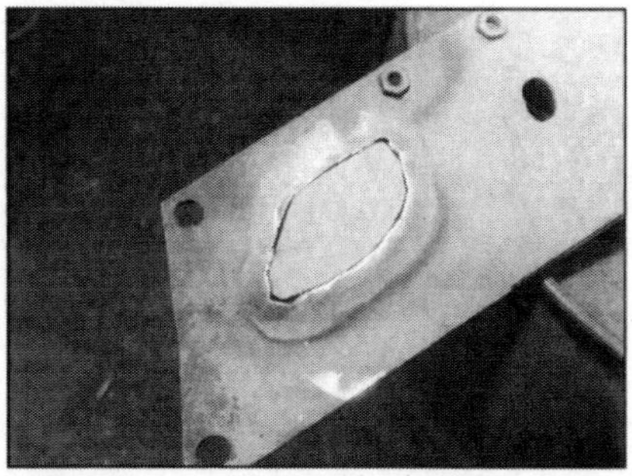

▲ FE25. Once more, a patch was cut to size ...

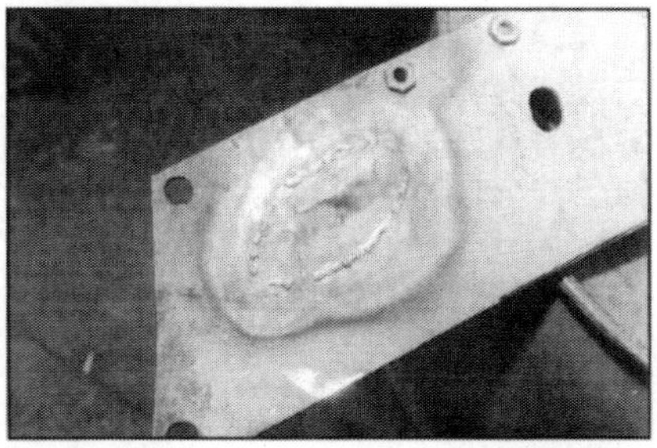

▲ FE26. ... and welded into place. Em ground the surface flush and sprayed the whole cross-member with primer.

▶ FE27. *It's impossible to believe that this is the same item that we saw just a few photos ago, but it is!*

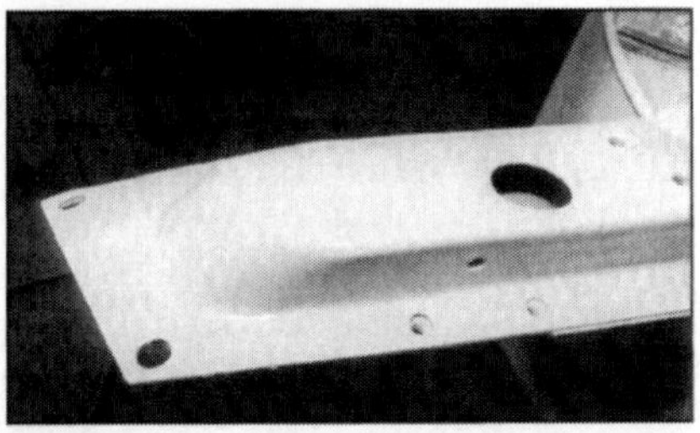

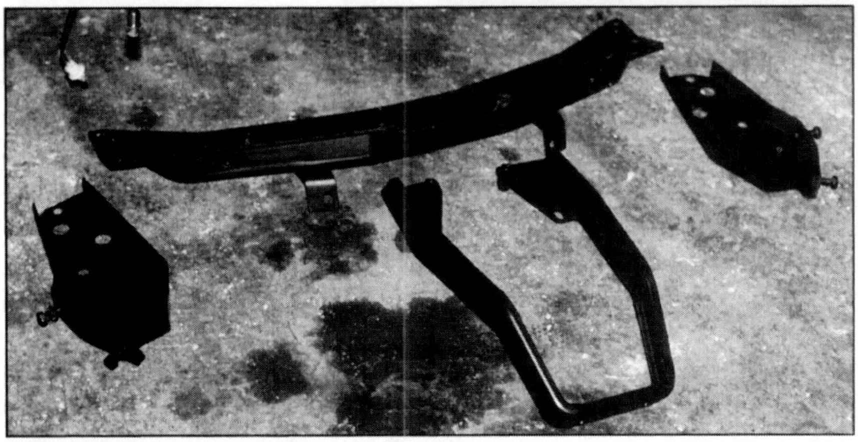

▲ FE28. *The other items in this particular sub-assembly were also thoroughly cleaned, shot-blasted and primed before being given a final coating of hard-wearing black paint. Left to right you can see: the passenger side bumper to frame extension, the revitalized cross-member (with brackets fitted), the centre front brace and the driver's side bumper-to-frame extension.*

▲ HA3. *You can see why so many owners remove the hood when they're working on their engines – there's so much more room! If you're working on your engine for extended periods of time (major overhaul, etc.) and don't want to remove the hood, then (as a safety measure) you should secure it by locking the support assembly using a bolt and nut through the existing hole to prevent accidental closure.*

HOOD ASSEMBLY

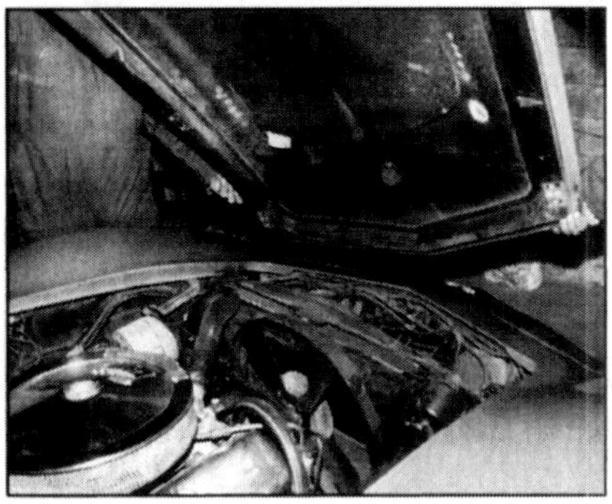

▲ HA1. *When the screws retaining the hinge straps and support assembly to the hood have been removed, the hood itself can be lifted away – it's a heavy item and really requires two people if you're going to avoid damaging your car or the hood. Make a note of how many shims (if any) are used. If you're just removing the hood for better engine access, scribe a line around the hood hinge upper straps to aid installation.*

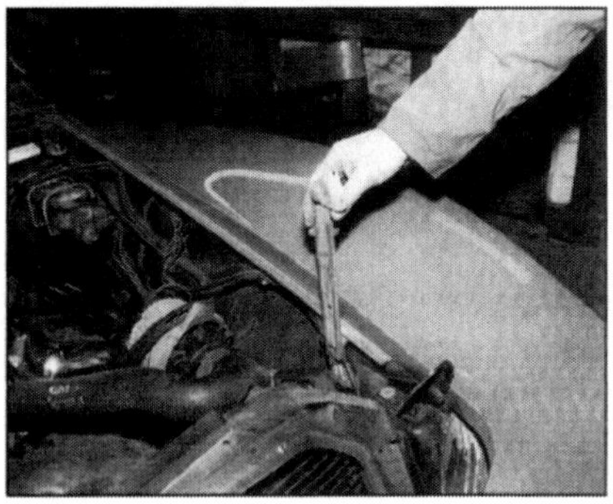

▲ HA2. *The hood support assembly is secured to the body by two screws.*

▲ HA4. *The hood bumper assembly incorporates the hood release mechanism and cable etc. Take great care to note what goes where at this point – make notes or take photos to aid reassembly.*

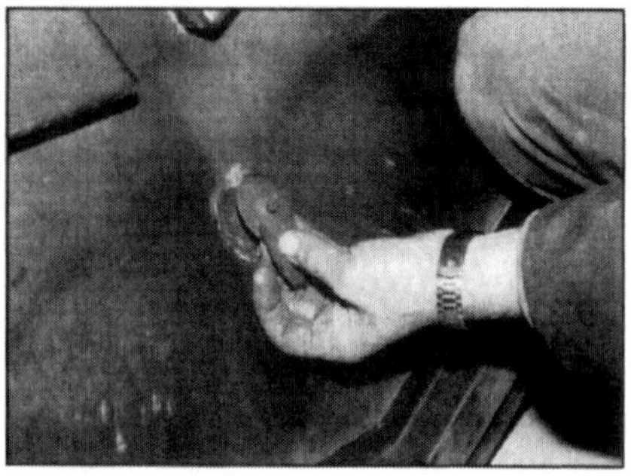

▲ HA5. *The hood badge is secured by a single screw through its centre …*

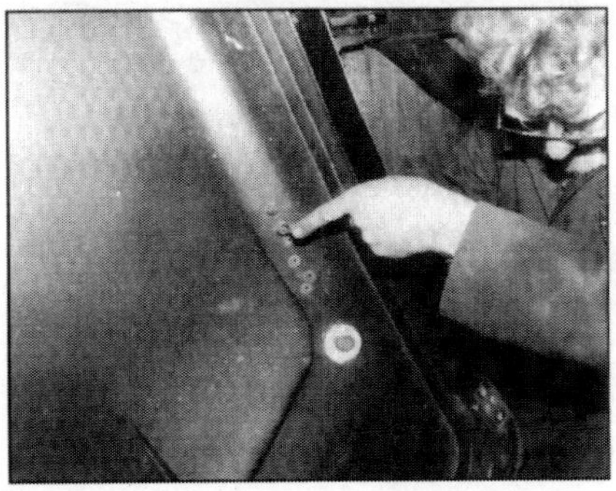

▲ HA6. ... and the hood badges, denoting the engine size are secured by a series of nuts.

Note: The front badge, including flags, numbers and moulding inserts, are retained in the front of the bodywork (i.e. not in the hood itself) by special nuts which may be reached by raising the hood and reaching down under it.

▶ HA9. ... which fits into the offside lock catchplate. As it is operated, the cable mechanism is released.

▶ HA10. This operates the lock on the nearside catchplate. It sounds complex, but it is relatively simple. The catchplates are held by four bolts and the cable passes through two clips to secure it to the hood.

▲ HA7. The weatherstrip was pulled away. Replacement weatherstrips are available from companies such as Auto Accessories of America, so there's no need to be too careful if yours is worn – you can buy a replacement. It helps to apply a little heat to soften the adhesive – a hair dryer will do the job, but don't soften the paint by holding the hair dryer too close or for too long!

▲ HA11. There's far more to the hood than meets the eye! All the metal components were found to be in good order and were placed in the pile for sandblasting and plating.

▲ HA8. The hood lock assembly with cable – again, take a long hard look at what goes where. The hood is operated by a cable from inside the car which pulls down a spring-loaded plunger ...

▶ HA12. Without a doubt, this part is fresh back from the platers! Here, Em is connecting the hood release cable to the nearside catchplate ...

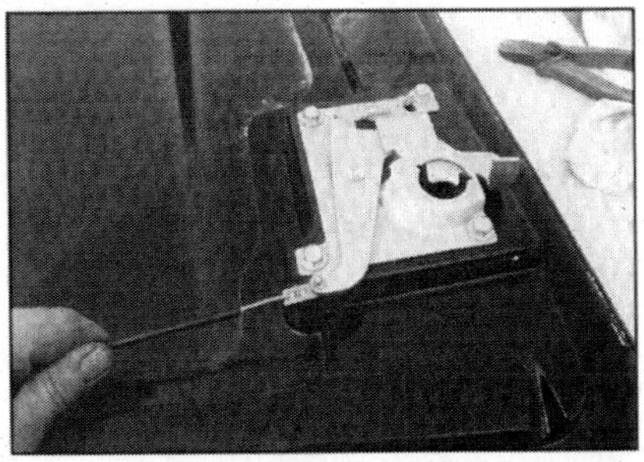

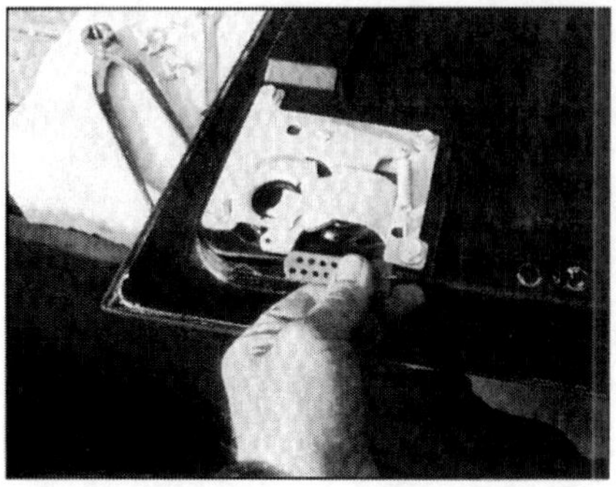

▲ *HA13. ... and then at the offside catchplate. He applied a smidgen of grease to all exposed parts of the cable and the moving parts of the various hinges and mechanisms. The position of the hood in relation to the hood body opening and fender surfaces is determined by the position of the hinges and position and adjustment of catch and bumpers. The fore and aft adjustment of the hood is achieved by loosening the lower hinge retaining bolts and moving the hood to the limit of the slotted holes in the hinge strap. The upper surface of the hood can be brought flush with adjacent surfaces by placing shims under the hinges and adjusting the bumper screws and catch bolt engagement.*

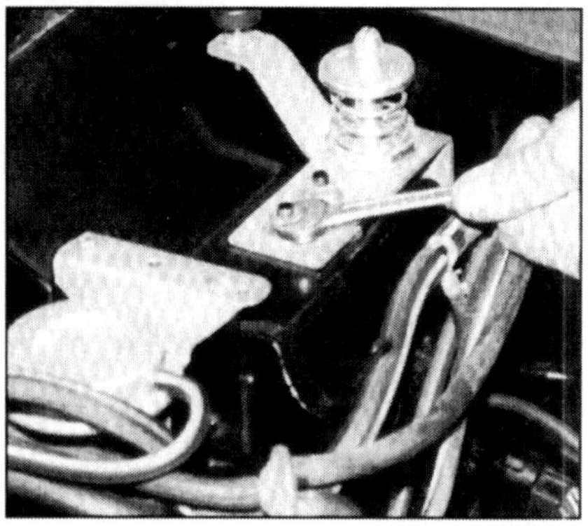

▲ *HA14. The entrance of the catch bolt into the plate may be adjusted by loosening two catch bolt mounting screws to allow movement of the bolt assembly.*

Synchronization of the catch release may be made by adjusting the cable retainer located at the offside catch bolt assembly. Once the hood mechanism has been reassembled correctly, the moving parts and cables should be given a light coating of grease to ensure they stay that way. It's a job to put on your three month tune-up list – it takes just a few minutes to complete and could save you endless hassle.

REAR END TEAR-DOWN

FUEL TANK, REAR LAMPS, BUMPERS AND TRIM

As the back of your Corvette is the only view that many motorists will see (as you drive by), it pays to take care of it! The '68 models were the last to feature four tail lights with separate reversing lights under the rear bumper. Our '69 car still has four lamps on the tail, but the inner two are combined reflectors and reversing lamps. The earlier set-up still has a great following, and watching those four lamps disappear down the boulevard on a warm, summer evening can still evoke some great feelings of nostalgia.

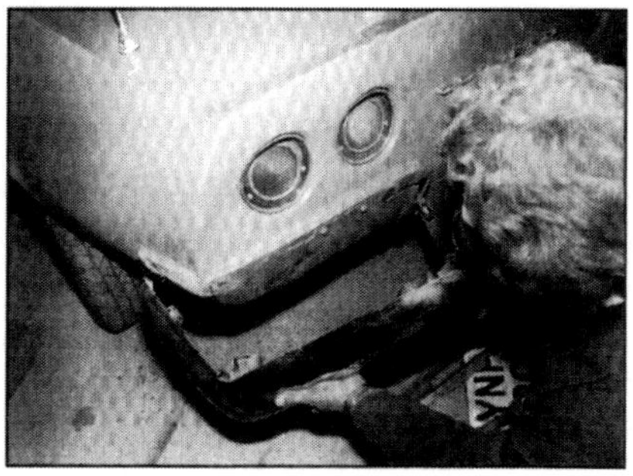

▲ *RE1. Removing the rear quarter bumpers on this model of Corvette should have been simple – in theory at least. However, as Em soon found, 20 years is more than enough time for bolts to rust solid in their respective threads. Plenty of penetrating fluid, applied each day for several days ahead of doing the job, will make all the difference where frozen threads are concerned.*

Both bumpers are bolted through the rear fender onto braces and struts, which are in turn fixed to the frame. Em started with the two outboard bolts, both of which are fixed to the same brace, the second inboard bolt fixes to a second brace and the furthest inboard bolt is fastened to a rod.

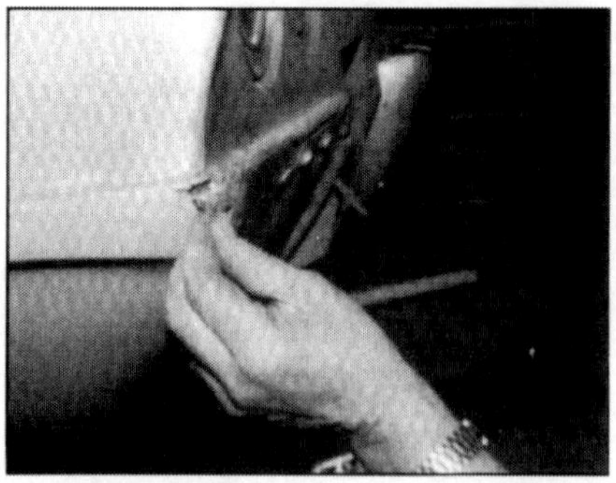

▲ *RE2. With the bumpers off, there's an opportunity to check the condition of the fiberglass. It tends to split and crack around the edges and the holes where the bolts pass through, especially if a previous owner has been a little careless at some point.*

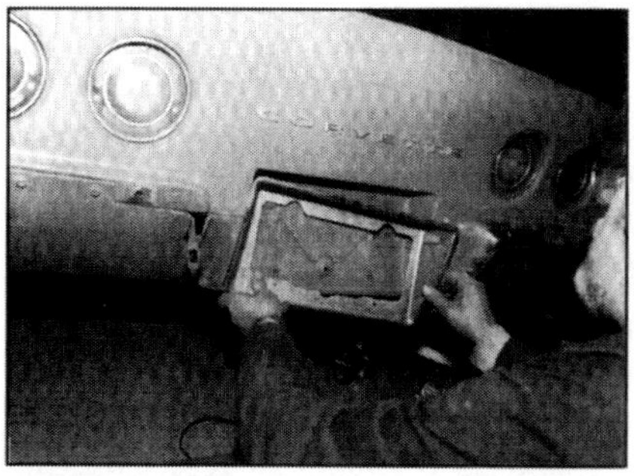

▲ *RE3. This license plate is secured by four screws, two horizontal and two vertical. With these removed, the plate can be pulled away.*

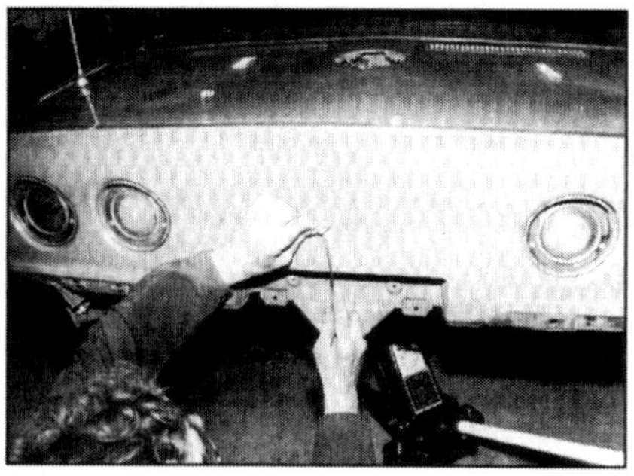

▲ *RE4. The name 'Corvette' has been denoted in various ways over the years – upper case letters, upper and lower case letters, script etc. On our car it is spelled out in upper case letters which push fit into their respective holes and are secured by washers and nuts at the rear. Em found it necessary to prize them free using a flat-bladed screwdriver – naturally, he took great care not to damage the fiberglass.*

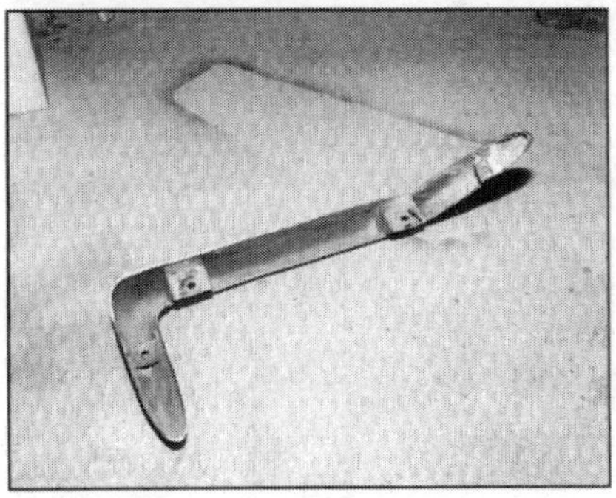

▲ RE5. Plenty of rust and the dirt of many years. The bumpers were placed ...

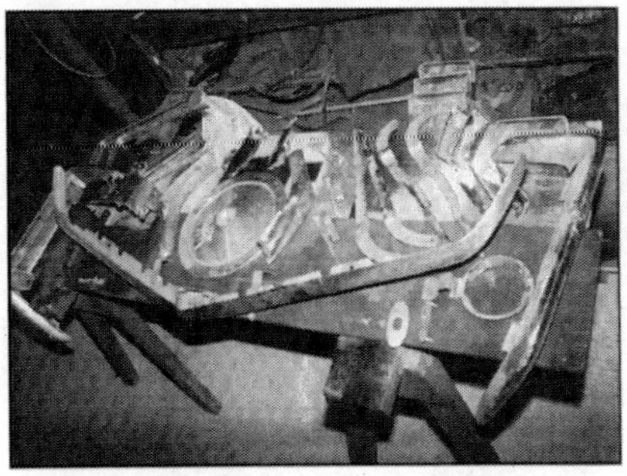

▲ RE6. ... in the ever-growing pile of ancillary equipment for rechroming. Most chrome parts are available as aftermarket items, though when you're dealing with large numbers of items (as here) it's generally more cost-effective to replate. A word of warning here! Only use a rechrome specialist who's recommended to you. A bad one may be more of a specialist in wrecking or losing parts.

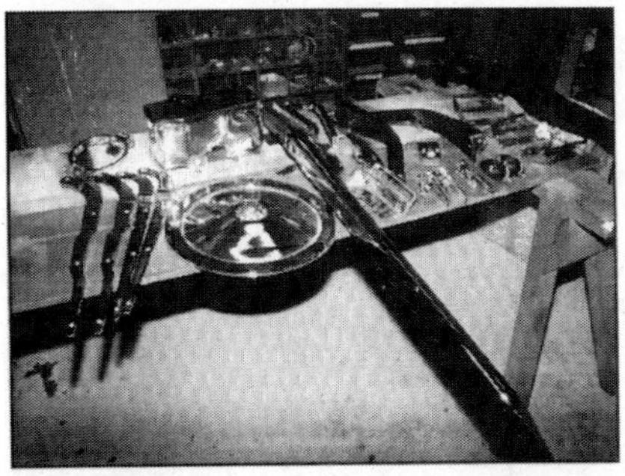

▲ RE7. A bit of difference in the 'shine' factor, here! Fresh back from the platers are several parts of the car, including the bumpers, the air cleaner top and the seat hinges.

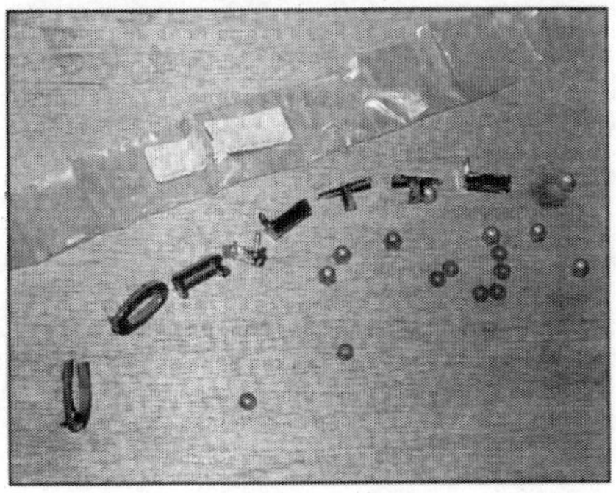

▲ RE8. Em found that the rear Corvette lettering was too far gone to bother with, so Auto Accessories of America supplied some shiny new letters. All you need is to be able to spell!

▲ RE9. The chromework around the fuel tank filler door and bezel assembly had suffered in the same way as most of the rest on the car.

▲ RE10. Em removed the tank itself prior to body removal. This demanded prior removal of the spare tire carrier bolted attachments. Once the assembly has been lowered, as here, it can be lifted away allowing access to the fuel tank.

FUEL TANK WARNING: Motor fuel is *extremely* explosive and flammable, so take extra precautions when you work on any part of the fuel system. Don't smoke or allow open flames, sparks or bare light bulbs near the work area, and don't work in a garage where a heating appliance of any type is present. If you spill any fuel on your skin, rinse it off immediately with soap and water. When you do any kind of work on the fuel system, wear impervious gloves and safety glasses, and have a suitable type of workshop fire extinguisher to hand. Disconnect the battery before working on the fuel tank.

▲ RE11. It's essential that the tank is empty, not just of fuel, but of the fumes as well – these can linger long after you think it's safe, and they are actually more dangerous than the fuel itself. A strict no-smoking ban was imposed and no naked lights of any description were allowed. The fuel tank is located between the frame rails to the rear of the rear axle. As a 1969 model, this car did not have the evaporation control system. The fuel level metering and fuel pick-up pipe assembly is located on the bottom surface of the tank. Before removing the tank, the fuel must be siphoned out into an approved motor fuel container. The boot from the filler neck has to be removed and the drain line disconnected. Note that we have already raised the rear of the car using a high quality trolley jack. This was left in place, even though the vehicle was also being supported by strong axle stands – you simply can't be too careful!

▶ *RE12. The tank is secured by two straps which pass around it, linking it to the frame via a fuel tank support member.*

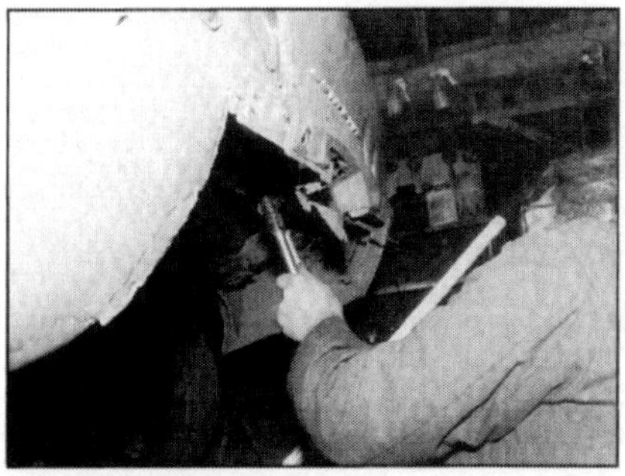

▶ *RE13. The support itself is held by a bolt at each end.*

▼ *RE14. With these removed the support and the rubber anti-squeak fitting can be drawn away.*

▶ *RE15. The anti squeak rubber is glued to the support, though it shouldn't be too difficult to remove. Given its position in life and the hassle involved to get to it, it's advisable to replace this as a matter of course.*

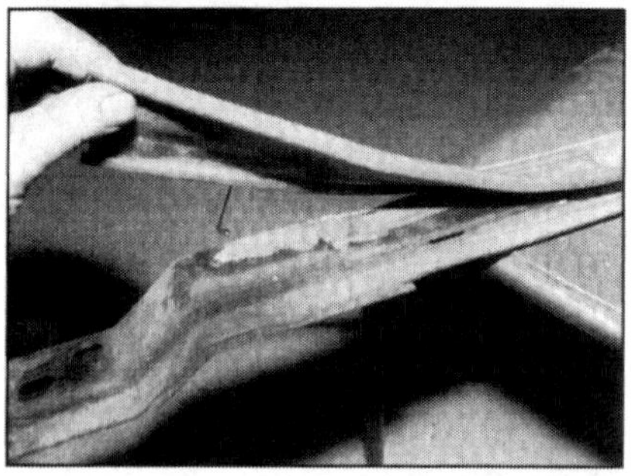

▶ *RE16. With the support out of the way, the fuel tank itself was removed. Before storing the tank – out of doors – seal off all the openings, so that neither moisture (or sparks!) can possibly get in, nor fumes get out.*

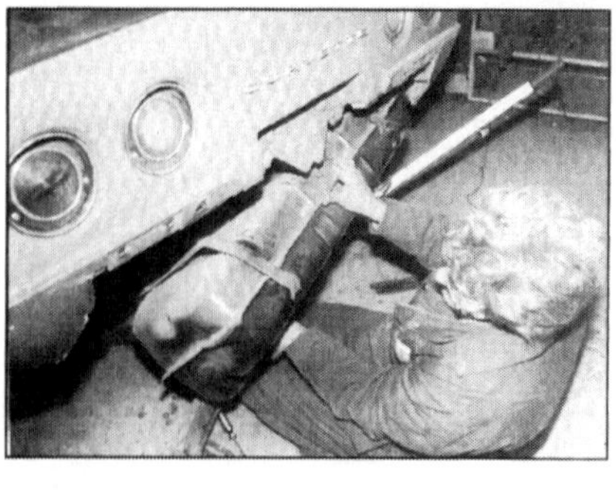

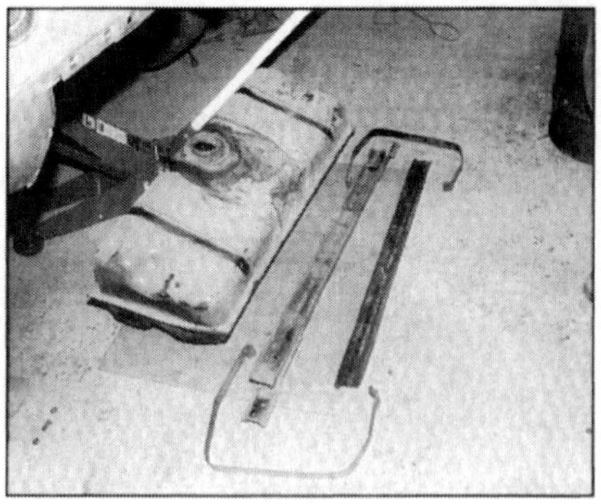

▲ *RE17. This photo shows more clearly the tank, straps, support and anti-squeak rubber. Note that removing the fuel metering unit demands the use of a special tool (GM Ref. J-23346) to undo the cam. It's hard and potentially dangerous without it. Pay a specialist or GM dealer to do it for you if you don't have access to the tool.*

▲ *RE18. It's important to make sure that the inner vent hole is clear before replacing the cap, otherwise a vacuum will form and no fuel will reach the engine. Replacement of the fuel tank is a reversal of the dismantling procedure, though we did take the opportunity to fit the new anti-squeak rubber mentioned earlier.*

▲ *RE19. Our car featured rear side marker lamps which are secured by two screws, accessed from behind the fenders.*

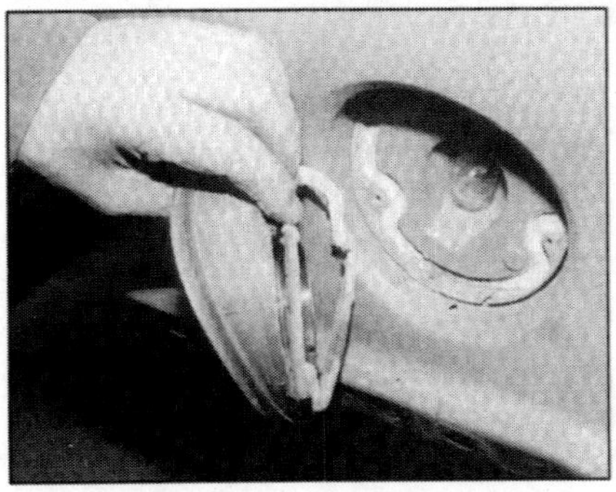

▲ RE20. The four rear lenses are secured by long screws which pass through the lens into the lamp assembly. With these removed, the lenses can be eased away along with the gasket. Sometimes the gasket will cause the lens to stick; in which case a gentle tap with the flat of your hand should dislodge it.

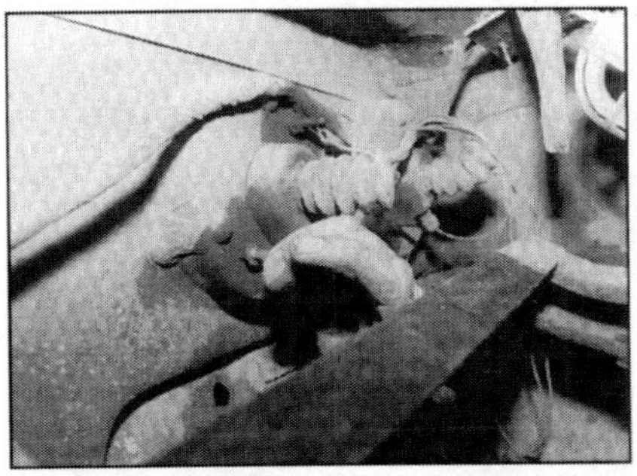

▲ RE21. The lamp assemblies are secured to the bodywork by means of nuts fitted over three captive studs in the assemblies. As with the fitments on the side marker lamp, you're likely to need plenty of penetrating oil and patience. Make sure you get all the encrusted dirt off first, and use a wire brush to clean up the visible threads.

▲ RE22. Having removed the lamp assemblies (including the license plate lamp), Em removed the wiring harness. Now is the time to use tags of masking tape and a biro to mark what goes where.

▲ RE23. The lamp assemblies were prepared for sandblasting. Note the masking effect of the gaskets which were installed on the production line before the bodies were sprayed.

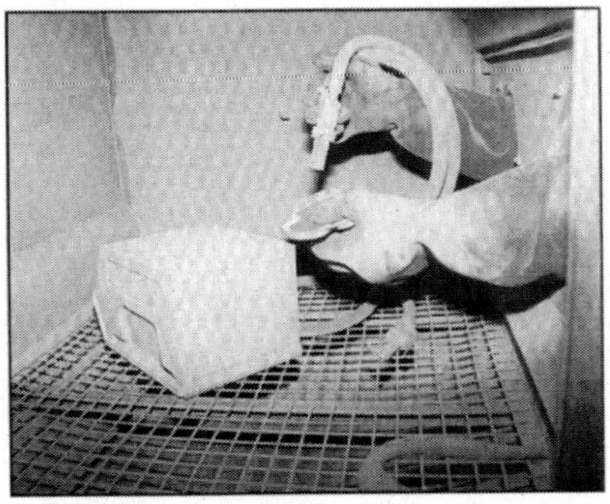

▲ RE24. However, the paint didn't last long once the assemblies were introduced to the joys of the sandblasting cabinet ...

▲ RE25. ... and when they emerged, and had been replated, the difference was obvious. Seen here are other items blasted at the same time, including various shims, bolts and the license plate lamp assembly.

▲ RE26. Cor, it's a 'Vette! Part way through the job of fitting the new chrome lettering.

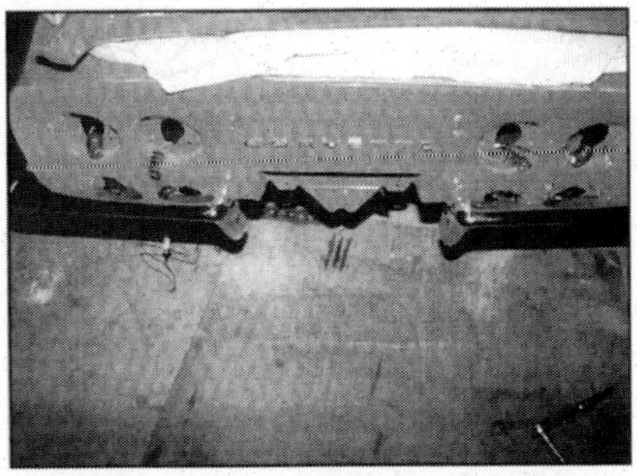

▲ RE27. Looking better already, with the newly chromed bumpers and 'Corvette' logo fitted.

▲ RE28. New gaskets are essential when you're replacing your rear lamps.

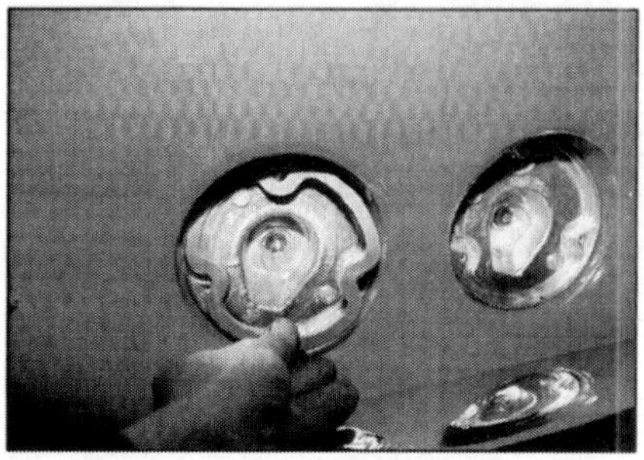

▲ *RE29. Once the lamp assemblies have been replaced, the gaskets follow. Em made sure they sat square and that they didn't foul the screw holes during the process of fitting the lenses, which in our case ...*

▲ *RE30. ... was a set of four replacement lenses. These tend to dull over the years (certainly over 20 years!) and it's usually a false economy to use the old ones. If you do re-use original lenses, clean the insides with soap and water and polish the outsides with a mildly abrasive paint polish.*

▲ *RE31. The finished job. With a recent paint job and the new chrome, it's just like it rolled off the production line.*

DOORS AND HINGES

The door hinges and springs have a stressful time in Corvettes – the doors are long and heavy, the hinges are close together and there's a lot of strain on them.

Your Corvette door hinges include springs designed to hold the doors open. When the springs fail to work, the door will close back on you – and it's irritating! While replacing door springs isn't the most difficult task, it is time-consuming and replacing the entire hinge assembly can be costly. It can be avoided altogether by taking just a few minutes a year to apply some preventive maintenance. Apply oil to the hinges and grease to the hinge spring rollers. Carry this out when you buy a Corvette, and annually thereafter, as mentioned.

Em planned to take off the doors and strip them down as part of the restoration project. The condition of the hinges and springs was checked and care was taken to identify and store the many and various fastenings so that there would be no difficulty in replacing them.

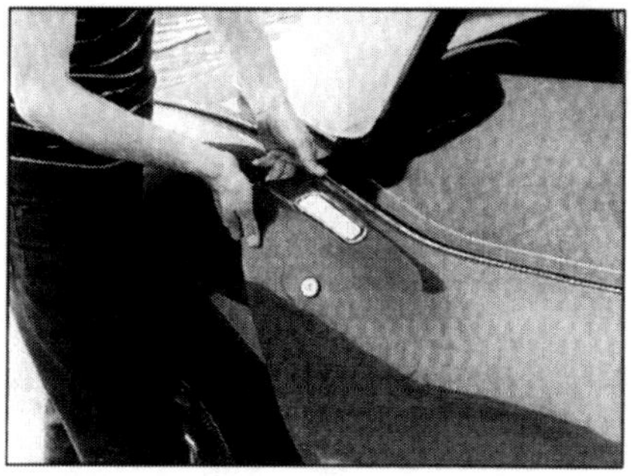

▲ *D1. First things first – find out whether your doors have got excessive wear in the hinges. Take hold of the door at the rear top, hold it just open and then pull up. You will probably feel excessive wear, and you will probably see it, because as the door sags down it will drag and bind when it shuts. Better yet, have an assistant lift and lower the door while you look at the hinge and see if there is evidence of wear there. Also look for any cracks in the hinge itself because that will also require replacement.*

▲ *D2. If the hinge has excessive play but is basically sound, you can get away with changing the brass bushing and the hinge pin. The pin can be tapped out of the hinge, but note carefully that the head of the hinge has a flat spot which allows the head of the hinge pin to clear the fiberglass as it comes out. Turn it if necessary.*

▲ *D3. Broken door hinge springs are common but they are easy to change. Don't be fooled by a hinge spring like this one that looks fine: look again and you will see that it is not held tight against the hinge and it is actually sitting up too far. Anyway, you will know if the spring isn't working as the door keeps shutting back on you! Just reach in and pull the spring up and out before replacing it with a new one.*

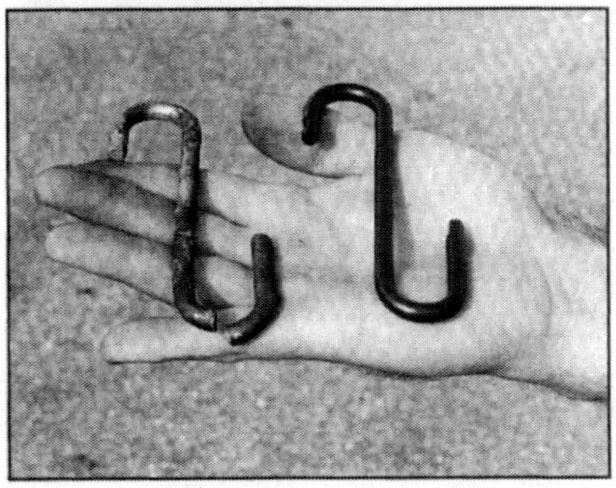

▲ D4. *You must always grease the rollers and the spring, as shown here, whether fitting a new spring or if the original spring is still OK. Lack of lubrication causes the spring to jam and bend the wrong way, typically breaking the bottom so the lower portion just falls off. A little grease saves a lot of hassle!*

IMPORTANT NOTE: You must always work with a careful assistant when disconnecting one or both hinges because the door must never be allowed to hang with its weight on one hinge only, or far worse damage could well occur.

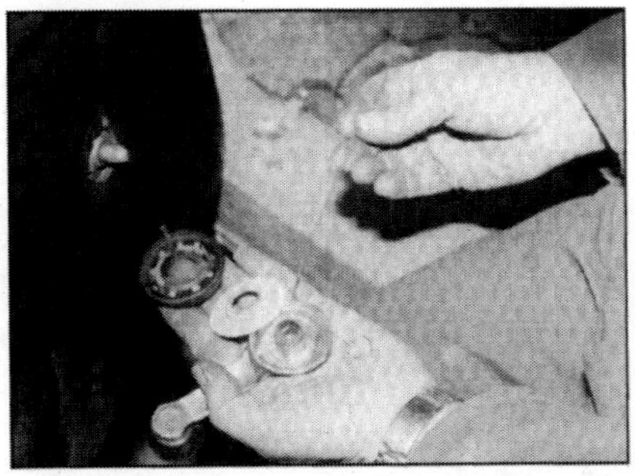

▲ D5. *We show removing the doors for a thorough overhaul and paint job, but even if you're just fixing springs, you'll still have to remove the trim panel in order to gain access to the hinge bolts. The GM special tool (Ref. J-7797) is required for removing the clips retaining the window crank and lock control, although this can be purchased from specialist companies, such as Auto Accessories of America. For what it costs, it's not worth trying to do the job any other way. Make sure that you store the various nuts, washers and springs carefully and in such a way that you know in which order to replace them. Just two screws secure the grab handle.*

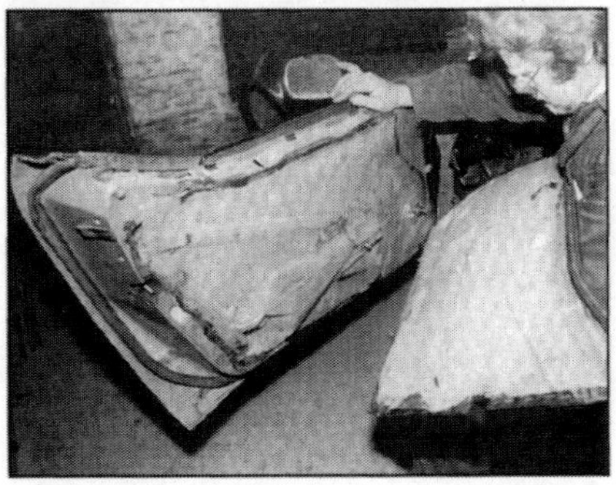

▲ D6. *We removed the four screws securing the trim panel, which are located at the inside corners of the inner door panel. After prizing out the plastic fastener clips located around the perimeter of the panel, the trim panel was eased away.*

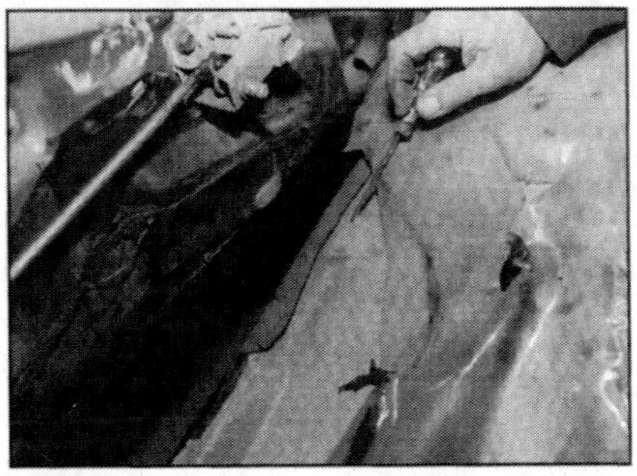

▲ D7. *This photo shows the fastener clips along the lower edge of the panel. Take care not to force them or they will break.*

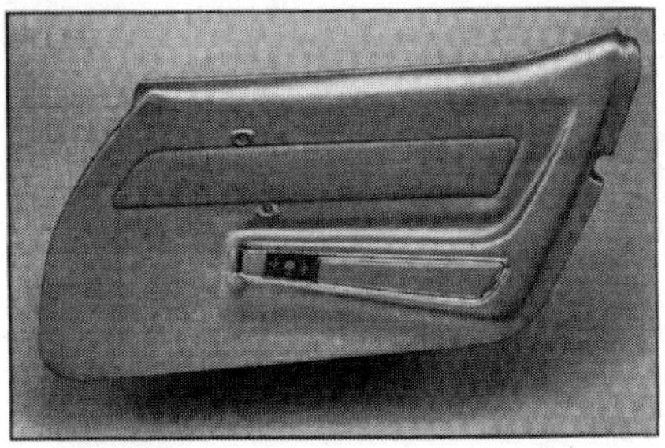

▲ D8. *The door panels on the 1969 cars were improved over those of the previous years, being redesigned to give more shoulder room. They were made from cardboard, foam and a heat-formed vinyl sandwich, but were not the most robust feature of the car. The vinyl was stretched thinnest at the point of maximum flexure and abrasion at the top of the arm rest. Consequently, you're unlikely to find a car of this vintage with good (or even reasonable) quality original door trims. However, this is not too much of a problem as they are readily available from aftermarket suppliers, such as Auto Accessories of America. They can be purchased as complete panels (with knobs and latches etc.) or as just the panel itself.* (Courtesy Auto Accessories of America)

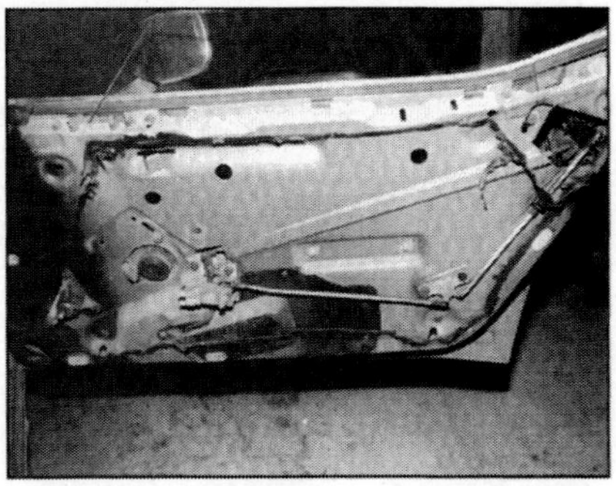

▲ D9. *After removing the protective plastic sheet in one piece, we were left with the door and the window/locking components. The mechanisms are complex and it's advisable to take your time and make plenty of drawings or take plenty of pictures so that you know exactly what goes where.*

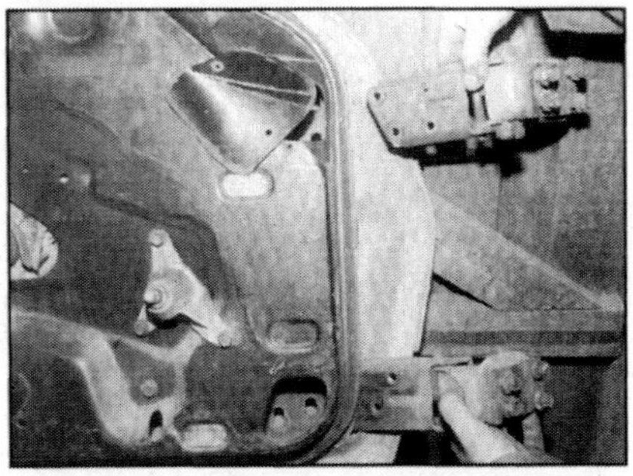

▲ D10. *For access to the lower hinge, remove the rocker panel and the cowl kick panel. For access to the upper hinge, remove the lower mast jacket cover and instrument panel pad (left side) or the dash panel pad (right side) and the air inlet ducts. (On power window models the wiring between the hinge pillar and the door has to be disconnected.)*

IMPORTANT NOTE: If the hinges are to be used again in the same positions, mark around the inner straps on the door panel and record the number of shims between the hinge strap and door panel. Examine the hinges carefully for signs of cracks or damage.

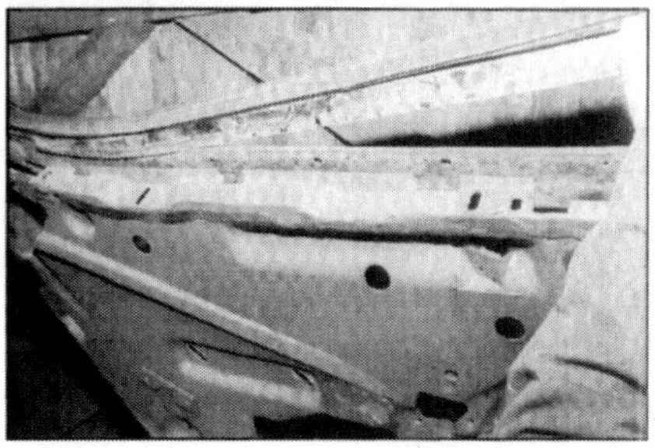

▲ *D11. The removal of the weather-strips, window seals and anti-rattle strips is simply a matter of easing them away from the door.*

▲ *D12. Em removed the window glass by lining up the two sash screws through the access holes. These were removed (holding the nuts to stop them turning) and the Phillips screw at the top of the channel assembly was taken out to disengage the stop. With the front and rear channels adjusted fully outward, the window was pulled carefully upward.*

▶ *D13. The window crank housing is secured by three screws. With this removed, we could then take out the crankshaft, pin and spring washer and remove the two retaining screws adjacent to the spring. With the linkage collapsed to the elongated position, the regulator can be slid back and forth to bring it out of the access hole (shown here).*

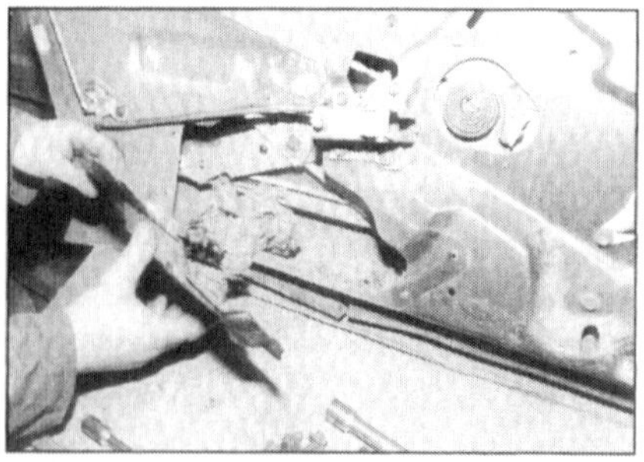

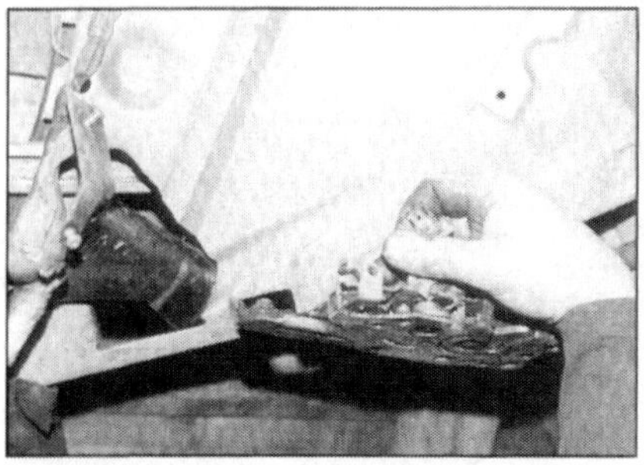

▲ *D14. The door latch and remote control mechanism is particularly complex. Refer to your Haynes manual for exact details of what goes where, and make sure that you don't lose any of those precious nuts, bolts and washers. This is a case for taking photos or even breaking out your video camera.*

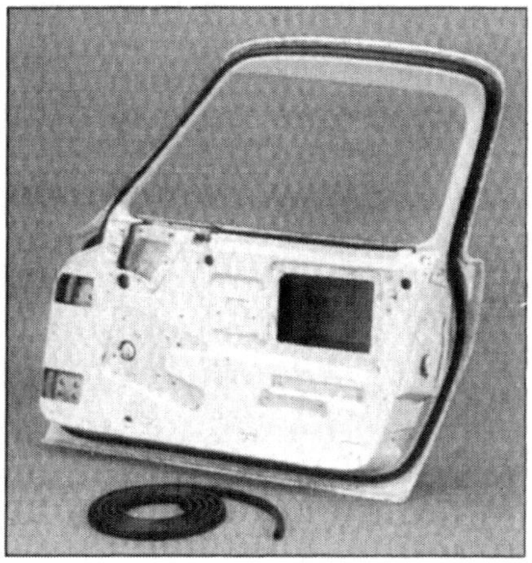

▲ *D15. If your car has had a good few years of use, it's unlikely that the weatherstrips around the doors will be worth saving. New weatherstrip sets are available to suit most years. In many cases, replacements are actually of better quality than the originals, and when you consider the damage that can be done to your Corvette's interior by leaking rainwater, the investment is well worthwhile. (Courtesy Auto Accessories of America)*

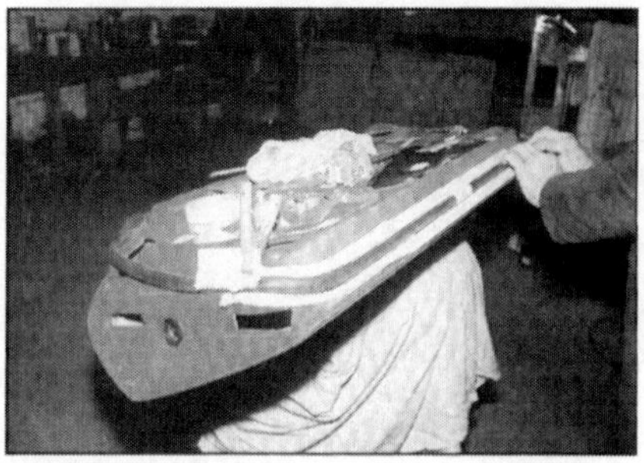

▲ *D16. A weatherstrip's effectiveness is only as good as its installer. Em took great care to get the strip exactly where it should be, holding it in place with clamps (with protection to stop them marking the fiberglass).*

▲ *D17. When the doors were refitted, both they and the body had been professionally painted and the finish was as good as new. So that the rear edges of the front fenders were not damaged during the fitting process, Em used masking tape to protect the vulnerable areas.*

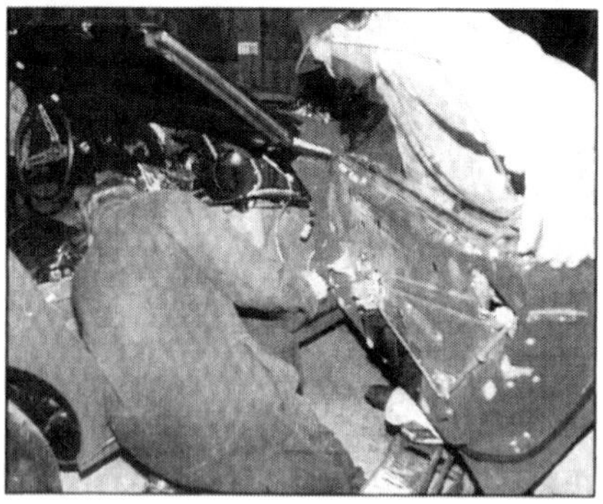

▲ *D18. It is essential that the door is not left to 'hang' on just one hinge or swing loose during hinge-fitting. Patience is required, especially if you have had the body off altogether and where the hinge positions and/or number of shims may have varied from the original.*

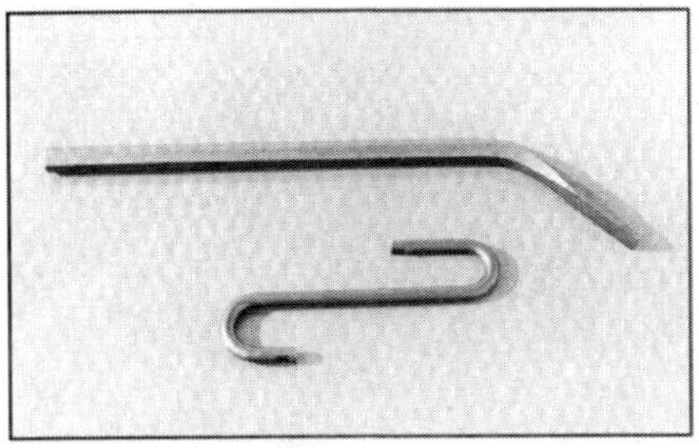

▲ *D19. Fitting a new spring requires an installation tool to twist it as it goes in. Space is tight when installing the spring and the tool makes the job a snap.*

Em put grease into the notches of the hinge instead of on the spring – this made it easier to hold the spring and guide it into place. He put the spring up behind the hinge until it rested in place in the two notches. This spring has to be inserted at a steep angle – don't be fooled into thinking it won't go in! Remembering that the springs are prone to breakage, Em greased the rollers after the new spring was installed. Then he opened and closed the door a few times before packing some more grease between the rollers and spring.

We have mentioned the hinge bushings before. There are two small brass bushings on each one, and it is important to get oil on both of them even though the top bushing is more difficult to see and to get oil to. You will need to squirt oil between the two sections of the hinge to get at this.

▲ *D20. Take your time and get the alignment spot-on. The same applies to the adjustment of the lock striker. Coat the roller with a layer of grease. Note that on this 1969 car, the depression plate triggered the mechanism, unlike the one-year-only thumb button/depression plate system of the previous year.*

ADJUSTING THE DOOR STRIKER

▲ *D21. The door striker is a single metal bolt/washer assembly threaded into a tapped floating cage plate in the body lock pillar. The door is secured in the closed position when the door lock fork-bolt snaps over and engages the striker bolt. The striker can be adjusted up, down, in or out by loosening the striker bolt and moving the striker as required before retightening it.*

Does your striker need adjustment? This is how to tell. When you're sure that the doors are correctly aligned, apply modelling clay (or body caulking compound) to the lock bolt opening, and close the door as far as necessary for the striker bolt to form an impression in the clay.

Don't completely close the door because this makes getting the clay out unnecessarily difficult.

The striker head should be centred fore and aft. Spacers are available from any GM dealer in 5/64 in., 5/32 in., 1/4 in. and 5/16 in. sizes and can be used individually or in specific combinations in order to make up any difference.

Fit any spacers required by using a pencil to mark the position of the striker on the body lock pillar and then remove the striker bolt with a 5/16 in. wrench. Remove the striker, install the spacer(s) and reverse the procedure to reinstall, making certain that the striker aligns with the pencil mark.

IMPORTANT NOTE: Never close a door completely when it has been removed and reinstalled until you have confirmed that the lock fork is engaging correctly with the striker.

FOLDING TOP LID

The folding top compartment lid is a convertible feature. When lifted up, it allows the folding top to emerge from hibernation in order to provide protection against inclement weather. When the sun shines, the top can be folded down again, totally out of sight in the rear compartment and thus not disturbing the smooth lines of the car.

The lid features a complex mechanism of linking rods, springs, and shims. It's very efficient and logical but it pays to take the time to study and record exactly how it works before you start taking it apart. Once more, a camera will help, and even more so, a video camera, which will be able to record what happens as the mechanism is operated. The use of masking tape to mark items as appropriate is often helpful.

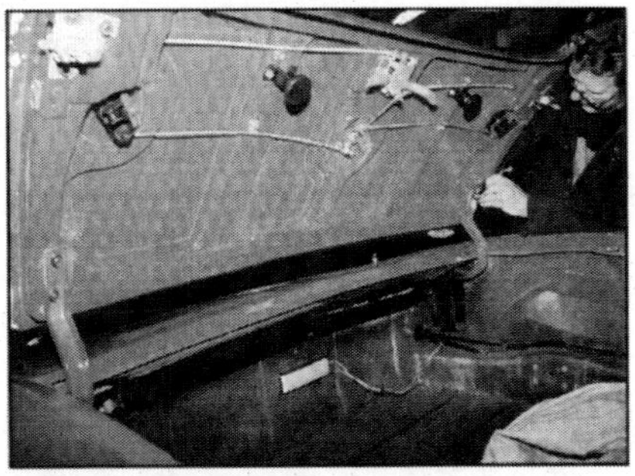

▲ *TL1. This photo of the lid* in situ *shows the lock/latch mechanisms and the various linking rods. The basic premise is that the two handles in the centre of the lid activate two rods which operate latches at either side of the lid and two more which operate locks, just below and slightly inboard of the latches.*

▲ *TL2. Start by removing the two bolts at either side which secure the lid to the hinge assembly. With these removed ...*

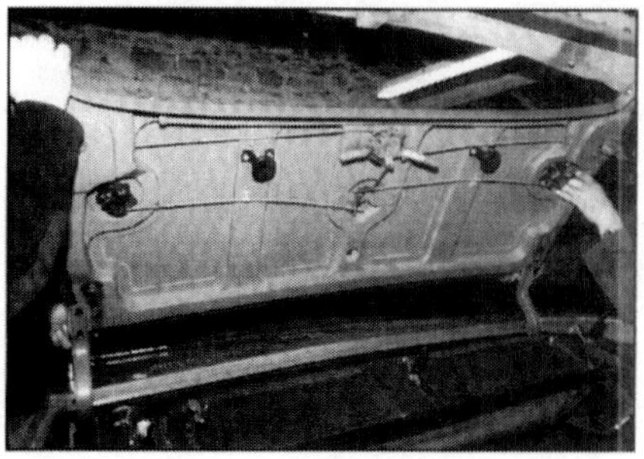

▲ *TL3. … the lid can be lifted off – note that Em had enlisted a second pair of hands for this part. Make sure that the weight of the lid is never taken solely on one hinge, otherwise fiberglass damage will surely occur.*

IMPORTANT NOTE:
Removal/installation of the hinge assembly requires the use of a special tool (Ref. J-9559) to restrain the spring which is very strong. If you don't have this tool, or lack the experience to use it properly, we recommend that this task should be carried out by a Chevrolet dealer service department or other qualified repair shop.

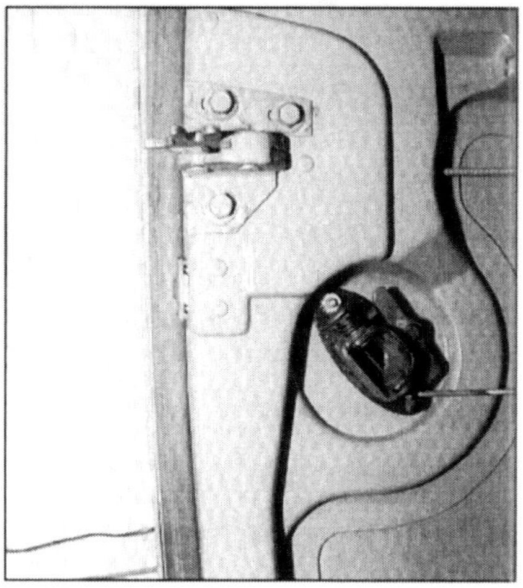

▲ *TL4. This photo shows the nearside top lock and latch mechanisms. To remove the top locks, take off the rod mounting clamps and remove the nut and washer assemblies. Ease the lock out and carefully remove the cover and gasket. The latch should also be disconnected from the operating rod and, if the latch is going to be replaced, a line should be marked around the contour of the latch assembly. Three retaining screws secure the latch and with these removed, it can be taken off the lid.*

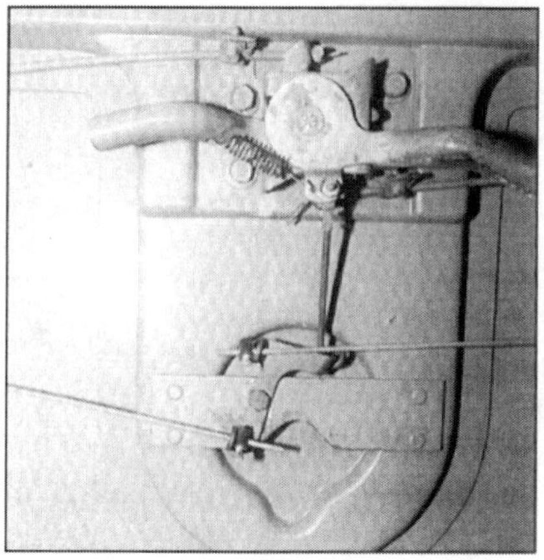

▲ *TL5. The locks and latches are operated by the compartment control. Remove the cable mounting clamps and disengage the rod from the control by removing the retainers. Remove the four screws and detach the control from the compartment lid. The weatherstrip around the lid is glued in place and can be pulled away – use a hot air gun or hair dryer to melt the adhesive where required. The weatherstrips are available as aftermarket products, so it doesn't matter if the strip is damaged in the process.*

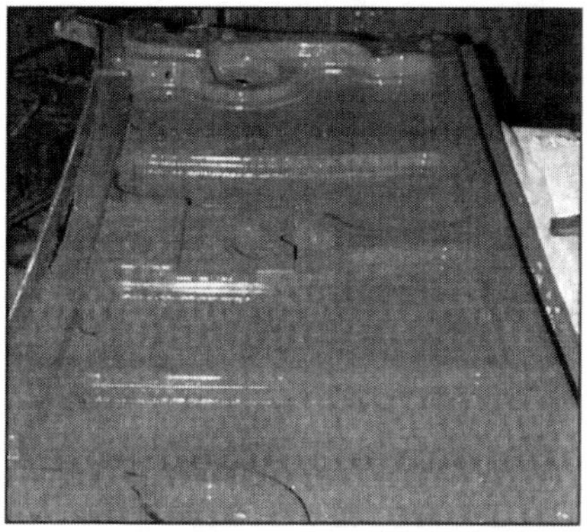

▲ *TL6. Fresh from the spray shop, the project car's 'naked' lid, ready for the refitting of the various components.*

▲ *TL7. Em first of all tried a 'dry run' with the weatherstrip.*

▲ *TL8. When he was happy with the fitment, he secured it in place using masking tape.*

▲ *TL9. The glue was applied and the masking tape left on until the adhesive set, to ensure that the weatherstrip was there to stay!*

▲ *TL10. The compartment lid is correctly adjusted when there is 1/16 in. to 3/16 in. clearance at the sides and 3/16 in. to 1/4 in. clearance at the rear; the surface should be flush with the surrounding panels. Add or remove shims as necessary to raise or lower the top surface of the lid.*

Adjust the screw and bumper to the fully down position. Close the lid, allowing the pin to engage in the bushing and locate the housing. Tighten the nuts.

Tighten the screw until the bumper is firmly seated against the lower surface of the lid assembly. Apply time and patience and the correct number of shims and you'll be rewarded with a flush-fitting lid. The latch striker can be adjusted in or out to the extent of the slotted mounting holes. The latch release is adjusted by loosening the latch assembly retaining bolts and utilizing the fore and aft adjustment slots. The folding top compartment locks are adjusted by loosening/tightening the rod mounting clamps to suit.

WINDSHIELD – REMOVAL AND REPLACEMENT

To remove the original windshield, lock the wiper door in the 'open' position and remove the wiper arms. Then remove the weatherstrips and various trim mouldings as described in the 'Windshield Rust and fiberglass Repair' section, on the following pages, dealing with the welding of the windshield frame.

It is necessary to cut through the original caulking material, the simplest method is to use a piece of piano wire, fitted at each end with pieces of wood to use as handles (as supplied in the GM 'home repairer' windshield replacement kit). You'll need a helper to produce the sawing action required to cut away the material, working up one side of the windshield, across the top, down the other side and finally, across the bottom. At this point, the old glass should simply lift out. Again, that helper can earn his day's pay. Naturally, when dealing with glass, you should take great care. Wear thick leather gloves; don't handle glass when it's wet; wear goggles in case of the risk of shattering glass; store glass safely, away from bare legs, children and animals.

INSTALLATION

The windshield on this car was replaced using a synthetic rubber adhesive caulking material which 'glues' the glass to the pinchweld frame (previously, a rubber channel was used). We had the windshield professionally fitted by Chris at Worcester Windscreens, but it is quite feasible to fit one yourself, given a little confidence. GM sell a kit for just this purpose, comprising a tube of adhesive caulking material/primer/applicator gun, a pressure sensitive adhesive sealing strip, piece of steel piano wire (0.02 in. diameter) and two pieces of wood for handles, black weatherstrip adhesive and rubber glass spacers.

It should be noted, as we said earlier, that whenever you are working with glass, heavy duty gloves must be worn to avoid the risk of injury. Chris is not wearing gloves, but then, it's his job and he's far less likely to make mistakes than the average home installer.

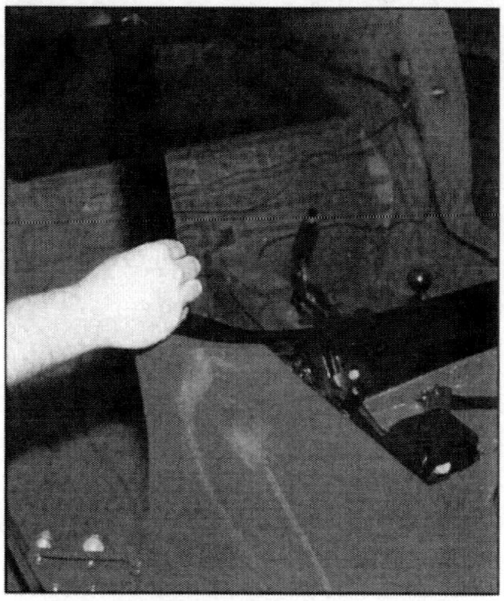

▲ *WS1. Naturally, our windshield frame was considerably cleaner than average – after all, it had just been restored! Even so, Chris made a thorough examination and managed to weed out quite a few bits of scrap material which had accumulated there during the restoration process. Where replacement follows removal, the old caulking material is likely to be the problem. It should be cleaned off as much as possible. It does not have to be perfect, but there should be no loose bits of caulking material around to get in the way. It's important to make sure that the four upper reveal moulding retaining clips are not damaged; if any are found to be bent or distorted, they should be replaced.*

Chris used weatherstrip adhesive to stick the rubber spacers around the window opening at the top, sides and bottom. There are two types of spacers – the flat type go at the top and the sides, and the stepped type go at the bottom.

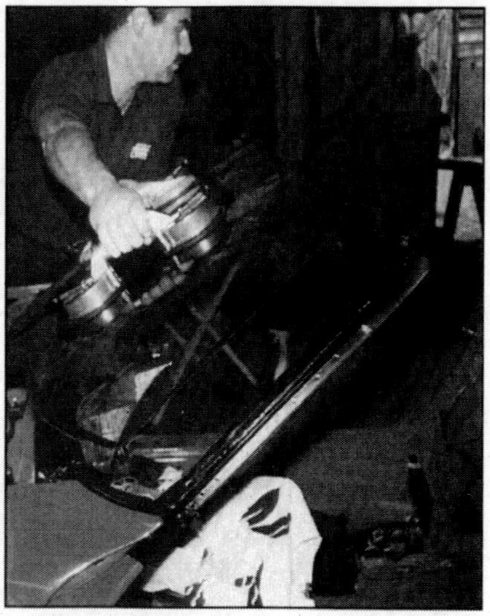

▲ *WS2. Chris used suction cup holders to position the new windshield in the body opening. Most home restorers won't have these, which makes that extra pair of (gloved) hands essential for safe working. He checked that the overlap of the glass to the body pinchweld/retaining flanges was equal and had a minimum of 3⁄16 in.*

In some cases it is necessary to place shims under the lower spacers to achieve the correct overlap. When he was happy, Chris marked the position with a grease pencil on both the windshield itself and the pillar.

▲ *WS3. He then removed the windshield and placed it on his (well-protected) jig. The jig not only served to protect the glass from damage but also to stop it sliding around while he was working on it. The inside edge of the glass has to be scrupulously clean and he ensured there were no traces of grease or oil. He then applied a film of silane primer to the inside edge around the glass and the inner edge of the windshield frame.*

▲ *WS4. Chris mixed the caulking material and accelerator, noting that there is only around 35 minutes working time before the mixture goes 'off' and becomes unworkable. The material was placed in the cartridge and a continuous bead of caulking material squeezed around the edge of the windshield opening. He was looking to make a 'pyramid' approximately ¼ in. wide at the base and ⅜ in. high. Any air bubbles were refilled and dispersed as he went along.*

▲ *WS5. As a professional, Chris had all the right equipment, and was able to fit the glass on his own. As a DIY project, it is most certainly best to have a helper on hand at this point. The glass has to be positioned carefully on the spacers with the marks on glass and frame being matched. The windshield only has to be pressed into position lightly to ensure a proper seal.*

▼ *WS6. Chris water-tested the windshield straight away using a cold water spray. In this case, the seal was good first time out, but if any leaks had been discovered they could have been sorted by using a flat-bladed screwdriver to paddle the (still malleable) caulking material into the defective area from inside the car. At this point, all that was left for Chris to do was give the screen a thorough clean to get rid of any traces of grease or dirt which could cause smearing of the glass.*

WINDSHIELD RUST AND fiberglass REPAIR

Wherever there is steel on your Corvette, you've got to look for rust – especially when the car is an older model Corvette. Though the front door pillar and windshield surround may look like it's fiberglass, it's actually a steel construction covered in fiberglass panels and trim mouldings. It has to be steel to give the car some strength and rigidity, especially important on the roofless models.

This whole area is ripe for rust, and you won't have to remove much of the trim to spot it. As you'll see here, John Fletcher's '69 model has suffered quite badly, which gave mechanic Em a chance to show off his welding prowess.

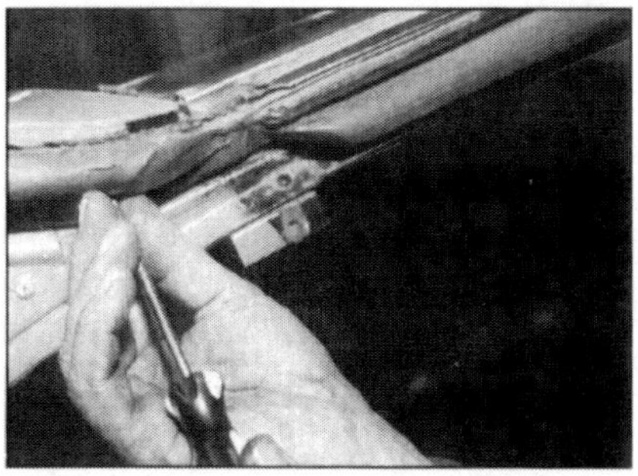

▲ *WR1. The pillar weatherstrips are secured by two self-tapping screws at their lower point. Remove them …*

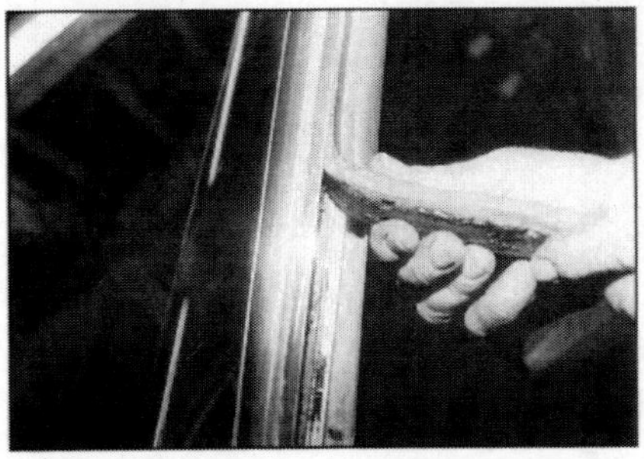

▲ WR2. … and the weatherstrip can be pulled away.

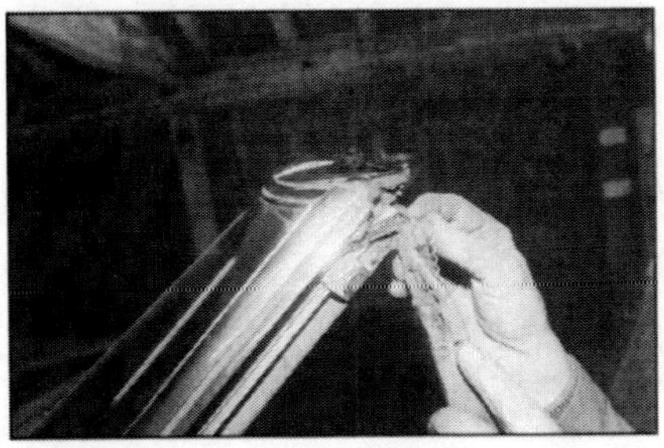

▲ WR3. It is also secured by a self-tapping screw at the top of the windshield pillar.

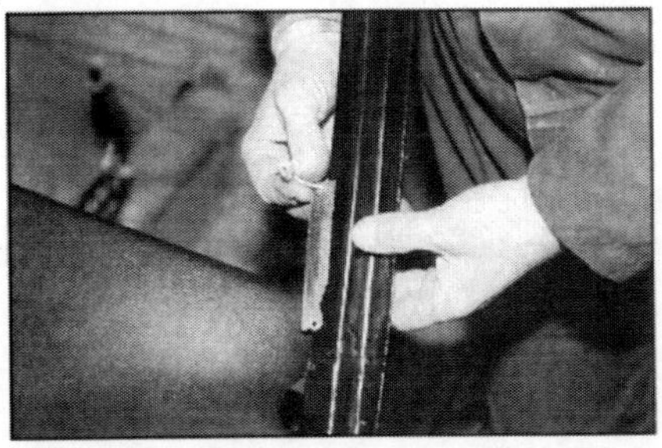

▲ WR4. While you're in this area, you can remove the Body Number plate, which is located just inside the windshield pillar on the driver's side of the car. It is a vital piece of your car's identification, so don't lose it!

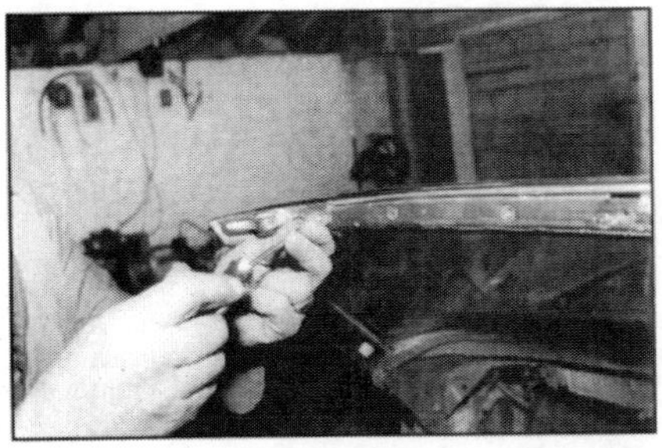

▲ WR5. Self-tappers also hold the upper reveal moulding …

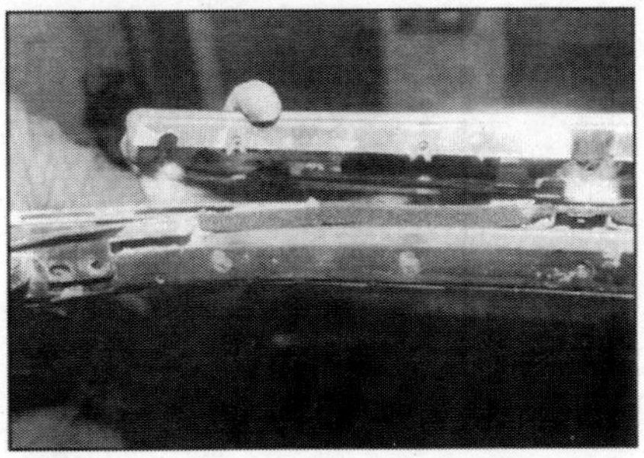

▲ WR6. … and the upper retainer moulding.

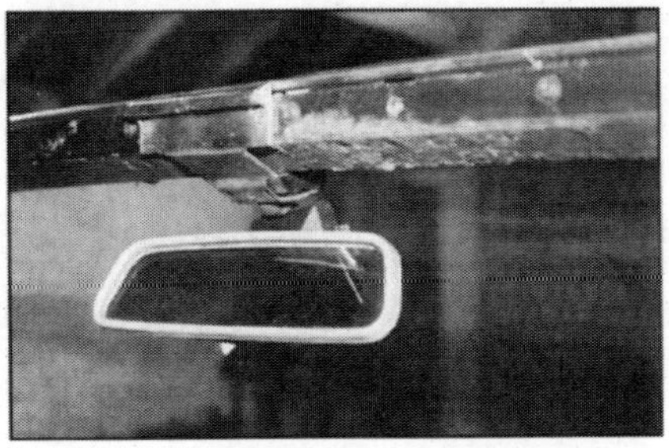

▲ WR7. With all the various trim and weatherstrips removed, it can be seen that rust has well and truly taken a hold. Even on this front rail, there's plenty of corrosion …

▲ WR8. … though the chief areas of concern are at the top and bottom of the windshield pillars.

▲ WR9. At the driver's side top, a little prodding soon uncovered a hole. Careful cutting back to sound steel revealed that the hole was actually a much more substantial section of 'nothing', where there should be metal. Em removed all the 'dead' metal and cut various pieces of mild steel to suit. Rather than try to create a single piece suitable for the whole damaged area, which would have been very difficult, given the many and various angles involved …

▲ WR10. … he applied two pieces as shown here, on the inside of the windshield frame and along the top …

▲ WR11. … and then Em cut a separate piece to fit into the corner. Notice how the steel has been snipped along its top edge so that it can follow the natural curve of the header rail.

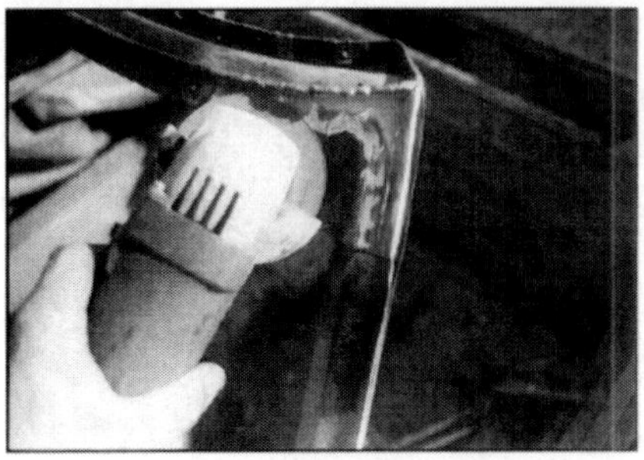

▲ *WR12. After welding it in place, the whole area was cleaned up using a power sander which left …*

▲ *WR13. … an impressively neat repair.*

▲ *WR14. The rust story was much the same at the other top corner of the frame.*

▲ *WR15. Em used the same repair process, removing all the rusty metal, before cutting and shaping a repair piece of mild steel then …*

▲ *WR16. … welding it into place before grinding it down to a smooth finish.*

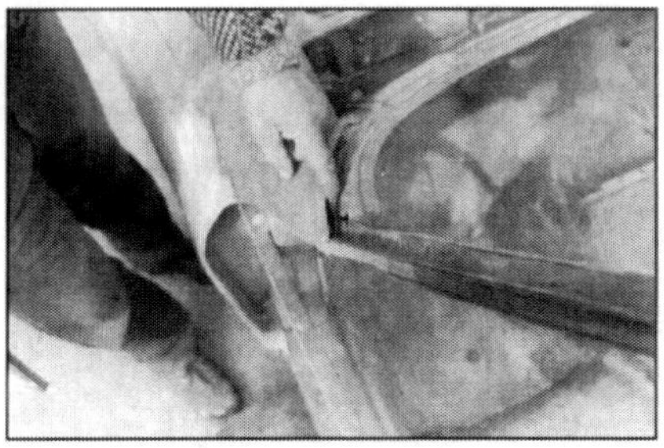

▲ *WR17. A far more difficult rust trap to access is at the bottom of the windshield frame.*

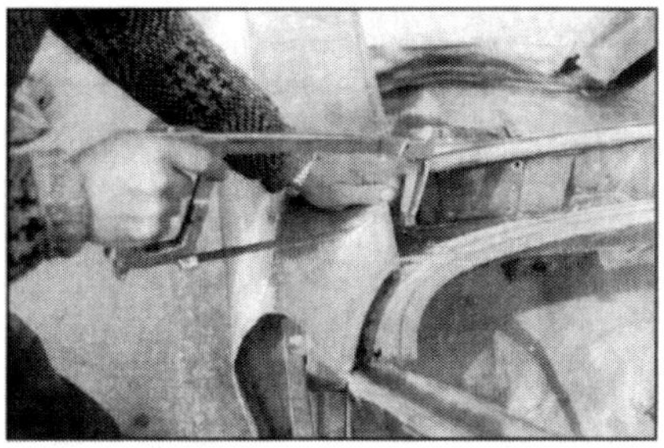

▲ *WR18. It's so difficult, in fact, that you have to cut away part of the front fender to get at it! This may seem sacrilegious, but it's not – left to its own devices, the rust will eventually rot this area completely, resulting in a very unsafe Corvette.*

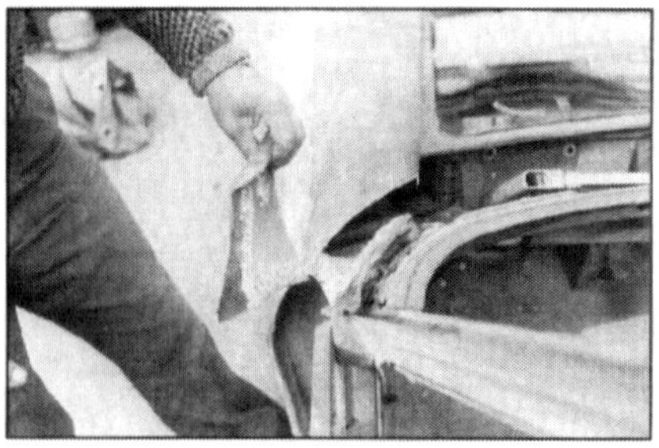

▲ *WR19. Em cut away just as much as necessary to gain access to the inner panel, which is riveted into place.*

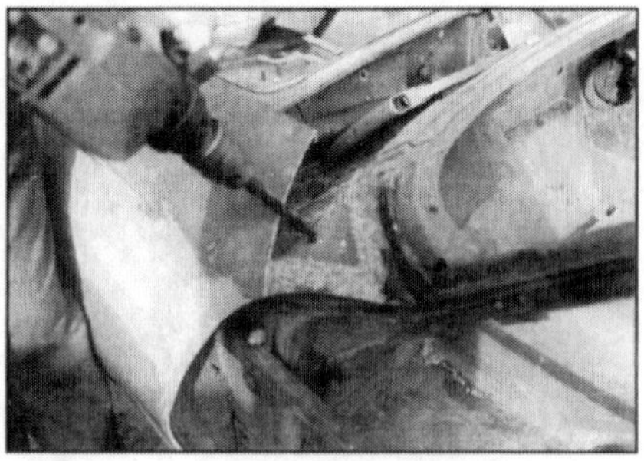

▲ *WR20. Removing the rivets was easy work with an electric drill …*

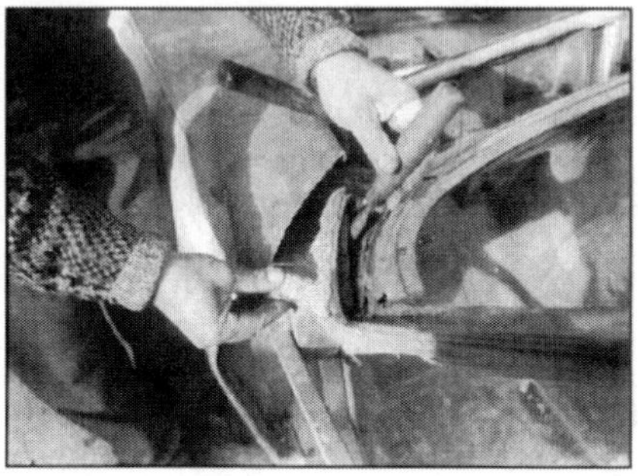

▲ *WR21. … and with a little persuasion from a flat chisel, the inner panel came out.*

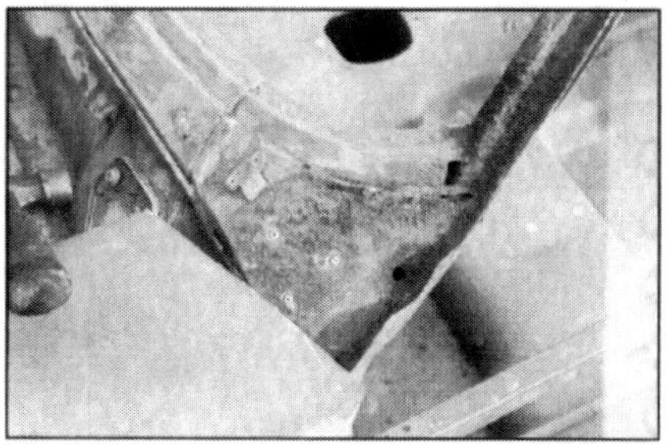

▲ *WR22. This left yet more rust and corrosion, particularly at the bottom of the windshield pillar.*

▲ *WR23. Once again, Em cut a piece of mild steel to shape …*

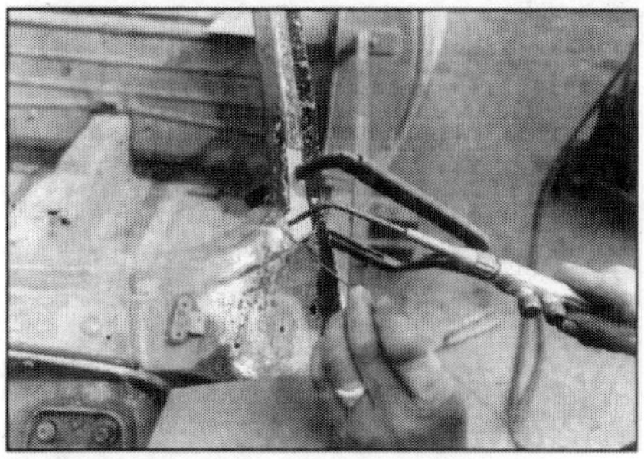

▲ *WR24. ... and welded it into position. The natural fire risk when working around fiberglass has to be uppermost in your mind when performing this kind of repair. At all times Em had a helper, with a fire extinguisher on hand, to check for sparks and possible fires starting whilst his attention was on the matter in hand. You must also check that no interior trim, underbody seal, electrical wires or anything else flammable is at risk on the other side of the area being welded.*

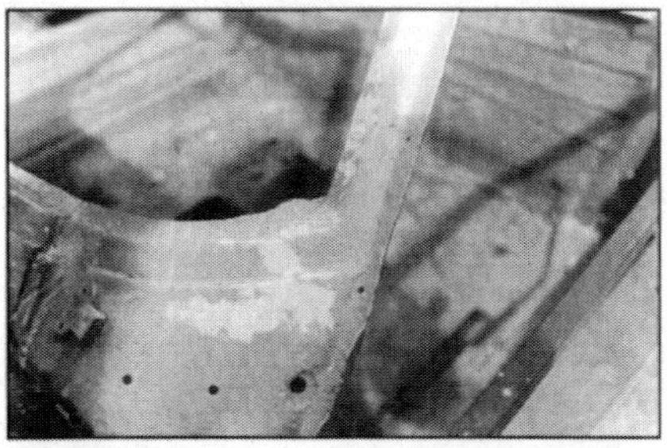

▲ *WR25. With the join cleaned up and the area primered, it was as good as new.*

▲ *WR26. The inner fender section that had been cut out earlier, was pop riveted back in place ...*

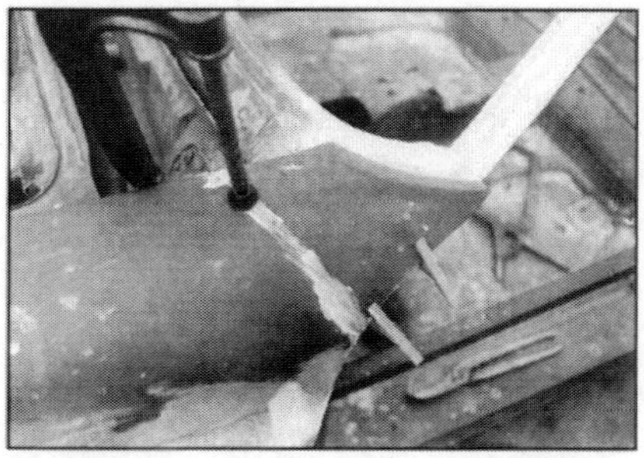

▲ *WR27. ... and the top section secured in place with clamps (and sprung clothes pegs!) and temporarily glued with GM fiberglass bonding adhesive.*

▲ *WR28. Once this had set, several strips of fiberglass were placed over the top of the join and treated to generous helpings of resin/hardener mixture. The joining edges were previously chamfered in the approved manner. Don't forget, gloves, goggles and a mask should always be worn when dealing with fiberglass and resin.*

▲ *WR29. When it was hard, the area was sanded down ...*

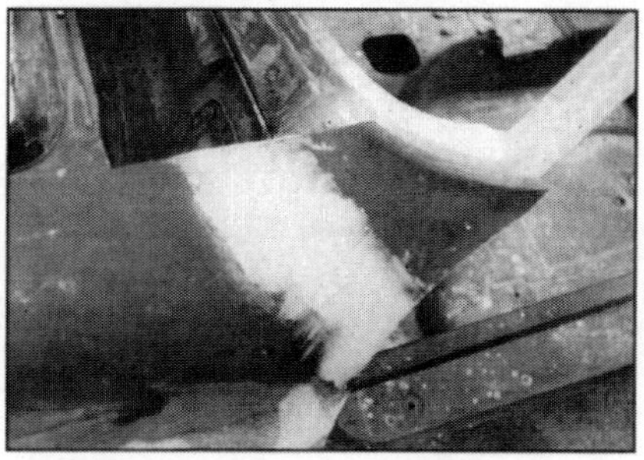

▲ *WR30. ... and bondo applied. Though the bond has to stand proud of the bodywork to a small extent, it's important not to apply too much, as it just makes extra unnecessary work sanding it smooth. Always remember that when sanding down filler or fiberglass, the air is filled with harmful particles – wear a suitable particle face mask at all times. Obtain one from your specialty auto-body repair parts supplier.*

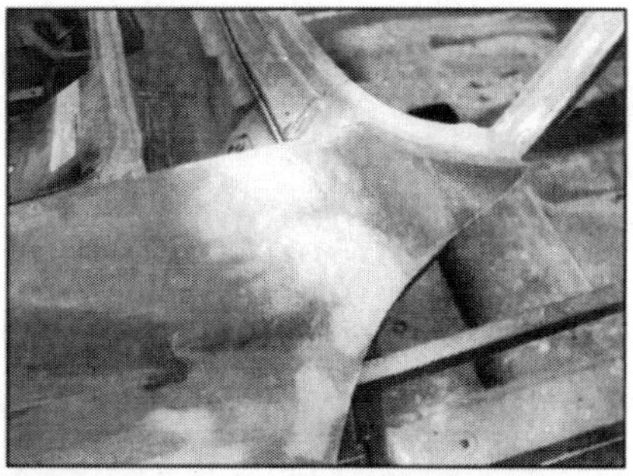

▲ *WR31. Using wet or dry paper, Em took his time to ensure that the join was perfectly smooth. Jobs like this are particularly difficult because the fender is curving in two directions. It's a wise move to practise on a scrap fender, or similar, before attempting this kind of work on your pride and joy if you are a beginner. We can't emphasize enough the need for time aplenty, and patience too, when working with fiberglass. It's the only way you'll achieve this kind of superb result. At this point it is ready for re-painting – in our case, a complete respray was on the schedule.*

fiberglass PANELS: REPLACEMENT AND REPAIR

CHOOSING fiberglass

Whether you are repairing accident damage or part of a restoration, you will want to choose the right type of fiberglass for your car. Different types of fiberglass have been used over the years and different types are available now, either from GM or as aftermarket panels – and you *can* tell the difference! Be aware that aftermarket fiberglass panels will affect the value of your car, but only if you intend putting it in for concourse judging where originality is of greater concern than anything else. For the everyday Corvette the next owner may not know enough about the subject to care, and for a car that is meant to be used and not pawed over by a judge, who cares anyway? In fact, hand-laid parts can be stronger than press-moulded parts.

Here's how the choices shape up:

▲ *FG1. Hand-laying fiberglass is the most cost-effective way of producing panels and, for that reason, most aftermarket fiberglass panels are made this way. However, only a small number of Corvettes from 1953 and early 1954 had panels made in this way. Hand-laid parts are easy to identify because you will see the traditional fiberglass texture on the back of the panel.*

▲ *FG3. Sheet moulded compound panels have been used from 1980, and these are known as SMC panels. If originality and retaining maximum value in your car is important to you, ensure that you use an original type of panel.*

▲ *FG2. From 1953 through 1980, a press-moulding system was employed, giving a smooth interior appearance.*

▲ *FG4. Panels are often replaced on a Corvette in order to give a custom appearance and many of them are self-coloured. Colours are not likely to match exactly but it is up to you how close you want to get.*

▲ *FG5. Original bumper covers were not made of fiberglass. The main problem is that the originals were flexible whereas fiberglass – as anyone who has driven into a gate post can testify – is not!*

▶ *FG6. As we mentioned in the Buying chapter, one of the best ways to spot damaged or poorly replaced fiberglass panels is to look down the car with the light shining down it. Wavy areas will show repairs where panels have been forced into position or where the replacement panel itself is of poor quality. You will also see evidence of overlapping, hand-laid fiberglass where patches have been let in or part panels fitted.*

▲ *FG7. The only way of knowing for certain what is going on beneath the paint is to inspect a panel with the paint off!*

▲ *FG8. Paint defects might not be paint defects at all, but problems with the fiberglass beneath. It goes without saying that it costs a whole lot less to rectify paint than it does to replace fiberglass.*

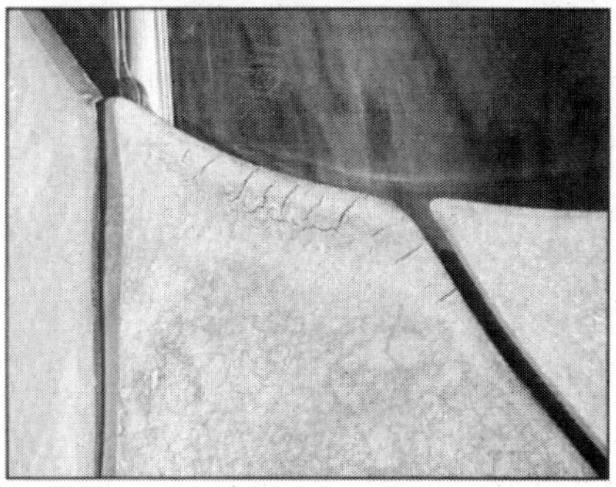

▲ *FG9. Unfortunately, on a number of 1973/74 Corvettes, cracking like this is common in the body panels. The only answer is to replace the defective panels, since filling will only give a very temporary repair.*

▲ *FG10. Here, adhesive is being applied to the rear of the panel before clamping the bonding strip in place – see Fitting fiberglass Panels section to follow. Some experts advise that if you want to give a perfect surface to the inside of hand-laid panels, see if the manufacturer will grind down the rear surface for you so you can fill and prep them yourself. But be warned: it's a hell of a big job and you probably still won't end up with exactly the right finish!*

SAFETY NOTE: Never sand, saw or grind fiberglass without wearing suitable breathing apparatus, and be sure to clear up the dust afterwards. The dust can be severely injurious to health.

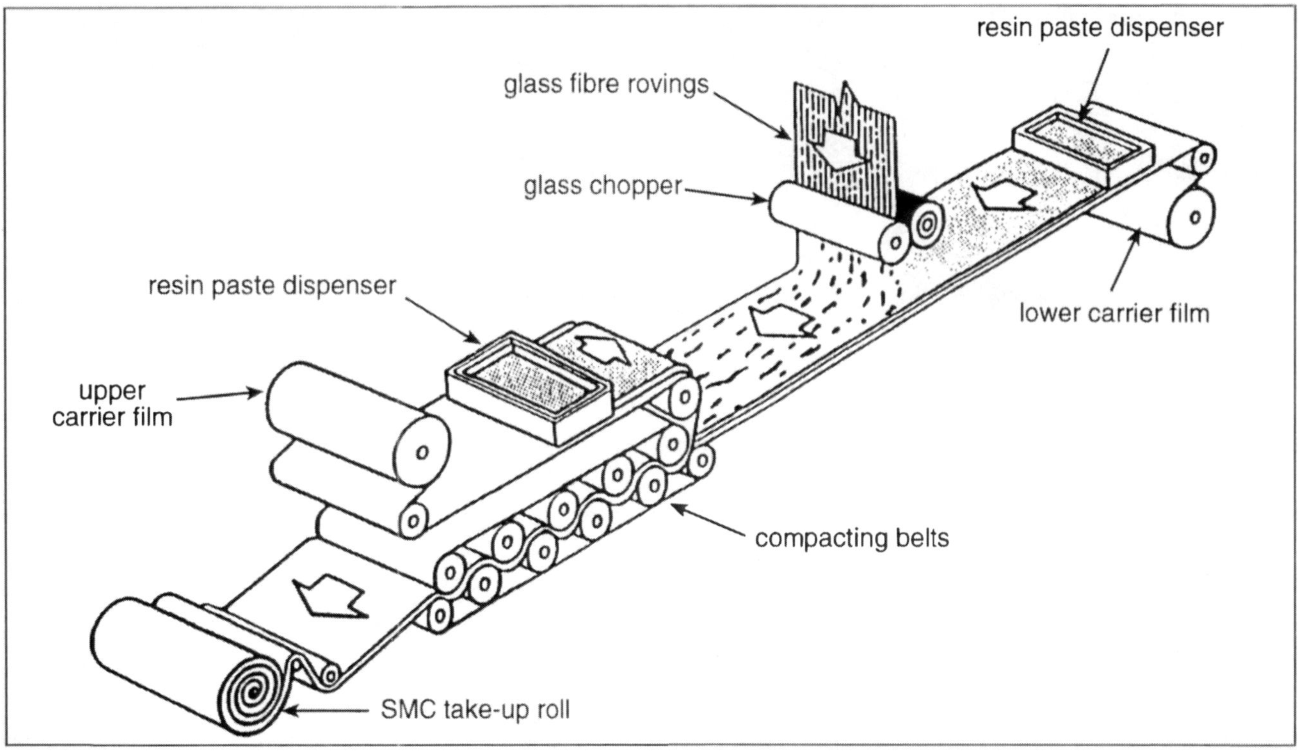

▲ *FG11. This is how the SMC sheet manufacturing process gives a smooth finish to both sides of the Corvette's body panels.* (Courtesy I-CAR)

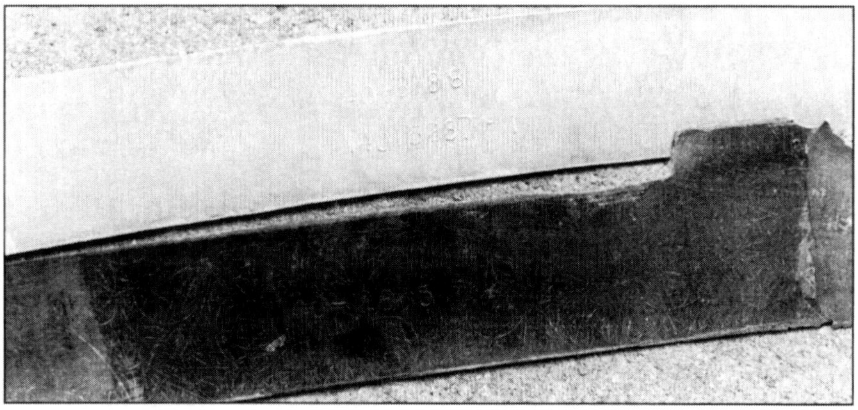

▶ *FG12. The SMC finish mentioned above can be identified by its light grey color and smooth surface. The original type of bonding strip, unlike the one shown here, is also fairly smooth but you can see the reinforcing fibres much more clearly on the surface.*

MINOR fiberglass REPAIRS

There's little doubt that Corvette owners have a real advantage over owners of steel-bodied cars when it comes to repairing bodywork. fiberglass is generally a much easier material to work with than mild steel or aluminium, especially after a little practice. We've covered major fiberglass work already – fitting a new fender etc. – but for the average owner most body work will take the form of more minor repairs to cracks and splits. Here, we show two typical areas of damage and how best to repair them. Remember the safety rules when working with fiberglass and filler; wear non-porous gloves at all times (although not those made of the kind of plastic dissolved by resin!) and wear a mask when there are dangerous fumes and/or dust or fibres in the air. Also, wear goggles whenever using resin or when power sanding.

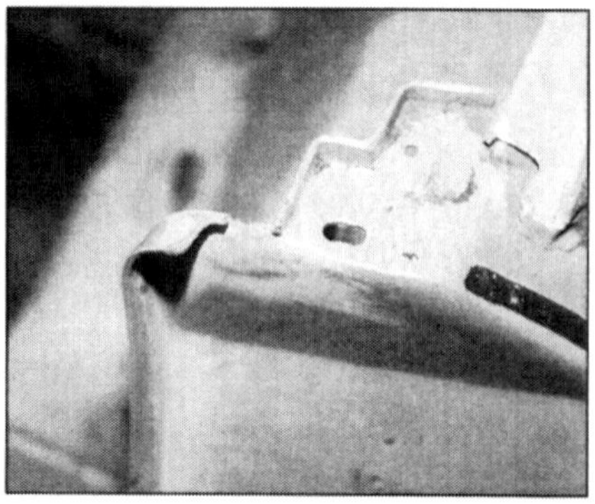

▶ *MR1. This is the sort of problem you're likely to come across – heavy-handed tightening of a screw, leading first to a cracking of the fiberglass and ultimately to a piece breaking off altogether. It's not a total disaster, though, as Em shows us here.*

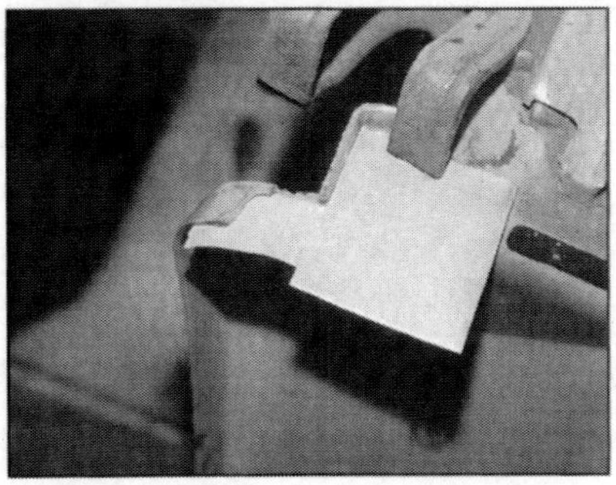

▲ MR2. First, he made up a mild steel template, or mould, to fit under the area to be repaired. This was held firmly in place by a pair of welding clamps. The bodywork was protected by the template on the top edge and by a soft packing piece underneath - otherwise, he'd soon have been making some more repairs!

▲ MR3. A fiberglass patch (no need to be too accurate at this stage) was placed on top of the steel plate in the area of the repair. Once dry, the template was removed and the new piece required only a little smoothing into shape, and a hole drilling ...

▲ MR4 .... to produce a perfect result. When working with fiberglass, wearing gloves will prevent your hands from the irritation –and possibly far worse health hazards –caused by close contact with fiberglass.

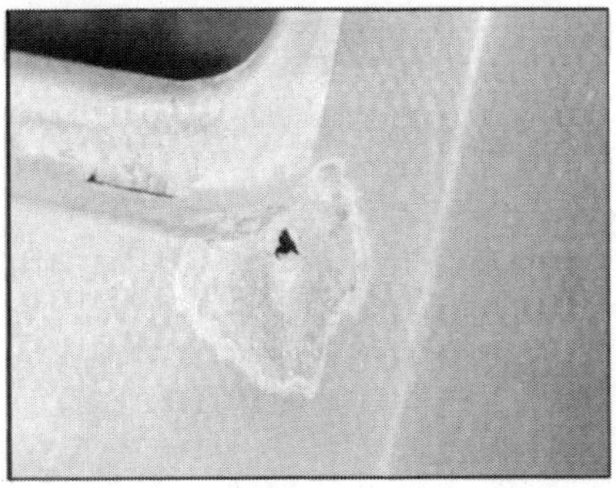

▲ MR5. Small holes present less of a problem for the home restorer compared with similar holes in a steel bodied car – holes in fiberglass don't rust! Em had noticed this small hole on the rear fender just behind the folding top compartment, and set about putting it right before the car was resprayed. Though it may seem a little odd to the untrained eye, he took off the paint layer for an inch around the hole –it's essential to repair to 'raw' fiberglass, not over paint!

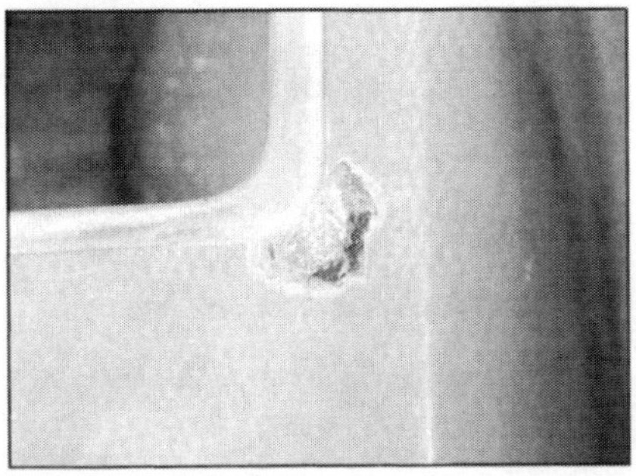

▲ MR6. As earlier, he cut a suitably-sized piece of fiberglass and stuck it to the affected area using the fiberglass resin/hardener mix. You must brush the edges of the area with the resin and then place the fiberglass mat on top. The resin effectively holds the mat in place and helps keep it there while more resin is stippled into it from above. The mat must be thoroughly 'wetted out' by the resin, otherwise the fiberglass will have little strength. As it becomes wetted out, the mat fibres become more malleable so that it can take on the shape required, and the white colouring almost disappears – looks more like 'glass' in fact! Don't forget, gloves should always be worn to prevent the resin and/or fiberglass causing skin irritation.

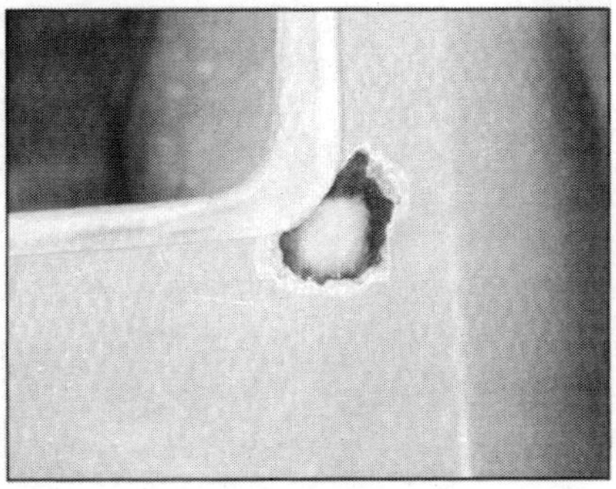

▲ MR7. Once it had set, Em sanded down the fiberglass patch to a smooth finish – as ever, wearing a mask for safety's sake.

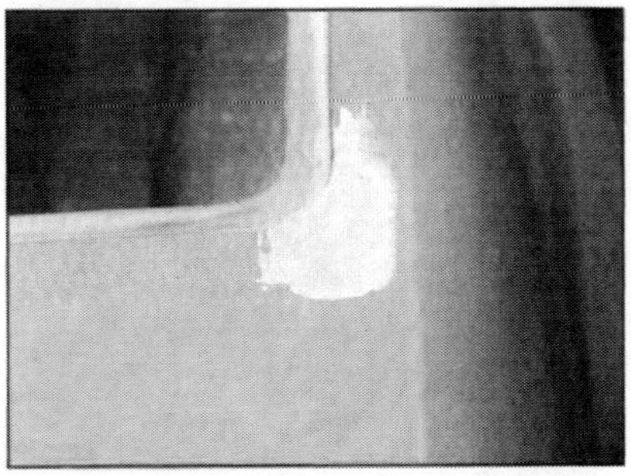

▲ MR8. In order to bring the repair area up to the same level as the surrounding paintwork, Em applied a thin layer of body stopper (a version of body filler used for just such small filling jobs).

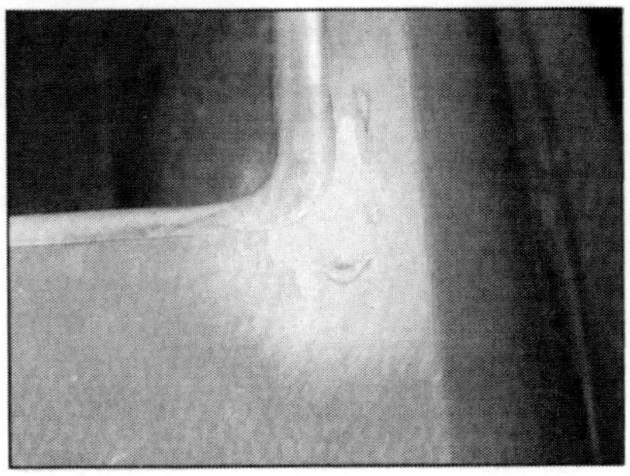

▲ MR9. It was then necessary to sand down the area to get it almost smooth. After this, a further, light layer of stopper was applied ...

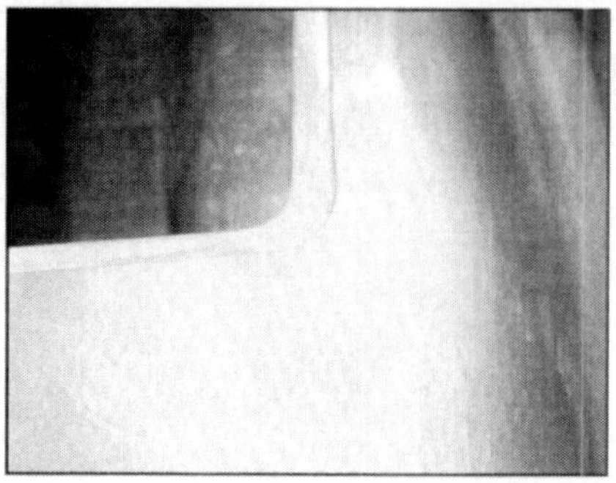

▲ *MR10. ... and this was sanded down, using wet-or-dry paper, to a perfectly smooth finish. As with most areas of fiberglass body repairs, this is well within the reach of most owners, as long as they are prepared to invest the time and patience required. Em made this look easy, but many less experienced owners would probably have had to have applied more stopper and sanded down another twice or even three times to achieve the same finish. The tyro is also likely to apply finish paint before the reality of a lousy finish becomes apparent. But we all have to start somewhere! And the more you practice, the easier it will become. We've mentioned it on numerous occasions, but it's vital to reiterate – don't get resin, filler or fiberglass on your skin or in your eyes, and don't breathe in the fumes or fibres – all are extremely dangerous. Read the safety notes in Appendix 1 of this book before carrying out any fiberglass repair work.*

FITTING fiberglass PANELS

Unless you are wearing X-ray specs, it can be difficult to see all the work that needs to be carried out to fit a new front fender to a Corvette. With the help of Tom Falconer at Claremont Corvette and Em in John Fletcher's workshop, we fitted a new front fender to a hood surround with both items off the car. That way, you get to see the inside story!

One of the huge advantages of fiberglass panels is that you can fit them without the need for highly developed special skills or special equipment, unlike traditional steel panels. You will need to take just as much care in fitting them, and the key to success is not to rush the job and not to move on to the next stage until you are completely sure that the work you have done has been done right.

▲ *FP1. The hood surround has been laid out by Em upside down on trestles – see for yourself where the headlamp aperture is situated (bottom right of this illustration), and Em also takes a look at the new fender.*

◀ *FP4. These bonding strips came in two parts, the front part taking on the complex shape of the front of the panels, as shown in the previous shot, and the rear part following the contours of the hood surround.*

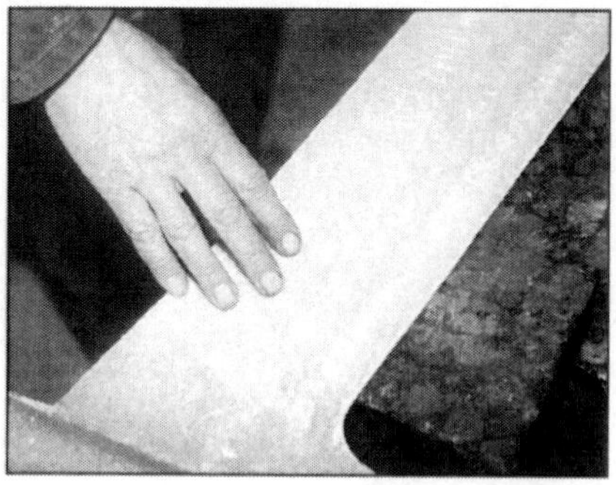

▲ *FP2. New panels produced by GM are almost smooth on their inner face, while some replacement panels just show raw fiberglass. Claremont's panels, used in this sequence, lie somewhere between the two: not as smooth as factory panels but by no means as rough as regular hand-laid fiberglass.*

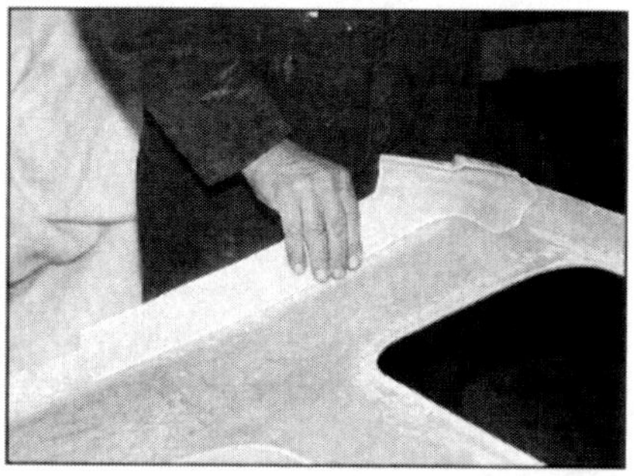

▲ *FP3. The correct, factory-approved way of fitting panels together is with the use of bonding strips. These are fiberglass strips that overlay the joint between the two panels being bonded together. Of course, they are specially shaped to fit exactly the contours of both panels. If you use non-factory parts, you run the risk of panels not fitting together properly and bonding strips fitting where they touch – which is not all that often!*

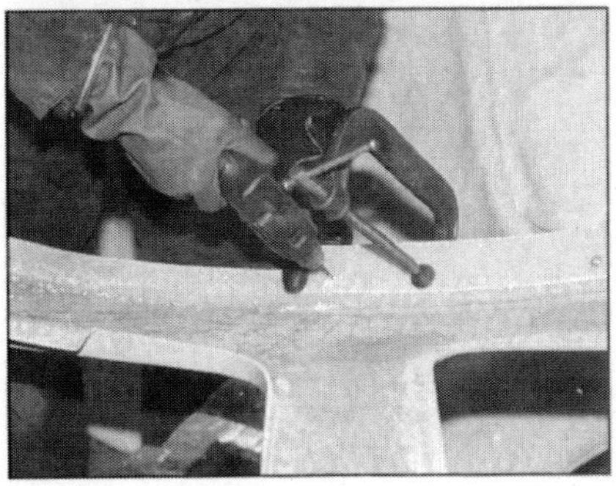

▲ FP5. As mentioned elsewhere in this book, you must make sure that you are wearing gloves, an efficient face mask and goggles when using a power tool on raw fiberglass. Here, the front part of the bonding strip has been clamped carefully in place – do not over-tighten the clamp because you could easily damage the top surface of the panel. In fact, you should use a soft pad beneath the head of the clamp. The bonding strip and hood surround are being drilled so they can be temporarily screwed together, with large flat washers beneath the screw heads and nuts.

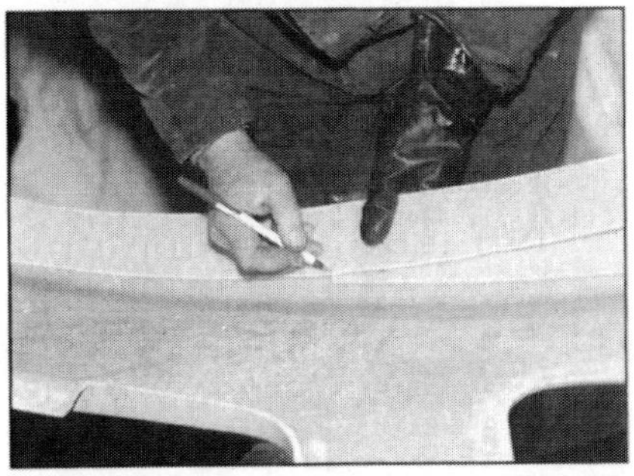

▲ FP6. Now Em marks out the over-long rear part of the bonding strip ...

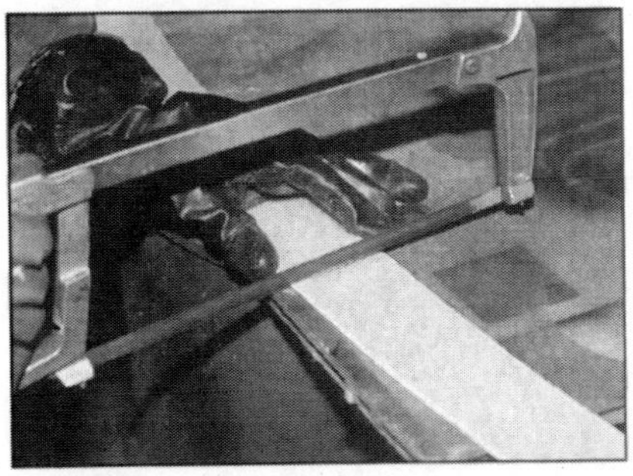

▲ FP7. ... and cuts off the excess.

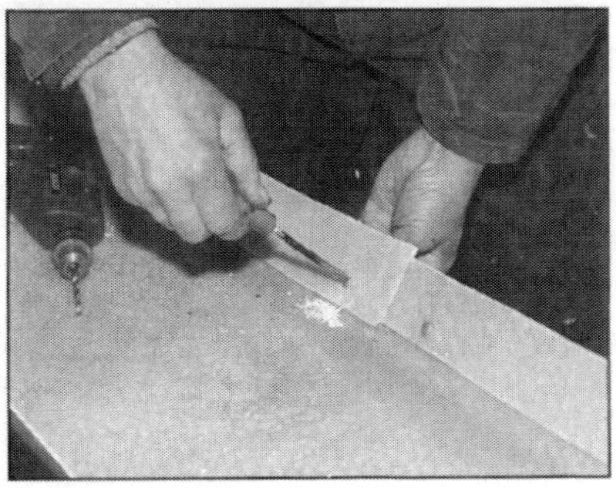

▲ FP8. Next, the rear of the bonding strip, going right the way to the back end of the hood surround, can be drilled and screwed into place. Note that half of the bonding strip sticks out over the edge of the hood surround.

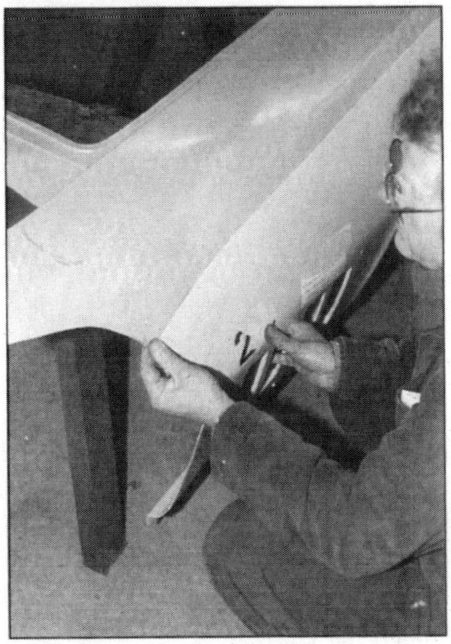

▲ FP9. You can now stop reading this while standing on your head! The hood surround has now been turned over so that it is the 'right' way up on the trestles. Em offers up the fender and compares it with the shape of the hood surround, resting the fender on the overlapping part of the bonding strip.

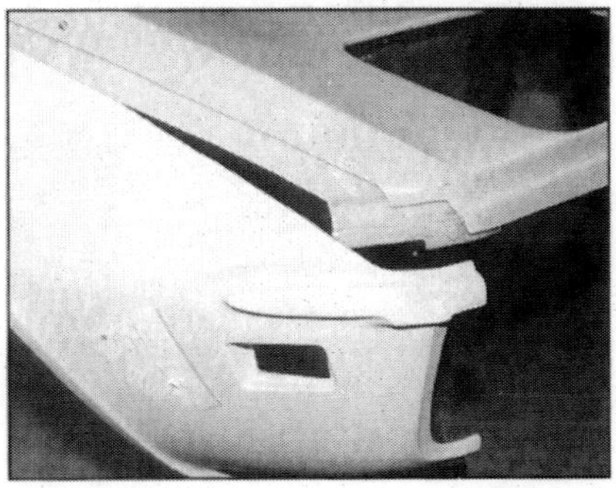

▲ FP10. Note how with non-factory panels you almost always have to carry out a little careful fitting to get the shapes to match exactly.

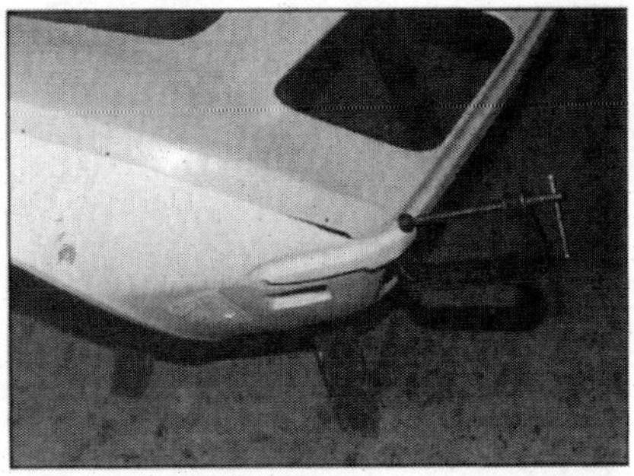

▲ FP11. After temporarily removing the front part of the bonding strip and trimming back the front corner of the fender surround, the front of the fender is held in place temporarily with a lightly tightened clamp. If you have several extra pairs of hands available in your workshop, you won't need the clamp at this stage!

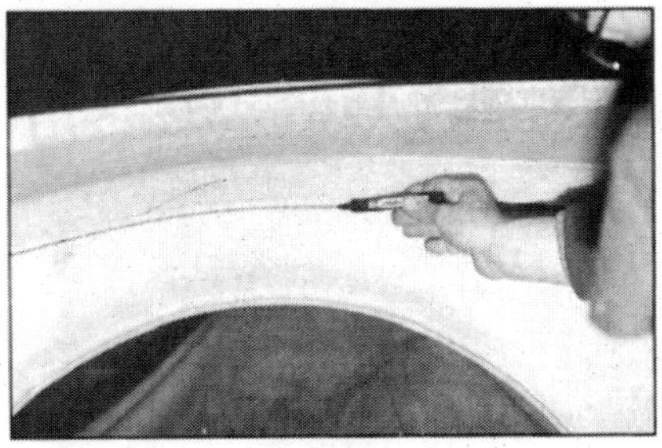

▲ FP12. The rear of the fender is also clamped into place and the shot you see now was taken from underneath the hood surround, the camera being approximately where the engine would be on a regular Corvette! The fender was found to overlap the hood surround to quite a considerable degree, so Em used a marker pen to draw along the joint ...

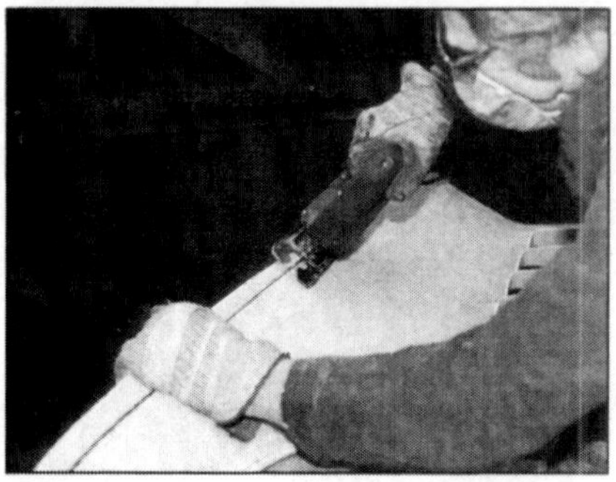

▲ *FP13. … showing just where the surplus fiberglass could be cut away. Note that Em uses a power jigsaw (complete with all correct protective gear!) and also note that this work is being done out of doors so the dangerous dust doesn't gather inside the garage. He cuts just a little way away from the marked line …*

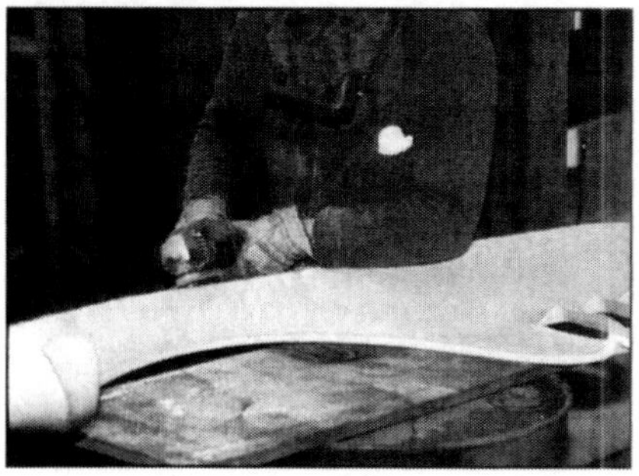

▲ *FP14. … then uses the sander to finish the edge of the panel to exactly the correct shape. Goggles are essential when using the sander because of the high rotational speeds involved.*

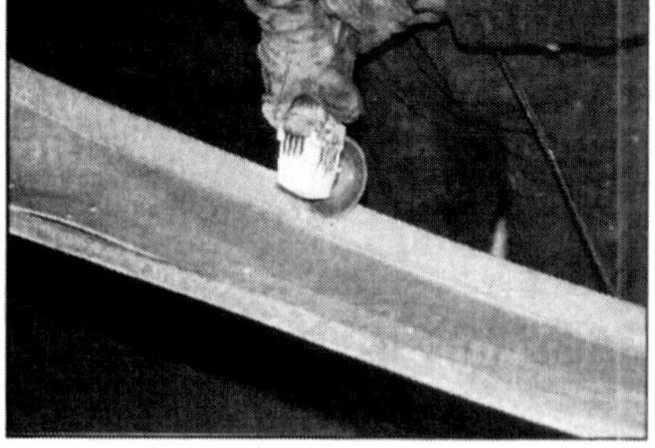

▲ *FP15. Now, on the inside edge of the fender where it will sit against the bonding strip, Em uses the sander to lightly sand off the surface of the fiberglass. You will see why in a moment!*

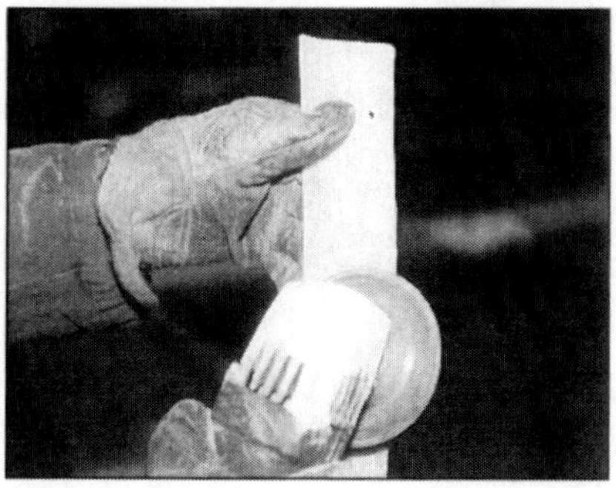

▲ *FP16. He also sands the inside of the bonding strip, once again very lightly and taking off a minimal amount of material.*

▲ *FP17. All this surface sanding was so that the bonding paste used for 'gluing' the surfaces of the fiberglass panels and the bonding strip together will adhere as well as possible. Em has carried out experiments with bonded together fiberglass and has found that bonding paste is actually stronger than the original laminate and that if you forcibly tear the two apart, the laminate destroys itself before the bonding paste quits.*

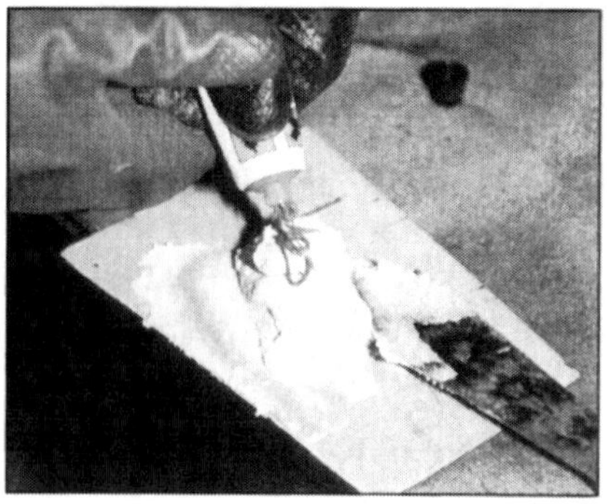

▲ *FP18. After scooping out an appropriate amount of the bonding paste onto a clean surface, hardener is added to the paste – read your instructions carefully to establish the appropriate amount of hardener to use.*

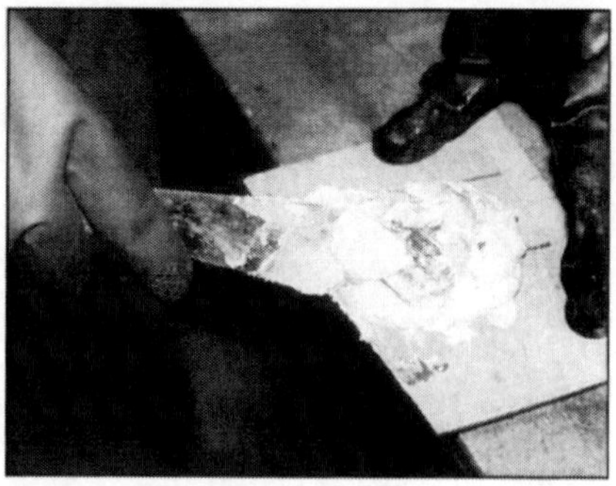

▲ *FP19. The whole lot must be mixed together very thoroughly. Always wear gloves as a safety precaution. Never mix the bonding paste on card. If you do you will pick up some of the fibres into the paste. Em used a scrap piece of steel plate; others choose to use rigid plastic.*

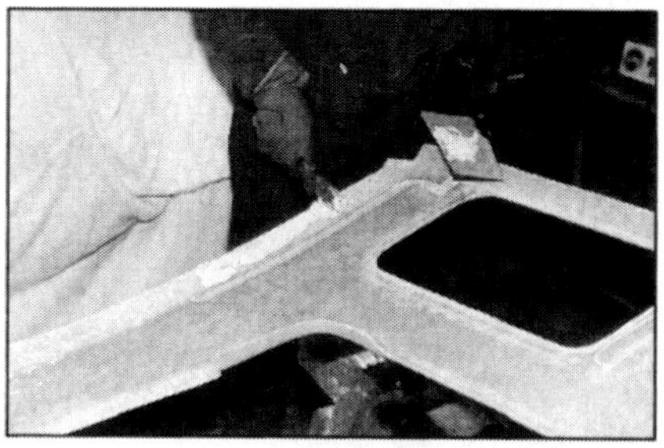

▲ *FP20. Em smeared on an even coat of the paste onto the hood surround – note that it is now upside down again on the trestles.*

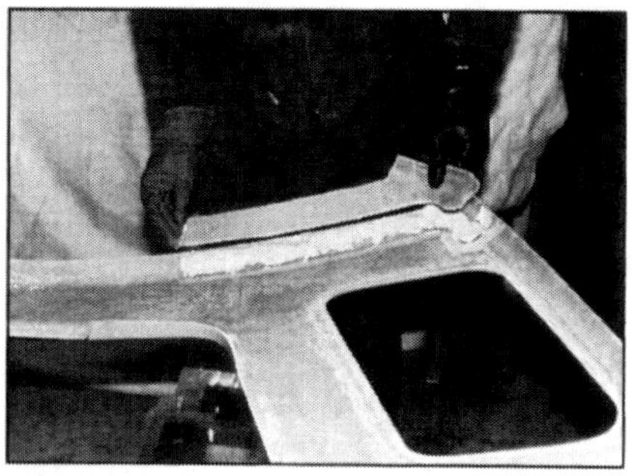

▲ *FP21. Then he lowered the bonding strip into place. Em experimented with masking tape placed around the area where the paste was to be applied, but after this one experiment, it was not found to be worthwhile. Better yet, the bonding paste can be added to the bonding strip rather than the panel to which it is fitted.*

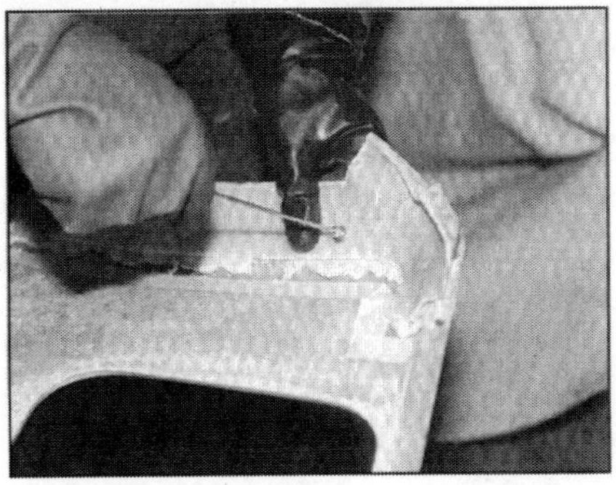

▲ FP22. The fixing screws and flat washers are refitted to hold the front of the bonding strip in place.

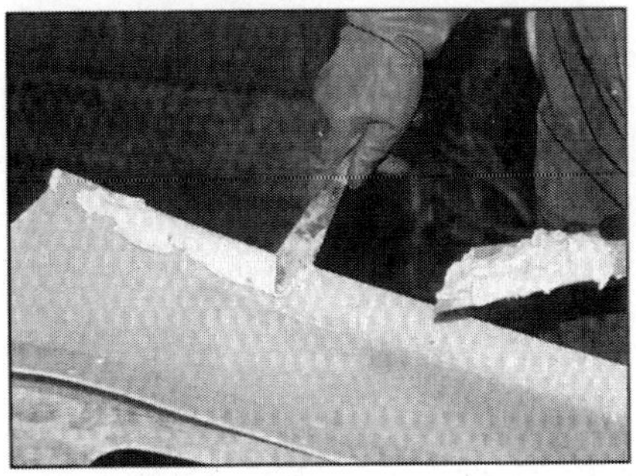

▲ FP23. At the back of the hood surround now – and do remember that it is upside down and laid out on a pair of trestles for demonstration purposes only – Em applies more bonding paste …

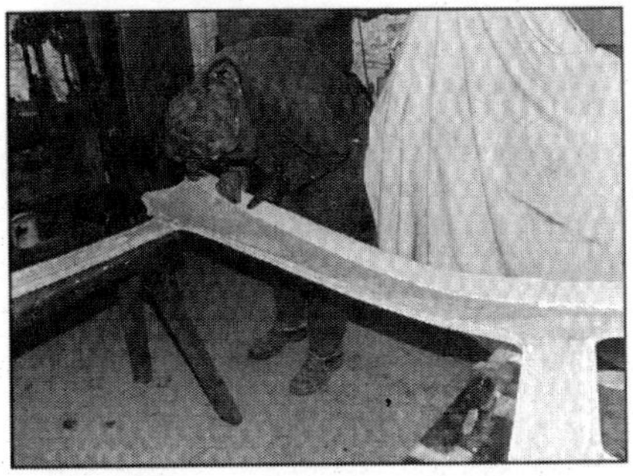

▲ FP24. … and the rear of the bonding strip is offered up into position. Note that Em continues to wear his protective gloves so that the bonding paste does not get onto his hands.

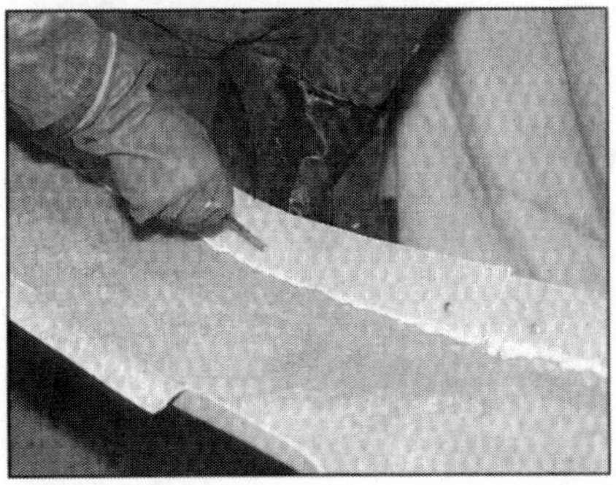

▲ FP25. All of the screws are refitted, holding the bonding strip exactly in place while the paste goes off.

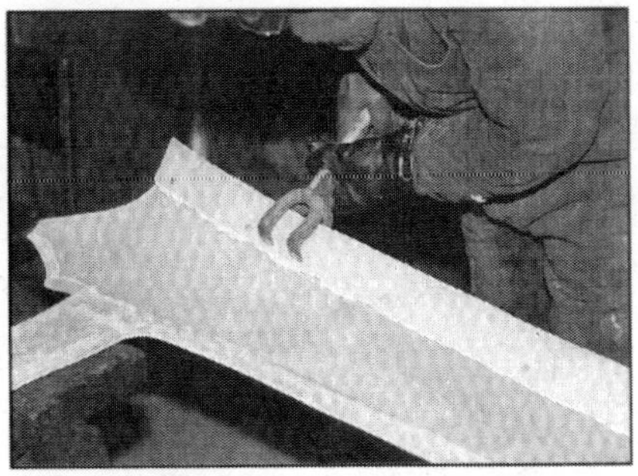

▲ FP26. Just to make quite certain that the bonding strip is pulled down tightly onto the hood surround in between each of the fixing screws, Em goes along the bonding strip with a self-grip wrench gently squeezing the two panels tightly together.

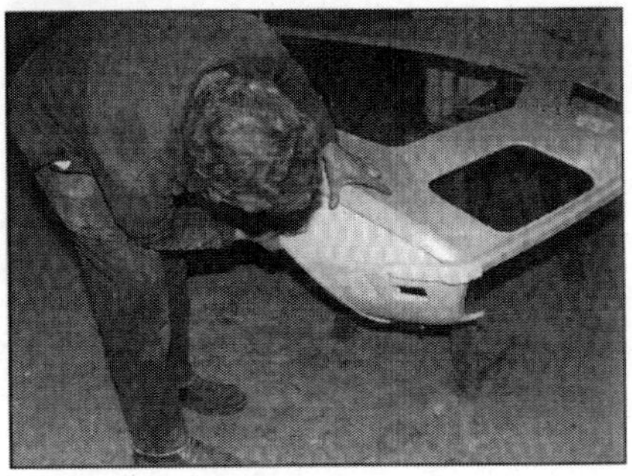

▲ FP27. Once more the hood surround is turned over on the trestles and now, right way up, the fender is again offered up.

▲ FP28. The drill is used once again to drill right through the fender and bonding strip …

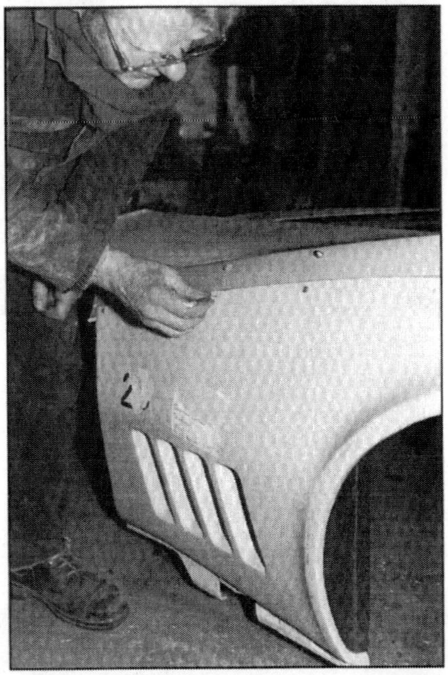

▲ FP29. … before temporarily screwing both panels together.

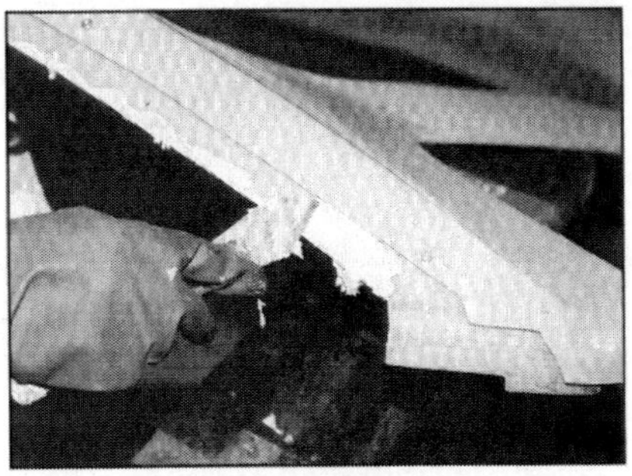

▲ FP30. Here you can see the bonding strips sticking out from beneath the hood surround. Em applies another layer of bonding paste.

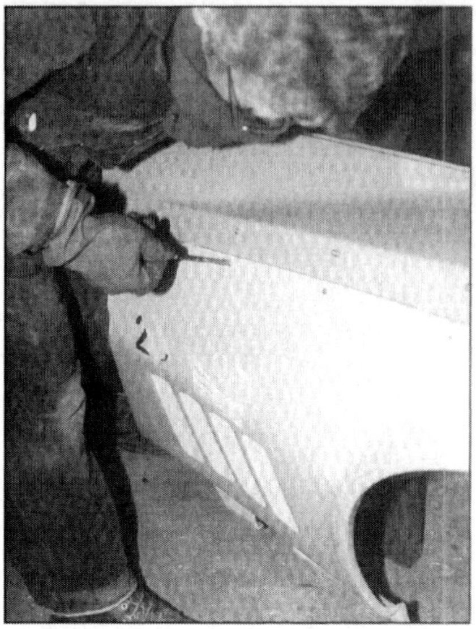

▲ FP31. The fender is offered up and screwed into place, squeezing out surplus bonding paste …

▲ FP32. … which shows up on the inside just as it did in the factory. Note that hood surround and fender are now upside down once again.

▲ FP33. Em uses a long-reach clamp to ensure that all the bonding surfaces are close together. In this case, if the outer surface of the fiberglass is damaged, it's not desirable but it is not the end of the world since it is going to be sanded away, as we will explain later.

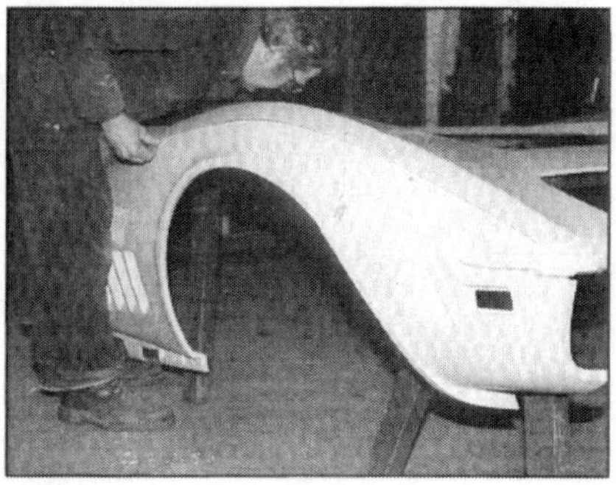

▲ FP34. Once the bonding paste has thoroughly gone off, all of the temporary mounting screws can be taken out.

▲ FP35. Em examines the joint between the two panels …

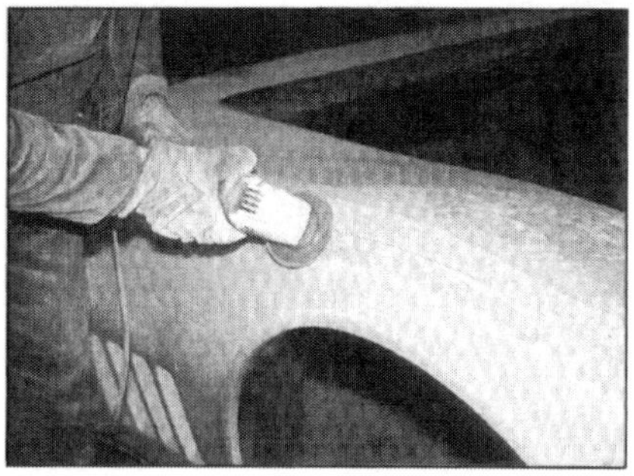

▲ FP36. … and then it is back on with protective gloves, face mask and goggles while the machine sander is used to sand a Vee, or shallow depression, all the way along the outer surface of the panel joint.

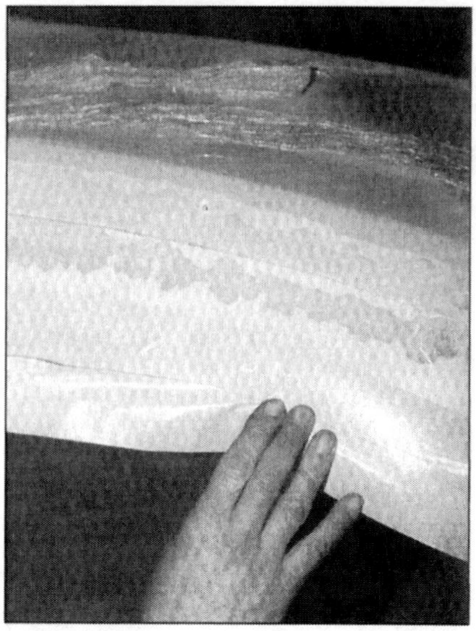

▲ FP37. Em used broad masking tape to protect the fender beneath the joint from any fiberglass resin that could run down as the next stage is carried out.

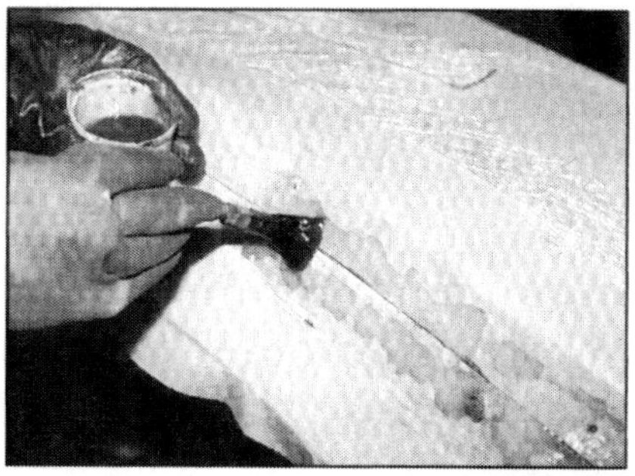

▲ FP38. Note the protective gloves once again. Resin, complete with hardener mixed in, was applied by Em in a really thick coat the full width of the area to be jointed.

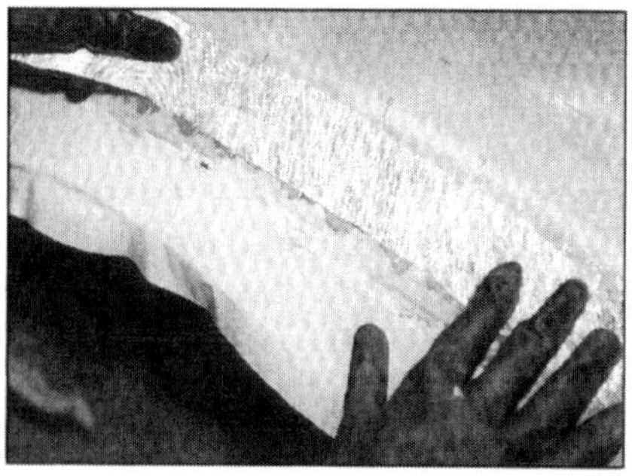

▲ FP39. A cut-out strip of fiberglass mat was placed onto the resin …

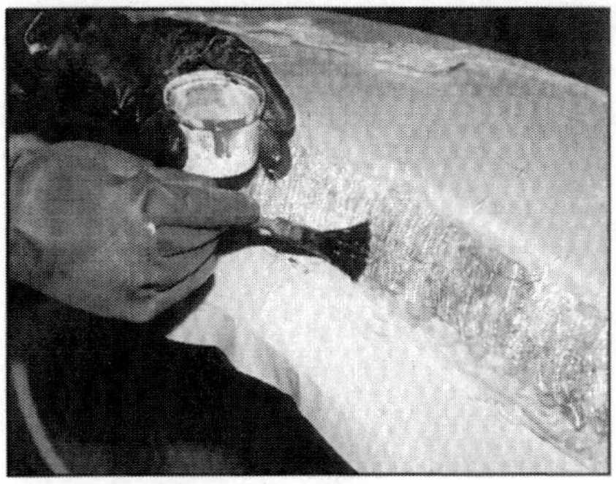

▲ FP40. … and more resin briskly stippled in until the mat became translucent and saturated with resin.

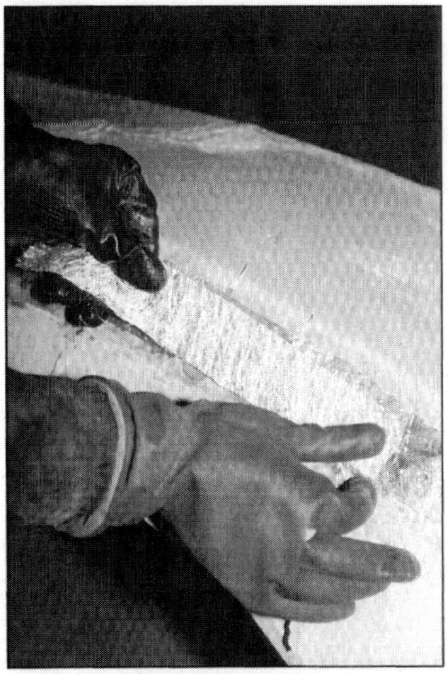

▲ FP41. Next, a second layer of mat was applied on top of the first …

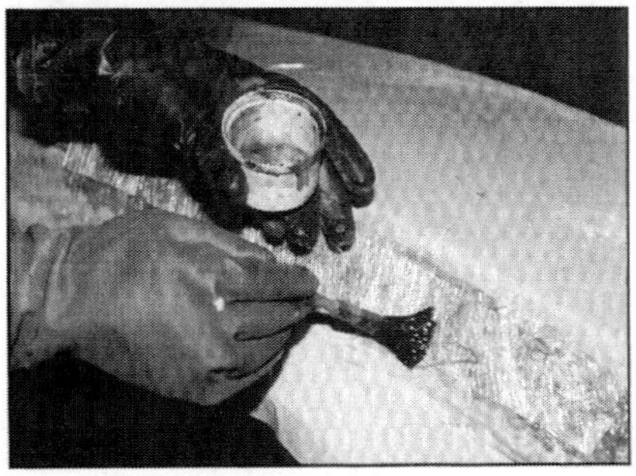

▲ FP42. … and more resin stippled in until the fresh layer of mat was similarly saturated.

▲ FP43. The correct way of getting all air out of the mat and resin and ensuring that everything is fully up to strength is to use this fibreglassing tool available from your specialty supplier. It is rubbed hard over the surface of the wet resin and mat, and ultimately produces a much stronger joint. Once the resin has gone off, you can sand down any high spots in the fiberglass and apply bondo to give an invisible joint.

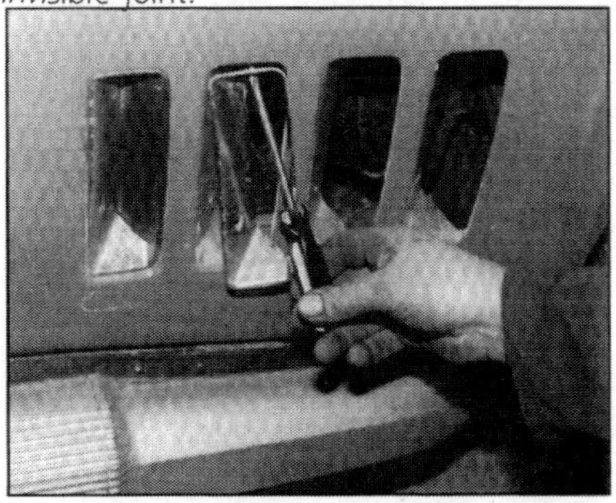

▲ FP44. Now we are going to switch our attention away from the demo. piece to work that Em carried out on John Fletcher's Corvette on which all four fenders were replaced with new. Before starting to remove a damaged fender from your car, make sure that all badges and body trim are removed and stored safely for later use.

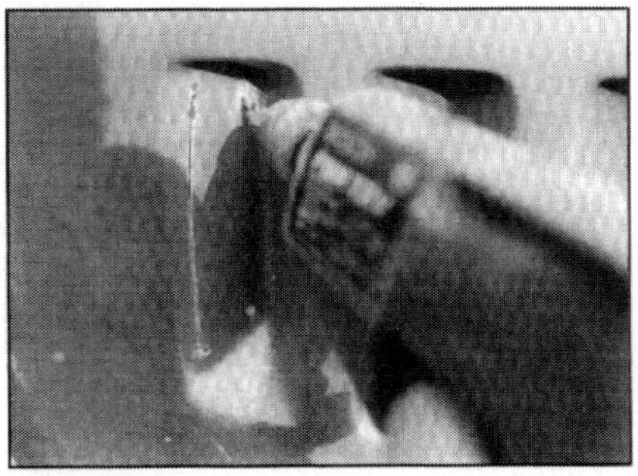

▲ FP45. The panels used on John's car, supplied by Auto Accessories of America were found not to have these correct cutaways in place. Em had to drill holes in the panels …

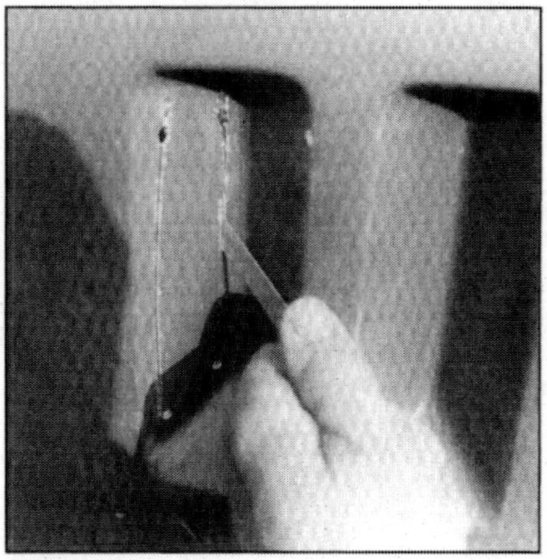

▲ FP46. … before inserting a saw blade …

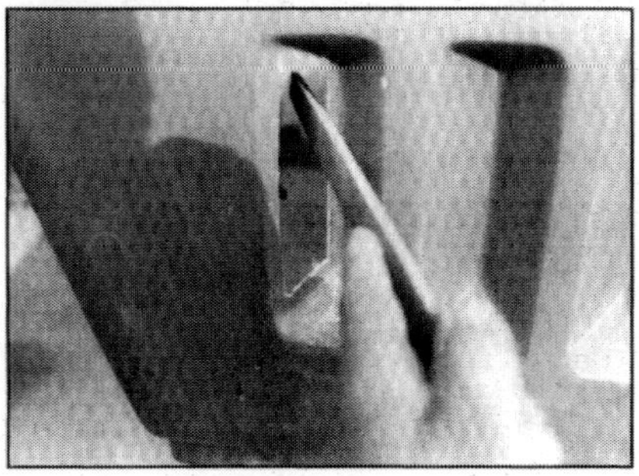

▲ FP47. … and filing out the holes to original spec.

▲ FP48. After some careful handiwork, everything looks exactly as it should.

▲ FP49. Here you can see Em working outdoors to remove part of the old, damaged front fender.

▲ FP50. You can see here how he removed the fenders in sections, working back to the part of the fenders where they join to the structure of the car …

▲ FP51. … until just the centre part of this rear fender, where it fits to its strengthening rib, is exposed and can carefully be cut away.

▲ FP52. The final stage was for Em to sand all of the remnants of the old panel away …

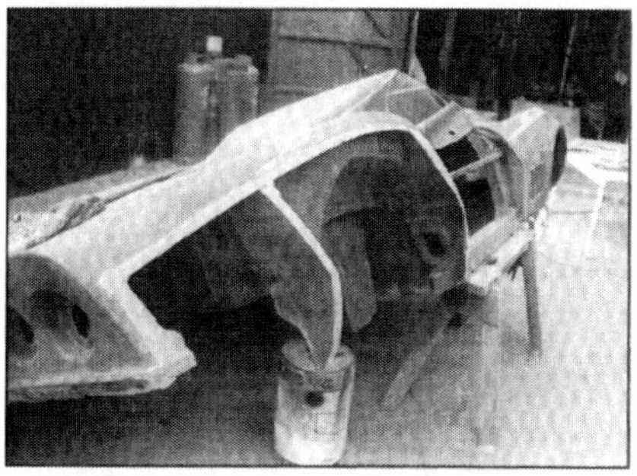

▲ FP53. … until, with reference to the size and shape of the new fender, the fender surround area and strengthening panel are ready for the bonding strip and panel to be fitted.

▲ FP54. Here, the bonding paste has gone off (hard) and Em is removing the screws holding the new fender in place.

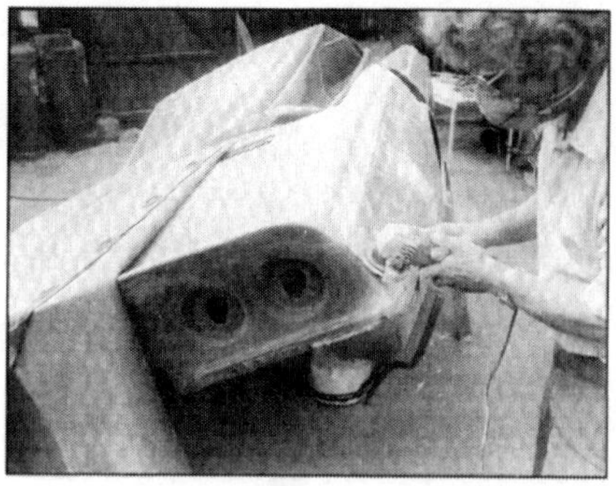

▲ FP55. Now you can see Em sanding back the join prior to applying the fiberglass and filler …

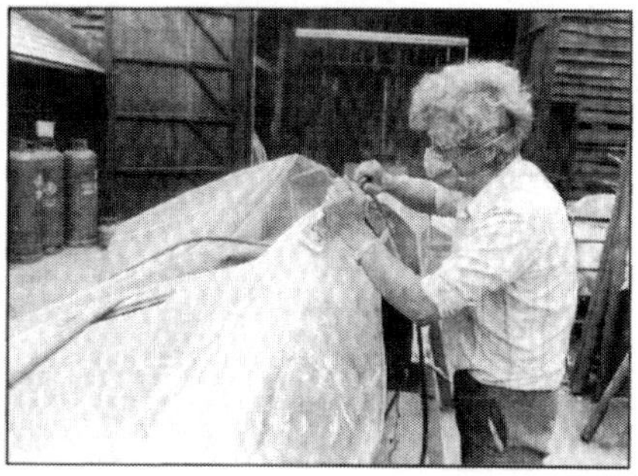

▲ FP56. … and the long process of sanding down the bondo to get a perfectly true surface is underway.

▲ FP57. This beautifully fitted front fender is an excellent testament to Em's skills, and serves as an inspiration to anyone who wants to fit new fenders to their own Corvette.

BODY REMOVAL

To lift or not to lift: that is the question:
Whether 'tis nobler in the mind to suffer
The slings and arrows of outrageous framework,
Or take arms against a sea of rust,
And by welding repair them?

It will take more than a little soliloquizing to help you to decide whether to lift your Corvette's bodywork from the frame. But, be warned: don't just rush into it because it seems a good idea at the time. You will need to lift that body if you aim to carry out a total restoration – but bear in mind that planned 'total restorations' that never work out are the rocks upon which many Corvettes have sunk. You will certainly need to lift the body if frame rust is severe. Otherwise, consider leaving things just as they are and not committing yourself to a great deal of work and expense before the car becomes driveable again. The point is, once you lift the body, there will be an awful lot of work to do – and an awful lot of work will force itself upon you because you just won't be able to put the bodywork back again without carrying out every repair you can find – and then some. For the great majority of Corvette restorations it would be better to ...

... bear those ills we have,
Than fly to others that we know not of

as the man said.

WHAT YOU WILL NEED

It goes without saying that you will need space for 'two vehicles' and room to work around them. You will need the willing help of plenty of assistance or a strap set and hoist of at least 2,000 lb capacity, for safety. You will then also need good roof trusses or a really strong temporary support. You will need plenty of time: a full week's work, including lots of overtime. You might need three times that long if you encounter lots of rust – which is highly likely. You can save time, though, by spraying every single fastener on suspension, brackets, body mountings and bumpers with penetrating oil several times during the week or so before you start. And that is just to get the body off! On soft-top cars you will need to consider supporting the body across the door openings to prevent flexing as the bodywork is lifted off.

OVERVIEW

Before looking in detail at how we removed the project car's body, we'll take an overview at what is involved.

SAFETY NOTE: Disconnect and *remove* the battery as your very first job. Put it somewhere safe so that nothing can short across the terminals. As soon as any fuel pipes or hoses are disconnected, plug or blank them off and ensure that no gas is allowed to run across the workshop floor or vent into your workshop space, because if it does there will be a real risk of explosion.

- All wires connected to the frame, engine and transmission will have to be disconnected. These will include the coil, temperature senders, alternator wires and an almost infinitely different number of connections, depending on model and year. Consult your manual carefully here, but also you must realize that you are bound to forget something. The way around this problem is to lift the body slightly, check again, lift a little more, check some more, and so on until you are confident everything is disconnected.
- You will also have to make sure that you unclip certain cables, such as the backup light switch lead which is fixed with body clips. Later models have another wire for smog equipment running to the side cover.
- You should by now have disconnected the battery, but don't forget the battery cable on 1968-up models to be disconnected from the frame under the battery compartment, and the other battery cable at the bellhousing bracket.
- You will see smaller wires to the starter solenoid and a thinner ground wire. (Incidentally, this is all too often connected to the large solenoid stud which is one of the most common reasons for heater blowers and windshield wipers not working.)
- As you disconnect each wire, make a tag on the end of it with masking tape and write down precisely where it came from. Also, put back all washers and nuts on the studs from which they came.
- Have a camera with flash gun handy so you can photograph every single item before you take it off, and also have a note pad to write down any idiosyncrasies you may come across.
- On models from 1974 onward, the oil pressure sender is electrical and all you have to do is pull off a wire connector. On earlier models there is a nylon tube which may solve the problem for you by breaking because of its old age and brittleness. Here's a chance to write in that notebook so you can obtain any spare parts you may need for later.

Here's a fairly random selection of items that you may need to disconnect.

- Power brake hoses.
- Fuel vapour canister hoses (where fitted).
- Control hoses for the heater and headlights. Tape up the ends of all hoses and connectors before leaving them.
- Engine heater hoses and upper and lower radiator hose. (Don't waste time trying to remove welded-on radiator hoses if you are carrying out a full restoration. Cut them off with a hacksaw, taking care to cut away from the radiator stub. You can then cut through the redundant stub of hose later and fit new ones when you rebuild.)
- Tachometer cables, speedometer cables, cruise control or automatic transmission cables, transmission interlock cable (disconnect it near the steering column by prizing out the 'Mickey Mouse' spring clip and removing the small cotter pin.
- You will invariably have breakages and damage, and you'll identify faulty and missing components at this stage – so,

once again, use your notebook.

- Air conditioning gas is a major source of damage to the ozone layer so you should have a specialist drain down the AC system for you if you have to disconnect it. You should also never disconnect the system yourself because of the pressures involved. Sometimes the air conditioning compressor can be unbolted from the engine and wired to the inner fender so you don't have to open up the system at all. Once your specialist has disconnected the AC lines, blank them off with tape to keep out dirt and moisture – the system's greatest enemies.
- Disconnect the brake lines to the master cylinder.
- Disconnect the clutch linkage by removing the spring clip and removing the clutch pedal rod at the upper end of the bellcrank.
- It is best to remove the chrome bumpers and the supporting brackets and any braces connecting the rear frame bar to the lower panel or license plate support.
- Also from the front of the car, unbolt and lift off the nose. After removing the front bumper brackets look for the supports used to attach the nose to the frame horns, then lift the nose free.
- Disconnect the ground strap for the antenna base plate and the safety cables that connect the seat belt brackets to the frame.
- On 1963 to 1966 cars the parking brake cable can be disconnected at each end of the front cable, but on later models you will have to disconnect the stud end. On 1967 to 1982 cars, remove the pulley bolt. Note the cable routing in all cases so that you can reassemble it correctly.
- Disconnect the gear shifter cable from auto transmissions and on four-speeds unscrew and remove the ball and backup lockout from inside the passenger compartment.

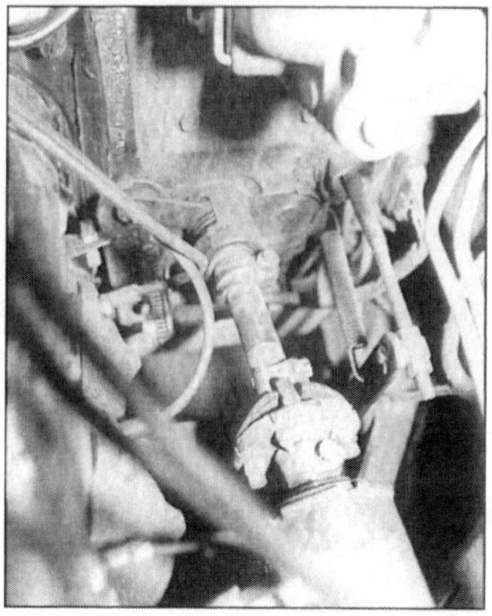

▲ *BR1. Now we are on to some real hardware. Disconnect the steering column from the steering box by removing the two bolts which pass through the flexible coupling joining the steering column to the steering shaft. This is known as the 'rag' joint. You may wish to remove the column by just undoing the bolt from the shaft clamp using a $\frac{7}{16}$ in. socket. Spread the clamp by tapping a screwdriver into it and this will make it easier to slide the column in or out. Remember to mark the alignment of the shaft to the clamp to save yourself having to set up from scratch on reassembly.*

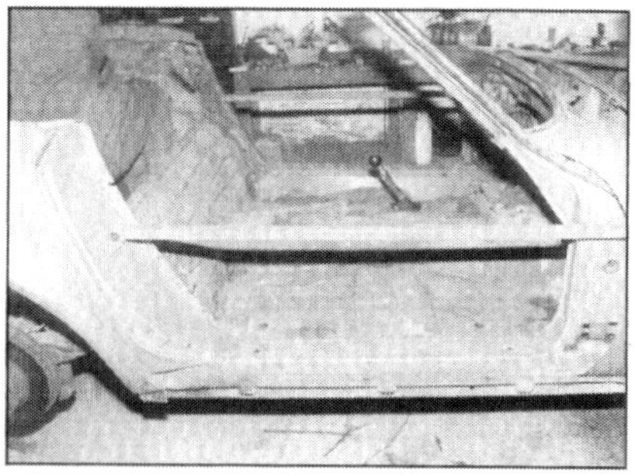

▲ *BR2. Because we had removed the doors from our car, Em wisely decided that a strong metal brace should be fitted at either side in order to retain the body's structural integrity.*

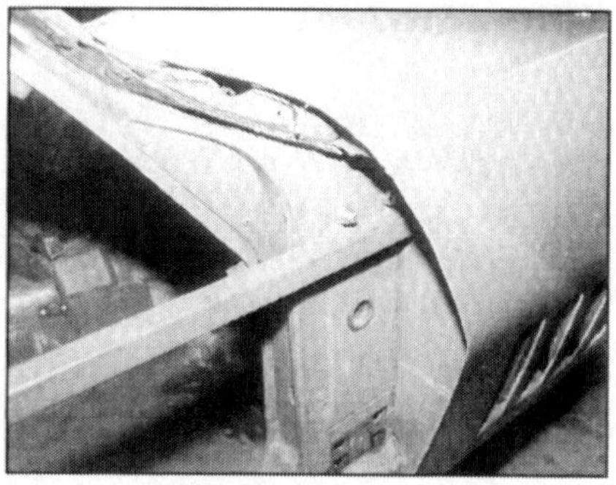

▲ *BR3. Em designed the brace specifically to bolt into existing holes for the door lock and front hinges respectively. As such, lifting the body and moving it around created none of the flexing problems (and possible resultant damage) which can so easily occur.*

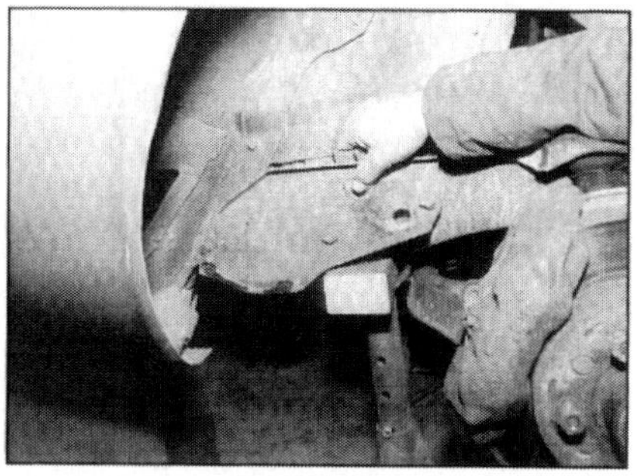

▲ *BR4. Corvettes from 1963 to 1982 have four body mounting bolts on each side, and the body is also connected to the frame horns at the radiator support mountings.*

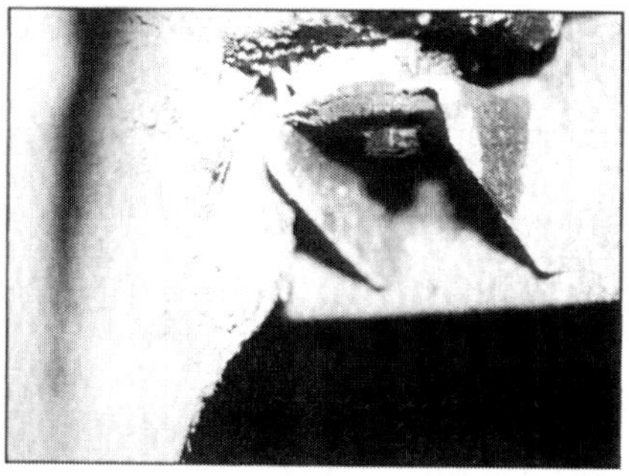

▲ *BR5. Take as much time as you need to ease out the body mounting bolts because most of them are screwed into hidden case nuts. The case nuts are prone to braking free and then the bolts will have to be cut out and, worse still, you will have to let in new case nuts or drill them out and let in non-original replacements from the top. Spray the case nuts even more generously than normal with*

releasing fluid before commencing work. As you can see here, the nuts/bolts are likely to covered in several years worth of road dirt. A little patience could save a lot of frustration at this point.

1969–82 Corvettes had four body mounts on each side of the frame.

1969–72 models were connected with metal spacers instead of the rubber cushions used with other years.

1963–67 Corvettes also have four body mounts per side on the frame. 1963 had an additional two 'mounts' per side that anchored the seat belts to the frame: these were replaced with cables on later models.

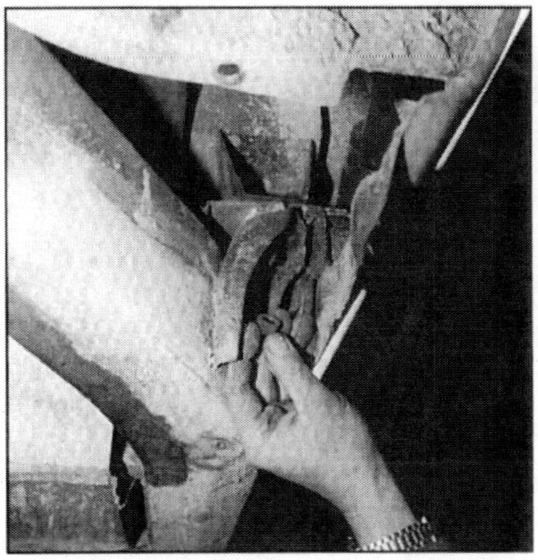

▲ *BR6. You are better off with the No. 1 body bolts because both nut and bolt head can be accessed. This picture shows the bolt being removed from the driver's side.*

▲ *BR7. Before you can get at the No. 2 body mounts, you have to remove the rocker panel moulding on 1963 to 1967 cars, and both the moulding and kick panels on 1968 models onward. The kick panels are fixed by a single Phillips head screw and are likely to be rusted in.*

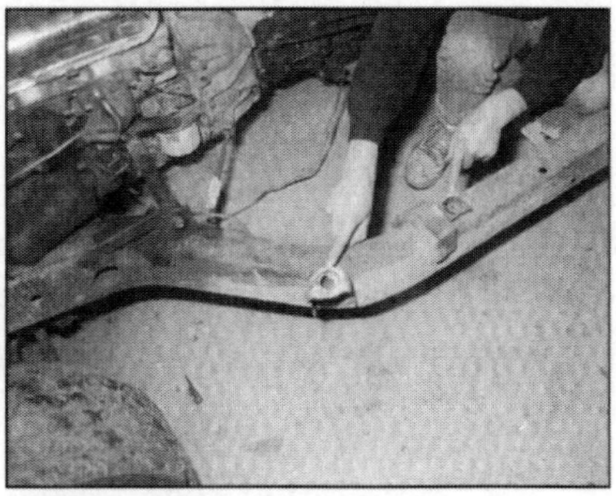

▲ *BR8. Here's another view of the positions of mounting bolts Nos 1 and 2.*

▲ *BR9. Before you can access the No. 3 bolts, you have to remove the plate located on the inner fender well in front of the rear tire. Four securing screws hold it in place ...*

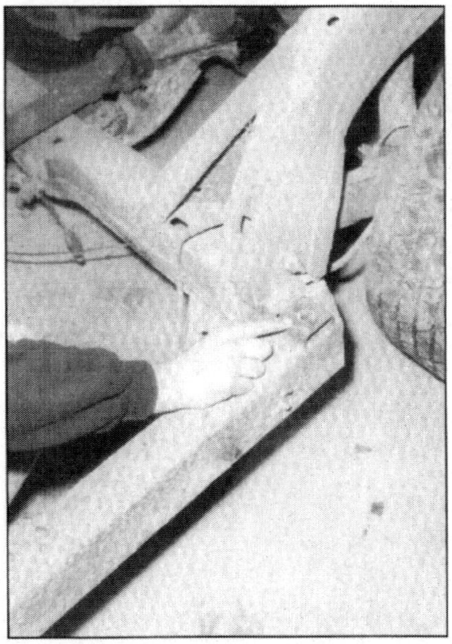

▲ *BR10. ... and once it's off, the mounting bolt can be reached.*

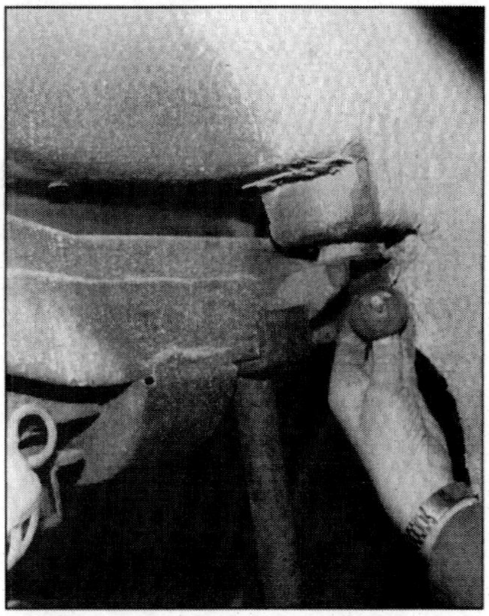

▲ *BR11. This gives you an idea of where the No. 4 body bolts are located, at the rear end of the fender well. The case nuts are inside the passenger compartment in the back corners.*

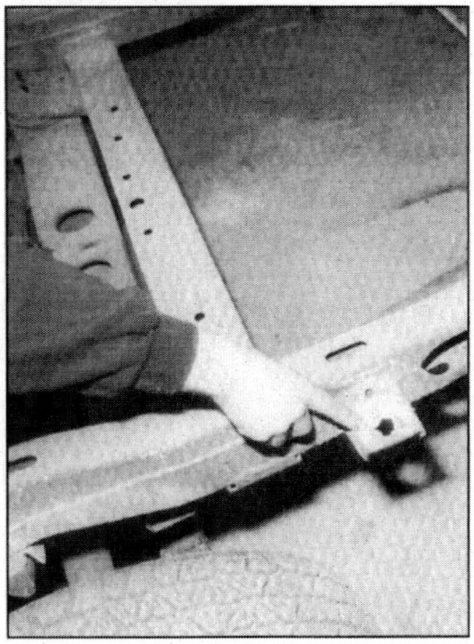

▲ *BR12. The position of the No. 4 body bolts is easier to see with the body removed!*

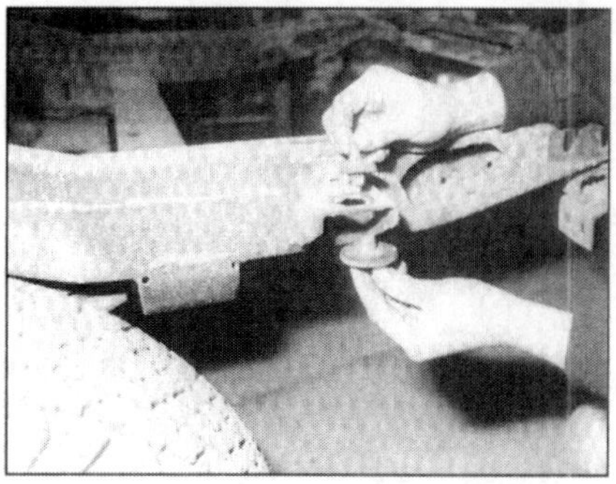

▲ *BR13. Things don't always go well – this bolt sheared clean in two …*

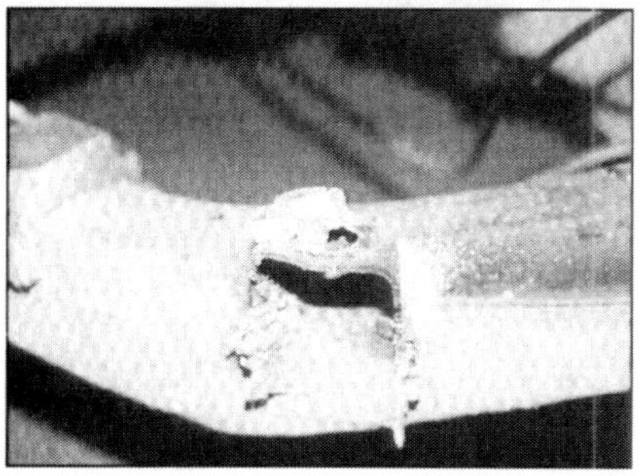

▲ *BR14. … and this spacer has seen better days!*

LABELLING SHIMS

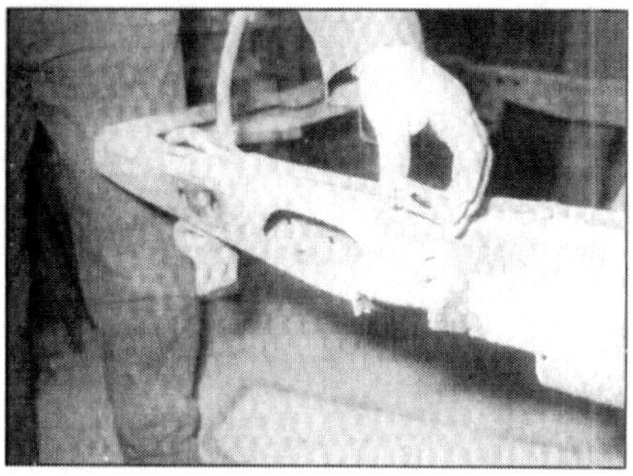

▲ *BR15. If you don't think about it beforehand, the shims will end up in an untidy pile and you won't have the faintest idea what went where. Put each set of shims in a separate plastic bag, write on a piece of paper where they came from and seal the end of the bag with a bag tie. Either that, or put each set in a glass jar with a note. If you have difficulty in getting the shims out, lift the body a little at a time packing it up with a block of wood between the body and frame, and get those shims out before they fall into an unidentifiable pile.*

LIFTING THE BODY

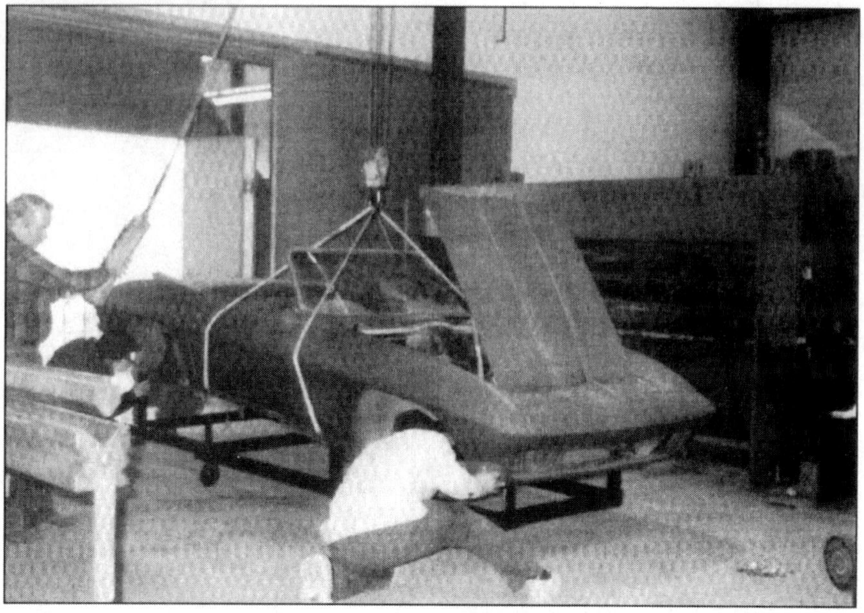

▲ *BR16. Using a strap and hoist arrangement is by far the simplest method of removing the Corvette's body. With its adjustable strap, such a set allows the body to be perfectly balanced and levelled so that removal and realignment is greatly facilitated. The hoist or chainfall should be rated at least 2,000 lb for safety, although a stripped Corvette body weighs considerably less than that. The weight carrying ability of the rafters, tripod or whatever is used to support the hoist should be examined carefully, too. Tighten up on the straps and recheck where they are attached to the body. Next, pull up with the hoist until the body begins to raise as weight is taken off the suspension.*

▼ *BR17. In this case Em decided to use friends and neighbours (bribed with a promise of a ride in the finished car!) to raise the body. Go around each of the four corners and use a bar or flat-ended tire lever (iron) to carefully prize the body off the frame. The body often becomes stuck to the frame at its mounting points and it's better to break it loose in this manner. Not only could the body suddenly snap off, the strain on the body and hoist also is greatly increased if the body is not broken loose first.*

▲ BR18. *As we said earlier, it is essential to check for those wires and hoses that you have not remembered to disconnect. If there are a number of people doing the lifting, nominate one of them to look and look again for any connections that need to be disconnected.*

Make sure you have enough headroom when lifting the body right off the frame. Remember that the frame will come up a little way as the weight of the body is removed from it. If all else fails, it may help if you remove the fuel tank. It sits quite high up on the frame and might just be the last straw in preventing you from lifting the body and moving it clear of the frame. The best way, of course, is to lift the body and roll out the frame as shown here.

Another tip is to remove the rear bumper bracket to give more clearance when the body is lifted off. As you can see, Em drew the short straw for pushing out the rolling chassis.

▲ BR19. *With the chassis out of the way, the lifting crew could gently lower the body down. Note that the bracing bars across the door apertures made ideal lifting points.*

There is plenty to think about when considering whether you should remove the body from your Corvette's frame. Plenty of food for thought, or as the man himself put it:

And thus the native hue of resolution
Is sicklied o'er with the pale cast of thought

... which is just as it should be.

PUTTING ON THE PAINT

It's not impossible for the keenest of enthusiasts to paint his or her own Corvette but, in all honesty, the only way the great majority of us are going to see a truly professional finish on the car is to have a professional body shop carry out the work. In any case, the best paint for the job – two-pack-paint – is emphatically not suitable for home workshop use since its fumes can be toxic.

There is a type of lacquer that is suitable for home spraying which must also be treated with the utmost respect because of the health and safety risks associated with it, but the finished paint job is not as resilient as it would have been if you had used the best two-pack paint. The safety provisions mentioned in this chapter and at the back of this book are for those who really do insist on carrying out the work themselves, but we strongly recommend that you leave the work to a specialist with a low-bake spray booth, capable of the highest standards of workmanship.

▲ PP1. *John Richards, proprietor of Avonvale Autobodies, examines John Fletcher's Corvette with great interest as it's off-loaded from the delivery truck outside his workshop premises late one evening.*

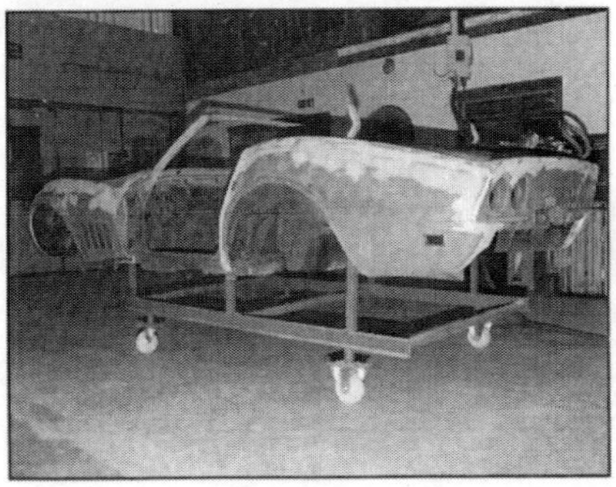

▲ PP2. *The only way you can carry out a full spray job to this extent is by having a completely stripped bodyshell in this state of undress. Never move a bare bodyshell around without its doors unless you've added the bracings shown in the bodywork repair section, or the bodyshell is mounted on a purpose built trolley like the one shown here.*

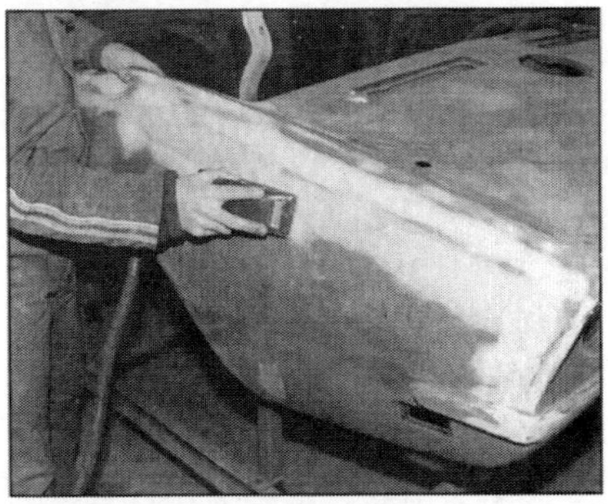

▲ PP3. *Any bodyshop worth its salt will want to take extra time and trouble eliminating even the most minute imperfections from the bodyshell. Avonvale examined the body panels, especially where Em had replaced fenders and smoothed out imperfections that couldn't be seen with the naked eye, using more bondo. Here a machine sander is used with an extractor tube to sand down the majority of the filler. When you are carrying out this work at home, it is ESSENTIAL that you wear an efficient face mask so that you are not breathing in dangerous dust.*

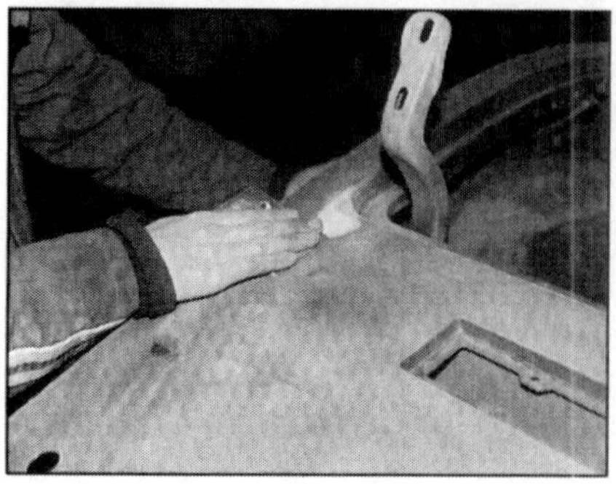

▲ PP4. Some small areas, particularly in concave curves and right on the top of convexes and ridges, have to be sanded by hand, because with a machine sander it's impossible to follow the contours.

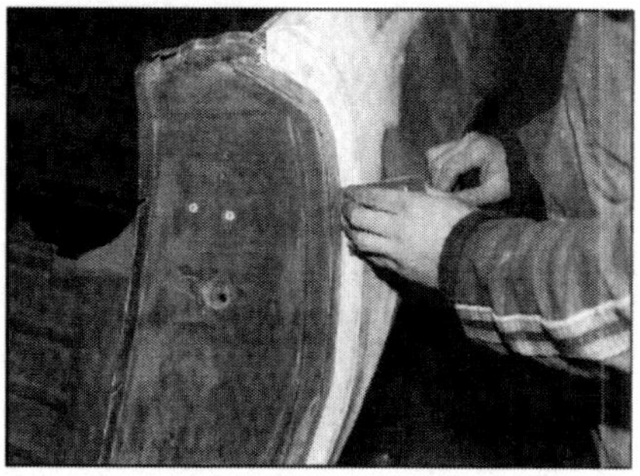

▲ PP5. Similarly, the insides of mouldings and the complex shapes around the door shut pillar also have to be sanded carefully by hand. It's essential to bear in mind that only an expert can hope to get rid of minor body imperfections at this stage. Because the body is not finished in shiny paint, you just won't be able to see the ripples that will make themselves apparent later. The expert runs his hand flat over each of the panels and, through experience, can feel any ripples in the surface. You can try it and you'll certainly pick out the worst in this way, but if you want to do your own spraying and you're not experienced, be prepared to put on a thin coat of gloss paint and then go back and fill and sand out all the ripples that it shows up.

▶ PP6. As we said before, there's been a lot of dust flying about, so the air line is used to blow every trace of it out of all the nooks and crannies.

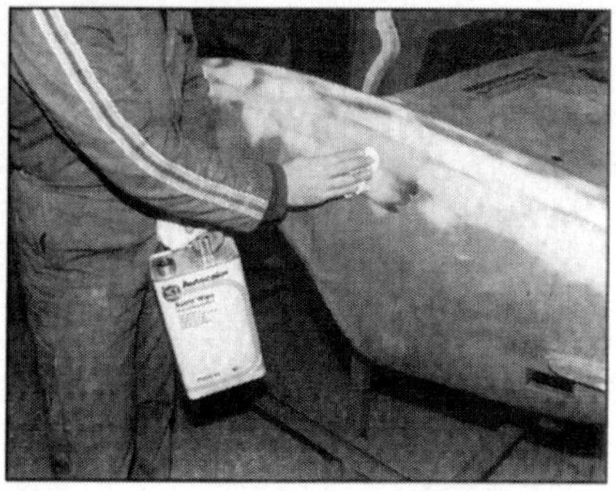

▲ PP7. You mustn't even try to apply paint to a car without the use of spirit wipe! Totally unseen, both grease and silicones will get onto the surface of the body and either of them can make a mess of the finished job. Silicones, in particular, are the painter's enemy, so do all you can to avoid them. Most domestic polishes and car polishes contain them, so even if you are just sanding down a small area of your car for touching-in, wipe it down first with spirit wipe so that you don't sand silicones into the surface. Silicones have the effect of repelling paint, and within seconds of applying it will leave small craters called 'fish-eyes' all over the painted job, often so bad that in the middle of the craters you can see the primer beneath. If that happens, the only answer is to strip off all of the paint and start again – a daunting thought! So use a clean rag (not one that might conceivably have been used before and so have polish on it!) and wipe every square inch of your car's body several times over with spirit wipe, turning and replacing the rag frequently. You must do this between each stage in the paint process: before applying filler; before applying primer; before applying finished paint.

▲ PP8. The Corvette's shell has now been pushed into the Avonvale spray booth, and Ben gets ready to mask off all of the car that isn't going to be sprayed. This is time-consuming work but essential if you want to avoid the second of the painter's worst nightmares: dust! The cockpit, engine bay and trunk area are all extremely difficult to clean totally of dust, and the only way is to mask them off before painting the car.

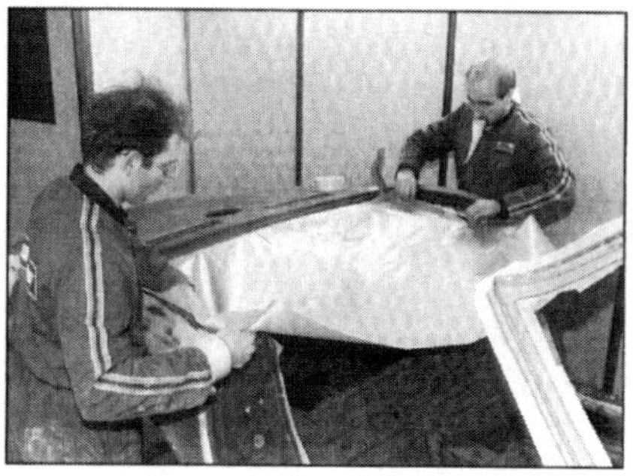

▲ PP9. Once a strip of masking tape has been placed all the way around the folding top bay, half of the width of the masking tape exposed, sticky side up, Ben and a helper use brown masking paper to cover over the gap.

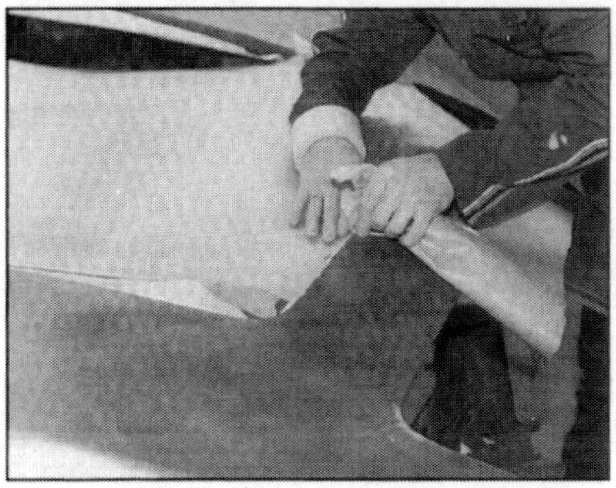

▲ PP10. Ben finds that the best way is to cut the paper oversize, stick it down to the masking tape and them tear it back so that it fits the aperture exactly. That way, paint can go right up to the top edge of the aperture. Some little time must be taken in ensuring that all of the edges are neat and tidy.

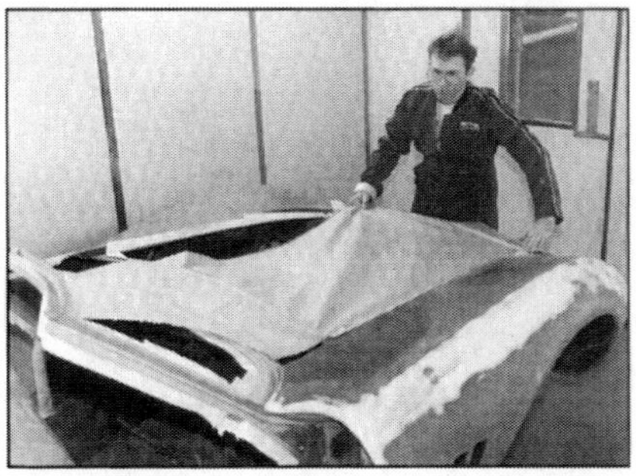

▲ PP11. Ben continues to construct a paper 'hood' for the car ...

▲ PP12. ... and continues to cover over the cockpit and folding top bay.

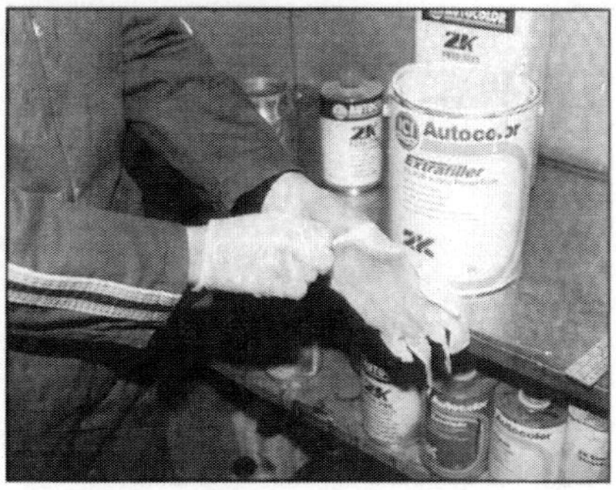

▲ PP13. We said earlier that these two-pack paints could be injurious to health, and so they are. Ben follows essential working practice and pulls on a pair of gloves.

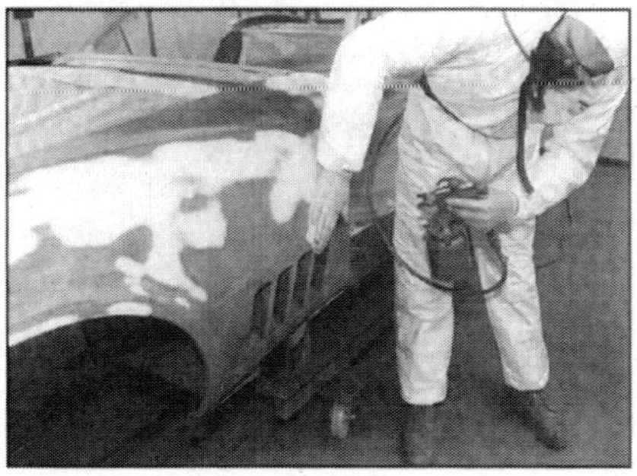

▲ PP14. The car is spirit wiped once again and, while Ben uses the air from his spray gun to help dislodge any stray bits, he wipes the surface of the whole bodyshell with a tak rag – a specially designed sticky cloth which takes hold of any loose dust rather than leaving it to fly about later. You can buy them from specialty auto-paint suppliers or some accessory stores.

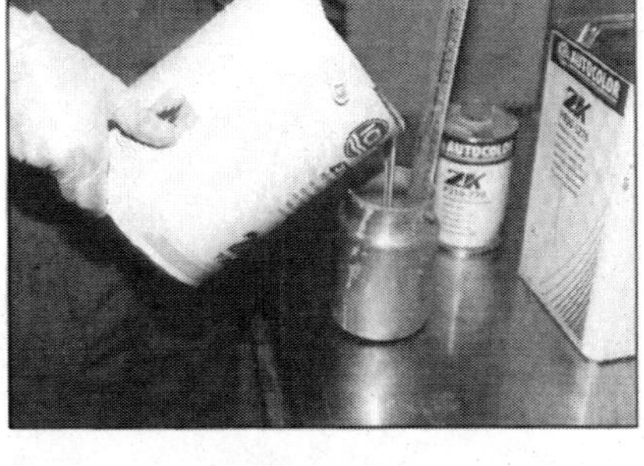

▲ PP15. Meanwhile, in the paint mixing bay, paint is poured into the spray gun pot in the right proportions ...

▶ PP16. ... followed by exactly the right amount of thinner ...

▲ PP17. ... and the chemical hardener that will set the whole thing off.

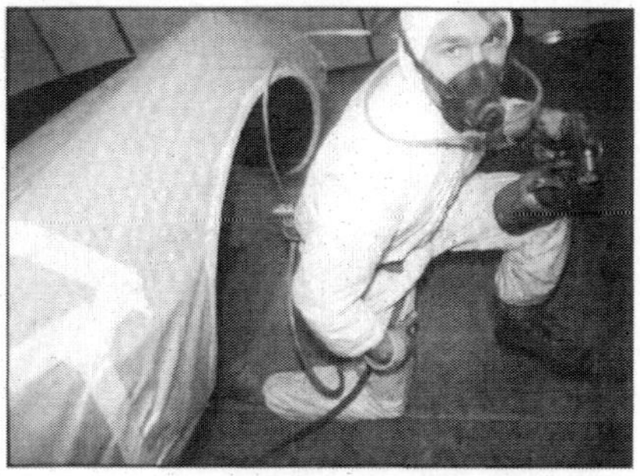

▲ PP18. No it's not a deep sea diver, it's a painter working with two-pack paint, and illustrates the steps that have to be taken to counter the safety hazards. (It also illustrates quite clearly why this stuff is not suitable for home use!) Ben is wearing a complete set of throw-away paper overalls with integral hood to stop the paint from getting on to his hair, plastic gloves to prevent the paint from coming into contact with his hands, and most important of all a full air-fed face mask which provides an air supply from outside the booth so that there is no chance whatsoever of Ben breathing in any of the fumes from this potentially toxic paint. Some painters prefer to wear a full face-covering mask so that the paint stands no chance of getting into the painter's eyes. We recommend it.

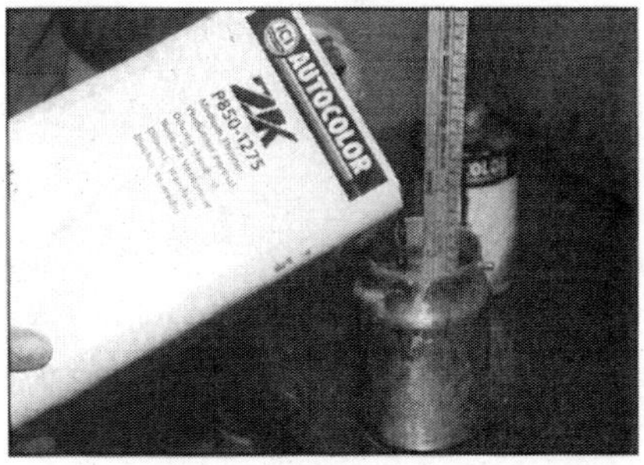

▲ PP19. Ben has been all round the car once, putting a thin coat of primer over each of the repaired areas and also over any tricky shaped areas where he would later have to delay with the spray gun as he goes over the whole job. Fiddling around with tricky areas when you're trying to paint the whole job is a sure fire way of achieving unwanted runs in the paint. So, Ben starts at one corner of the car, working along the length of the rear fender from end to end in 50 per cent overlapping strips.

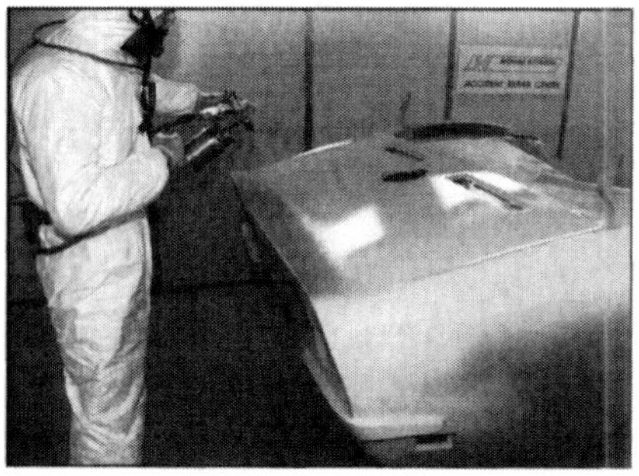

▲ PP20. Then he turns his attention to the trunk lid area, starting right at the top, and again working in overlapping strips …

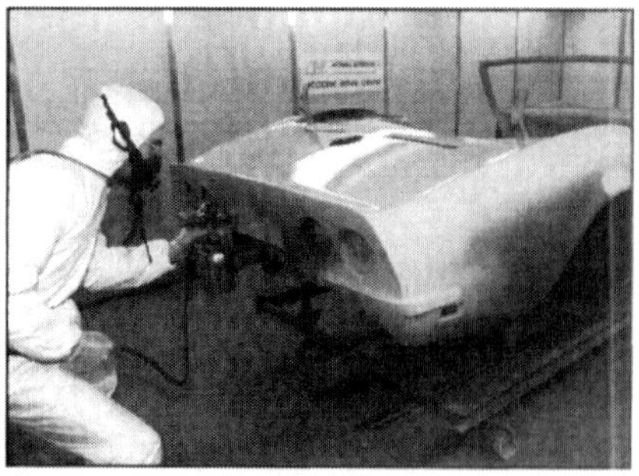

▲ PP21. … and working right down the rear panel to the bottom of the rear valance.

▲ PP22. Then it's the turn of the passenger side rear fender …

▲ PP23. … followed by the front fender …

▲ PP24. … followed by the front of the bodyshell, the hood surround …

▲ PP25. … the other front fender, working once again in continuous strips until he gets to the wheel arch …

▲ PP26. … when he switches his attention to one side of the wheel arch at a time. Then he works back down the body rocker panel area and back to the point where he first started. The beauty of two-pack primer is that it can be put on thicker than traditional lacquer primer and shrinks back far less, giving better coverage of minor imperfections in the surface of the body. It can later be sanded back even further, it stops repairs showing through and it gives a more durable finish. Unfortunately you do have to have the equipment and dress of a passenger in the space shuttle to be able to use it safely!

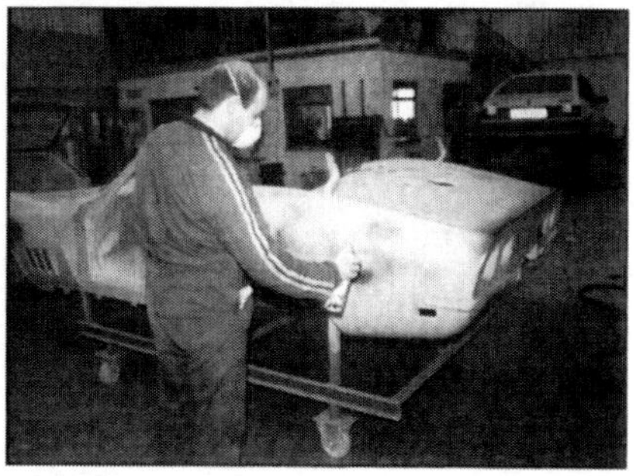

▲ PP27. The low-bake oven has done its job, the primer has gone off and the bodyshell is pushed back out into the Avonvale workshop. Whatever you might think to the contrary, Avonvale have not now decided to paint the car in aerosol paint! They are using black aerosol lacquer to put a very fine misting coat of black paint over the whole of the shell.

▲ PP28. Every bit of the shell is covered and where you think there might be minor blemishes, a little more paint is applied – but note that it's still not a complete covering coat – it's there for a special purpose!

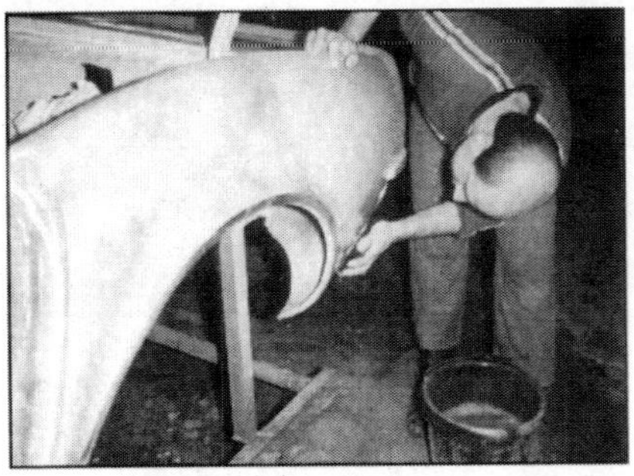

▲ PP29. As the whole shell is wet-sanded with a fine grit of wet-or-dry paper, tiny scratches and imperfections will be revealed as the black misting coat is left lying in such scratches while the sanding block takes the black paint off the flat surfaces that surround it. The body finisher can then decide whether just a little more sanding will take out the minor blemish – making the most of the value of two-pack primers – or whether it will be necessary to apply a little two-pack stopper (like ordinary bondo but much thinner and finer and therefore more suitable for small blemishes) before sanding down and re-priming that bit of the car.

The perfectionists at Avonvale wouldn't let us into the spray and bake oven while the top coat was applied because of the risk of our cameraman stirring up dust and causing imperfections in the surface of the paint. The principles that were followed during the application of primer are exactly the same for top coat – and the results?

▲ PP30. Well, the results speak for themselves and the fully finished bodyshell looked nothing less than stunning when it came out of the spray booth.

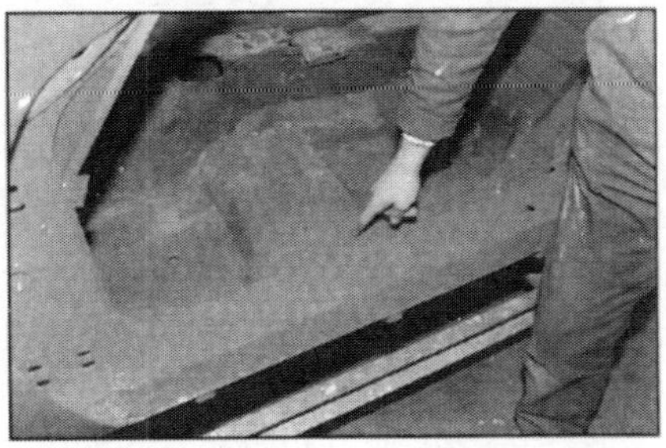

▲ PP31. When these bodyshells were painted in the factory, there was always a little overspray on the floor. For that reason, the car's owner, John Fletcher, stipulated that the overspray had to be recreated and the paint allowed to harden off before the cockpit area could be masked off once again and the whole car painted as described earlier.

▲ PP32. The depth of shine that you get from two-pack paint is difficult to emulate with lacquer except by a process of cutting back and polishing for hours at a time. Two-pack paint also has the advantage that it retains its shine for years to come, whereas lacquer needs the elbow grease treatment at regular intervals.

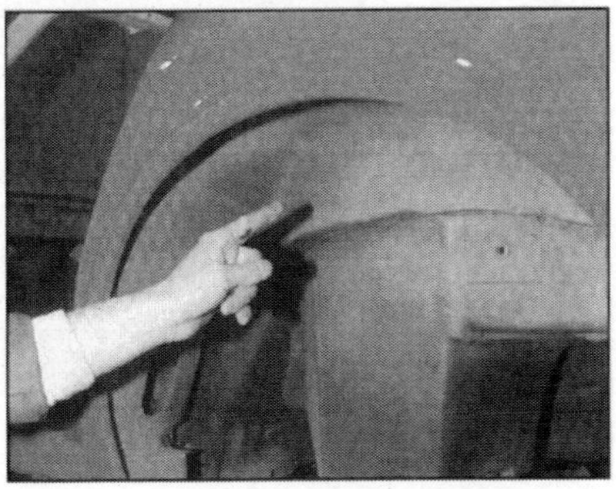

▲ PP33. Bearing in mind that John Fletcher's car had been stripped down to the barest of bodyshells, it was now time to paint the insides of the wheel arches, the inside of the engine bay and the trunk.

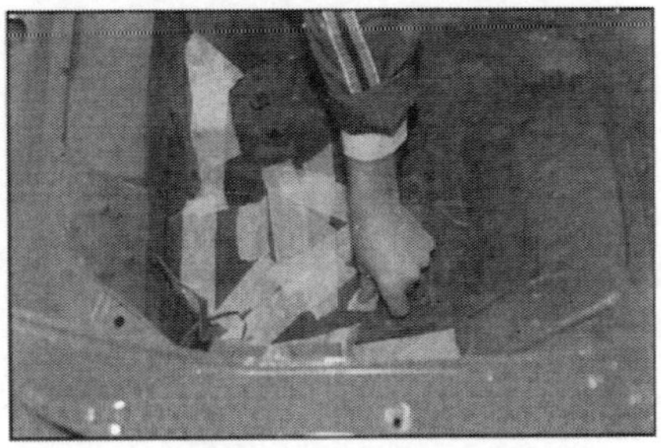

▲ PP34. Ben ensures that every small mounting hole and access hole is blanked off with masking tape …

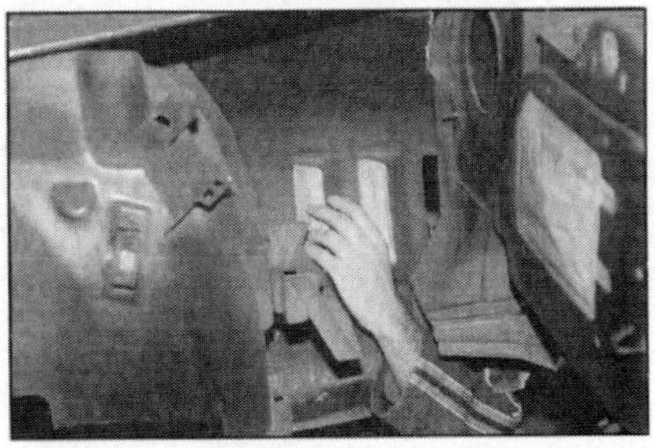

▲ PP35. … not forgetting the all-important vents in the front fenders!

▲ PP36. Headlamp apertures receive the treatment …

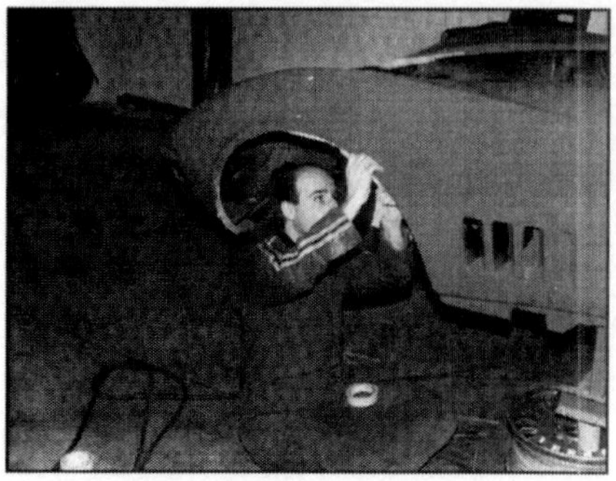

▲ PP37. ... as do the newly painted edges of the wheel arches ...

▲ PP38. ... and the edges of the hood surround.

▲ PP39. As a kind of reverse of what had gone before, Ben concentrates on covering every square inch of the newly painted bodywork in masking tape and masking paper. A little creative thinking was needed here to ensure that all the bits he wanted to paint got painted while none of the bits he didn't want to paint didn't get painted!

▲ PP40. Once again, it was time to apply the spirit wipe, the blast of air from the spray gun and the tak rag ...

▲ PP41. ... before painting beneath the wheel arches ...

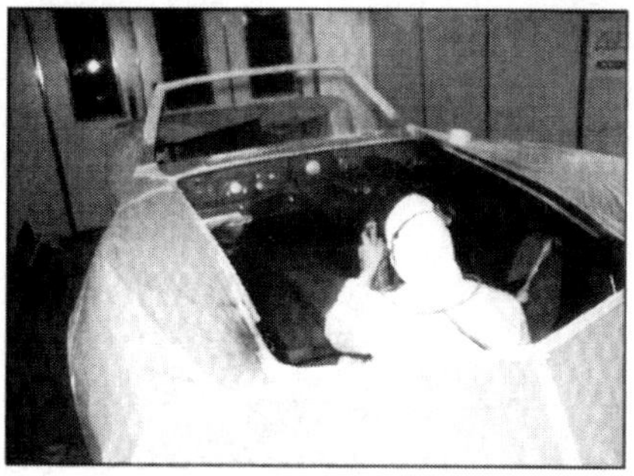

▲ PP42. ... inside the engine bay ...

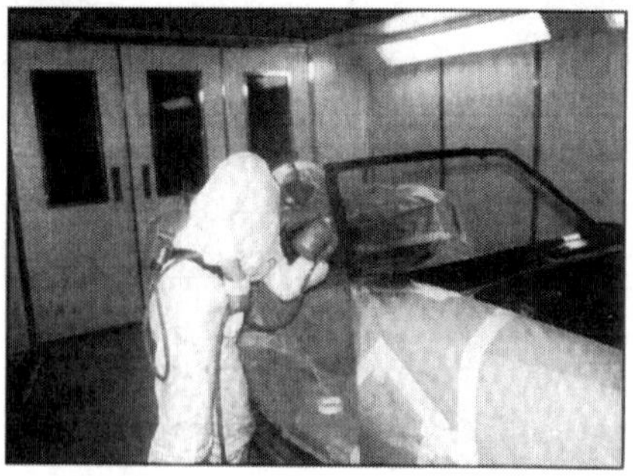

▲ PP43. ... and the windshield surround with exactly the correct satin finish (not high gloss) black paint.

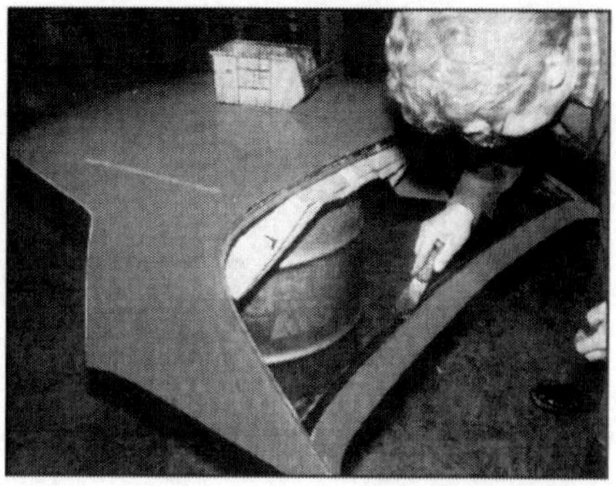

▲ PP44. Meanwhile, the car's hardtop was being cleaned up back in John Fletcher's workshop.

▲ PP45. Em pours on some of the special glassfibre paint stripper that he used to get the old paint off the car. Note that you absolutely must not use regular paint stripper, since this will also strip off the fiberglass gel coat. Better still than pouring the stripper onto the bodywork is to pour it into a jar and apply it with a brush. Be sure always to wear suitable rubber gloves, goggles and protective clothing so that none of the stripper can get onto your skin. If any of it does get onto your skin or into your eyes, stop immediately, wash off with copious amounts of cold water and seek medical advice.

▲ PP46. You have to take the utmost care when scraping and stripping paint off fiberglass since it is all too easy to damage the surface beneath which will inevitably be softened to some degree by *the stripper. It takes far longer than doing the same job on steel bodywork and you should allow sufficient time to do the job properly and realize that it takes lots of patience! Note that Em has spread sheets of old newspaper all over the floor to catch the drippings and scrapings as the paint is removed. A good idea! Take care that children and animals' paws can't come into contact with the paint stripper or the residue as the paint is stripped away, since it is extremely caustic.*

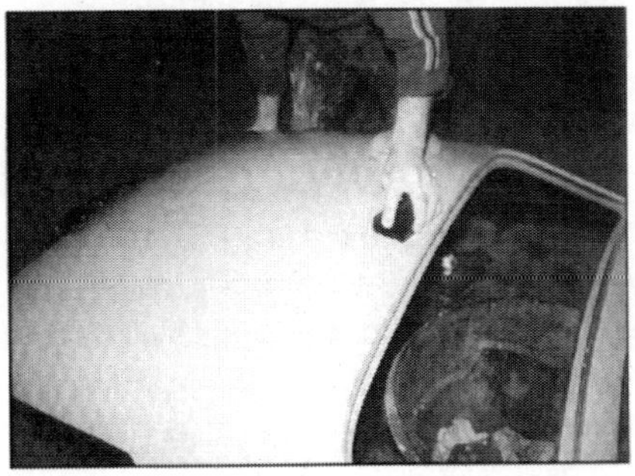

▲ *PP47. At Avonvale, the hardtop was primed, sanded and painted in exactly the same way as the rest of the car.*

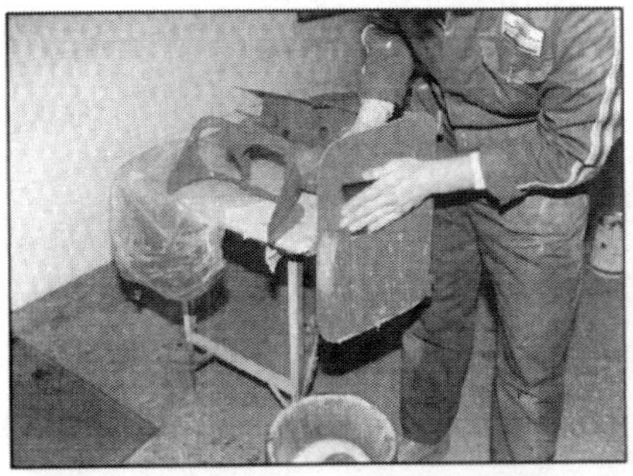

▲ *PP48. Painting all of the minor components in a restoration of this size is in itself a major undertaking. Ben sands one of the headlamp covers with wet-or-dry having first given it a single coat of finish paint.*

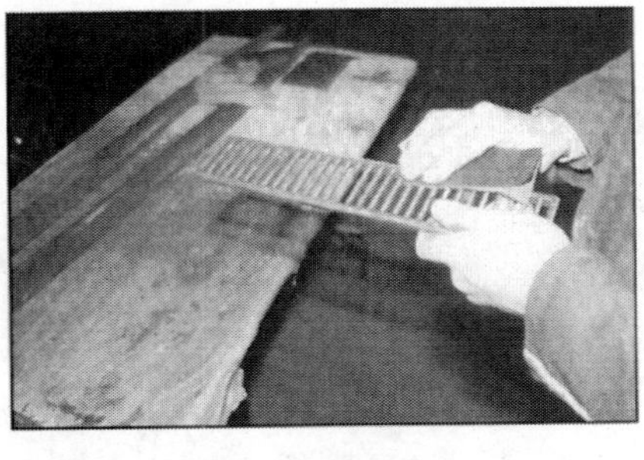

▲ *PP49. Small areas such as this grille just have to be sanded by hand and the whole process will take several hours.*

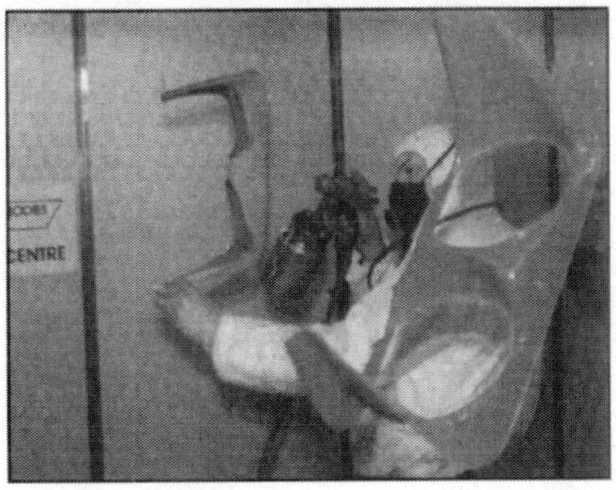

▲ *PP50. The only way of painting such small components is to hang them up, and even then you need a lot of space to ensure that you don't get overspray on one component while painting another. Professional spray booths have vertically moving streams of air that take the paint overspray down into a washing and filtration plant beneath the floor and well away from other components hanging up in the booth.*

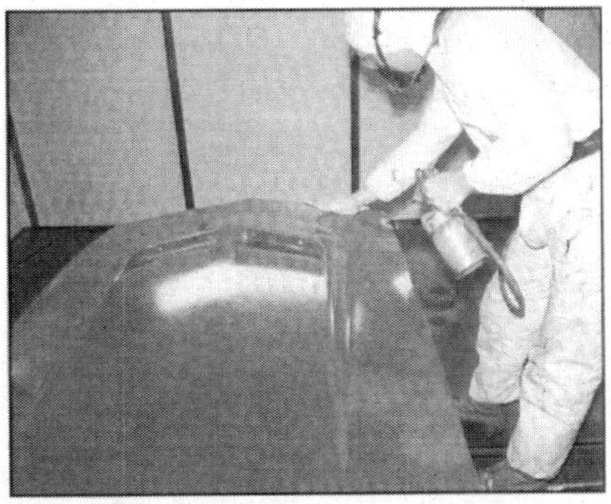

▲ *PP51. However careful you are, in a job of this magnitude, there will usually be the odd blemish. After giving the hood its first coat of paint, Ben noticed that one corner was not as he wished it to be, so he sanded and stoppered it once again …*

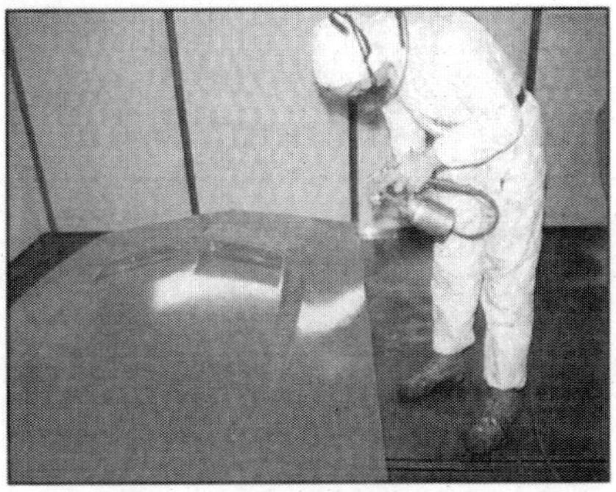

▲ *PP52. … and gave that small area a coat of paint before giving the hood its final two coats.*

▲ *PP53. Two-pack paint builds much more heavily than lacquer and you should rarely need more than three coats. The finished hood and components look magnificent, even under the searching gaze of the workshop's overhead lights.*

▼ *PP54. Back at John Fletcher's home, the painted bodyshell had a magnificent as-new air about it. It's when you reach this stage in a restoration that you gain the extra impetus that you need to go on and finish the job to a high standard.*

REPAIRING FRAME RUST

In this section, we show you how to go about repairing rusted-out frames, but before getting into the how-to there are a couple of important pointers to consider.

Number One. Whenever you carry out anything more than the simplest rot repair, and in particular whenever you get near suspension mounting points, always have the frame alignment checked by a specialist repairer with an alignment check jig. If all of the pick-up points are not exactly right, your Corvette's handling could become downright dangerous. If fact, why not have the checks carried out before you commence work and then you will be able to find out if some ham-fisted bozo caused accident damage or repair damage before you even owned the car.

Number Two. It is not just originality for originality's sake, it is also a practical point: ensure that you don't obliterate the frame ID number when repairing the frame. We'll discuss frame replacement later, but the same applies to frame replacement as it does to cutting out and repairing the part of the frame with the ID number on it – make sure that you transfer the old ID number to the new frame or part. MIG-weld in the repair then grind down the weld, finishing off with bondo and paint so that the weld is invisible and the ID number looks original.

In America, many States demand that at least two ID numbers out of the body tag number, engine number and frame number match on the car. If they don't, the car cannot be registered, so it is essential that you cling on to those ID numbers for dear life! If your engine completely blows up at some time in the future you may have no option but to replace it, and then where would you be if you had also lost your original frame number?

REPAIR OR REPLACE?

In almost all cases, frame repair is greatly preferable to frame replacement. For one thing, you will be extremely fortunate if you come across exactly the right frame for your car in excellent condition – and bear in mind that frame types differ quite considerably between models – and even a GM part, still being made for the later cars, is bound to have the latest features added to the tooling during its production life and so not be 'correct' for your 'Vette's model year.

The second problem is that frame replacement can be horrendously expensive, not just because of the high costs of the frames themselves but because of all the stripping out and inevitable renewal that you will have to carry out.

The third reason is that frame replacement can be extremely time consuming. All of those fixtures and fittings on your old frame will have become kind of comfortable and won't take kindly to being asked to move. You will encounter frozen threads, snapping studs and rounding over hexagons, and you may find it difficult to locate replacements for some of the more esoteric bits and pieces that don't normally need replacement.

Corvette expert John Pfanstiehl points out that quite a few repairs to the frame can be carried out with the body in place, but if you do so, be aware of the following:

SAFETY NOTE: There are a number of highly flammable parts on your Corvette, and welding the frame with the body in place can be very dangerous. fiberglass is flammable, and you don't need telling about motor fuel – and not just fuel in the tank, but fuel in the pipelines leading to the front of the car. Beware, too, that hydraulic lines can contain flammable fluids, and interior trim can catch fire from overheating from beneath. Also, any rust-proofing materials injected into the frame, and underbody seal and oil thrown onto the frame will be prone to catch fire. ALWAYS have a fire extinguisher to hand, and stick to MIG-welding rather than oxy-acetylene welding for frame repairs. Not only is fire less likely but you will also induce far less heat distortion into the frame itself. Even though the body may need to come off for a complete restoration project, or for frame repairs, it's almost always better to leave it where it is and work around it. Read the safety information in Appendix 1 before starting work.

FRAME DESCRIPTION

FD1(Below left). Though the stunning bodywork of the '68 series cars was all new, the frame was inherited without changes from the 1963/67 model Corvette. Basically, it is a strong steel box frame, strengthened by five cross-members. The transmission cross-member has two holes through which the twin exhaust pipes are directed under the differential in order to avoid contact with the fully extended transverse spring. These can be seen clearly in the centre of this photo. On manual transmission cars, the cross-member is welded in situ *whereas, on vehicles with auto transmission, it is removable so that the Turbo Hydramatic can be removed without the necessity to displace the engine. At the rear of the car the perimeter rail angles up vertically and then continues back in order to support the rear bumper and fuel tank. For 1969 extra reinforcement fillets were introduced to brace this by the front of the trailing arm.*

The differential can be found on its own cross-member which is itself mounted on a higher part of the frame. It has two bushes which are extremely effective at absorbing vibration.

At the front, the suspension comes with unequal length wishbones (or A-arms) similar to the Chevrolet sedan cars of the era.

REPLACEMENT PARTS

The first thing to consider is whether the replacement sections you are purchasing are exactly the same as those on your Corvette. As we said earlier, there are many variations, so talk things over with your suppliers and even send them a photograph or sketch to ensure that you are getting the right part. Small repair parts can be made up in your own workshop but it can be very troublesome and time consuming to make the complex stuff and get it right. Obtain from your supplier the full list of the parts they have available once you have established what needs replacing on your car.

FRAME REPAIR – GENERAL NOTES

If you are working on a stripped-down frame or even one where the body has been taken from the frame, you will need to ensure that it is level and true before commencing work. Most workshop floors are not true, so you will have to make up floor bearers checked with a spirit level in every direction to give yourself a true surface to work from.

When replacing frame rails, you will have to cut and separate the welds that fix the cross-member to the side rails. Cut back from the rails, taking care not to damage the cross-member. Be sure not to cut into the cross-member at all, otherwise you won't have anything to weld to! A useful tip from John Pfanstiehl is to cut an access hole in the vertical side of the frame section which is being removed. Then the welds can be cut from the inside of the frame with less risk of cutting into the cross-member. Alternatively, cut the side rail away from the weld and grind back to the weld joint.

Before cutting the cross-member from the side rail, be sure to have taken accurate measurements so that you can ensure that it is precisely aligned when welding everything back together again. Also measure the exact location of all clip holes, such as for the fuel and brake lines and drill new holes in the replacement section.

With the old side rail out of the way, the six-inch sleeve insert can be slid into the frame rail, leaving three inches exposed. Tack weld it into place, then slide the replacement section over the sleeve. You can now weld the section to the side rail and then to the cross-member. On 1963 to 1967 models the rocker panel moulding brackets must now be welded on.

SAFETY NOTE: You should only carry out this work if your welding is of a high standard. The weld needs to be a good, strong, penetrative weld providing maximum strength, and you need to know enough about when to make a single or double pass with the welder and whether your welding is sufficiently strong. If your welding skills are not of a sufficiently high standard, leave it to an expert.

▲ *FD2. You are sure to need to use a cutting disk, so it is important (very important if you use a cutting torch, as shown here) to make sure that the entire workshop area is cleared of flammable materials before commencing work.*

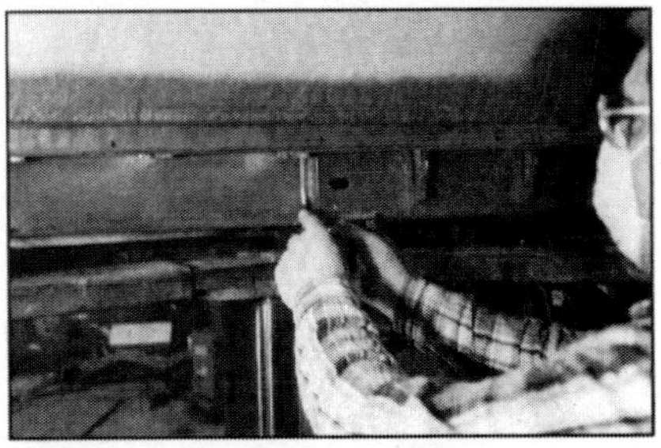

▲ *FD3. You can purchase frame sections in different lengths, but do remember that you can work off the repair section and cut it to the best lengths for your own purposes. You don't have to use the whole thing if it doesn't suit you!*

▲ *FD4. This is the frame of the '69 car we show being worked on. Despite being over 25 years old at the time, it showed no signs of previous accident damage. And, though the caked-on mud hid a few rot-spots, in general the frame was satisfyingly solid and was certainly repairable.*

Corvette rust problems usually begin at the end of the side rails where accumulated dirt can trap moisture. As you will see in this section, this is where most of our troubles were. However, Em applied his considerable metalworking skills to repair these and other areas that needed repair.

▲ *FD5. He started work at the front of the car, carefully checking all sections for signs of obvious rust or damage. The engine cross-member leads a hard life; not only is it used for jacking the car up at the front, but any minor driving errors (misjudging the height of a curb, for example) are likely to lead to damage.*

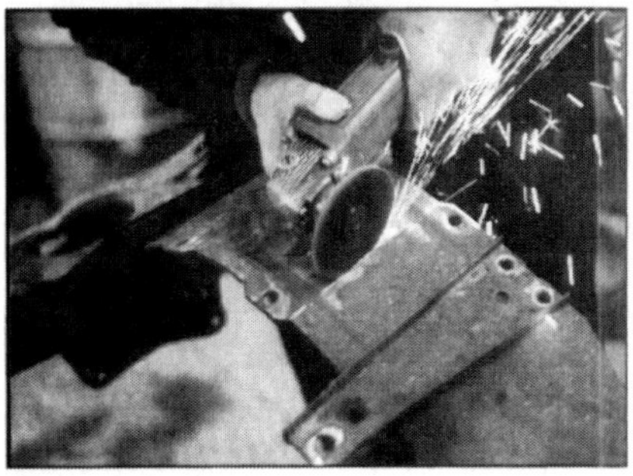

▲ *FD6. It was decided to cut away the damaged section, the simplest method being to use a grinder. Em had carefully marked the line to cut before starting.*

▲ *FD7. However, it wasn't as simple as that; having cut the cross-member, the various spot welds had to be drilled out.*

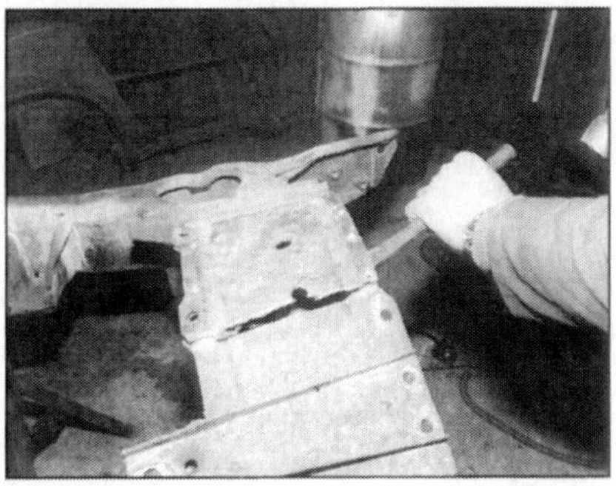

▲ *FD8. Even then, the section required a little more 'persuasion' before it would leave its happy home!*

▲ *FD9. Eventually it could be pulled away. This photo shows that this particular plate is shaped in two planes and that it would require skill in large amounts to recreate the exact shape.*

▲ *FD10. Em's first task was to cut a piece of steel to the approximate size and shape and bend it to the curvature required. Here, it has been laid in place to ensure that the basic elements of the replacement panel are correct.*

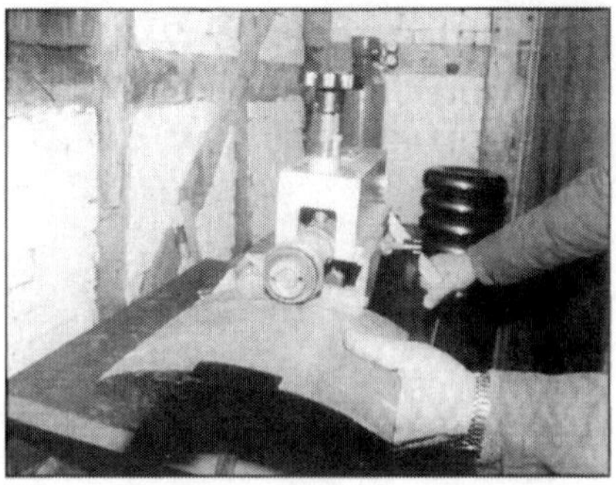

▲ *FD11. The panel had to be put through the roller again with a suitable mandrel in order to create just the right angle for the 'lips' at either side. Though this is a skilled operation, it's not beyond the reach of the enthusiast, though the machinery itself may be better hired than bought because of the high cost for a one-off application.*

▲ *FD12. After having drilled the holes in the right places, Em tried one last mock fitting ...*

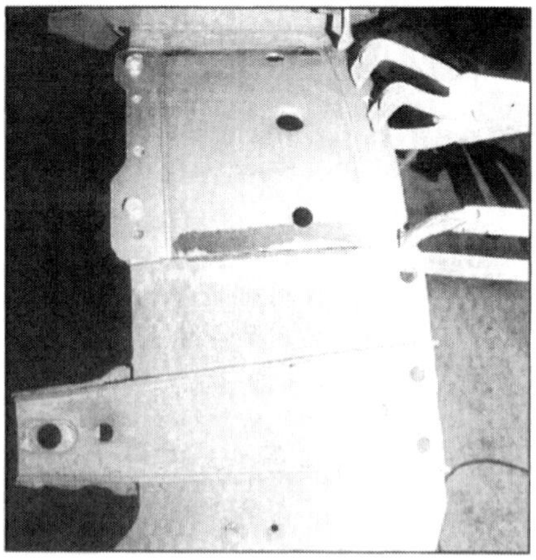

▲ *FD13. ... before the panel was bolted into place and the welding clamps applied.*

▲ FD14. Part way through the MIG-welding process, the panel has been skip welded around its outer edge.

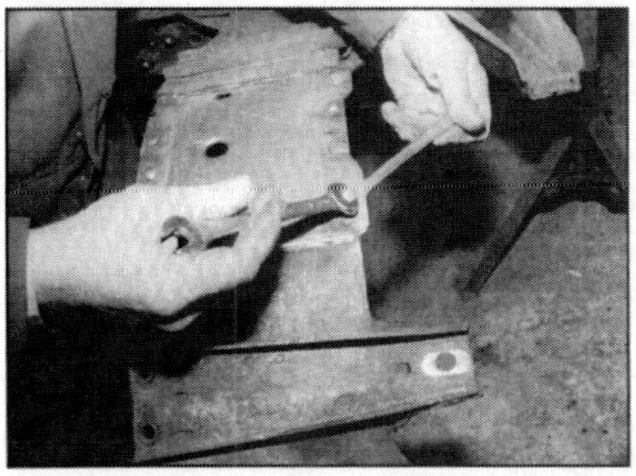

▲ FD15. Em then seam welded it down the two 'cut' edges. Once the bolts were removed ...

▲ FD16. ... the rough edges of the welds were ground down flush so that, once the frame respray was complete, there would be no trace of any repairs.

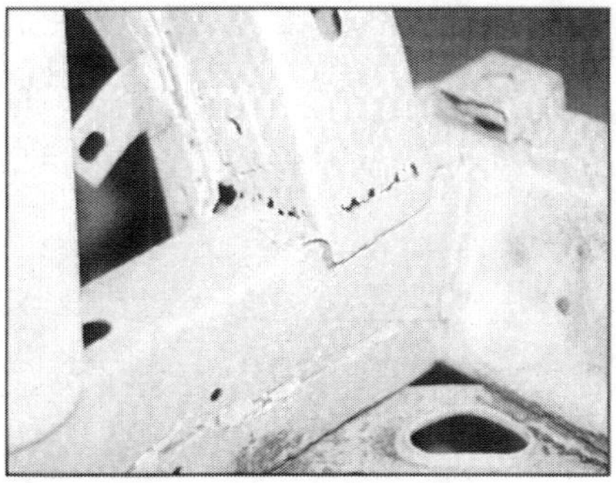

▲ FD17. The rear section of the side rail on either side is a common rust trap on '63–'82 Corvettes, as you can see here after the area had been shot blasted.

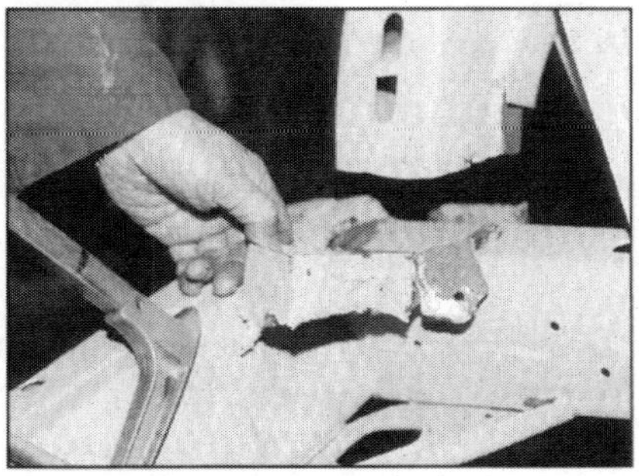

▲ FD18. Whenever you're removing the 'dead' metal from any part of the frame, it's essential to cut much further back than the immediate rotten area – rust spreads further back into the good metal than you would think.

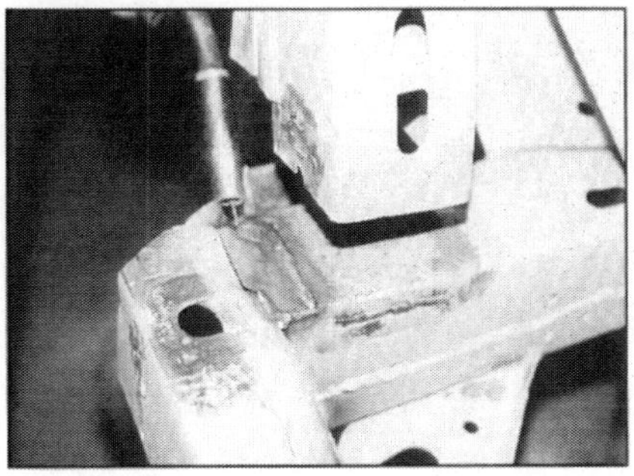

▲ FD19. Em also discovered a minor rotten area in the lower frame rail, which was cut away, a suitable replacement piece formed and then MIG-welded into place.

◀ FD20. Em then created a 'patch' to fit exactly the front of the area in question. Time and patience is required here to get both the size and the shape correct.

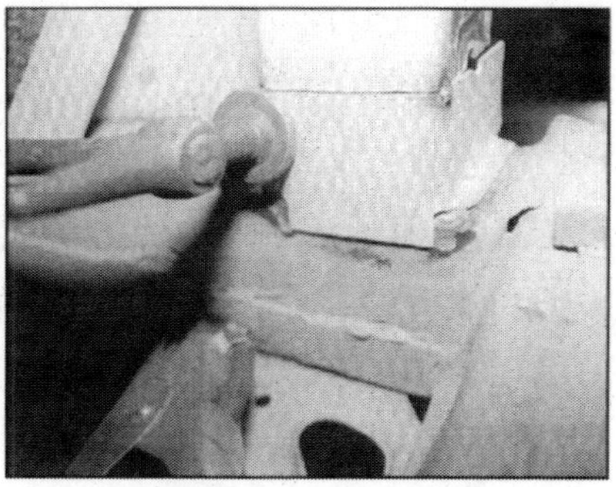

▲ FD21. He skip welded one edge into place, the heat from which operation caused the patch to distort slightly. This is something which you have to be watching for constantly. A gentle tapping back into place is all that is required.

▲ FD22. Having skip welded all around the patch, Em then performed a neat seam weld. Having completed the front part of the patchwork ...

▲ FD23. ... he then cut two more patches to fit to the rear of the affected area. These were welded in place ...

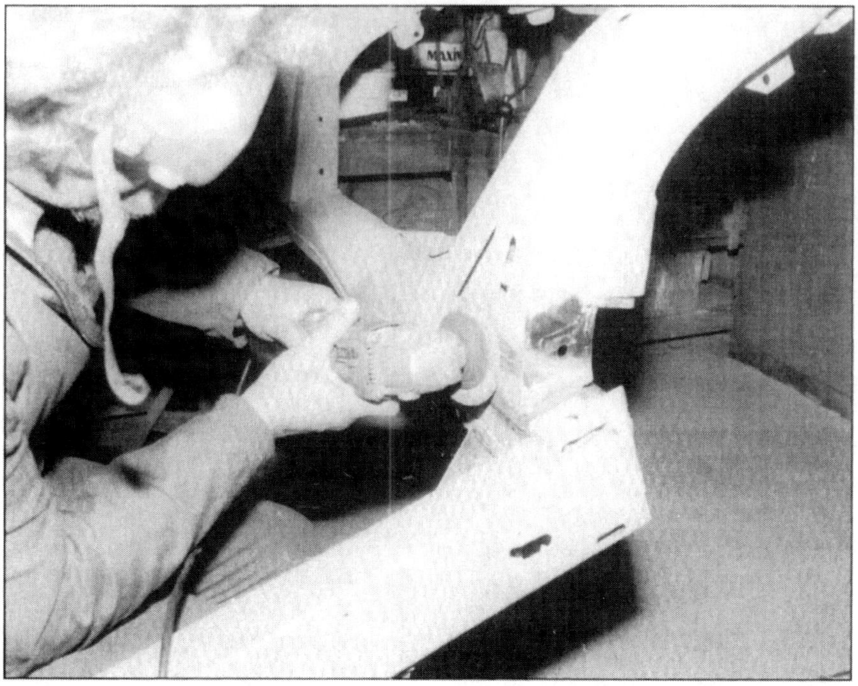

▲ FD24. … and once completed, the various welded sections were ground down to a smooth finish.

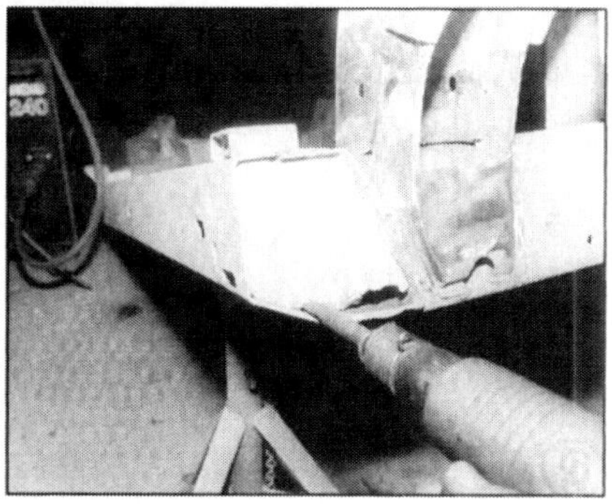

▲ FD25. The end of the frame rail is another area where water and dirt can collect to create the perfect environment for rust. After the years have passed, there's almost bound to be some repair work required! Em took a pneumatic chisel to it. Again, not a usual DIY tool, but they can be purchased quite cheaply provided you have a suitable air compressor to drive it. Naturally, it's vital when using this kind of equipment to protect your hands and your eyes with gloves and goggles in case of flying lumps of debris and rusted metal.

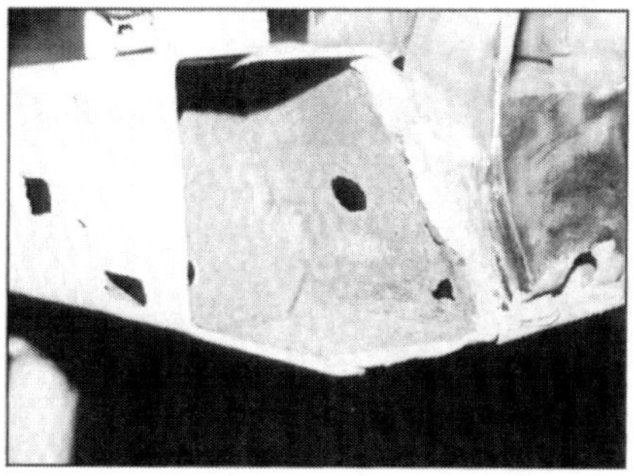

▲ FD26. Once again, Em cut back well beyond the area of obvious damage.

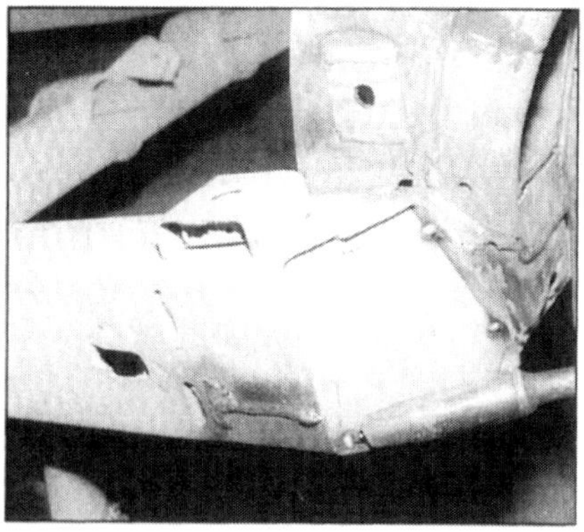

▲ FD27. The same repair principle was applied as previously – a steel patch was cut to suit the cut-out area in question and then MIG-welded into place. The patch was skip welded into position then seam welded for neatness, originality and to prevent water getting into this particular box section. Note, also, the extra welded section along the bottom of the rail. It was found that this edge had rotted as well as the back, so it was replaced as a matter of course.

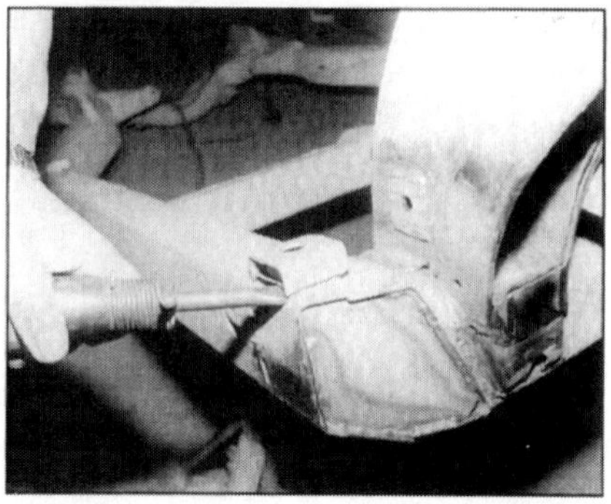

▲ FD28. It was found that the No. 3 body mounting bracket had suffered from the dreaded rust bug. Once more, the air chisel made light work of removing it …

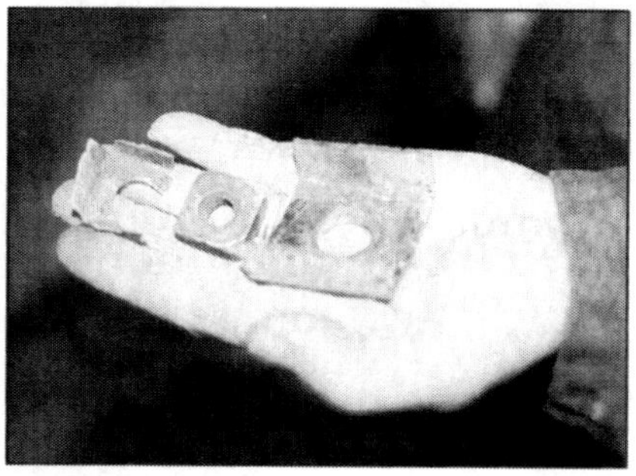

▲ FD29. … whereupon Em was left with these sections, the rot being most evident on the section to the far left. It was this section which needed to be replaced.

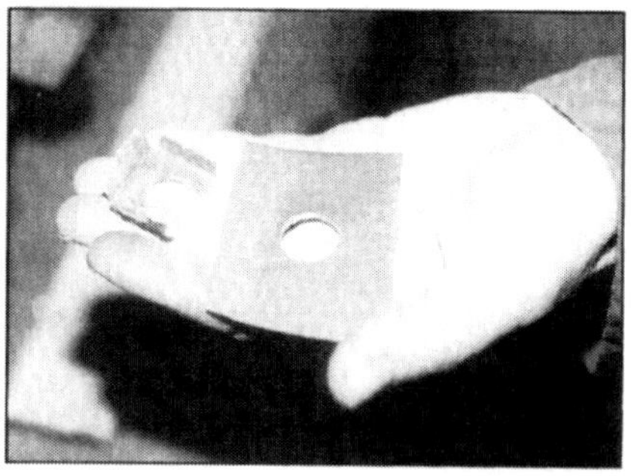

▲ FD30. Em cut a flat piece of steel and started by drilling a suitably sized hole in its centre. At this stage it's hard to see how it could become anything like the right shape.

▲ *FD31. However, after a little accurate measuring, followed by some equally impressive work with the tin snips, topped off by ...*

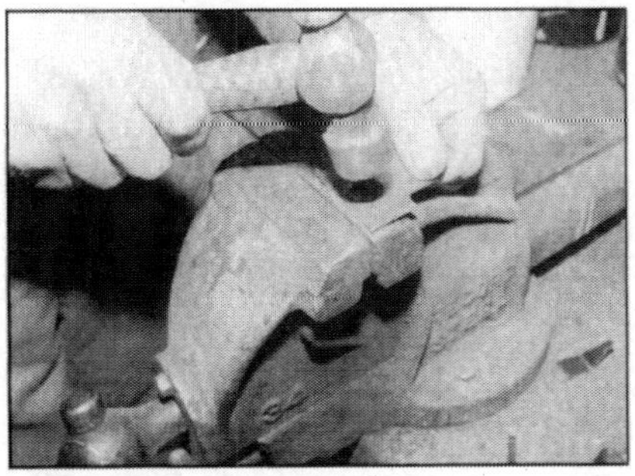

▲ *FD32. ... some judicious work with a hammer and Em had produced ...*

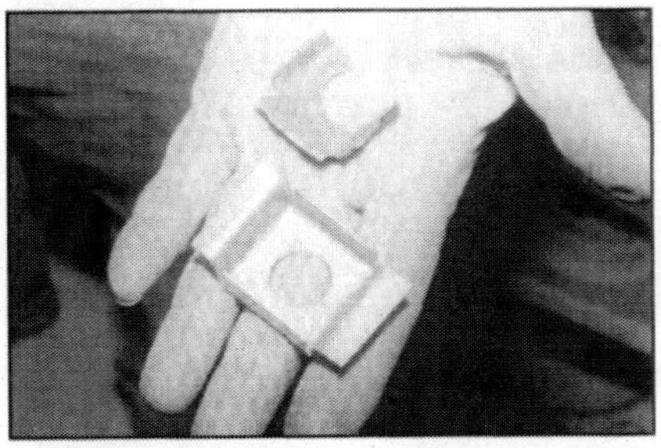

▲ *FD33. ... a perfect replica of the original – except for the rust, of course!*

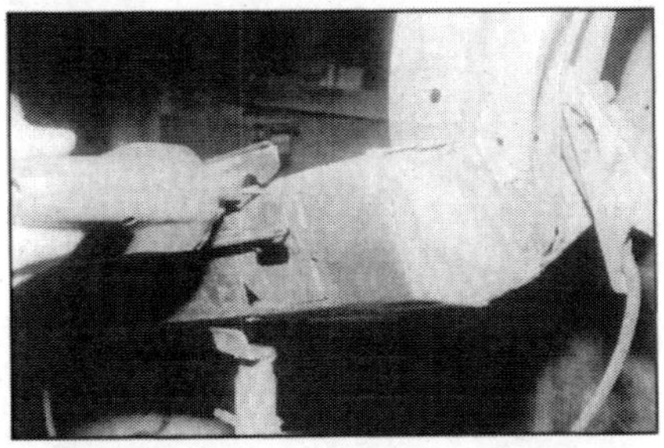

▲ *FD34. This area on the outer edge of the side rail was found to have a little too much rust for comfort. Em used the existing hole as a starting point for his jigsaw ...*

▲ *FD35. ... and cut away a square panel, as seen here.*

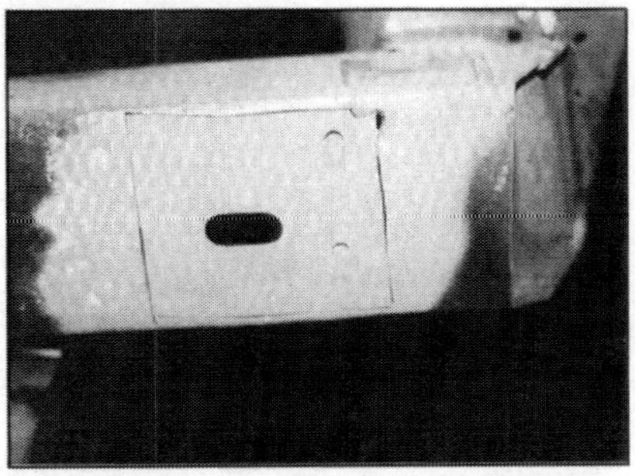

▲ *FD36. A newly fabricated panel was put into place ...*

▲ *FD37. ... skip welded ...*

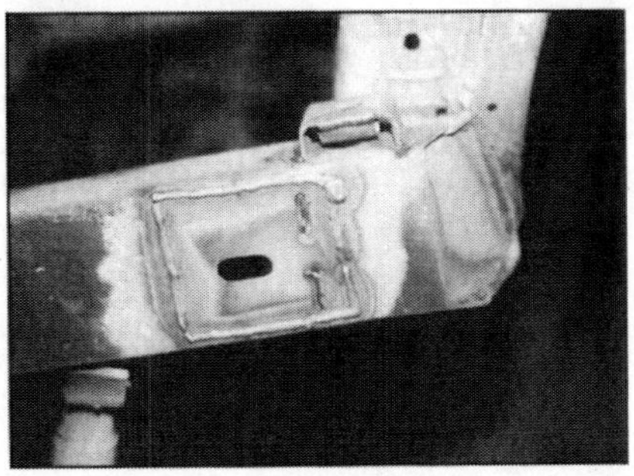

▲ *FD38. ... and seam welded.*

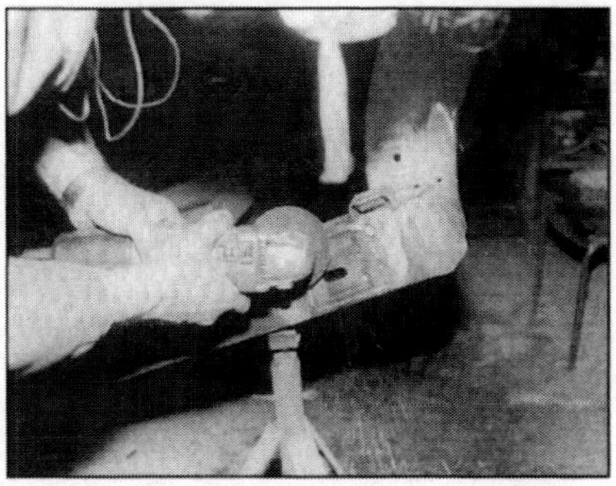

▲ *FD39. Grinding the welds brought the surface flush once more. You might think that this particular corner had its fair share of troubles, but they weren't over yet!*

FRAME RAIL GUSSET REPAIR

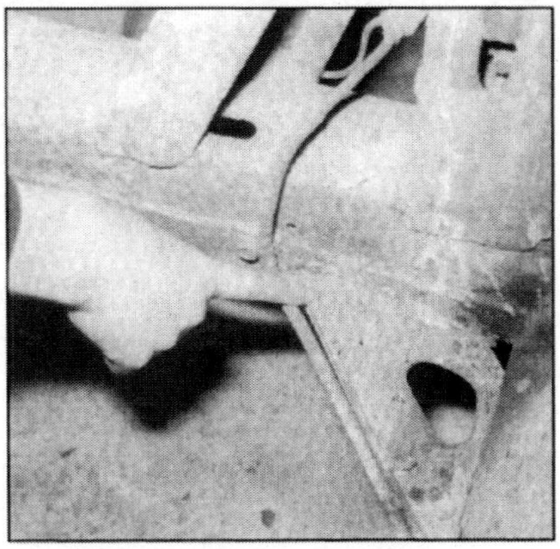

▲ *FD40. The frame rail reinforcing gusset has nothing more to do in life than sit there collecting moisture and dirt in roughly equal quantities. Unless it's been superbly rustproofed on a regular basis, the odds are you'll find that you've got trouble in this area.*

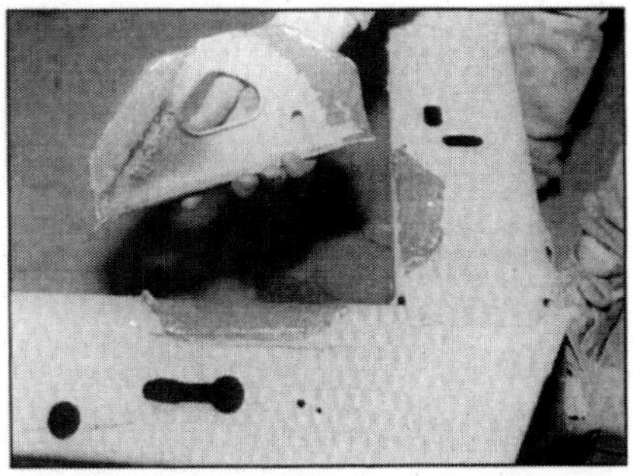

▲ *FD41. With the frame upended, it may look as if the gusset is just a little world-weary, but there's often plenty of rot beneath the surface.*

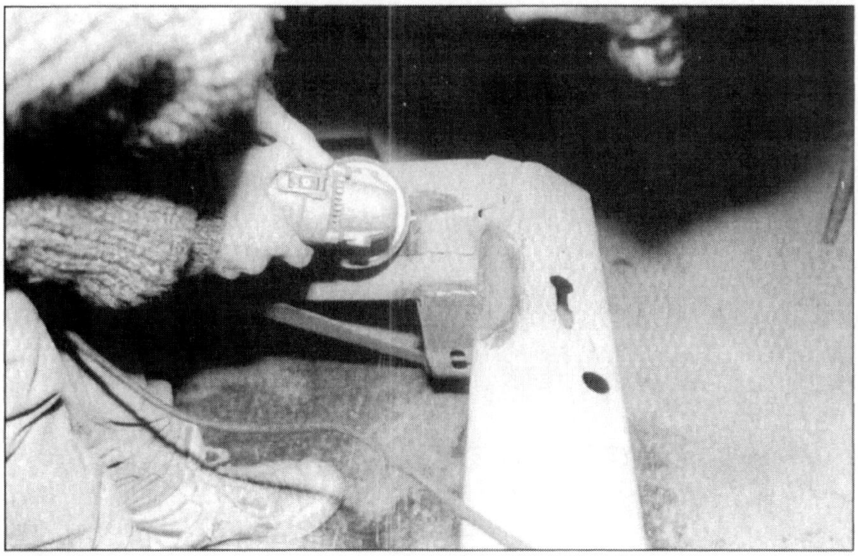

▲ FD42. There was none on this particular section of the side rail, so all that was required was a good clean-up with the grinder to remove all traces of the previous welds.

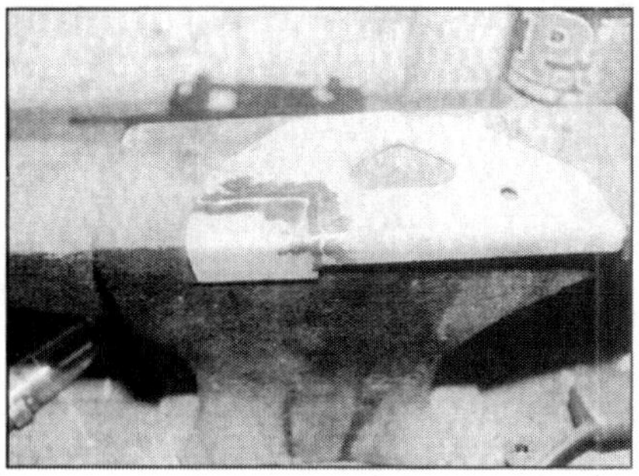

▲ FD43. As mentioned, there was plenty of rotten metal on the gusset, especially on the left-hand corner which, as you can see here, had to be cut away and a repair plate fabricated. It was made a little too big so that after it was welded in, it could be cut exactly to size.

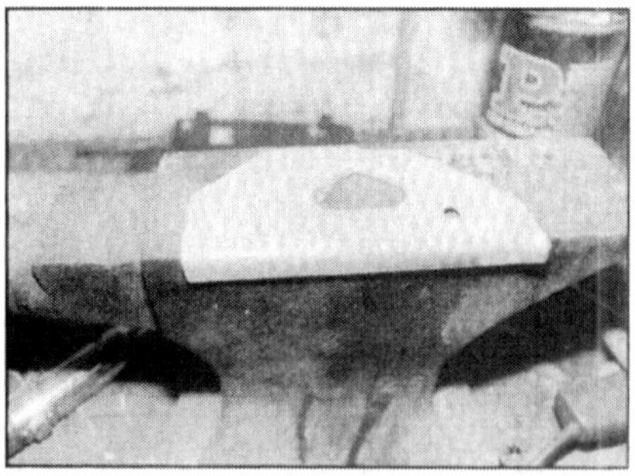

▲ FD44. After the grinding process, it was impossible to see where Em had been! Note that there is still one hole to be drilled at this point.

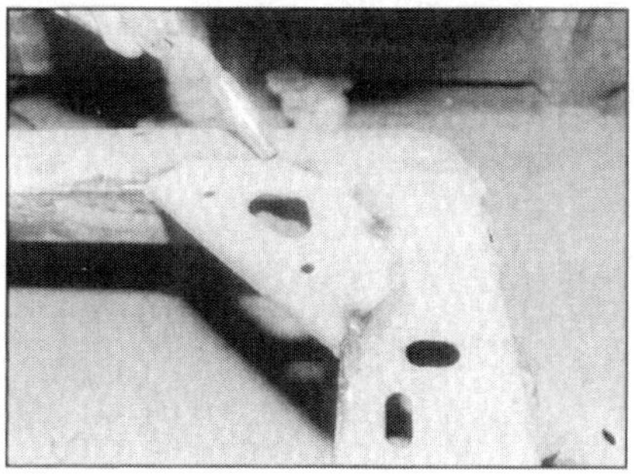

▲ FD45. The gusset was skip welded and then seam welded into place. The 'missing' hole has now been drilled.

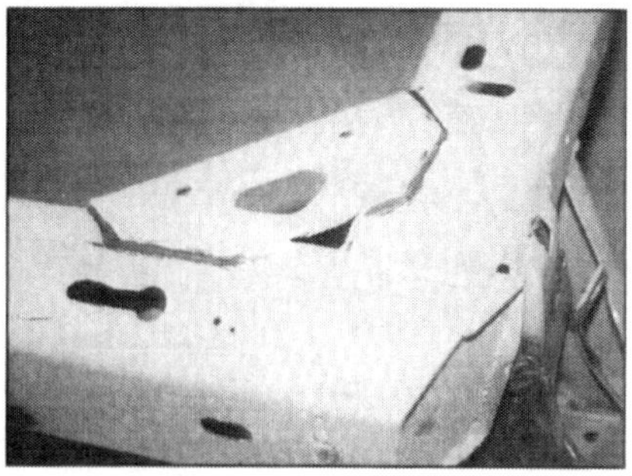

▲ FD46. Once more, a perfect piece of workmanship, neatly cleaned up and ready for some paint and anti-rust treatment to prevent an early recurrence.

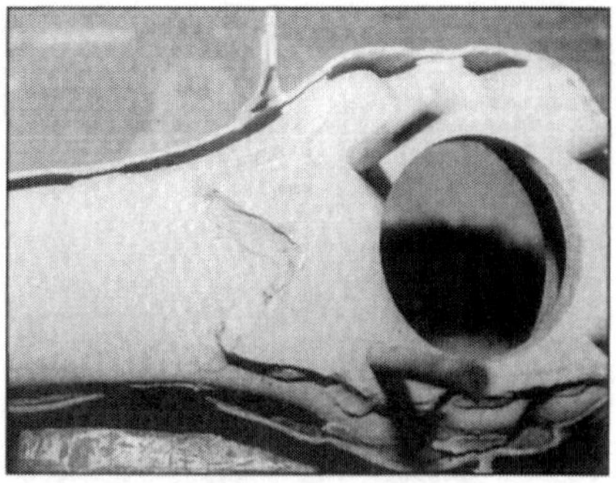

▲ FD47. Looking from the inside, you can see the basic design flaws of the trailing arms fitted to Corvettes from 1963 to 1982. They are made up from two sections skip welded together, and they really are bad rot spots. Look out for bulging of the steel around the skip welds and for lumpiness on the flat areas of the arms (see also Figure CH25).

▲ FD48. In this case it was possible to push a screwdriver down in between the join. Not good!

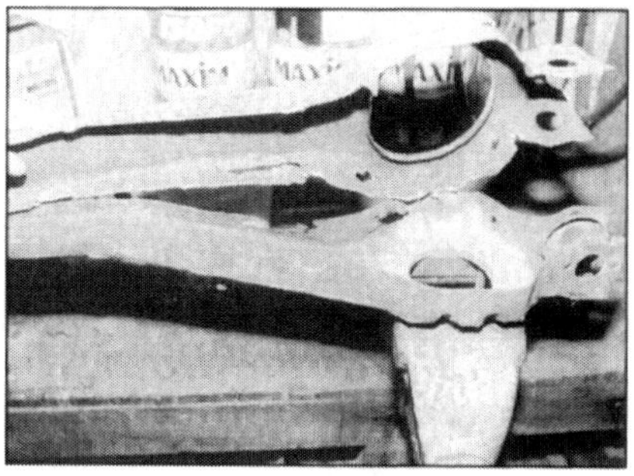

▲ FD49. It took very little effort to actually split the trailing arm section into two halves. As you can see, the whole assembly is incredibly rotten and well beyond repair, even for someone of Em's talents! The answer? Expensive new replacements, unfortunately.

RADIATOR SUPPORT

The radiator support serves unseen and unthought of; it's very much a case of being out of sight, out of mind. But its role in life means that it is going to be more susceptible to rusting than many parts of the car. It heats up and cools down with the radiator and is in the firing line for road spray. Not only that, its box section construction allows the rot to work its way from the inside out. It's definitely an area to keep an eye on every time you're working under your car.

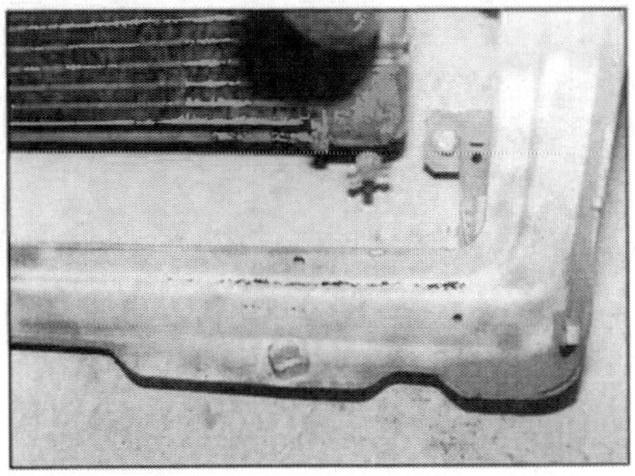

▲ *FD50. When Em sandblasted the radiator support on the '69 restoration car, we weren't too surprised to find that there were patches of rot. Fortunately, they took the form of a series of small holes which meant that ...*

▲ *FD51. ... Em could break out the MIG-welder once more and weld-in patch repairs, although forming all of these complex shapes in the repair sections took above average skills. Most home restorers would buy new (or good-as-new) replacement sections.*

FRAME PREPARATION AND PAINTING

Frames can be cleaned and prepared for painting by sandblasting (the method we used for most of our metal components, including the frame) or chemical dipping. The stripping can be done by hand with chemicals and paint strippers. Sandblasting does an excellent job in removing surface rust and preparing for painting. However, three cautions should be noted:

1 The inner surfaces and overlapped areas of the frame are not cleaned.
2 It is nearly impossible to remove all the sand from inner areas of the frame (it will continue to seep out, particularly when paint is being sprayed).
3 The surface finish is noticeably rougher than it was originally (this is important for serious restorations) – although a surface-rusted surface will be rougher anyway and spray-on primer-filler can be sanded to restore a 'new' finish.

Chemical dipping has none of the above problems but does have the alternative problems of higher cost and the difficulty of taking the frame to and from the stripper. It generally takes two to four people to lift the frame and a truck or large trailer for transportation.

Hand methods have the obvious advantages of low cost and convenience, but like sandblasting, only reach the exposed areas. If chemical paint strippers are used, remember that they should be very thoroughly rinsed off.

▲ *FW1. However you get the dirt and rust off, it's vital to get plenty of primer back onto the virgin metal as soon as possible. Every minute it is left unpainted is another minute that rust can get a hold. For such a large area as the frame, it's best to use a compressor-driven spray gun. You won't need an expensive one if all you're spraying is the frame and small components – in fact, you'll find one invaluable.*

SAFETY NOTE: Don't forget that when using spray paint of any type, you don't want it in your lungs – always wear the correct type of breathing mask.

▼ *FW2. Em used a hard-wearing black paint finish for the final coat. Unlike spraying bodywork panels, which will be on show the whole time, the usual rules with regard to dust etc., aren't so important. Note that Em has the workshop door open for better ventilation and so that the paint fumes will disperse quickly.*

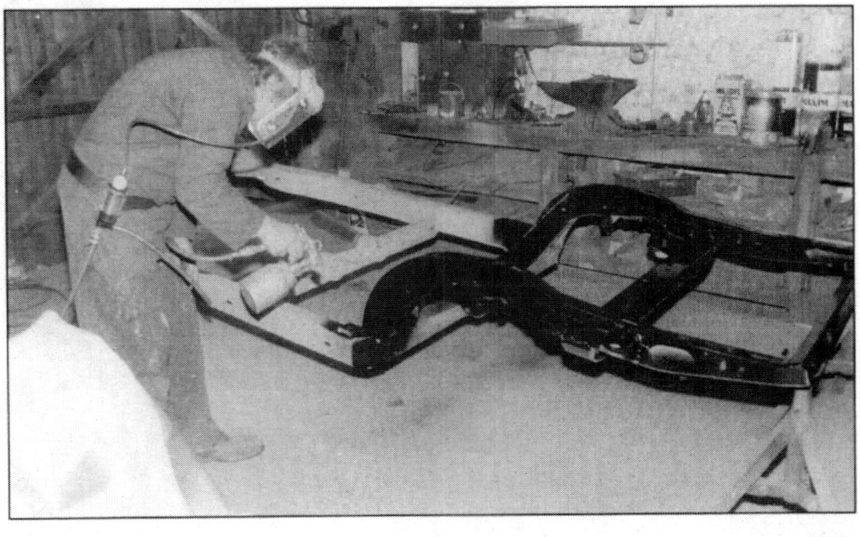

RUSTPROOFING THE FRAME

We used a wax-based fluid to inject and 'paint' the frame in an effort to prevent (or at least, delay) the onset of further rust problems. It's important to note that this should not be allowed to get onto the manual shift clutch, any braking surfaces (rotors, friction parts, etc.), onto 'hot' areas such as the exhaust, or into devices such as the carburettor. Cover anything of this nature with plastic bags tied or taped up tight.

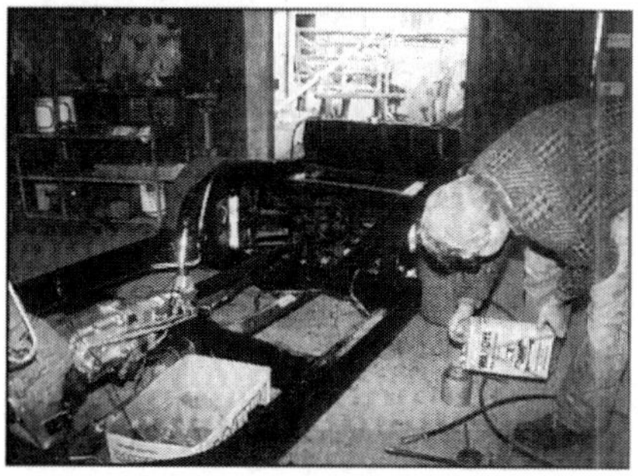

▲ *FW3. Here, Em is loading up a spray gun in order to 'force feed' the rustproofing into the various hollow sections to prevent them rotting from the inside out.*

▲ *FW4. As we've already seen, the rear of the side rails is a spot favoured by rust, so we started there. Get that nozzle in, on and around the whole area, not forgetting …*

▲ *FW5. … this area which gave Em so much grief on restoration. Remember that this kind of rustproofing fluid can safely be used to protect brake lines. For this, Em used a brush, to ensure that the wax was worked into all the nooks and crannies.*

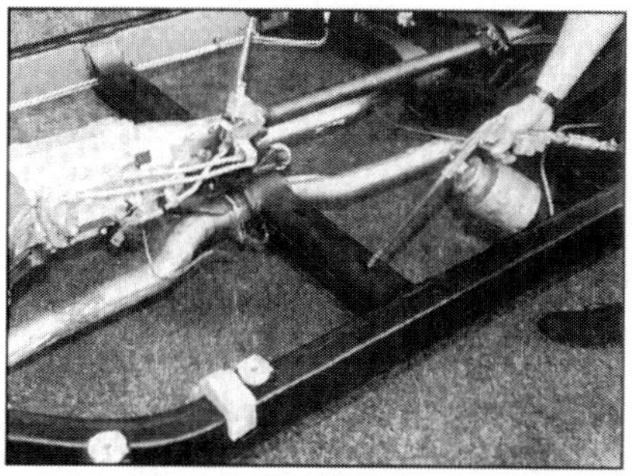

▲ *FW6. Don't forget to inject the cross-members for they, too, are hollow and provide ideal breeding conditions for rust bugs. Many fluids 'creep', and once inside such box sections they will tend to find their way into the more inaccessible places.*

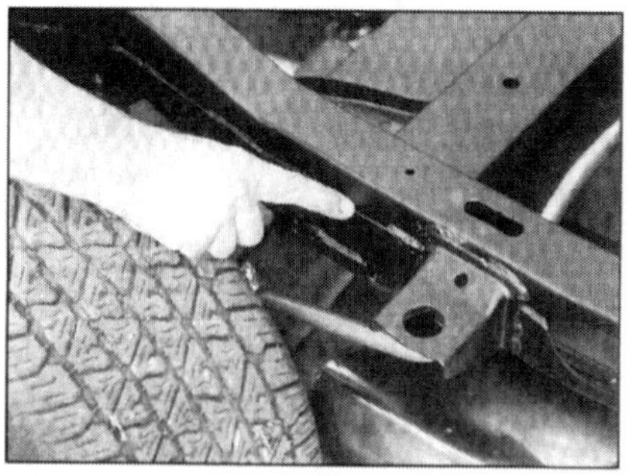

▲ *FW7. The side frame rails feature lots of places for rust to start. The way the rail is constructed at this point allows water to get in easily – make sure you get plenty of rustproofer in here.*

RESTORING FACTORY FINISHES

Original frame paints were usually the cheapest ones available at the time, and most parts were not even primed. It's not surprising that the finishes didn't last. If durability and a glossy appearance are desired, the frame should be primed and painted with two-part epoxy paint – although this must only be applied by a professional bodyshop, for health and safety reasons. If you're working on a restoration project to factory original condition, take care to thoroughly research not only what type of paint was used, but also what type of stencilling of part numbers or other markings were found on the frame. It's quite difficult to redo this work later.

One of the best sources of such information is the National Corvette Restorers Society (NCRS). Most questions on frame finishes and detailing for individual years can be answered through its restoration articles, judging manuals, or members' expertise.

The availability of replacement frame sections now has enabled Corvette owners to repair rotted frames at low cost, avoiding the problems of frame replacement which often costs a small fortune.

REFITTING THE BODY TO THE FRAME

In this section we will show how to replace the body on the frame while causing minimal grief later on. For instance, you must make sure that the gear shifter, which has to go through the inner rubber boot first, doesn't rip the boots – take time to ease it carefully through them. Also, be aware that wires and cables can easily get trapped between the body and frame and not be noticed until everything is bolted down. An then you must make sure that bolts and fasteners for ancillary components are oriented correctly so that they can be removed when necessary when the body is on. Examples include the parking brake cable pulley bolt and the rear shock upper bolt. They can go on either way with the body off, but with it on again it is a different story!

▶ *BB1. At some point in a full restoration project, you'll end up with a Corvette that looks something like this – not a bit like a Corvette, in fact! However, it wasn't long before …*

▼ *BB2. … the newly painted body was standing by, waiting to be returned to the frame. Note the body stand which makes life much easier when working on the body while it's off the car. As much as possible had been left off in order to make the body more wieldy as it was manoeuvred into place.*

▲ *BB5. Any experienced restorer will tell you that threads in the frame can often become clogged during the restoration process. Finding out after the body has been put into place is no fun, so the time to run an appropriate bolt through them is now. If there is any resistance, clean the threads and try again. It's a good idea to put a good helping of grease on the threads to prevent problems for future restorers. If you don't have thread taps, invest in a set, or at least cut a slot down the length of a scrap bolt and run that through the thread.*

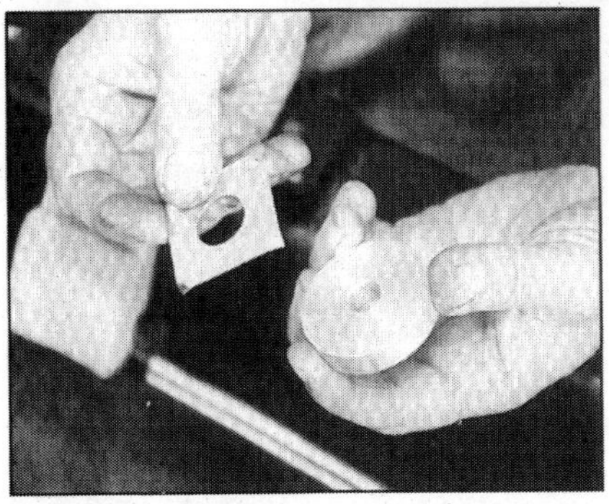

▲ *BB3. You'll need to amass the correct number of shims (or rubber buffers, depending on model) and spacers before you start. Somewhere in your well-organized tear down log you have the details. As part of our total restoration, we'd replaced all the spacers and washers with new ones – the cost is not great, and it's not something we're planning to repeat on a regular basis!*

▲ *BB4. It's important to make sure that the mounting pads are in place.*

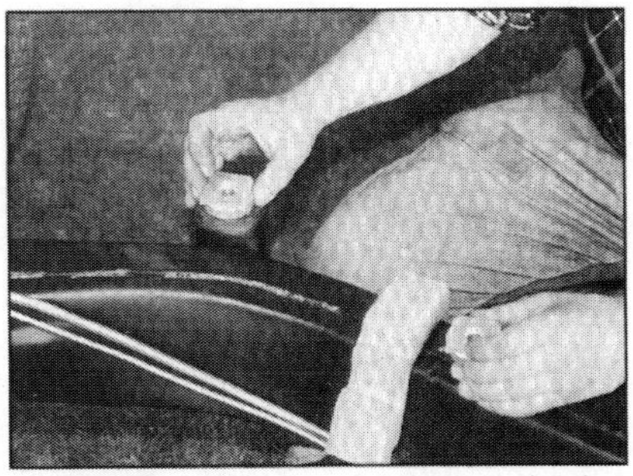

▲ *BB6. Em checked his records and allocated the shims/spacers to the eight body mounting points. These are the passenger side No. 1 and No. 2 points.*

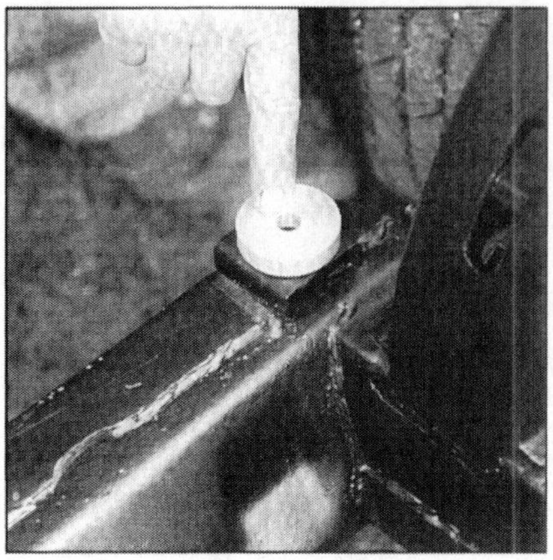

▲ BB7. This is the No. 3 body mounting point and …

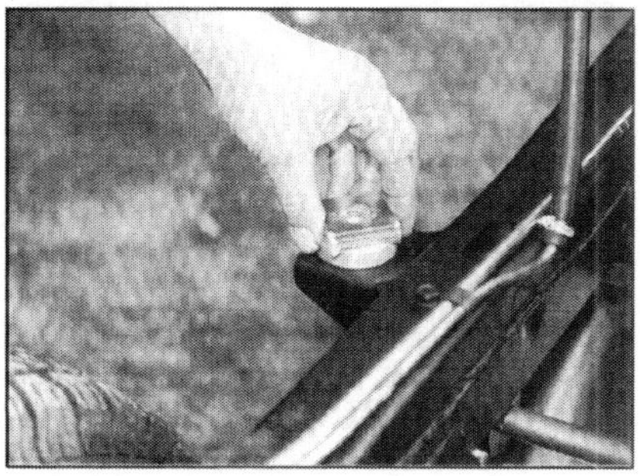

▲ BB8. … this is the No. 4 point, requiring six shims in this case. Masking tape was used to hold the shims in their respective mounting locations on the frame to prevent them being knocked out of place as the body was lowered. For detailed restorations like this, it's wise to use the same type of shims and spacers the factory used.

▲ BB9. It's a good idea to complete tasks such as fitting this exhaust-to-frame earthing strap whilst the body is off. It's far easier and allows much better access. It's a good idea to coat such connections – as well as all bolt threads – with a layer of copper grease to prevent corrosion problems in the future.

▲ BB10. All together now – lift! Once more, friends and neighbours provided the lifting power for our body fitting section. The cameraman got the best job!

▲ BB11. The body was lifted oh so carefully over the car from the rear then …

▲ BB12. … a little further. This shot illustrates why you need at least six lifters, because of the height you need to raise the heavy body.

▲ BB13. It was in exactly the right position to be lowered straight down …

▲ BB14. … to the point where it's almost there. This is where extreme care has to be taken to avoid trapping cables and pipes and, of course, that no errant fingers are in the way. Always have an extra helper on hand to look and check while everyone else lifts and lowers.

▲ BB15. That's it, body down. Em then took some time to work his way all round the car to ensure that everything was exactly in the right place ...

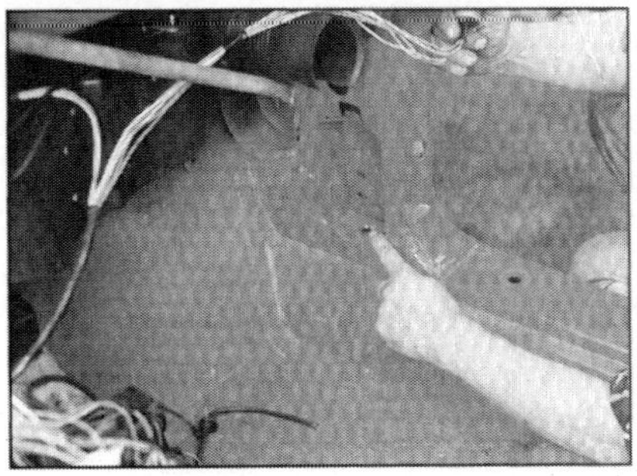

▲ BB16. ... notably that the body mounting holes lined up. This is the passenger-side No. 2 mounting hole, easily accessed from the footwell.

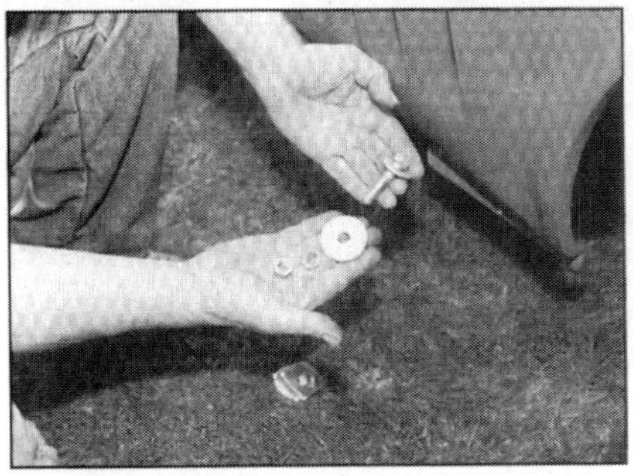

▲ BB17. He collected the new bolts, washers and lockwashers so that they were ready to be fitted in the correct order. Note that the bolts for No. 1, No. 2 and No. 3 mounts are inserted from the top; only No. 1 body mount requires a securing nut – all the other bolts screw into captive threads; the shims are fitted with their tabs pointed downward on No. 1 and No. 4 mounting points, and with them pointing upward on No. 2 and No. 3.

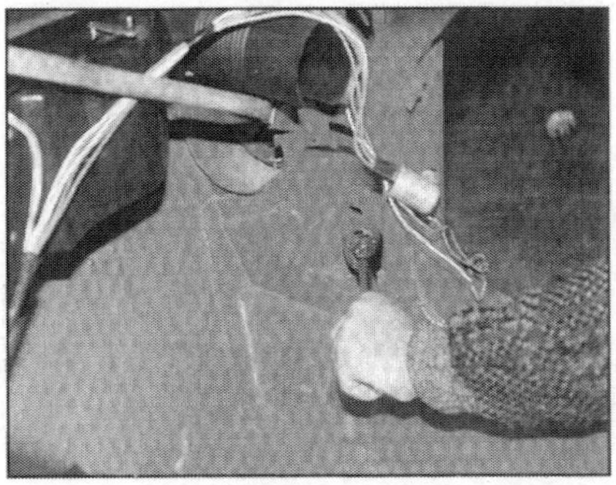

▲ BB18. The No. 2 mounting bolt is always the easiest to get to, especially with the body in its stripped down state.

▲ BB19. The No. 1 bolt is not quite as much fun, though with the engine ancillaries missing accessibility was improved. Note that Em has gained access with a wrench through the fender louvres.

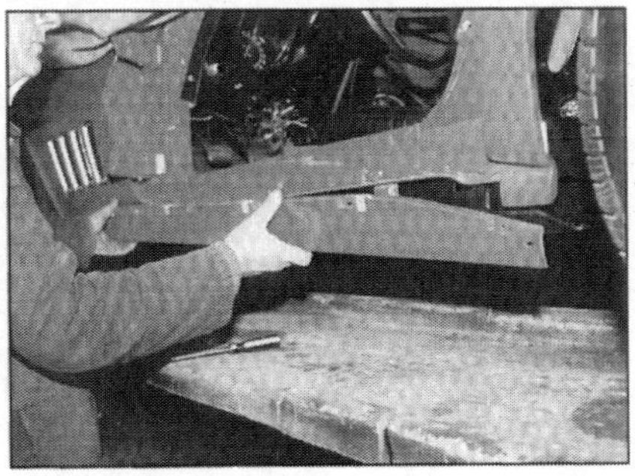

▲ BB20. Once the body was securely attached, Em fitted the two body rocker panel mouldings. Having drilled the required holes ...

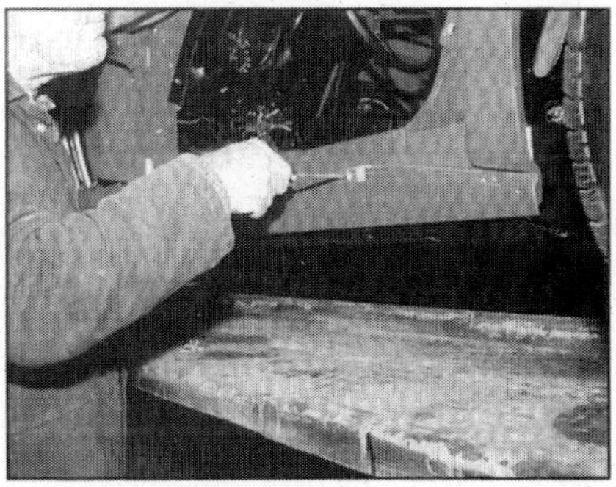

▲ BB21. ... the mouldings are held by simple self-tapping screws.

▲ BB22. You may find (as we did) that the aftermarket decorative chrome strip has to be trimmed a little at one end in order to fit the moulding exactly.

▲ BB23. Again, it is secured by self-tappers.

All body mounting bolts should be torqued up to around 40/50 lb ft. Make a note to check them after a week or so to ensure that they have not worked loose. Further checks should be made every time your Corvette needs a major service. After the body was reinstalled onto the frame, Em tightened all the body alignments. These alignments can be adjusted by changing shims, particularly the shims on the No. 1 and No. 2 body mounts. Don't forget to have the car resting on its wheels for this adjustment.

Remember that the fiberglass body is sandwiched between the chrome bumper and the supporting brackets and if you overtighten the bolts, you will crack the fiberglass.

One final tip from John Pfanstiehl is to take the trouble to double check the connection on the oil pressure line both at the engine end and at the gauge end. Leave either of them off, and your engine bay or – worse still – your beautiful car interior will be sprayed with oil when the engine starts up again.

HARD TOP OVERHAUL

Note: Please refer to the fitting windshield section for information on how to fit glass to hard top rear screen.

The hard top (or auxiliary top, to use its official terminology) is a useful option for owners who live in states (or even countries) where the climate is less well disposed to be sunny. For those who like to use their hardtop, the rakish design of the rear window glass offers a feeling of greater interior space than can be found in the vertical back windowed coupé. Naturally, this is something of an illusion, especially if the folding top is still folded down inside the car. (The latter can easily be removed and stored for the winter by undoing four bolts from their cage nuts.)

The hard top is secured in no fewer than eight positions to ensure a snug, leak-proof fit. Moreover, it has polished metal caps at its extreme rear corners to enable it to be stored on a timber or concrete floor or against a wall without sliding. The top is a one-piece moulding which has a hard plastic headlining fitted inside. The headlining is designed to fit exactly along the top and down the inner rear three-quarter panels.

The hardtop should be cleaned and cared for in the same manner as the rest of the car. However, during storage, it is advisable to cover it to prevent the build up of dirt and grit which could cause scratches on the paintwork.

Our major project car's hard top was not in particularly bad condition, simply showing its age a little, particularly in terms of the exterior paintwork.

Once again, dismantling the hard top called for patience and penetrating oil in equal amounts, and a number of labelled plastic bags in which to store the numerous fittings and fixings.

▲ *HT1. If it comes to the worst, you can always replace your hard top altogether. Auto Accessories of America have a selection for all model years, including this vinyl covered version suitable for this car. Normally the tops would be supplied in a ready-to-paint primered finish.* (Courtesy Auto Accessories of America)

▼ *HT2. It's usually a two (careful) person task to remove a hard top. However, if you really want to be sure you don't damage either top or car, you could invest in a hard top lift. And if you're feeling really good about your 'Vette, you could even have an electric version!* (Courtesy Auto Accessories of America)

▲ HT3. Ready and waiting, John Fletcher's hard top sits upside down on the bench. Note the thick cardboard to prevent damage to the fiberglass. Even though it was due for a complete going over, we didn't want to be repairing any more cracks than were necessary.

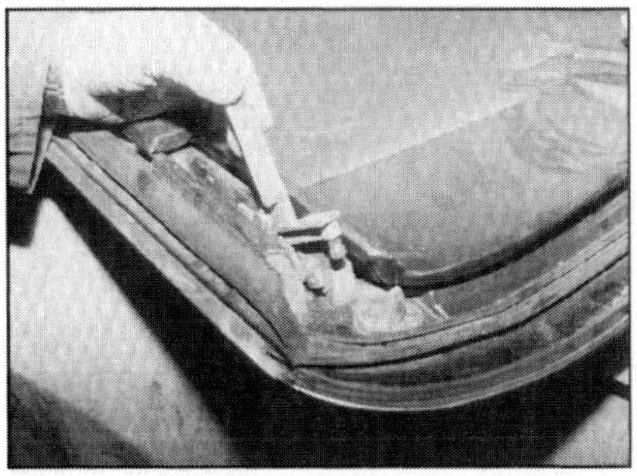

▲ HT4. The hard top is held at the front by a lock assembly at each side and one in the centre. The top locates in the windshield frame by means of a header guide at each side, shown here in front of the lock assembly.

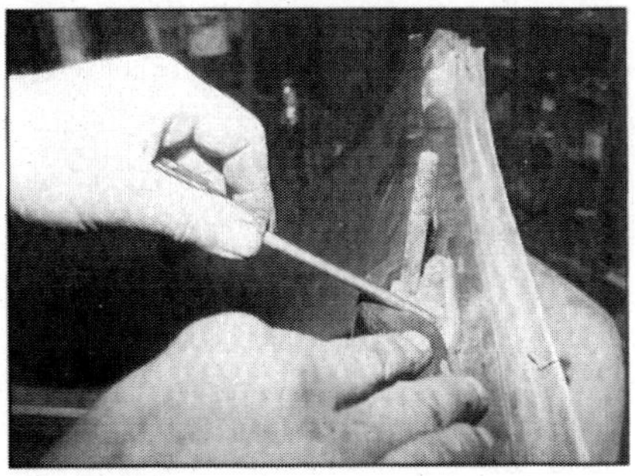

▲ HT5. At the rear, the hard top locates in the bodywork by means of bolts at either side of the top. They are secured by two Phillips head screws.

▲ HT6. A little further forward is a bracket, which is secured by similar screws.

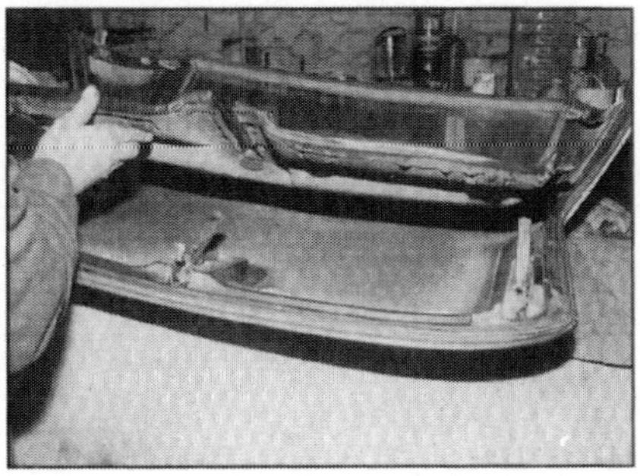

▲ HT7. The plastic headlining assembly is secured to the top by a series of plastic fasteners around the edge of the hard top inner panel. Disengage the fasteners by prizing outward with fingers or a flat-bladed instrument. After the headlining has been eased away from the hard top, it can be broken down into three parts (the roof section and the two which cover the rear threequarter panels) by removing retaining nuts at the inside surface of headlining. When reinstalling, the headlining has to be aligned to the top before engaging the plastic fasteners.

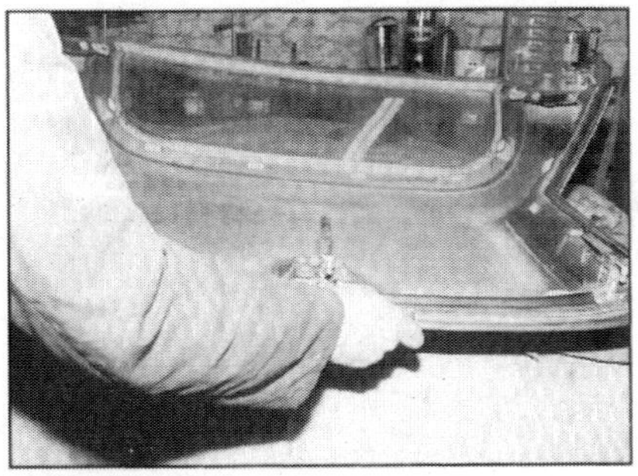

▲ HT8. With the headlining removed, access to the various fittings is easy, though penetrating oil will probably be called for, as the various fasteners may have corroded in place.

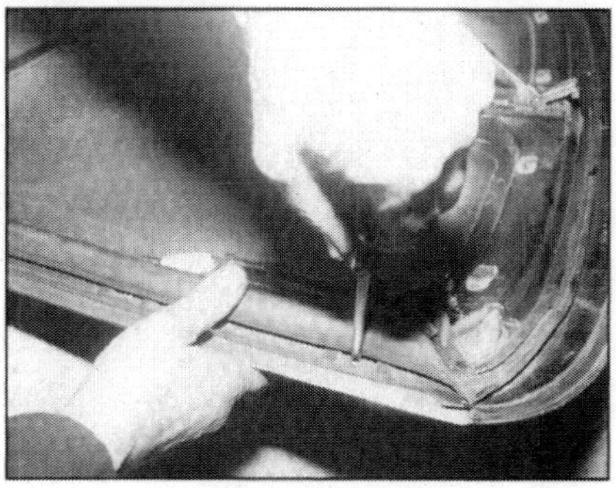

▲ HT9. The door window weatherstrip and drip moulding are retained by screws to the top assembly.

▲ HT10. Once the screws have been removed, the weatherstrip can be pulled away, as shown here.

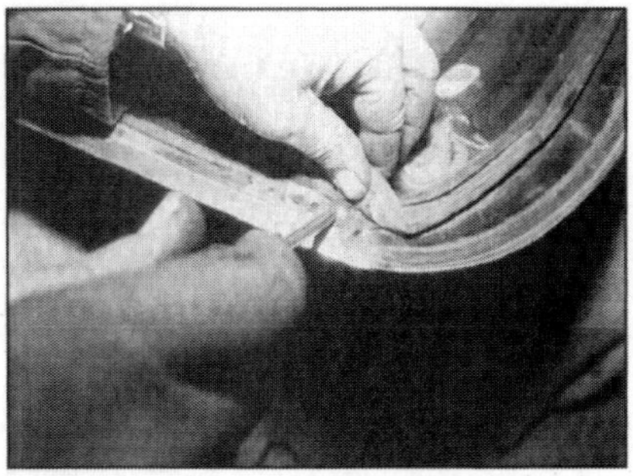

▲ HT11. The side outer and rear outer weatherstrips are removed by undoing the screws or drilling out the pop rivets at each end ...

▲ HT12. ... and easing out the special plastic retaining clips.

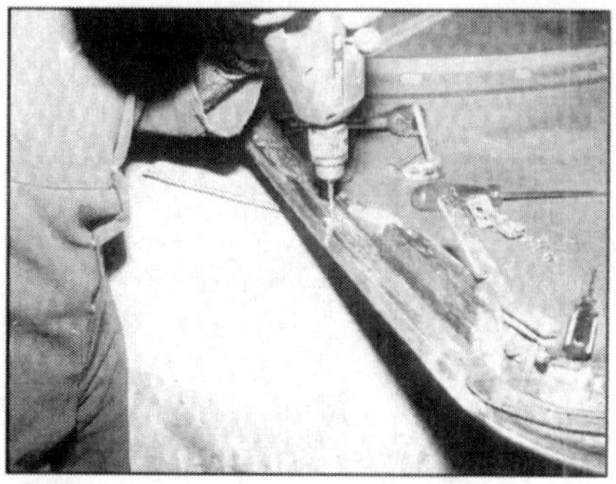

▲ *HT13. You'll possibly find, as we did when removing this moulding, that some pop rivet shanks insist on holding their ground and have to be drilled out. Naturally, lots of care is required to avoid damaging the fiberglass top.*

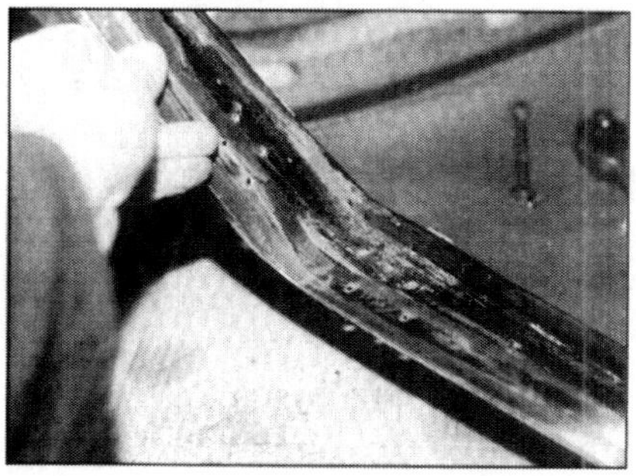

▲ *HT14. With all the pop rivets or screws removed (by whatever method!), the moulding can be eased away as shown here. Take care during the removal process, for though some mouldings are available on the aftermarket, others aren't.*

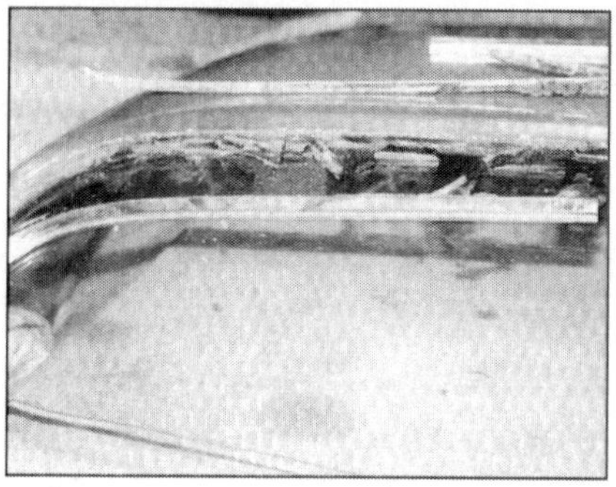

▲ *HT15. To remove the rear glass, first mark the position of the right hand upper reveal moulding end and prize out from the retaining clips. Repeat this operation for the left-hand and lower moulding assemblies. Remove the lower (inside) garnish moulding by removing four nuts from fixed studs and carefully pulling the moulding outward. Take out the glass – as usual, you are well advised to wear gloves when handling glass. Note how the hardened old caulking material has stuck to certain parts of the mouldings.*

▼ *HT16. You can now assess just what you are going to need in the way of replacement parts. Essentials will be weatherstrip kits, and it's good to know that almost all hardtop parts are available, even the rear and side glass for earlier models no longer available from Chevrolet dealerships. Auto Accessories of America can even add the correct date stamp to the glass – ask for it when you order it. Other parts you can buy include all the chrome hardware for a number of model years and hardtop mount kits.* (Courtesy Auto Accessories of America)

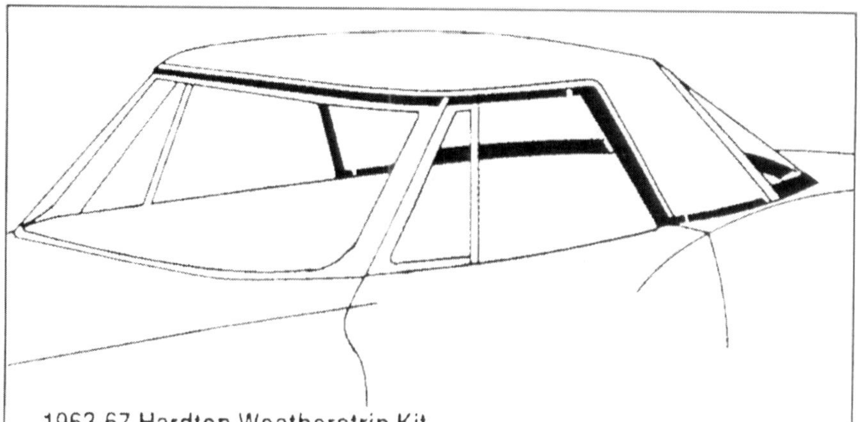

1963-67 Hardtop Weatherstrip Kit

REASSEMBLY

You may be dismantling your hard top to replace the ancillary parts or, like us, you may be completely restoring it, filling in all those tiny nicks that have appeared over the years and treating it to a top-class respray. Whichever it is, the reassembly process is simply a reverse of the earlier instructions, though they are usually carried out with a little more care, so as not to damage the fresh, gleaming paintwork.

INSTALLATION OF REAR WINDOW GLASS

When replacing the rear window glass, do a 'dry run' first, positioning the glass in the opening and checking that the overlap of the glass is equal all round with a minimum overlap of 3/16 in. In some cases it will be necessary to place shims under the lower spacers to obtain the correct fitting. When you're happy that the positioning is just right, use a grease pencil to mark the glass and top surface. The installation from here on mirrors that of the windshield, already detailed elsewhere in this book. Once the installation is complete, the mouldings can be replaced in a reversal of the dismantling process.

HARDTOP ADJUSTMENT

For fore/aft adjustment of left-hand and right-hand locks, mark the original position, loosen the two mounting bolts and move in the direction required. Be sure that the lock is engaged in its serrations before tightening the mounting screws. The adjustment for up and down is achieved by turning the latch bolts clockwise and counter-clockwise respectively. The header guide pins and rear pin housings are retained by mounting screws located at the flanges. The centre trim plate is removed by first taking out the centre lock and then the four retaining screws.

Chapter 4
Interior

INTERIOR TRIM AND CARPET

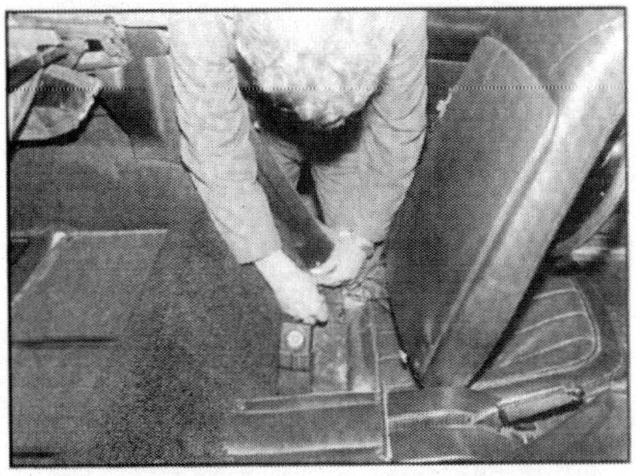

▲ *RT1. To get at the edges of the carpet around the underbody storage compartment area, we had to remove the rear quarter trim panels. These are secured by self-tapping screws both on the inside …*

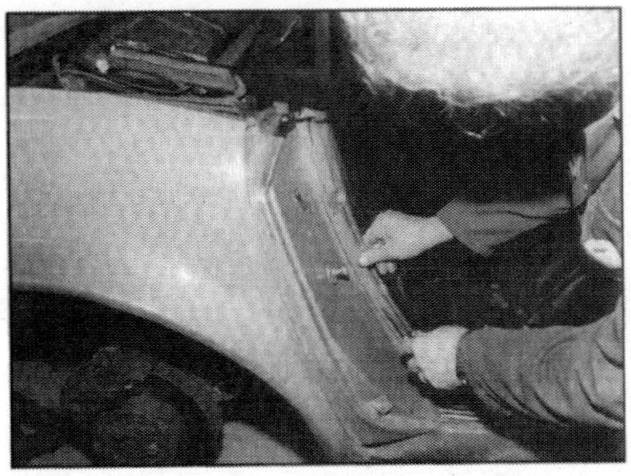

▲ *RT2. … and the outside. When all have been removed …*

▲ *RT3. … the quarter trim panel can be eased away.*

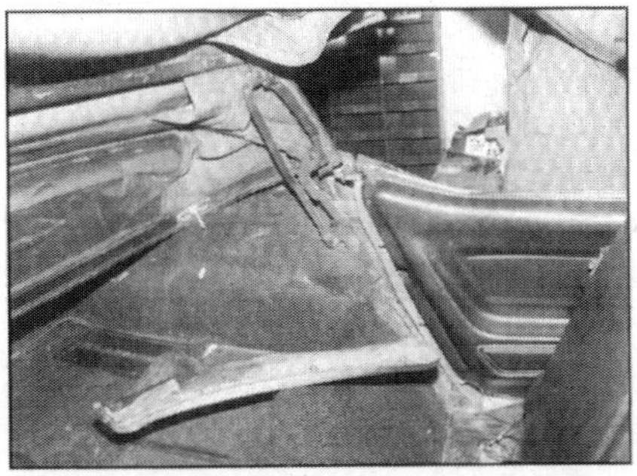

▲ *RT4. It also allows access to the lower folding top mounting brackets.*

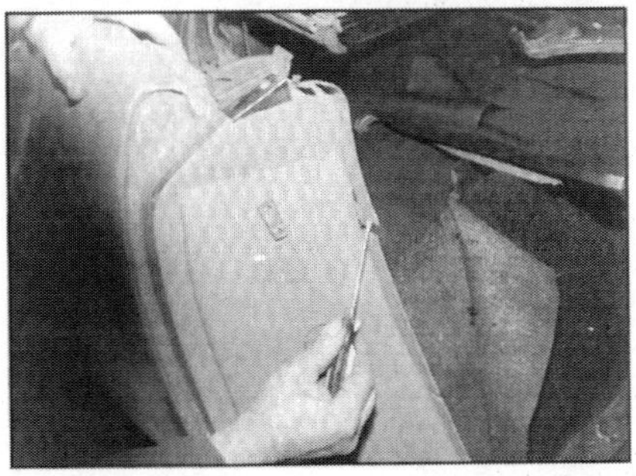

▲ *RT5. In this case, it also revealed that the fiberglass holes had cracked and would have to be repaired. This illustrates only too well how easy it is to damage fiberglass by inadvertently over-tightening trim screws.*

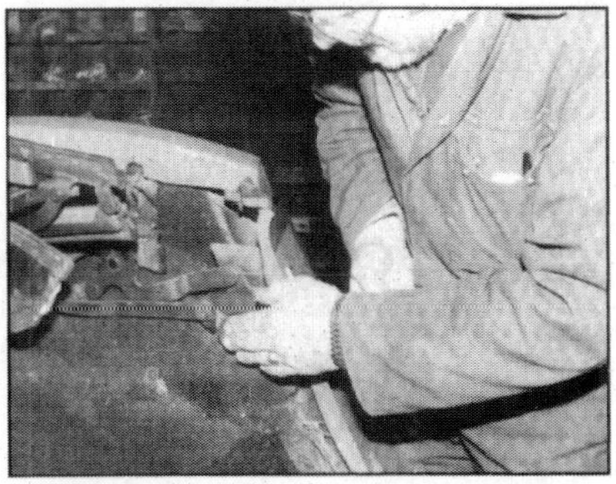

▲ *RT6. Em took off the complete folding top by removing the four bolts at each side, as described elsewhere in this book. You'll need a helper at this stage so that no undue strain is placed on the top or its mountings.*

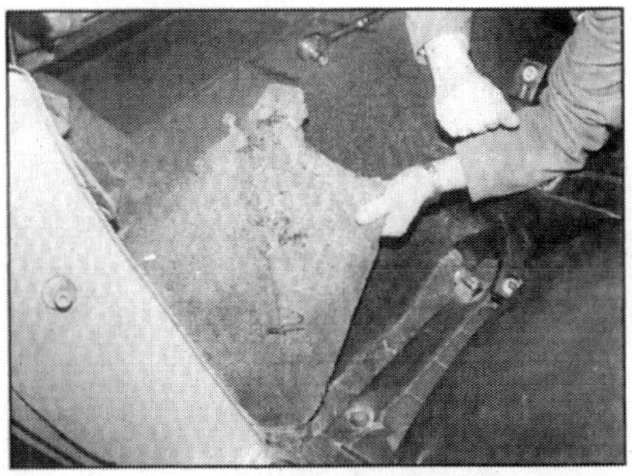

▲ *RT7. Part of our restoration project was to replace the carpets, meaning that the old ones had to be removed. Because of its location, the vertical riser of the rear carpets often gets worn or torn and suffers fading by the sun.*

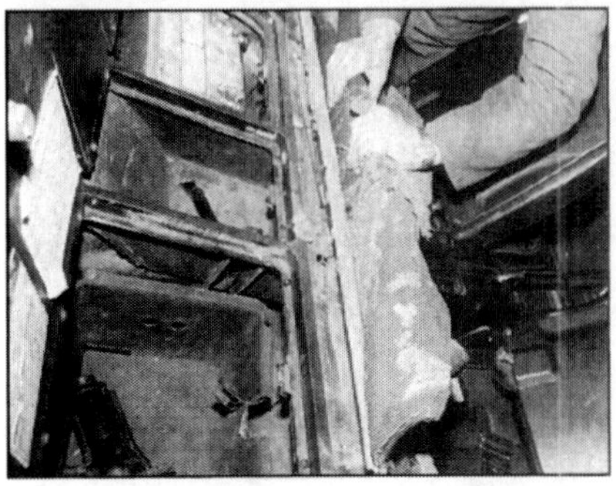

▲ *RT8. It's a good idea to use a heat gun in order to warm the glue and make removal easier. When you're replacing carpets, use the right kind of spray glue; 3M's P/N 8080 and Mar-Hyde's P/N 3410 work well, according to John Pfanstiehl.*

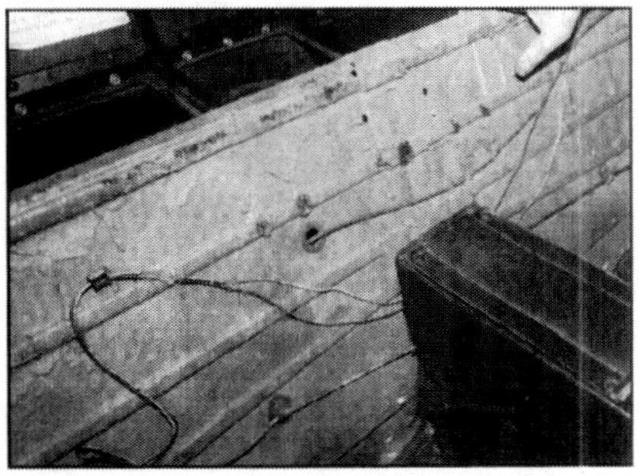

▲ *RT9. Removing the carpet exposes the wiring for the alarm, seat, belt and door-ajar switches. These harnesses make noticeable lumps under a new carpet – one option is to add insulation to reduce heat and noise, and to smooth out the lumps.*

▲ *RT10. The rest of the rear carpeting can be removed the same way.*

▲ *RT11. The rear fenders are covered with sound deadening mats which are removed in much the same way as the carpeting.*

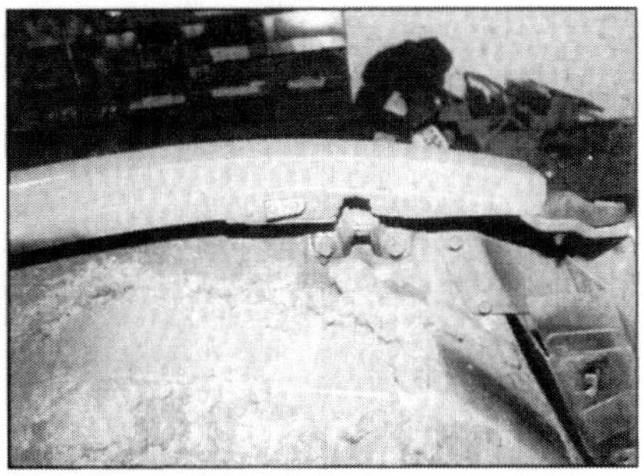

▲ *RT12. As with other parts of the car, minor fiberglass damage was in evidence, such as here around the folding top lid bumper ...*

▲ *RT13. ... and here, behind the underbody storage compartments where more cracks were found.*

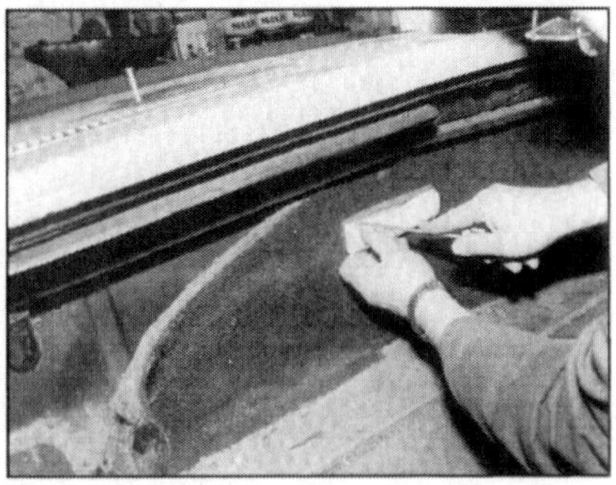

▲ *RT14. In order to pull out the final piece of rear carpeting it was necessary to remove the luggage lamp. Access to the backplate mounting screws was gained by prizing off the cover. Note the plastic carpet fixing lugs above the luggage lamp – care must be taken when removing these to avoid damaging the soft plastic.*

▲ *RT15. There are just two wires to the luggage lamp. Remove and tag, so you'll know where to replace them.*

▲ *RT16. Remove the right-hand box by opening the door and lifting the box upward. This photo shows the right-hand door complete with stickers relating to tire pressures, jacking instructions and limited slip diff. To maintain originality during restoration, replacement stickers can be obtained from a number of manufacturers.*

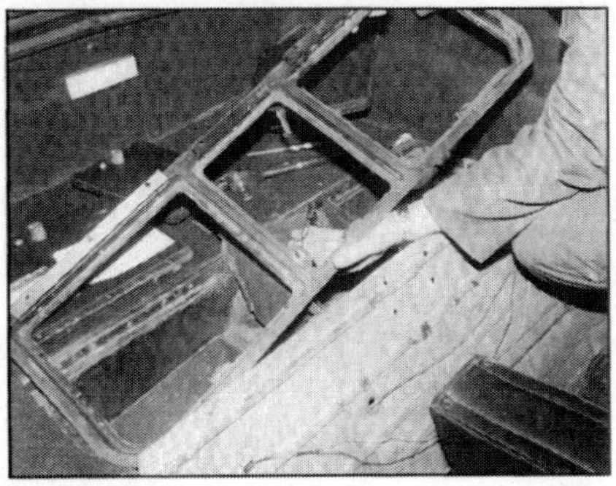

▲ *RT17. To release the storage compartment assembly, eight screws along the inner flange of the frame and the lower door hinges have to be removed. It can then be lifted out. To remove the doors, remove three screws (outer doors) and two screws (inner doors) at the upper hinge strap.*

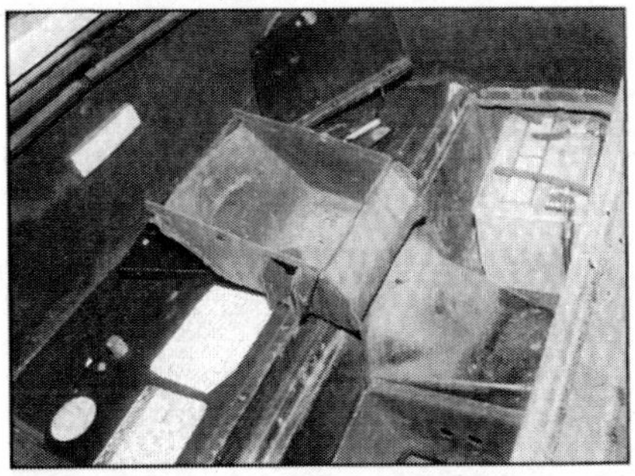

▲ *RT18. The centre storage compartment box can be removed by taking out the seven screws from around the top inside edge of the box.*

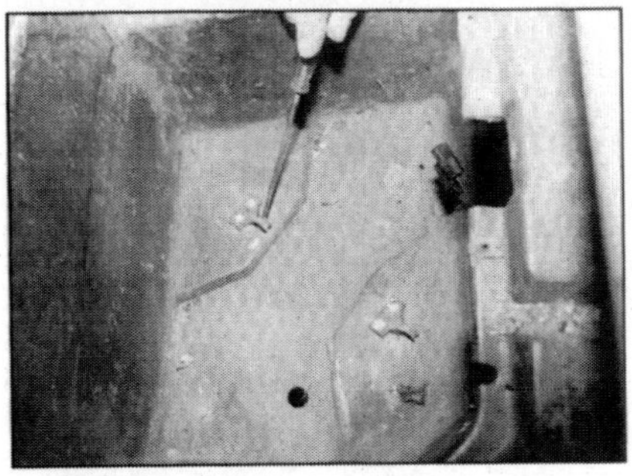

▲ *RT19. When removing the battery, always disconnect the negative cable first (and hook it up last on replacement). Then disconnect the positive cable and remove the battery hold-down clamp/strap.*

▶ *RT20. The old and some of the new – at top is the frame and the right-hand and centre storage compartments. Both the latter are available as new replacement parts from Auto Accessories of America.*

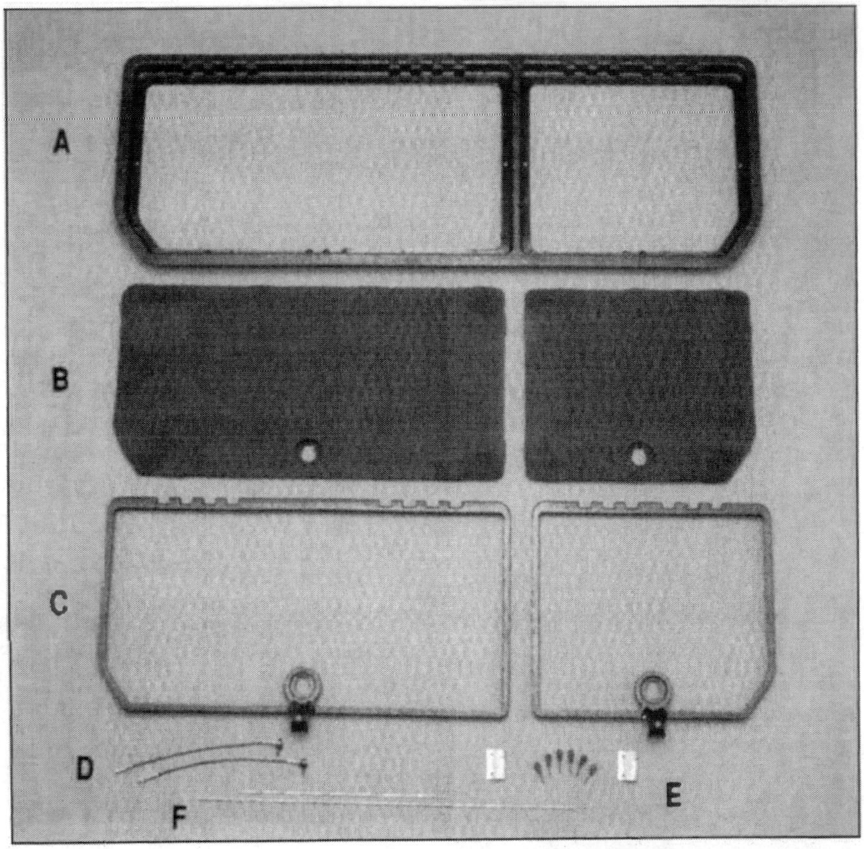

▲ *RT21. In 1979 the Corvette moved to two rear compartments (though early models were fitted with three). The components for these versions, too, are available.* (Courtesy Auto Accessories of America)

▶ *RT22. To remove the rear exhaust doors, the vacuum valve attaching screws should be removed. Then disconnect the vacuum valve at the door actuating rod, detach the hose and remove the valve assembly.*

▲ *RT23. With the doors closed, remove the screws and detach the door assemblies from the rear plenum.*

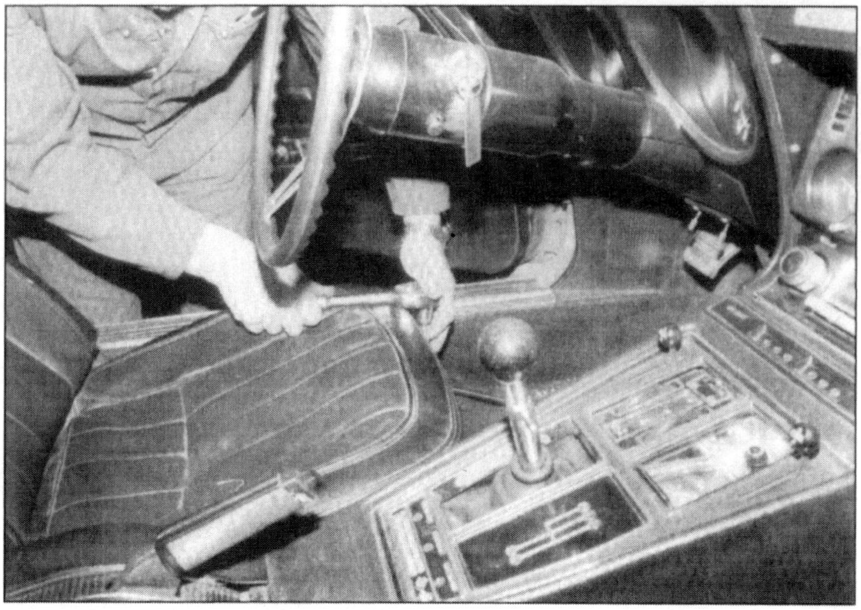

▶ *RT24. Working inside your 'Vette is made much simpler by removing the seats; it's not a difficult task and there's so much more room to move around in. If the car has a seat belt warning system, do not forget to unplug the sensor leads before seat removal.*

In this case, there were no lap belt and shoulder harness type seat belts to contend with, as they were fitted to 1972 and later models. Over the years, a variety of seat types and fitments were installed, so it's wise to check with your Haynes manual for the exact details, though they are all similar in essence. Those in our '69 model are notable in that they are raked at an angle of 33 degrees – the most severe in the Corvette's history. This was necessary to accommodate the lower roof line introduced with the new body style in 1968. The runners were held by four screws, two at the front ...

▶ *RT25. ... and two at the rear.*

▶ *RT26. And that's it – out it comes. If you're not working on the seats, take care to store them sympathetically; somewhere where they won't be damaged as you trip over them, or have oil spilled on them. And don't forget the rodent problem – seats left for long periods of time in a dark, secluded part of the workshop or loft can prove irresistible to mice looking for a new home!*

At this point, the seat can be removed and stored. In our case, we had a seat renovation in mind, which is covered in its own section.

▲ *RT27. The aluminium kick-plate serves to protect the high rocker panel from scuffs and general damage as well as holding the edge of the carpet in place. It is held by four self-tapping screws ...*

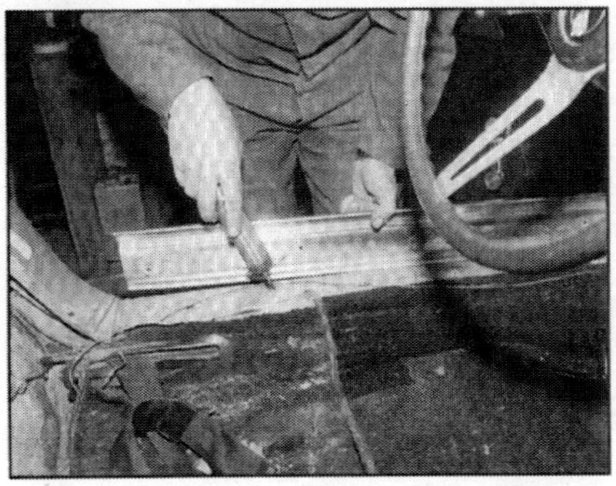

▲ RT28. … and with them removed, the plate can be lifted up and stored or trashed as appropriate. Note the wiring revealed by its removal. It's important that throughout any trim removal process, attention is paid to the possibility of what lies underneath any specific piece of trim. Though there won't be any immediate problems – with the battery disconnected – even the slightest damage to the wiring could end disastrously with overheating or even a fire.

▲ RT29. These seat belts are of the non-inertia type. They are secured to the floor by a single large bolt as shown here. The guiding cover is held by three screws. How do you know if your belts are still OK? Well, the only way to check them is by performing a destruction test, in which case they definitely won't be fit for use! We would recommend that any belts over 10 years old should be replaced as a matter of course – if you use your car regularly, consider swapping them in half that. They're not expensive or difficult to fit and they could save lives – one might be yours!

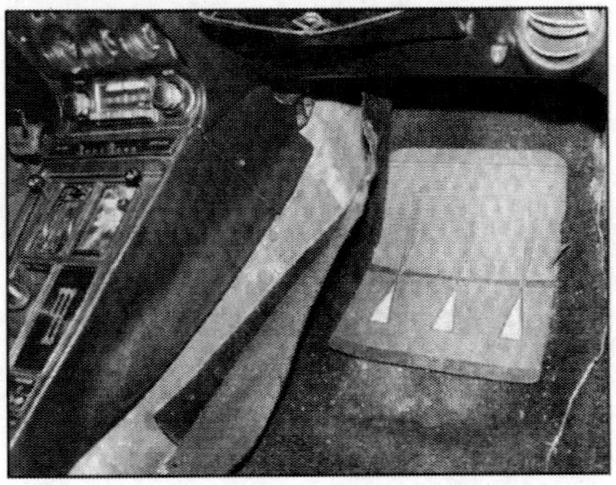

▲ RT30. Having removed the seats and belts, it's a good idea to oust the front carpets in much the same way as the rear carpets.

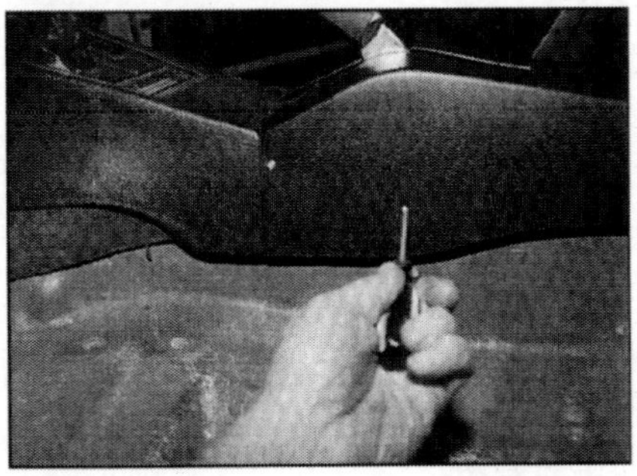

▲ RT31. Removing the centre console by taking out the two Phillips head screws on each side. Use a long screwdriver and hold the seat cushion down – with the seats already out and on their way to be renovated, we had it easy at this stage.

If any of the screws are loose or missing, make sure you inspect, tighten and align their screw clips after the console is removed. The one bolt remaining must have its position selected by an engineer who enjoyed setting folk problems to solve! Open the passenger side compartment door and lift out the storage 'bucket'. Now contort your body and reach down through the door opening in front of the glove box material to the rear centre area abutting the console. Unless it already has been removed, you'll find a ⅜ in. head bolt. Use a ¼ in. drive ratchet to unscrew this bolt.

▲ RT32. Raise the back of the console. It may have to be pushed slightly forward to allow this. Pull the parking brake handle up to facilitate removal of the console. The plastic slider strip will stay with the handle. Power window lifter switches can be unplugged by reaching into the front of the console after the rear is lifted. Then transfer the rear support and the front reinforcement from old console to new. Take care while tightening the nuts. If they start into the studs cross-threaded or bind on the studs while tightening, the studs will break loose inside the plastic. What to do about damaged or worn consoles is covered in a separate section in this chapter.

▲ RT33. To remove the upper centre cluster trim plate screws, tip the trim plate forward so that the windshield wiper switch connector can be reached and disconnected, and withdraw the trim plate. It is important not to bend the trim plate excessively as it is designed to collapse under impact.

REMOVING THE STEERING COLUMN

The interior of the '68 model Corvettes was more modern and practical than its predecessors. A much more convenient column-mounted ignition and steering lock switch was fitted and a tilt/telescopic steering wheel was available as an option. The 1963/68 models featured a large, 16 in. steering wheel, but for 1969, a smaller, 15 in. wheel was fitted.

▲ *RT34. Remove the pinch bolt from the intermediate steering shaft coupling and separate the coupling from the lower end of the steering shaft. Disconnect the linkage from the base of the steering column and the steering column harness at the electrical connector.*

Disconnect the neutral start switch and back-up lamp switch connectors (if so equipped), then remove the floorpan trim cover screws and remove the cover. Remove the two nuts that secure the floorpan bracket to the mounting studs and then the two nuts that hold the column bracket to the instrument panel. At this point the column is ready for removal – it may be handy to have a helper standing by.

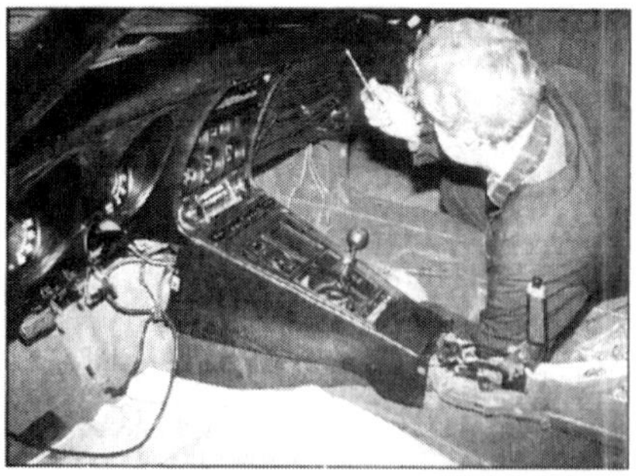

▲ *RT35. After removing the two screws on the outboard flange and the two screws at the inner flange of the centre cluster of the right hand pad ...*

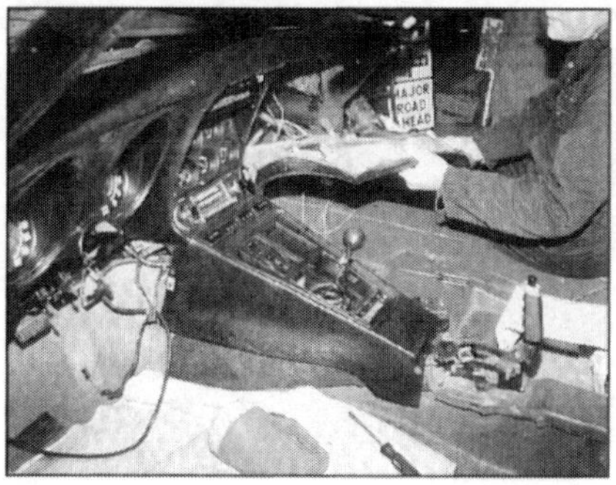

▲ *RT36. ... the pad can be pulled outward to release the retaining clips along the top of the trim side panel.*

▲ *RT37. As you can see, there's plenty of wiring down there. It's important when removing trim panels to ensure that no wiring is trapped or snagged in such a way that it could be damaged. Where appropriate, mark wiring and/or plugs by writing on tags of masking tape. The body rocker panel cover restrains some of the force of the kick panel and, had we not already removed it, four of the rocker panel screws would have had to be loosened in order to effect removal of the panel.*

The kick panel is held by a single screw at its centre; once removed, it can be pulled away rearward. The side trim panels then lift out by pulling them forward and upward to release the two spring clips at the front console assembly.

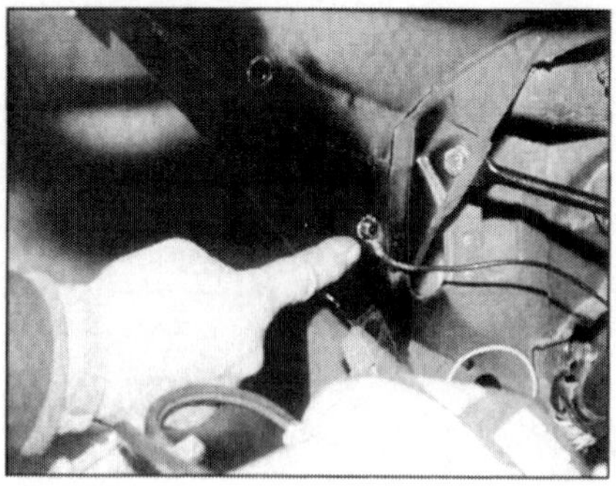

▲ *RT38. Always make sure that any ground connections make good solid contact with metal. If you're using eyelet connectors fitted under existing screws, make sure that the screws actually go into some metal behind the fiberglass – otherwise you're wasting your time!*

▲ *RT39. The instrument panel pad is secured by two screws on the outer flange ...*

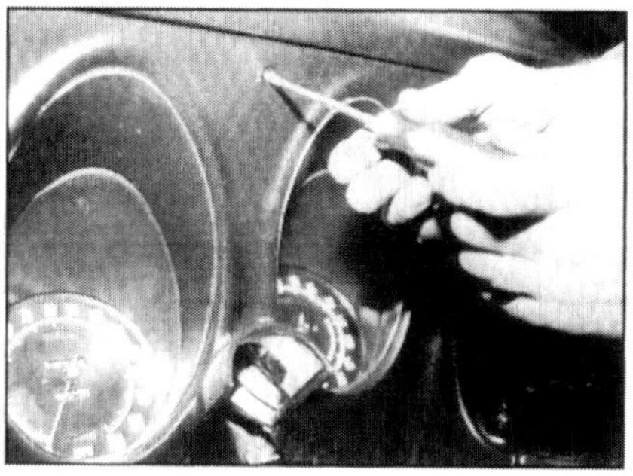

▲ *RT40. ... and three across the top of the pad. When pulled outward, the retaining clips are released ...*

◀ *RT41. ... after which the whole assembly will fall backward. Obviously, at this point it is still attached to the car via many wires and cables. Take great care to disconnect them all properly and tag them for later identification, where appropriate.*

▲ *RT42. The speedometer and tachometer drive cables have to be unscrewed along with the warning light bulbs from the instrument clusters.*

▲ *RT43. With the instrument panel pad removed, you're left with a whole bunch of wires – you'd better know where they go! Make good use of masking tape to label them as you work through. Alternatively, have someone take pictures, or use a video camera. It's very easy to become lax with regard to noting which wires go where and just as easy to believe that you will remember individual points – after several months and many hundreds of disconnections and dismantling procedures, it's extremely unlikely you'll remember everything. The camera, however, not only doesn't lie, but it doesn't forget, either!*

Console and instrument panel removal are covered in a separate section in this chapter.

▲ *RT44. The windshield pillar trim can be removed as described in the section dealing with windshield pillar rust repair.*

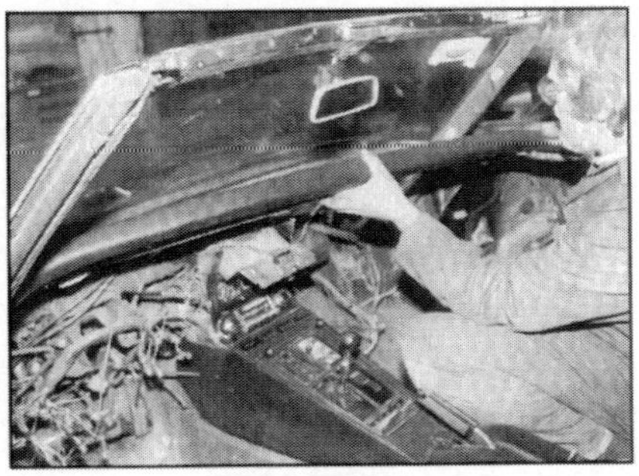

▲ *RT45. With the dash pad removed there is access to the reinforcing strut, which is actually two pieces.*

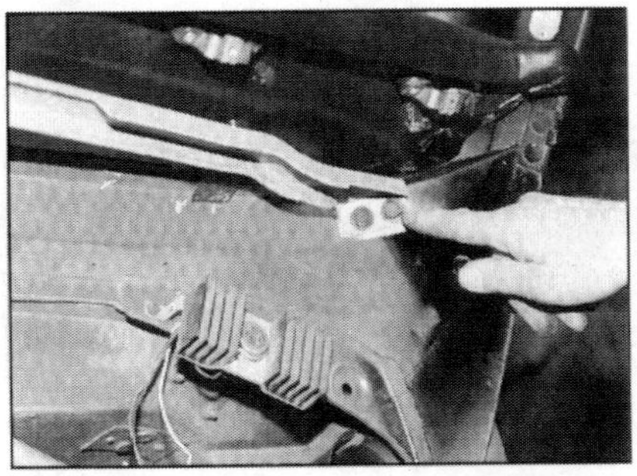

▲ *RT46. It is held by two bolts at the right …*

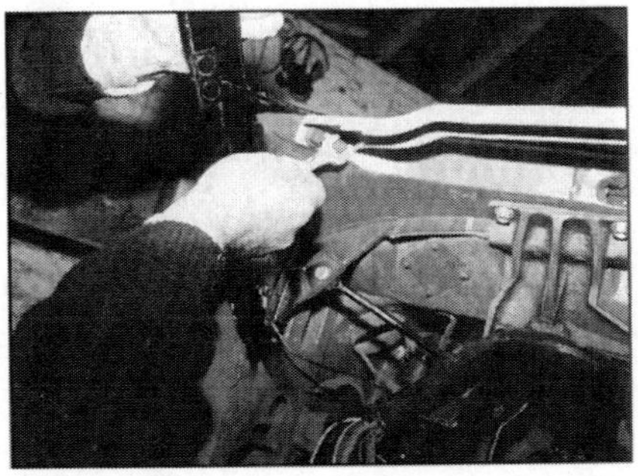

▲ *RT47. … two at the left …*

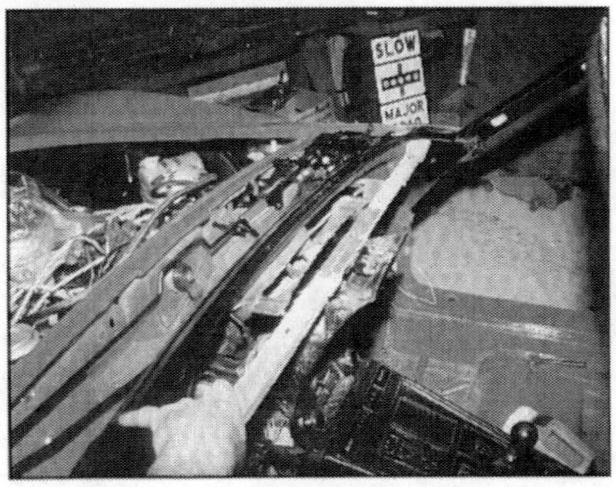

▲ *RT48. … and again, by a further two bolts in the centre. It can be removed and installed all in one piece.*

▲ *RT49. Convertible-style bodies have added reinforcements beneath the dash to help arrest scuttle shake and to offer more rigidity. There are five altogether, two of which are bolted from the cowl area towards the centre of the car. Two more can be found here, on the driver's side …*

▲ *RT50. … while the remaining brace is here on the right-hand side of the vehicle. Note that it is directly in front of the heater core and has to be disconnected …*

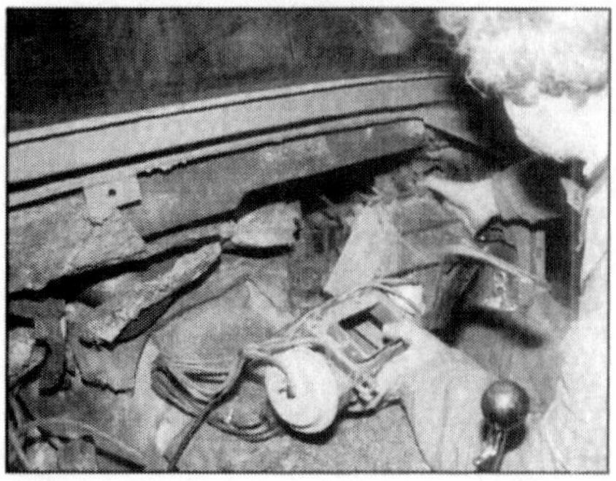

▲ *RT51. ... before the heater core can be removed. (If you want to remove the radio, this brace will also get in the way of that.)*

▲ *RT53. There's no need to remove the existing skin – just remove dirt, grime and grease from the surface and fill any gouging with sealant foam.* (Courtesy Auto Accessories of America)

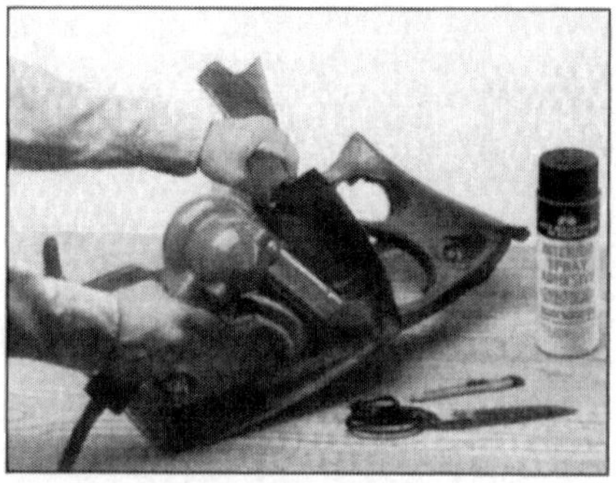

▲ *RT54. Having applied interior spray adhesive over the clean, dry surface of the original skin, place the new skin in position. A hair dryer or heat gun can be used to 'shrink' the skin into place.* (Courtesy Auto Accessories of America)

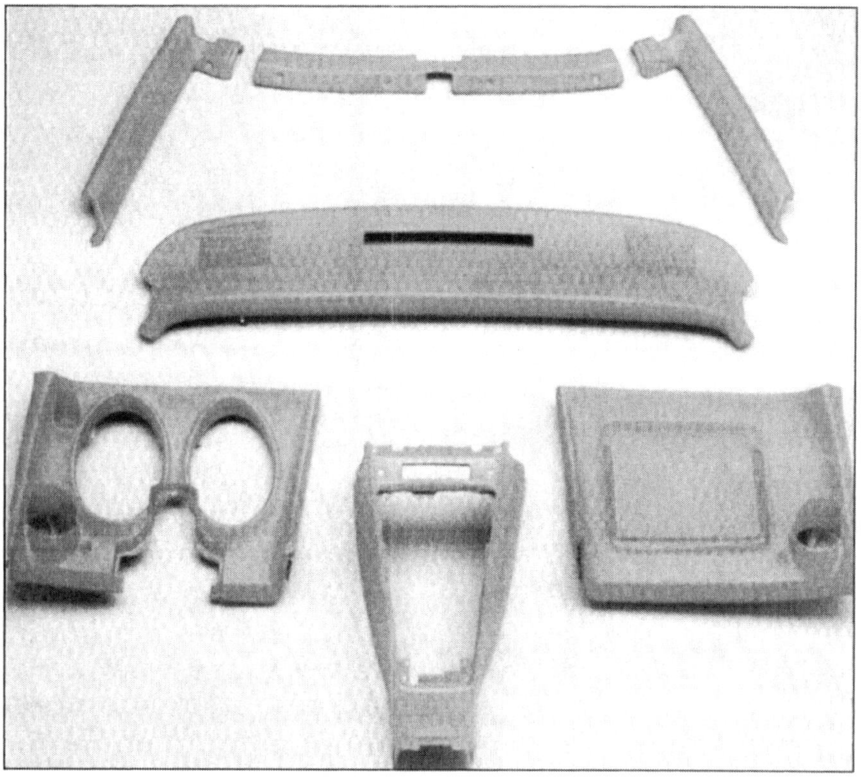

▲ *RT52. What you do with your interior trim parts once they're all on the workshop floor depends on their condition. Take the dash pads, for example. They get a whole load of wear and tear over the years and take a real battering from the sun which, magnified by the windshield and side glass, will eventually crack the pads. One way around these problems is to fit new interior skins. Using an accurate recreation of the ABS surfaces found in the Corvette when it was new, Auto Accessories of America can provide skins with the same texture, grain, stitching and color as the original items. Each skin is made slightly larger than the original so that it fits like a glove over the existing surface.* (Courtesy Auto Accessories of America)

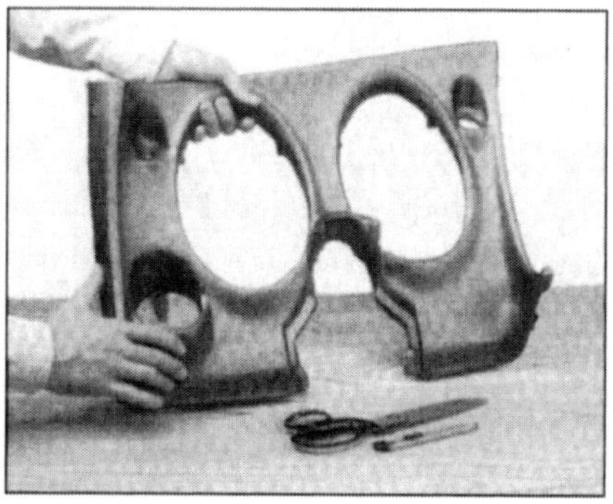

▲ *RT55. A cheaper way to restore a great look to your 'Vette. It won't be totally original, of course, but it'll be a whole lot better than a disgraceful dash!* (Courtesy Auto Accessories of America)

REFITTING INTERIOR TRIM

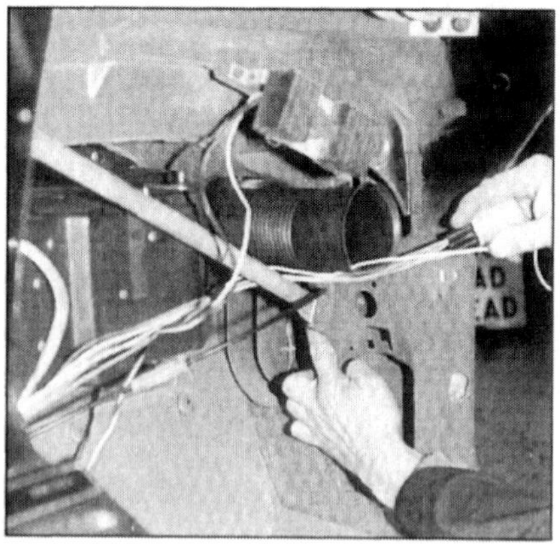

▲ *RT56. How easy it is to put your car's interior back together again will depend how well you documented the dismantling procedure. Certainly, it is not a job to be rushed, for a simple mistake*

at an early stage could result in another tear-down, later on, to put it right. In particular, take great pains to get the electrical wiring correct, especially the earthing points; they're a cause of great distress on most cars, but more than ever when the body is fiberglass.

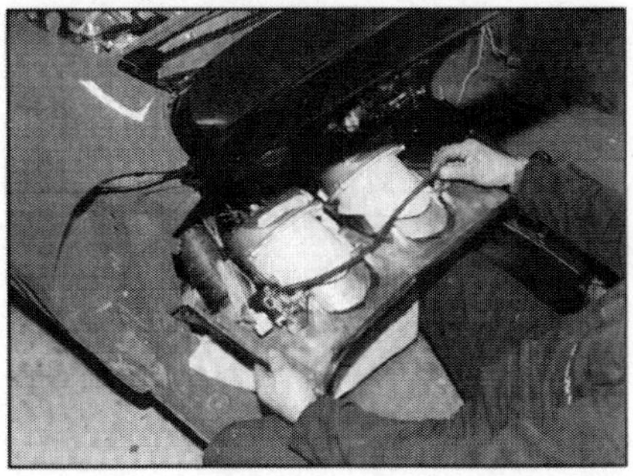

▲ *RT57. Another point to bear in mind is that the routing of the rubber tubes for the windshield and headlamp mechanisms must be perfect; if you trap a tube, or run it where it can be trapped later on, it could result in only partial operation of the item or a total failure of the system.*

▲ *RT58. When you get to this stage, you can allow yourself a tentative sigh of relief. If nothing else, it at least looks like you're making progress!*

REMOVING AND RESTORING THE SEATS

As you'll have seen elsewhere, John Fletcher decided that his seats were definitely in need of some restorative work. Many years of hard labour had taken their toll, so the seats were totally dismantled, the metal parts sandblasted and repainted and new seat covers fitted. This is how it was done with the able assistance of trim and upholstery expert, Bernie Lewis.

▲ *ST1. The runners were removed from the seats – another four screws – stripped down and prepared for blasting and replating. You'll probably agree they were about ready for it!*

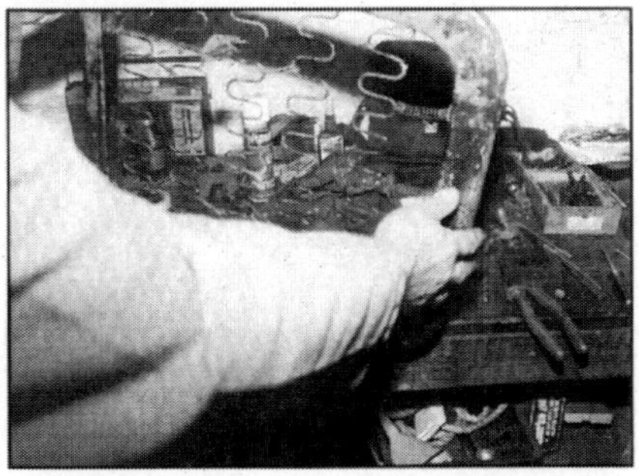

▲ *ST2. Em also renovated the frame and cushion wires.*

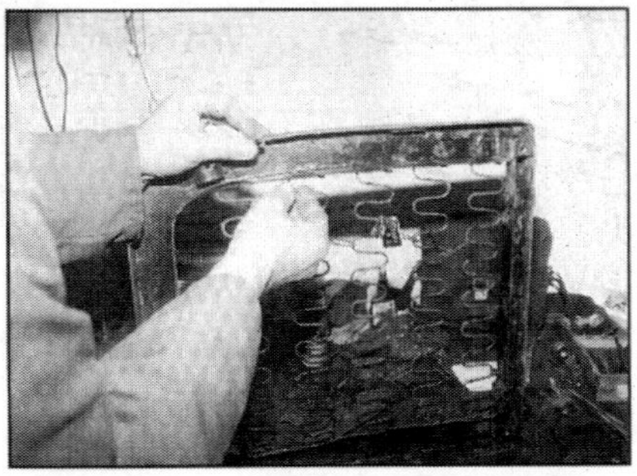

▲ *ST3. The latter were unhooked after the positions of the inter-wire hooks had been noted. Again, taking photos or a video is the simplest way to make sure that everything goes back the way it came out. Care must be taken when removing the springs – they can easily slip and cause personal injury, especially to unprotected eyes.*

▲ *ST4. The frame and wires were sand-blasted and then sprayed with primer. The seat back frame/wires were given the same treatment. Naturally, there's not much sense in putting old seat covers back on these superb frames ...*

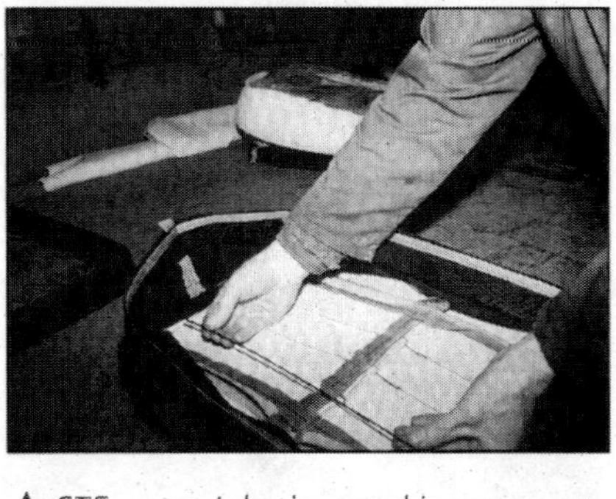

▲ *ST5. ... so John invested in a new set of covers. These are readily available from companies such as Auto Accessories of America. Most models and years are catered for. Depending on how you purchase them, you could be getting the whole kit, as here, but you may have to purchase minor components separately.*

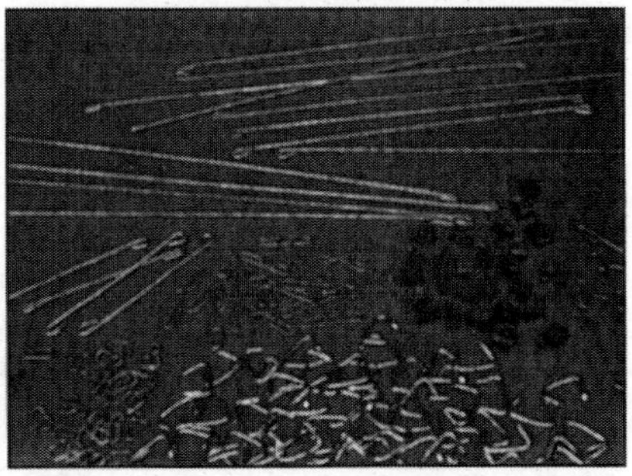

▲ *ST6. New hooks, wires, clips and hog rings all come complete with a 'Triple A' kit – the originals are likely to be rusty as well as being bent out of shape as you take them out – but you can buy them separately if you have to. (Courtesy Auto Accessories of America)*

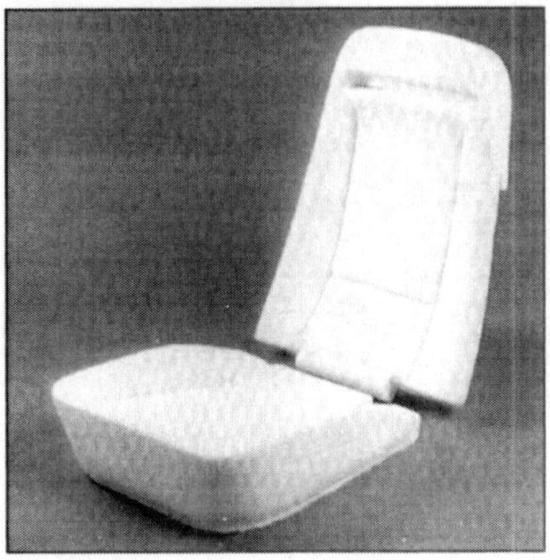

▲ ST7. If your seat cushions are really showing signs of wear, you can get new top and/or bottom seat foam to suit most models. (Courtesy Auto Accessories of America)

▲ ST8. At this stage Em gave the task to expert trimmer, Bernie Lewis. Bernie made the point that, although his years of experience gave him a very definite speed advantage, the home restorer should have no trouble fitting such covers to a Corvette as long as enough time is allowed. As we've said before, rush jobs are seldom completed correctly. Our cushions were in reasonably good order and were replaced, although they needed a little extra padding ...

▲ ST9. ... in order to be a snug fit inside the new covers. Bernie fitted standard seating foam around the edge of the seat cushion ...

▲ ST10. ... and then glued it into position. Remember the safety rules when you're working with glue – work only in a well-ventilated area and take regular breaks. It's a good idea to wear plastic gloves – that glue is very tricky to get off and can be an irritant. Always read the notes on the packaging.

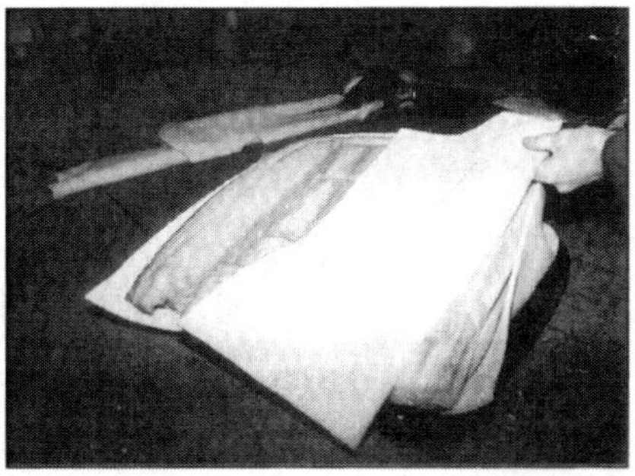

▲ ST11. The curves of the top of the seat cushion were mimicked perfectly ...

▲ ST12. ... and stuck into place in the same way.

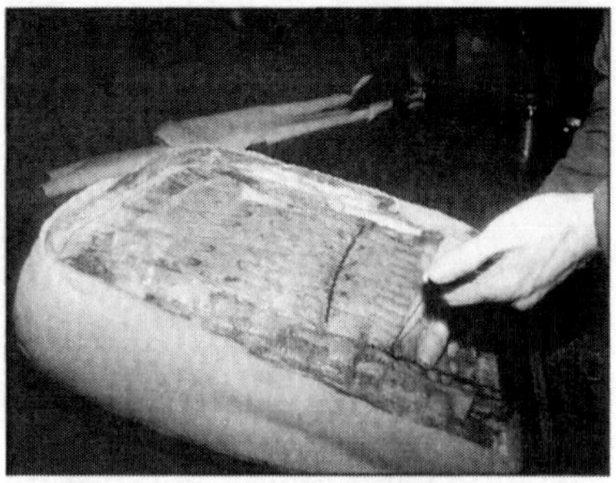

▲ ST13. To prevent the seat cover moving around, it has to be secured to the cushion. Gluing is not the way – it has to be 'sewn', which requires the use of this professional hook to pull the strong thread through and onto these wires on the underside of the cushion. Naturally, it's vitally important that the cover is perfectly placed before doing this – time and patience are required.

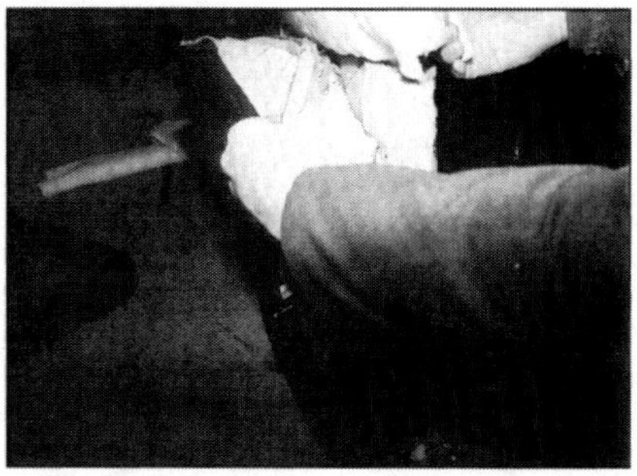

▲ ST14. Here you can see Bernie using the hook to pull the thread through.

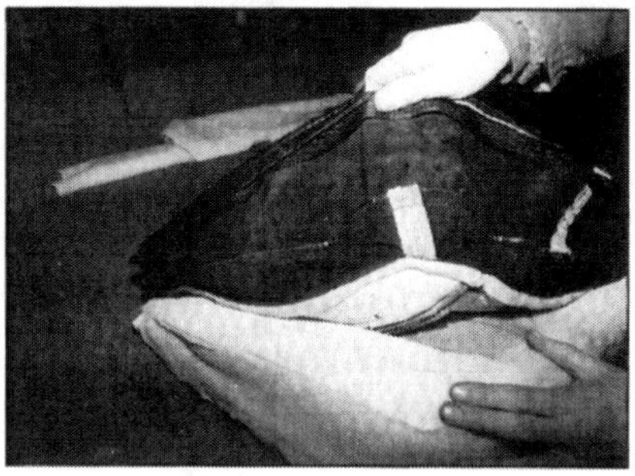

▲ ST15. The seat cover has to be placed down over the cushion very firmly ...

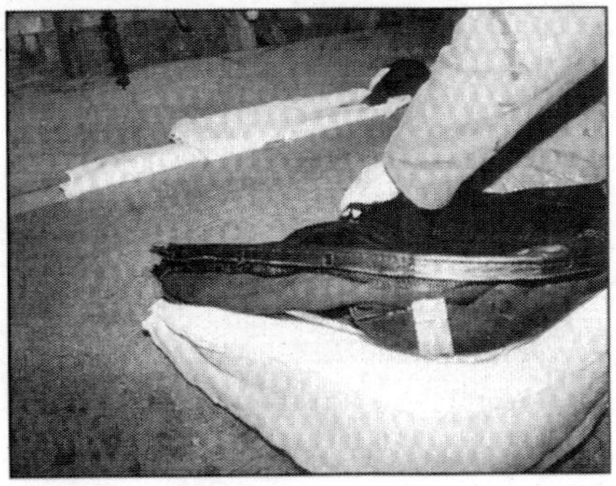

▲ ST16. ... and the fit has to be tight and snug all around the edges.

▲ ST17. When the cover is perfectly 'square' on the cushion, it can be upended and the edges pulled around the frame edges ...

▲ ST18. ... so that the securing clips can be installed.

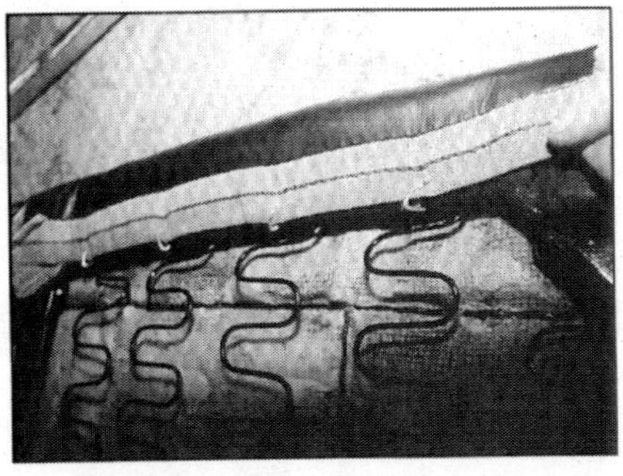

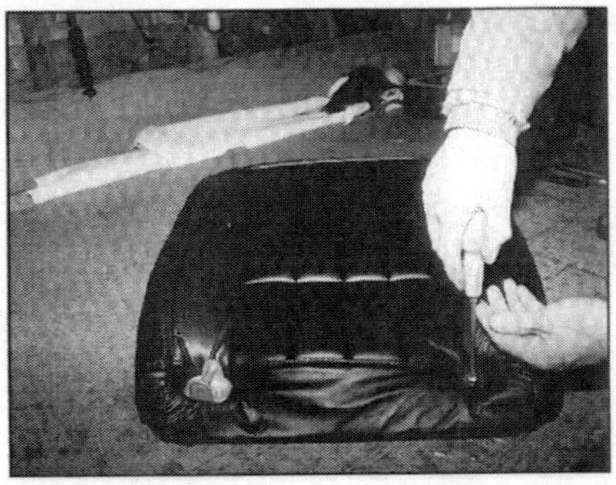

▲ ST19. An impressive result and, though this one was completed by a professional, there's no reason why the home restorer shouldn't be able to achieve the same result, albeit not as quickly. Here, Bernie is refitting the seat bumper bolts.

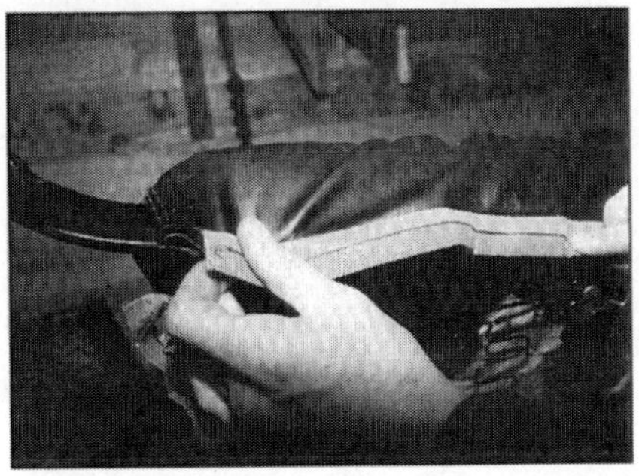

▲ ST20. He then progressed to the seat back cushion and started by placing the covers in position ...

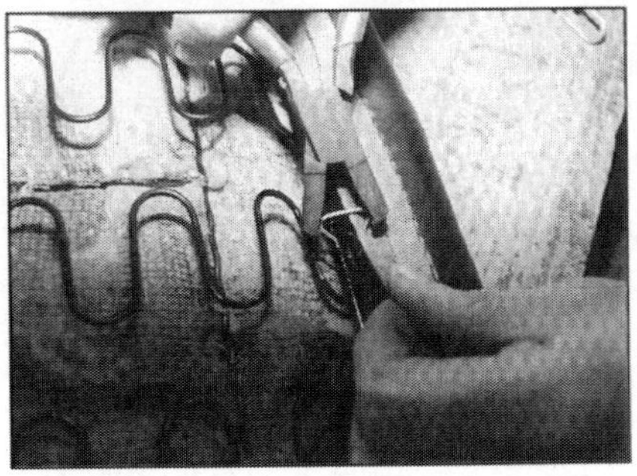

▲ ST21. ... and then new hog rings were installed around the edges. Note that hog ring pliers are required to do this operation safely – using standard pliers is a sure way to end up with scraped knuckles, or worse!

◀ ST22. It's important that there are no sharp edges left when the hog ringing is complete.

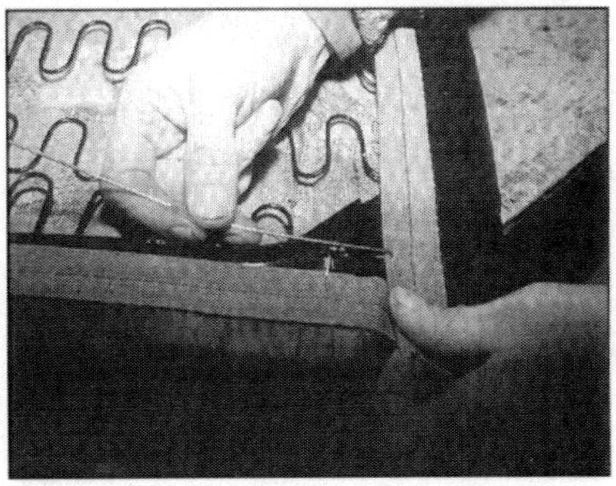

▲ ST23. Special wires are used to 'tension' the seat cover across the seat back ...

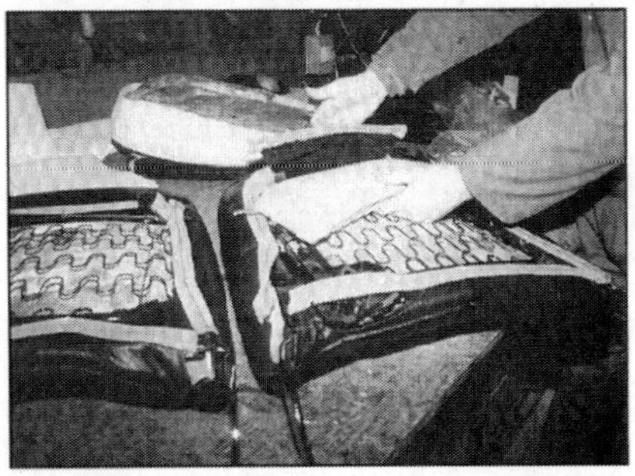

▲ ST24. ... and a foam padding piece is inserted at the top for the headrest.

▲ ST25. This is covered over by the top of the cover and the edge secured ...

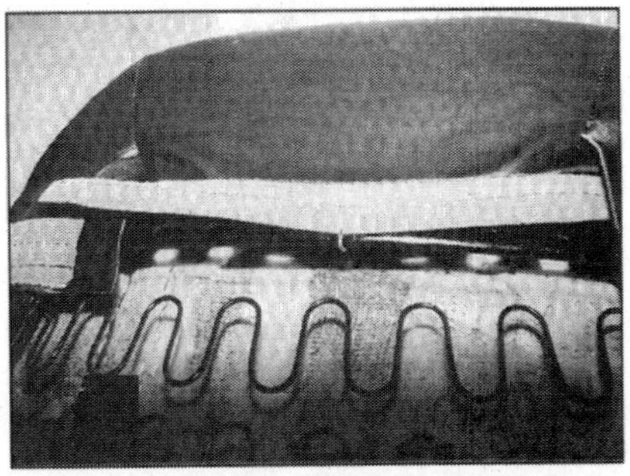

▲ ST26. ... using hog rings, as previously.

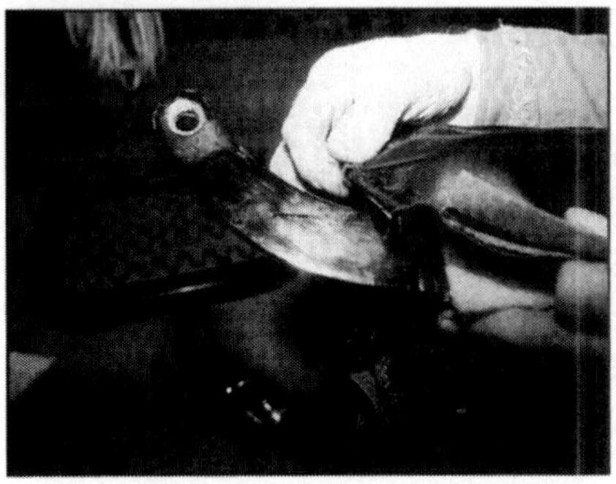

▲ *ST27. It's important that the edges of the seat cover fit perfectly ...*

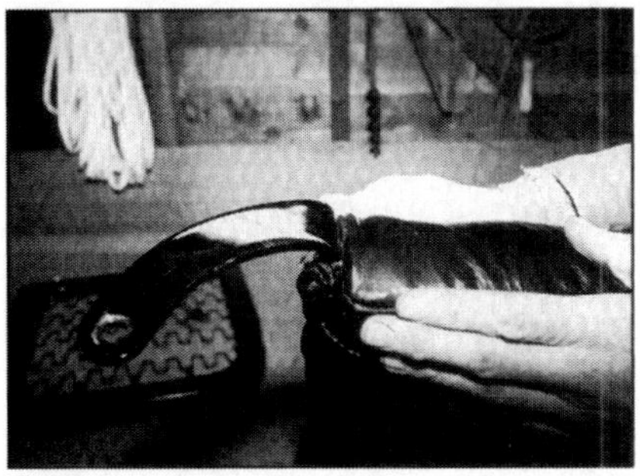

▲ *ST28. ... around the lower corners of the seat back where the hinge is situated. Once Bernie was satisfied with the covering job ...*

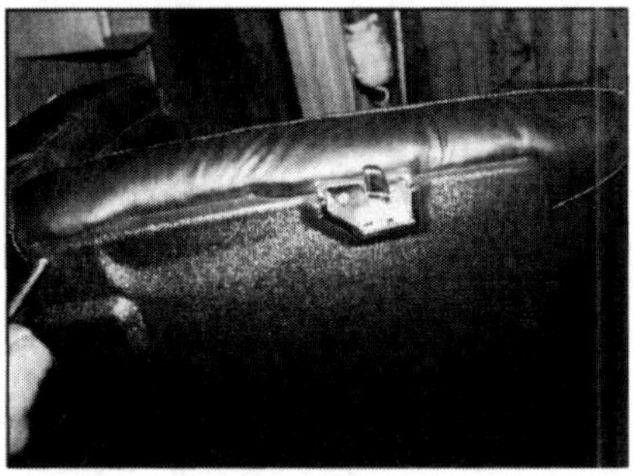

▲ *ST29. ... the seat back and squab were refitted and the seat back replaced.*

▲ *ST30. As good as new and, just as important, they look like the originals.*

▲ *ST31. If re-covering a seat is way beyond your capabilities, specialists such as 'Triple A' can supply covers ready mounted on new foam and ready to fit. You could even customize to a new or optional style, or convert to leather in this way. This is an example of the expert 1978 Pace and 1979–82 leather seat bases and covers.* (Courtesy Auto Accessories of America)

▲ *ST32. When you're dismantling any part of your car, keep a weather eye open for interesting stamps and marks. This one on one of the seat back cushions tells us that it first saw service on 21 October 1969.*

SPEAKER REPLACEMENT AND DASH REPAIR

A good sound system will add plenty to the enjoyment you get from your Corvette, but what's the use of a high-specification head unit if the speakers sound as if there's an angry bee trapped inside? The years will take their toll of any speakers, especially if you've been trying to keep the volume above that of the exhaust note, and there are few cars of any age which wouldn't benefit from new speakers.

To change the speakers you have to remove several trim pieces, so take this opportunity to check them out and replace as necessary.

▲ *SR1. While you're removing the dash pad in order to get at the speakers, consider renewing the pad itself, especially if the effects of hot sunshine or just plain old age have caused this sort of deterioration to take place. Both visors will have to be removed – mark them left-hand and right-hand – and you should also take this opportunity to clean up the chrome attachment rods. Use masking tape to fix the screws temporarily to the visors so that you know which go where.*

▲ *SR2. The attachment screws for the upper windshield trim piece are located on its bottom surface. In this case, the mirror does not have to be removed. Take off the trim piece with extreme care because it tends to go brittle after years of sitting there.*

▲ *SR3. The windshield post trim pieces can be gently pulled away from the post at the top from the velcro-type plastic snaps that hold them there. Original-style replacement parts will have the bottom snap fitted, but aftermarket replacements may not come with this bottom snap in place. However, the trim will be held satisfactorily with only the one snap in place.*

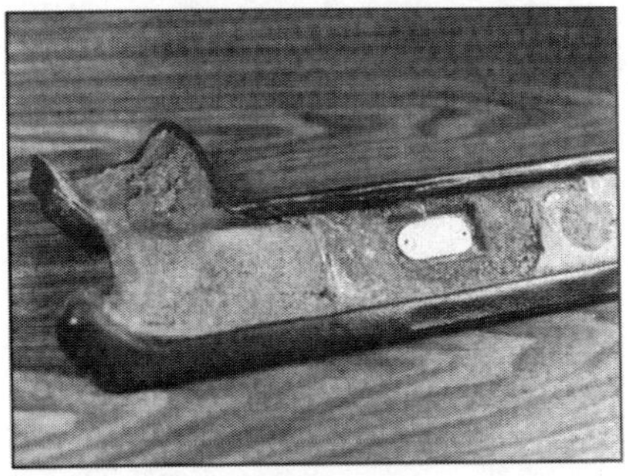

▲ *SR4. After scraping off any loose rust from the back of the trim piece steel backing, apply a light coating of oil or, better yet, wax chassis treatment – the thinner type rather than the heavier black goo!*

▲ *SR5. Two Phillips screws hold the right dash pad. After removing them, you can pull the lower corner rearward, which makes removal of the kick panel easier.*

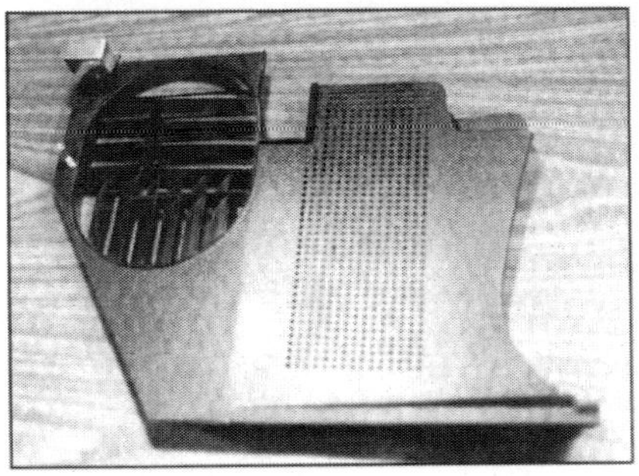

▲ *SR6. The right kick panel should be held at the top with a screw that you'll find way up under the dash, near the top of the carpet. On the other hand, you may not find it there because some owners fail to put it back properly, in which case the hole that's left is prone to letting water into the car. You'll have to pull the kick panel down and rearward in order to remove it. Unfortunately, if air conditioning is fitted, any replacement kick panel that you fit here may need to be shaped with a fine-toothed saw if the correct panel is not available.*

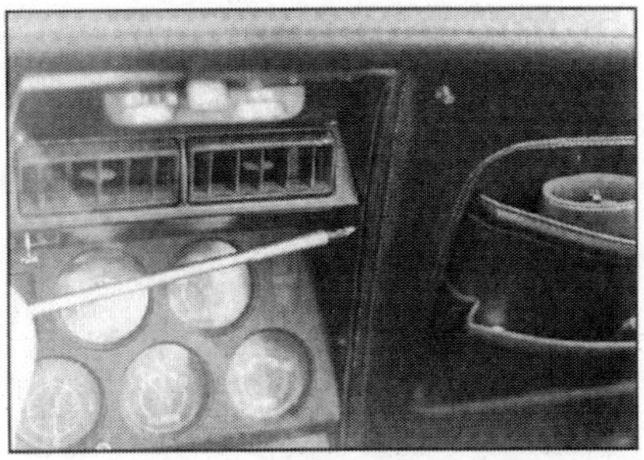

▲ *SR7. The two inside screws which go into the right dash pod will now have to be removed, using a long screwdriver. These screws are shorter than most so, once again, attach them with masking tape so you can identify them later. Also, when reassembling, only tighten them very gently because they are threaded into plastic and this strips out easily.*

Now the small trim piece at the front of the console can be removed. The right dash pad can be prized down and out, after the three holding screws have been removed, and it will then come out from under the top pad. But don't pull it straight out! Peer down behind the top edge to look for any wires attached to clips on the back of the panel so you can disconnect them rather than wrecking their connections. The three screws also have to be removed on the left dash pad.

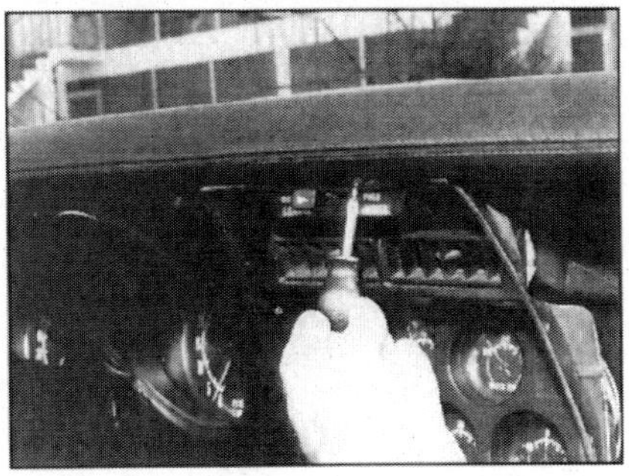

▲ *SR8. Up at the top centre of the gauge housing, there's another little screw to be removed. The housing is sometimes broken; if so, its top portion will fall down after you have taken this screw out. This is yet another screw that absolutely has to go back into the same place. If you use too long a screw you'll leave an ugly dimple, or even a hole in the dash top pad.*

▲ *SR9. On the passenger side, there's one last screw, attaching the top pad to its steady bracket, to be removed.*

▲ *SR10. For this job you are recommended to have an assistant on hand. Push up on the centre of the top pad while pulling down on the ends to bow the pad and enable it to clear the windshield posts.*

▲ *SR11. At long, long last, you can get at the speaker mount, taking it from the old top pad and fitting it to a new one that you may be fitting. If the new speaker has a different shape from the old one, use adhesive-backed foam to block off the air gaps created by the different shape of the new one. If you don't, the speaker will not be as efficient as it should be. Ensure that all the correct clips are installed on the front of the pad, and if you are re-using the pad, ensure that it isn't broken where the clips are to be attached.*

▲ *SR12. The sleeves which go onto the top screws of the left and right dash pads are often missed and fall out or stay hidden behind the vinyl covering of the dash pads. If the sleeve is missing the screw will go in way too far and on past the vinyl. You'll also have to ensure that the screw clips don't fall off or slide too far on their brackets during reassembly, and if any of the speaker screws seem loose, make sure they won't cause vibration by putting some glue on and around them.*

Before going any further, turn on the radio and make sure the speakers work correctly. Just remember how long all that dismantling took last time – you won't want to put everything back together just to find that one of the speakers isn't working!

It's also imperative to ensure that the speaker wires are connected the correct way round, and to make absolutely sure you should always mark the wires and the connections with a tag as you disconnect them. If you don't, the speakers will work against each other and there will be a huge drop in the volume and quality of the sound you hear. This is yet another reason for testing them out before reassembling the dash. If the volume or the sound quality seems poor, try switching the leads on one of the speakers and see if it improves things. Front speakers and any rear speakers fitted should also be in phase for best results.

PARKING BRAKE CONSOLE

You have two options when it comes to replacing your Corvette console. You can either use the original GM part or the lower cost non-original parts available from a number of Corvette specialists. It has to be said that the GM parts are better quality, but you are paying for it. On the other hand, aftermarket replacements provide you with custom features such as non-standard colours or a raised armrest pad.

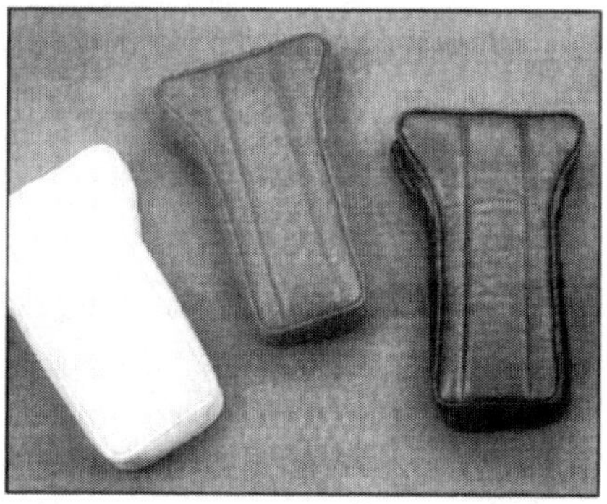

▲ *HC1. The padded armrest cushion is seen in many Corvettes, yet was never offered by GM. It is a useful accessory, comfortable for the driver, and it can even serve as a cushion for shorter passengers. The cushions can be ordered with a leather top surface, as shown here.* (Courtesy Auto Accessories of America)

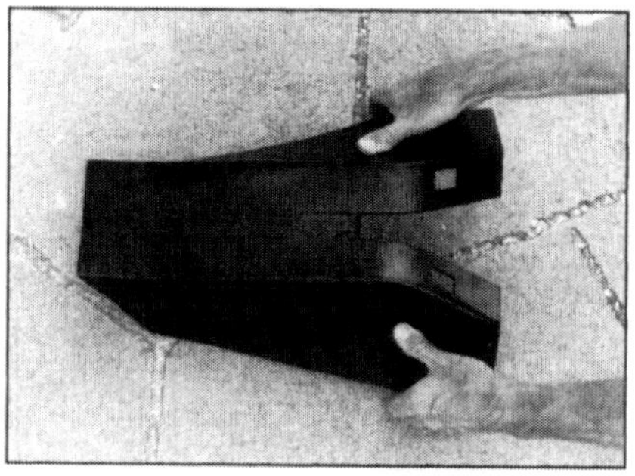

▲ *HC2. The non-GM replacement console is much more flexible than the original although the grain pattern is slightly different. It is made of self-coloured plastic, so that means that there is no coloured finish to scratch off.*

▲ *HC3. The front console is secured by two screws which pass through the sides of the unit towards the front. With these removed, the heater/side vent controls, gear lever (shifter) knob assembly, boot assembly and trim plate have to be removed in order to pull the console up and to the rear.*

Also, with console removed, this is a good time to check and replace the ribbed plastic slider.

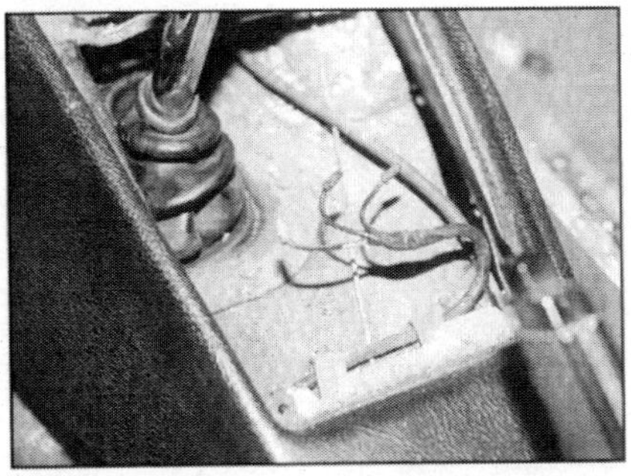

▲ *HC4. Whenever you're removing trim, be wary of the wiring behind it. It's easy to tug away happily and not realize that a wiring connector is attached to a particular panel until it's too late and you hear the awful sound of snapping plastic! Also, of course, you need to use masking tape to mark every piece of wiring you come across – use ball pen and make sure you understand what you've written!*

▶ *HC5. After so many years service, it wasn't surprising that this one was looking a little the worse for wear. Apart from the general tattiness, the vinyl was split in several places.*

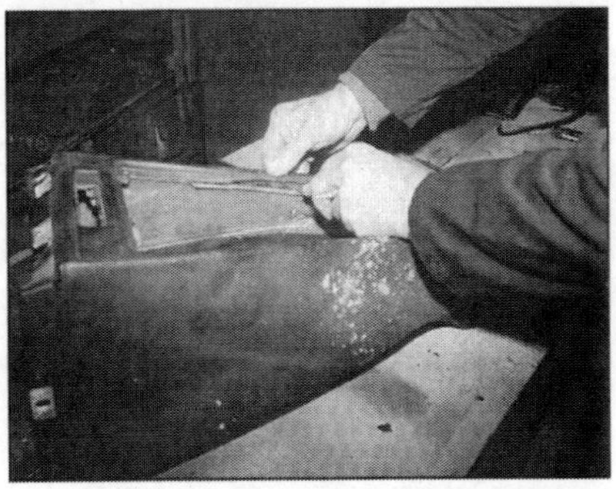

▼ *HC6. This is an aftermarket console and slider from Auto Accessories of America. Wherever you get your replacement body and interior parts from, it always pays to make sure you've got exactly the right part for the job – with so many decades of Corvette spares to sort through, it's easy enough to end up with a slightly different product from the one you need. Em took careful measurements before he fitted the new items.*

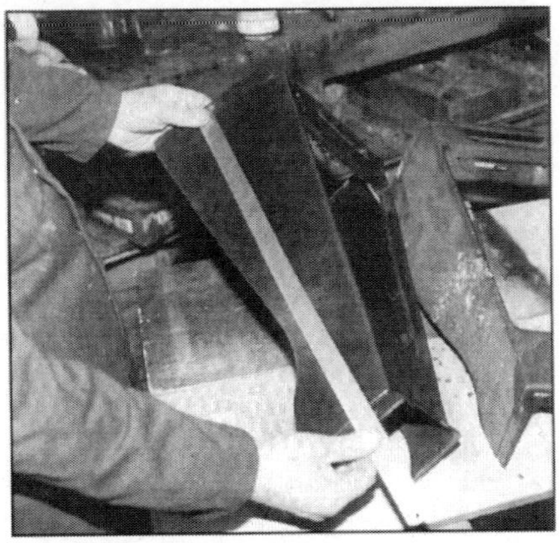

▼ *HC7. Fitting is a straightforward process, requiring just the two screws to be fitted at the front. The rear console is refitted in a reversal of the removal process.*

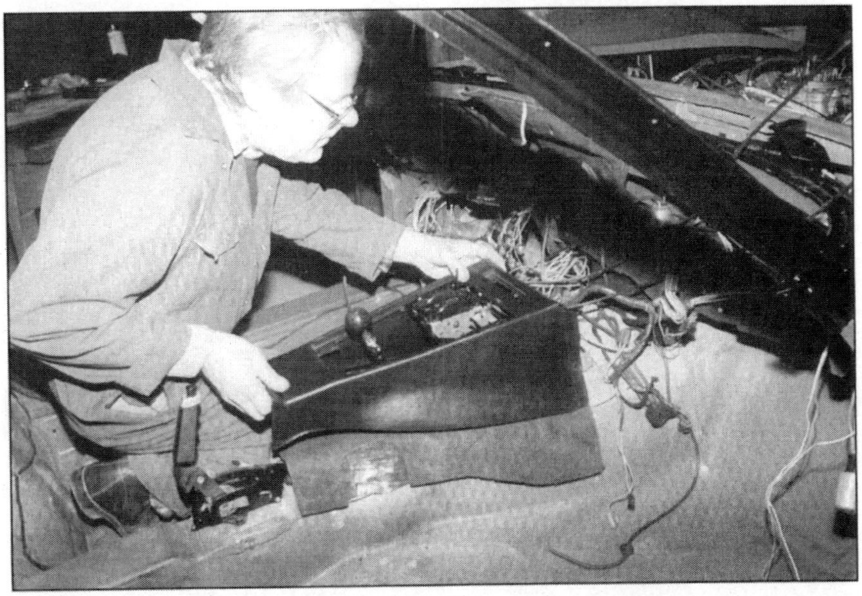

FITTING TIPS

- You may have to cut away some carpet so that it won't get between the rear support of the console and the compartment wall.
- It is difficult to locate the screw clips, so use a flash light to look in through the console screw holes and then use a piece of thick welding wire, or something similar, to manoeuvre the clip into position. Don't tighten up any of the screws until they are all in place so that you can still shuffle the console around, as necessary.
- Still before you tighten down the screws, move the brake handle fully up and down to test for binding of the slider. If adjustment doesn't do the trick, remove the console and investigate the problem.
- Don't overtighten the screws, because there is nothing more disheartening than fitting a new console and then straight away breaking it!

FOLDING TOP

Despite the fact that the open-air Corvette disappeared altogether from 1975 to 1986, it has to be remembered that it started life as a convertible long before anyone thought of sticking a roof permanently in place. Indeed, to many, the convertible model is the only Corvette to own. The convertible Corvette features a totally concealed soft top which can easily be lowered with one hand. Though the coupé of similar vintage offers a reasonable fresh air compromise, with its detachable fiberglass roof panels, only the folding top car offers true 'wind-in-the-hair' motoring, comparable only to riding a powerful motorcycle – but without the inconvenience of having to gear-up for bad weather and without the possibility of falling off!

CARE OF THE FOLDING TOP

To ensure a long service life for the folding top, never stow it when it is damp; allow it dry out first. Every three months (or more frequently if required), wash with a solution of lukewarm water and a neutral soap using a soft bristle brush. Then rinse with plenty of clean water. If additional cleaning is needed, a mild foaming cleaner can be used together with a little lukewarm water and a soft bristle brush. Remove the foam with a sponge or cloth, then repeat the operation if necessary. Rinse with plenty of clean water afterwards. After washing, allow the top to dry out thoroughly in the raised position before folding.

CARE OF REAR WINDOW

The texture of the large plastic rear window makes it easy to mark it, so some care should be taken when cleaning. Never clean it with a dry cloth; if the window has a layer of superficial dust, use a damp cotton cloth. Wipe across the window. When washing the window, use only cold (or lukewarm) water and a neutral soap solution. Rinse off with clear water and prevent streaks by using a damp cotton cloth to wipe off the excess. Many aftermarket suppliers offer a quality plastic window cleaner.

NEVER use a scraper to remove ice, frost or snow from the rear window. In addition, do not stick labels on the rear window or use chemical cleaners and/or de-icers, because they may damage the surface.

REMOVAL OF FOLDING TOP ASSEMBLY

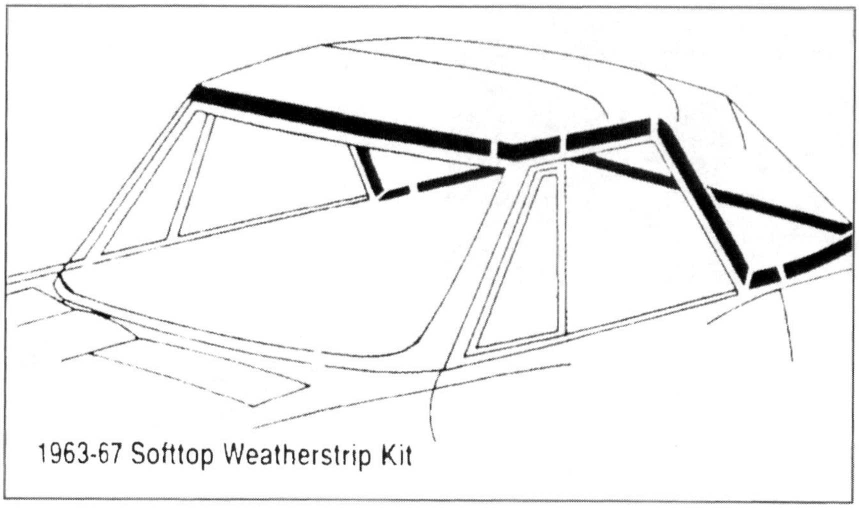

▲ *FT1. An examination of John Fletcher's car's folding top assembly told him what to expect and parts to order before work commenced. It's not just a matter of ordering a new folding top – you'll also need new weatherstrip.* (Courtesy Auto Accessories of America)

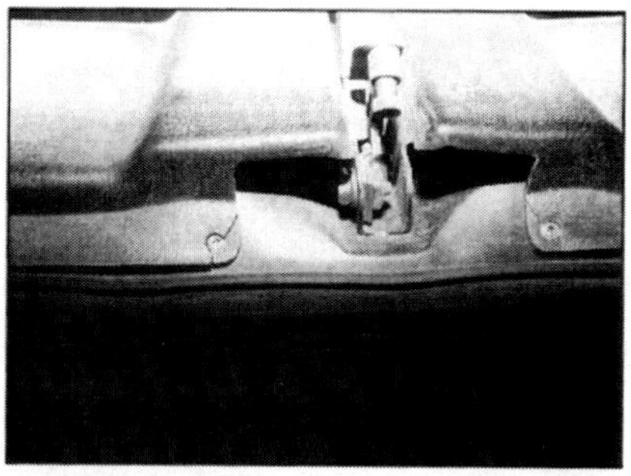

▶ *FT2. The plastic trim covering the front header panel is secured by screws and if, as here, the owner gets carried away while tightening, this is the kind of damage that can occur.*

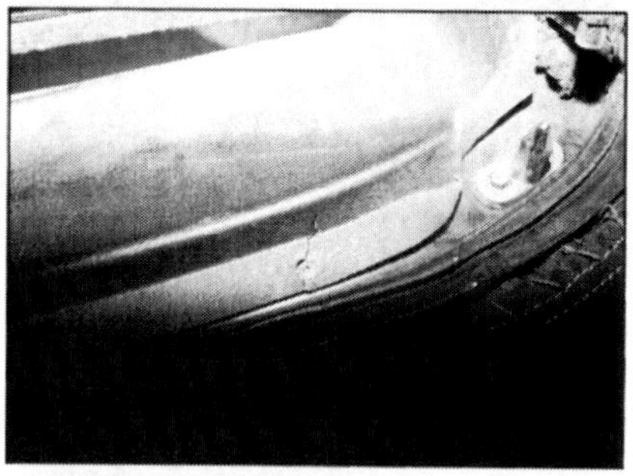

▶ *FT3. The same problem repeats itself at the end of the header panel – in this case, John had Em replace the trim panel complete with a new part from Auto Accessories of America.*

▶ *FT4. This is the rear bow locating pin which fits through the hole in the rear compartment lid. Note the slotted holes for the securing screws, necessary to allow adjustment.*

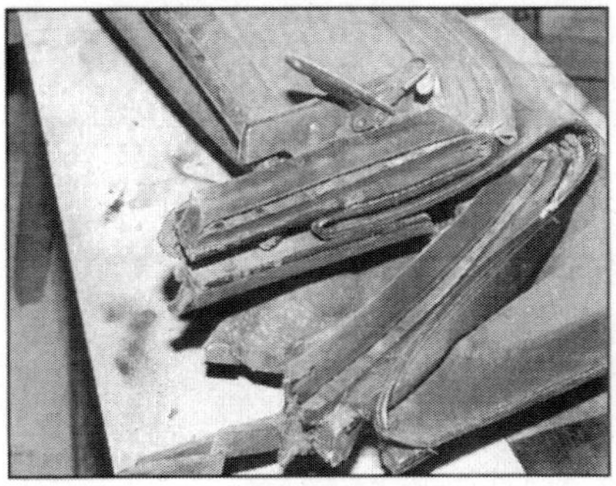

▲ FT5. The soft top itself was looking somewhat the worse for wear. A combination of old age and a lack of care was responsible. The folding top can be removed as a complete assembly. To do this, the top should be raised but with neither the header nor rear bow locks engaged. Removing the plastic trim at the rear of the lock pillar reveals the two frame hinge securing screws. (If the frame is just being removed as part of normal sun/rain driving, then the hinge outline should be marked on the lock pillar surface.) With these removed, the folding top can be lifted out of the car.

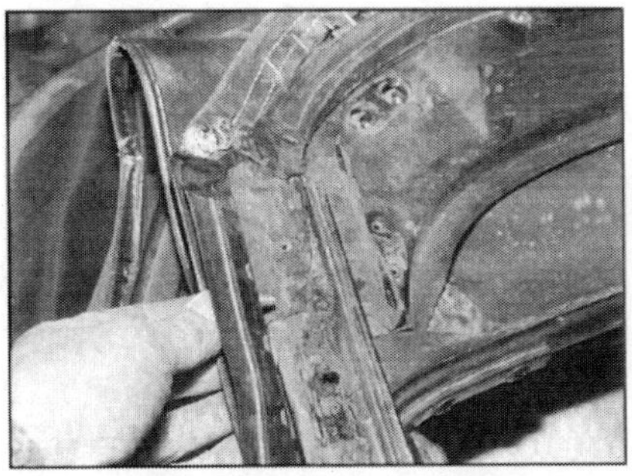

▲ FT6. Remove the side rail weatherstrip, which acts to seal the window opening. The weatherstrip is held in place by loose screws which are part of the assembly and by neoprene base cement which is applied between the weatherstrip and side rail surface.

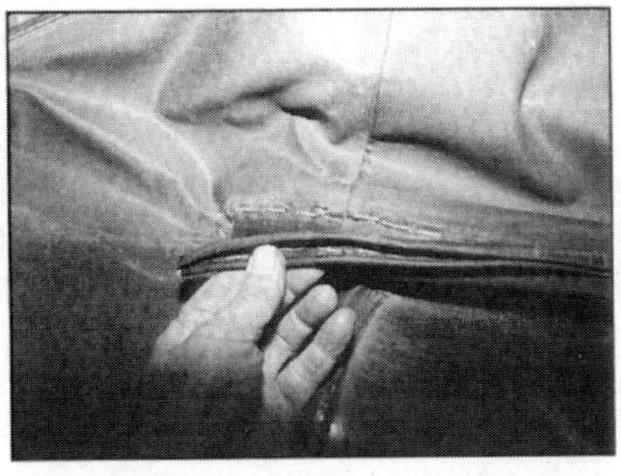

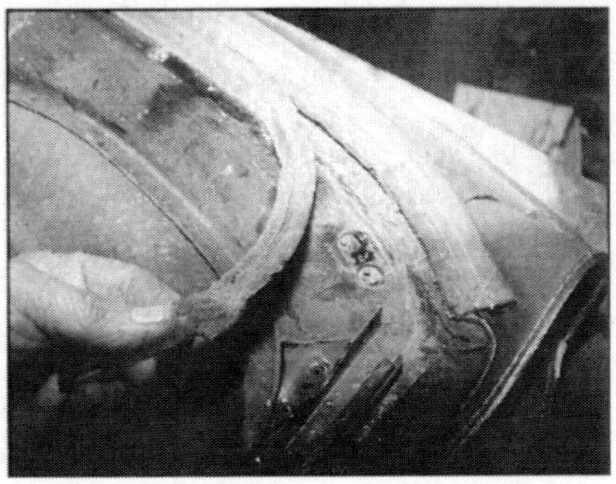

▲ FT7. The weatherstrip across the header panel is retained in a similar manner. When we removed it and pulled back the folding top material ...

▲ FT8. ... it was not a pretty sight – just look at all that rust!

▲ FT9. Em continued dismantling the whole folding top assembly by pulling the cemented trim from the rear side rail ...

◀ FT10. ... and then removed the end caps from the wire-on binding and the tacks securing the binding to the No. 3 bow.

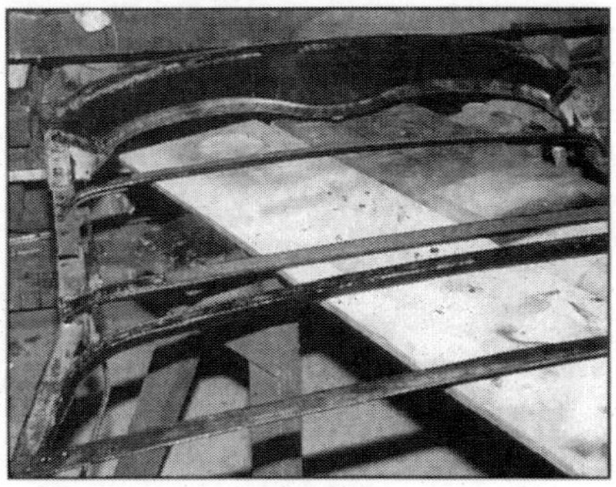

▲ FT11. With all the other various staples and tacks removed, he was left with what you see here – a bare frame/header panel, definitely showing the ravages of time. Note that the bows are counted from the front, with No. 1 being immediately after the header panel and the last one being the rear bow (rather than being called No. 4 bow).

▲ FT12. After the dismantling process, Em collected the various pieces of hardware for despatch to the sandblasters and platers. Here you see the three locks, various screws and plates and the locating pins for front and rear.

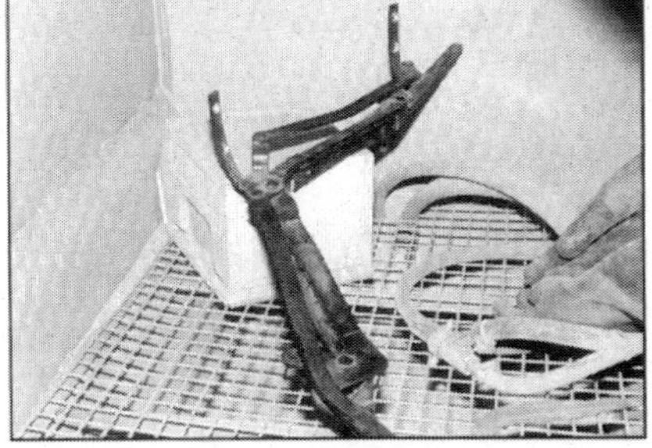

▲ FT13. Also in need of a little sandblasting – waiting here in the sandblasting cabinet – was the hinge assembly.

▲ *FT14. Compare the hinge assembly already blasted (on the right) to the one waiting its turn.*

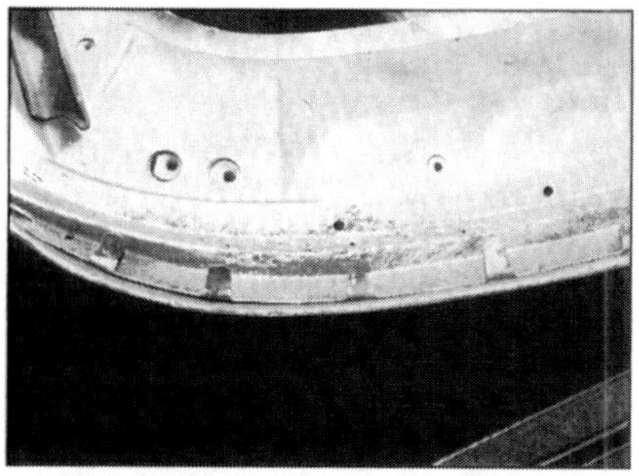

▲ *FT15. The header panel, too, was blasted and this showed up several areas where corrosion had pock-marked and holed the metal.*

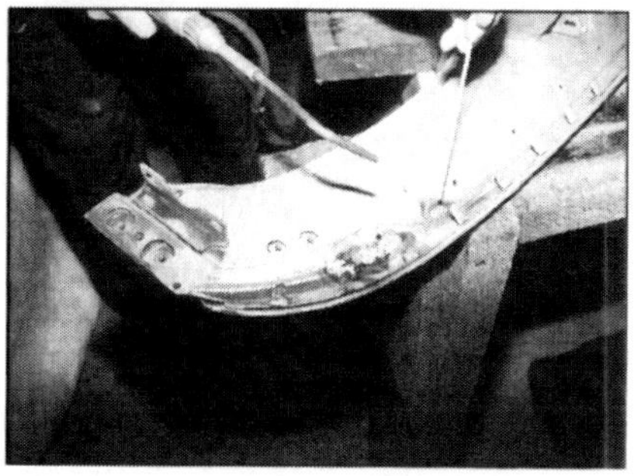

▲ *FT16. Once more it was time for Em to break out his braze welding equipment to patch up the damaged panel. You have to be certain you can fill the header rail with water and drain it so all the braze flux can be washed out, otherwise it causes more corrosion.*

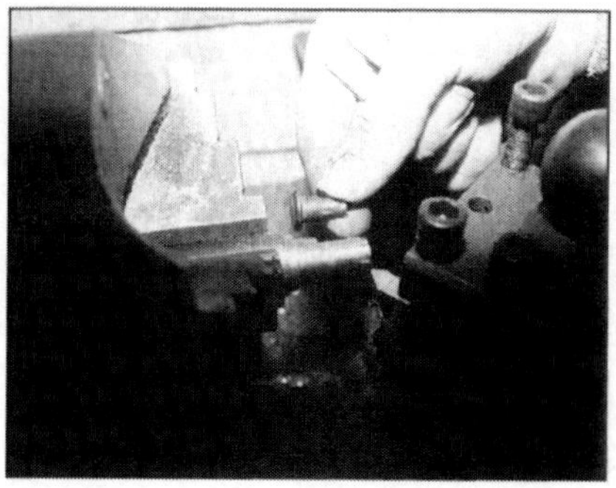

▲ *FT17. In removing the hinge assembly from the frame, the pivot pins were damaged. By careful measuring and some skilful work on the lathe …*

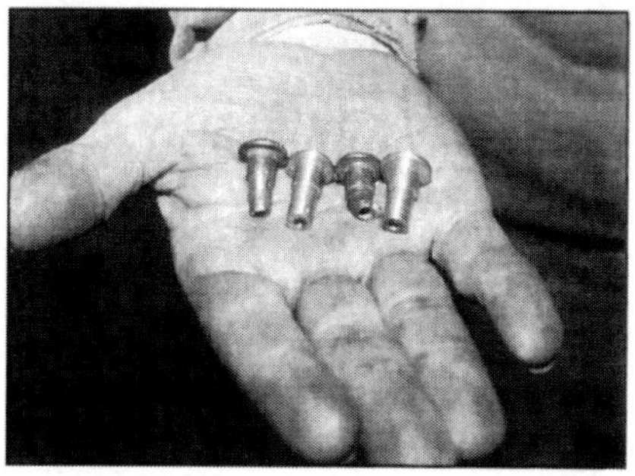

▲ *FT18. … Em was able make his own replacements …*

▲ *FT19. … which worked every bit as good as new. He put on a drop of light machine oil, a process you should copy twice every year. Of course, with the trim and top material fitted, care has to be taken to ensure that the oil does not stain it.*

▼ *FT20. If your workshop isn't as well equipped as this one, and if your frame is really bad news, then you can buy complete frames ready painted and assembled.* (Courtesy Auto Accessories of America)

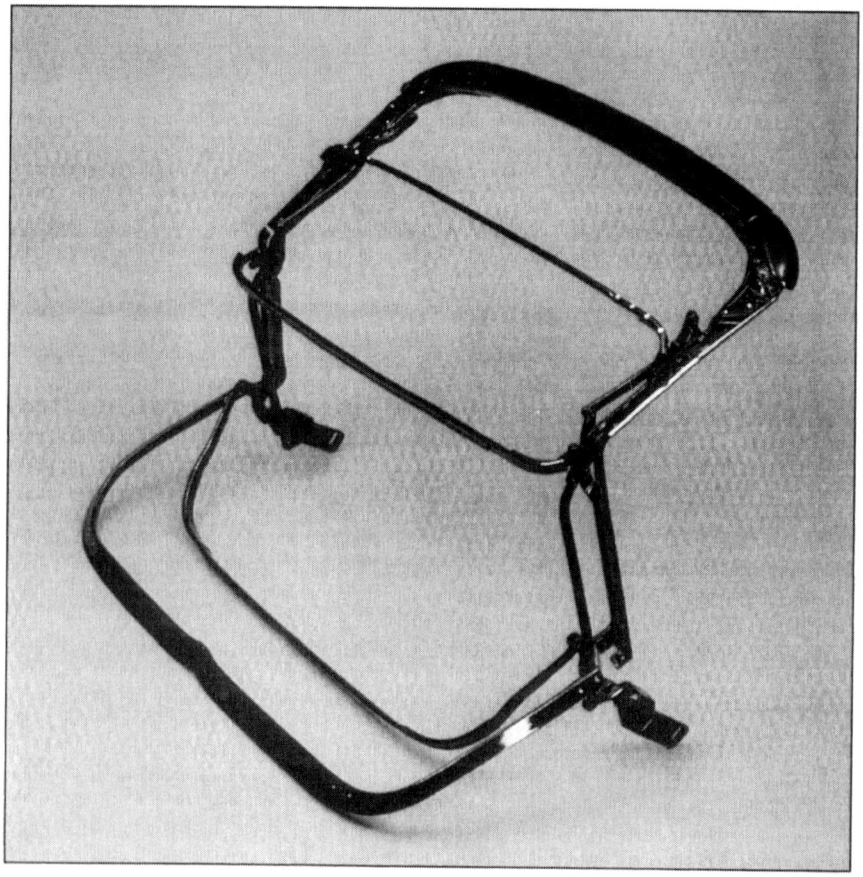

▲ FT21. All the newly plated and chromed fittings were replaced. This is the rear lock locating pin. The screws were not done up tight, but left with some slack to allow for final adjustments later on.

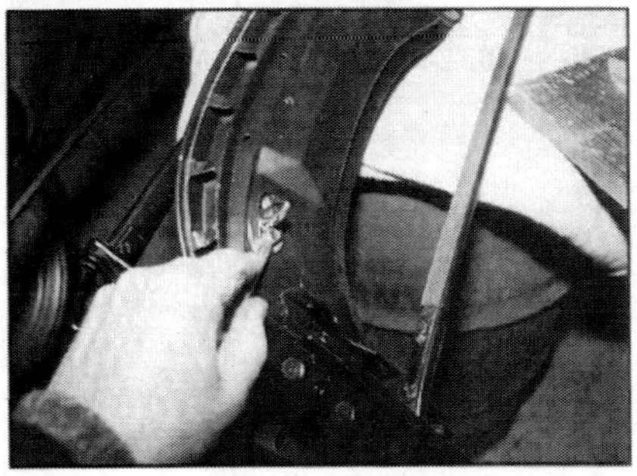

▲ FT22. Once the header panel had been fitted to the frame, the latch guide pins were fitted ...

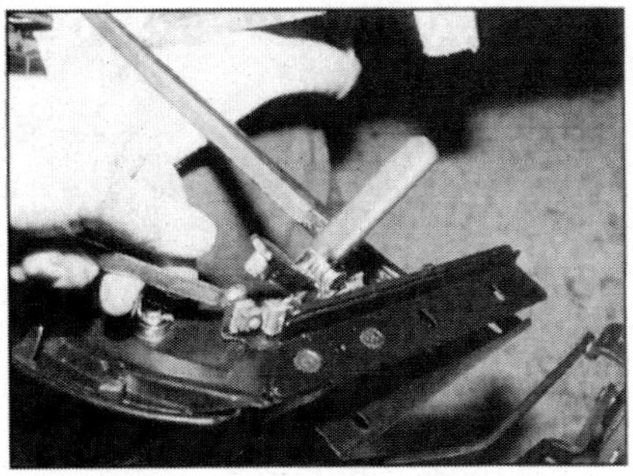

▲ FT23. ... followed by the three lock assemblies – one at each side and one in the middle.

▲ FT24. The complete header and frame assembly with all the fittings. Having been sandblasted and sprayed, the overall effect was that of being better than new.

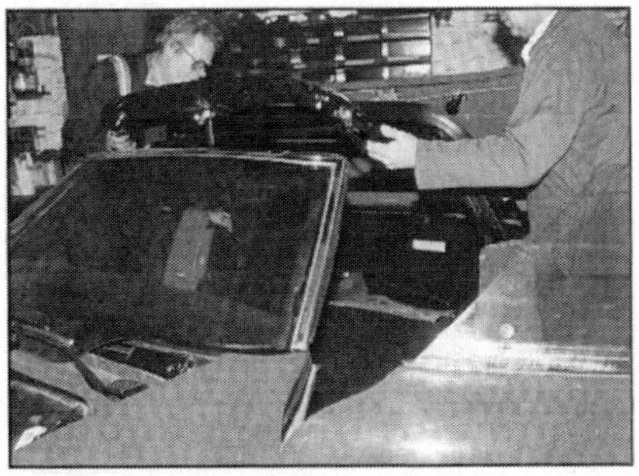

▲ FT25. Fitting the frame (as with removing it) is a two-person job, especially if, you have just treated your Corvette to a top-class paint job.

▲ FT26. The hinge assembly bolts to the car here at the lock pillar and at the corresponding position on the other side of the car. It's as well to leave the replacement of the plastic trim covers until last, in case you have to remove the frame again.

▲ FT27. Before progressing to trimming the frame, it's a very good idea to ensure that it operates exactly as it should and doesn't stick or jam. Note that the rear compartment lid is up at this point ...

▲ FT28. ... so that it can be dropped down into the folded position.

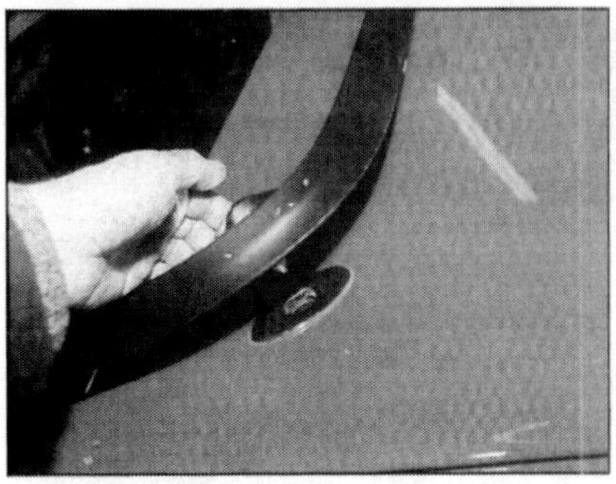

▲ *FT29. It's easier to align the rear locating pins before the folding top fabric is fitted. Once the frame was given the thumbs up, Em fitted the top trim and weatherstrip to the rear bow – the filler strip locks this all together and has to go in last.*

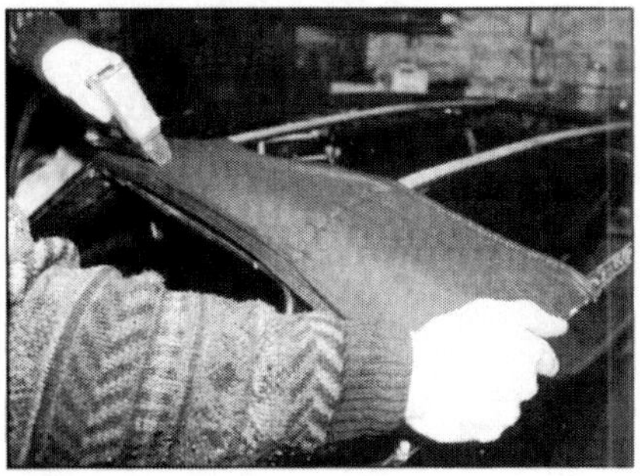

▲ *FT30. Em then installed the pads which run along the top of the frame from the header to the No. 3 bow …*

▲ *FT31. … like this. Note the top trim ready to be fitted sitting behind the No. 3 bow. The stay straps were then fitted using tacks.*

▲ *FT32. It's important to measure the centre of the header, the No. 3 bow, the rear bow and the leading/trailing edges of top trim and mark them accordingly. Keep an eye on these marks during the installation process, making sure that they remain in alignment as the job progresses. This is particularly critical when you come to start tacking or stapling – not quite the point of no return, but going back is difficult.*

With the rear bow locked down in the 'top-up' position, Em pulled up the stay straps and tacked them to the No. 3 bow.

With a piece of mechanic's wire, he pulled the trim hold-down cable assembly through the top pocket and attached the spring cable assembly to the header with a screw. With the leading edge of the trim pulled right up to the header (and aligned as mentioned previously) he stapled the top trim temporarily to the header.

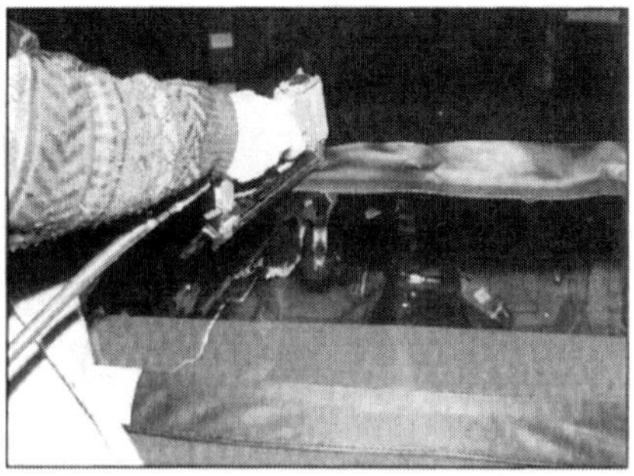

▲ *FT33. He drew the window extension up to the No. 3 bow, as always, maintaining the correct alignment. He allowed a generous overlap and stapled to the bow.*

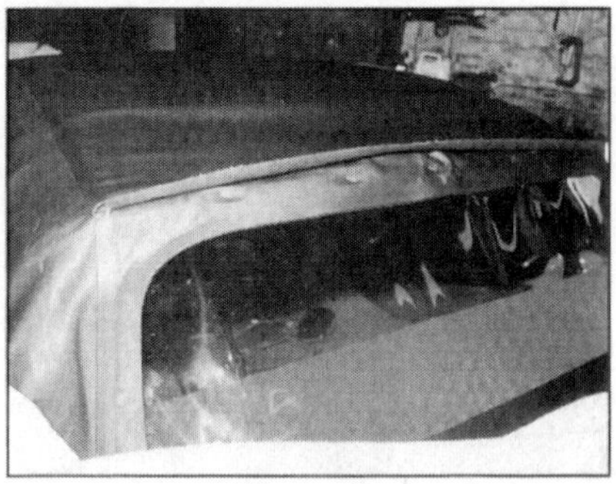

▲ *FT34. He then pulled down a portion of the top trim over the window extension …*

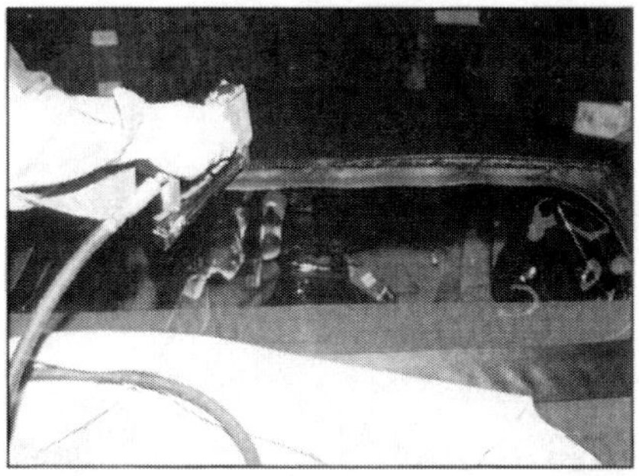

▲ *FT35. … and stapled that, too, to the No. 3 bow.*

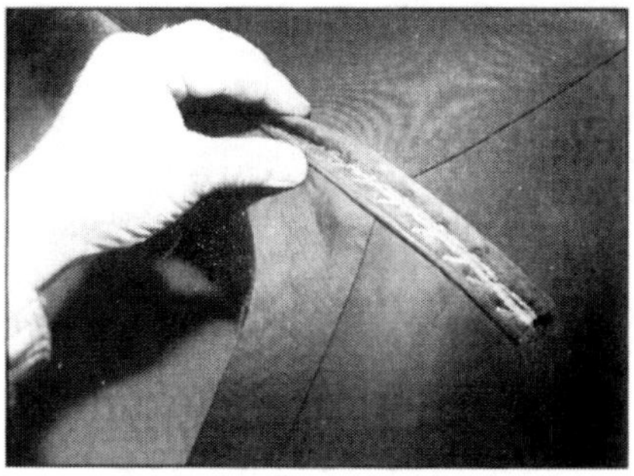

▲ *FT36. After the excess material had been neatly trimmed, the wire-on binding section was installed, folded over …*

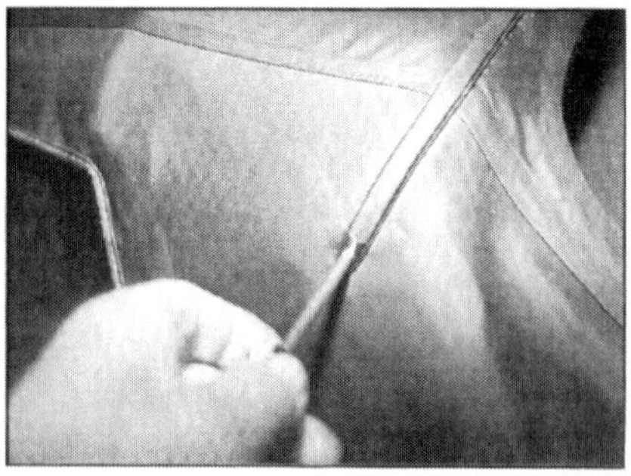

▲ *FT37. … and the binding caps fitted as shown here.*

▲ *FT38. Em then removed the temporary staples securing the top trim to the front header, having first locked it in place. He pulled the trim assembly up tight and aligned it ready for the final installation. The header weatherstrip is secured to the header by means of special plastic fasteners and neoprene glue.*

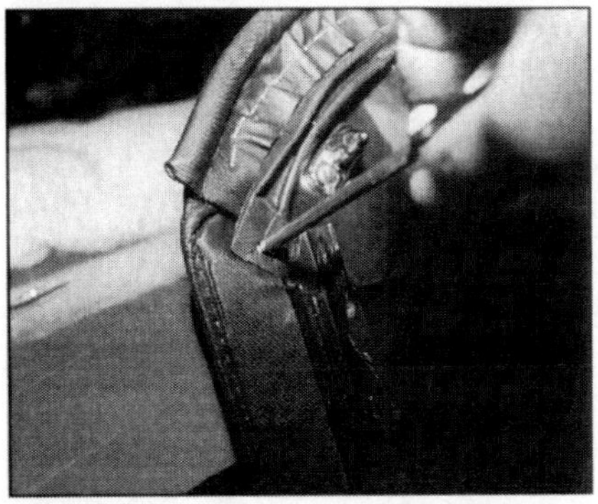

▲ *FT39. The header weatherstrip is secured at each end by a screw, as shown here. The side rail weatherstrip (which seals window opening) is secured by neoprene cement and by screws which are part of the assembly. As always, all surfaces to be glued must be rust, dirt and grease-free.*

▲ *FT40. The finished product. It's not an easy task, demanding plenty of time and patience and a willing helper on standby much of the time. However, if you're to keep the elements out, it's essential.*

ADJUSTMENTS

To correct variations in the top fit, adjustments are made at three locations, namely the header, the hinges and the rear bow hold-down locks. A combination of adjustments may be necessary to correct any given problem, plus you may need to carry out door and window positional adjustments, which are covered elsewhere.

HEADER

After removing header trim panel (or in this case, before fitting it) the header assembly can be moved fore and aft after loosening the two clamping screws. Loose top trim (when the top is up and locked) and excessive effort required to engage the header locks are both indications that there is some misalignment. To gain access to the header outer clamping screw, the side roof rail weatherstrip has to be removed.

HINGES

The hinges can be moved both up and down, and fore and aft in slotted plates in order to offer some degree of adjustment. The hinges live behind plastic trim plates which have to be removed for adjustment purposes. The following conditions indicate that the hinges need some adjustment:

- Poor top fit at the upper and rear edges of windows.
- Faulty fore and aft engagement of rear bow hold-down pins in the lock assemblies.
- Loose or over-tight top rear panel when the top is raised and properly locked in position.

REAR BOW HOLD-DOWN LOCKS

Again, slotted holes in the lock housing offer the opportunity for fore/aft adjustment, and turning the locating pins adjusts the amount of force that holds the seal to the folding top lid. The hold-down pins can be moved to left or right to centre the top on the body in the raised position, thus correcting poor alignment at the windows and difficult entry of pins into locks.

Chapter 5

Mechanical Rebuild

PART 1 ENGINE

There's only one good reason for overhauling your Corvette engine, and that's because you want to do it, although a supplementary reason might be that you are highly experienced and believe you can do a better job than a rebuild shop. Saving money is not a reason for rebuilding your own Corvette engine because crated units, often with a guarantee, usually work out cheaper than buying all the separate bits and pieces, to say nothing of the amount of time you will spend. If you do buy a crated engine, make sure that you buy from a reputable supplier, such as those mentioned in this book, and don't necessarily go for the cheapest engine that you can find. Cheapest is not always best when it comes to engine rebuilds and you may be unlucky enough to encounter the sort of engine rebuild shark who does the bare minimum, or less, and hopes that you'll be many miles and many months away before you discover the worst. Remember that one painted up engine looks pretty much as good as the next and, without your X-ray specs, you'll have no evidence as to what has gone on, out of view.

Also, a crated unit is probably the best way to go if you don't want to spend more time and money than is necessary, but just want to have an up and running 'Vette as quickly as possible. But do beware of the danger that we have mentioned several times before. If your car is a really good, original example, you won't want to lose the correct, original engine number, so you may want to arrange to have your engine reconditioned instead of buying one off the shelf. And then there's the added complication in America that in some States, at least two out of the three ID numbers on your car must match before you can register it. If you have the original engine ID, it would be extremely unwise to sacrifice it needlessly.

HOW FAR DO YOU WANT TO GO?

Even when it comes to rebuilding your own engine, there is a range of choices on how far to go in carrying out the work yourself. Unless you are a fully experienced mechanic with loads of equipment – in which case you'll hardly need the advice of this chapter – you would be advised to carry out your rebuild in association with a reliable and local machine shop. Use your *Yellow Pages*, or better still, a recommendation from another enthusiast who has had good work done, to find a machine shop where an excellent standard of work will be carried out and where they will be prepared to spend time talking with you and discussing what your requirements are. For a start, you'll want to decide whether you really do need to do a complete engine tear-down or whether, say, cylinder head overhaul will do the job. This is something that a good machine shop will be able to determine for you, but equip yourself with as much information as you can find for yourself so you've got a good idea of what you are talking about.

BUYING PARTS

One of the many pleasures of owning a Corvette is the wonderful availability of parts, mainly because of the excellent work done by a number of Corvette specialists. All of the engine parts used in the project car restored by John Fletcher were supplied by Claremont Corvette and Auto Accessories of America – for details see the listings at the end of this book. They both come highly recommended!

ENGINE TESTS

From your auto-accessory shop you'll be able to buy a cylinder compression tester at a reasonable price. This screws into each spark plug hole in turn and measures the compression in that cylinder. Here's how to prepare for using it:

Before running a cylinder compression test, the engine must be at its normal operating temperature and the battery must be fully charged.

Now clean the area around each of the spark plugs so that no dirt can get into the cylinders during the test, and remove all the spark plugs.

Fit the compression gauge to the No. 1 spark plug hole. (Some testers just have a tapered rubber tip which must be held down firmly over the hole) and ensure that hands and clothing, hair and anything else loose is kept away from moving engine parts. When the engine is cranked over six or eight times, the gauge will rise rapidly to its maximum reading.

On a brand new engine, the cylinder compression should be about 150 psi, although the precise figure is not important provided that it's not a heck of a lot lower or higher than this figure. If it's higher, it is almost certain that the compression ratio has been significantly raised on your engine at some time in the past, and this could be one of those times when you need to talk things over with your specialist. Another reason for the compression reading being too high could be that the combustion chambers are very heavily coated with carbon deposits. Modern fuel doesn't readily allow this to happen and this could be another indication that a lot of oil is being blown up past the piston rings.

If any individual cylinder or cylinders give a low reading, carry out the following extra test. Pour about a fluid ounce, or three or four squirts from the oil can, of engine oil into the cylinder and repeat the testing procedure. If the piston rings are worn, the oil will provide a temporary seal and you should see the cylinder compression rise. If the compression reading does not rise, this is quite good news because it means that the valves are worn, not the cylinders or piston rings, so you may be able to get away with just a cylinder head overhaul, which is far simpler to carry out and much less expensive. Leakage past the valves may be caused by badly worn valve guides or burned valve seats (most likely).

Other clues to back up the evidence of your compression test: an engine with badly worn cylinder bores will pump out black smoke when you rev the engine. Try having an assistant rev the engine quite hard while you watch from behind the car. Carry out this test with the engine hot. A quick test for worn valve guides is that after you've been travelling down an incline with the gas pedal closed, you then hit the gas pedal again and from behind, if things have got really bad, you'll see a big whoosh of black smoke. A symptom of burned valve seats or faces is that, if you're travelling down the self same hill with the throttle closed, you'll hear popping and banging through the exhaust pipe as unburnt gases travel down the headers and make little explosions in the exhaust system. In the very worst cases, you can end up with a hole in your muffler.

As we have said before, if compressions are way down, or there are any other queries, you are well advised to have further tests performed by a good quality automotive repair shop, but by now you'll have something to go on.

READING THE READINGS

It's most important that all the readings should be quite close to one another. If they're more than 18 or 20 psi different from one another, this could be an indication of wear in the cylinders with the lowest readings.

By themselves, the compression readings don't provide you with much information, apart from the relative levels mentioned earlier. Like a professor of English you'll need to deconstruct the text. If two adjacent cylinders give equally low readings, it is highly likely that the head gasket between them is blown. Look out for evidence of engine coolant in the oil, creamy sludging around the oil filler cap, oil in the water, and the possibility of spots of water being thrown out of the exhaust when the engine is hot, engine overheating and excessive coolant consumption.

OTHER CHECKS

If you have an oil pressure gauge, you're looking for 40 to 45 psi at 2000 rpm. with the engine hot, and the further below that figure you have, the greater the indication that the bearings, and undoubtedly the oil pump are shot. If you have a warning light rather than a gauge, have your repair shop connect a gauge temporarily across the sender terminals to give a pressure reading.

One word of warning is not to be fooled by rough running, a black exhaust and way-high fuel consumption, as symptoms of this sort can often be cured by a thorough tune-up. For instance, a very badly blocked air filter will make it seem like your engine is fit for cast-iron heaven, when all it needs is a good work out at the gym and a complete set of service components and correct settings. Make a point of always checking the simple things first!

CARBURETTOR REMOVAL

There are various types and combinations of carburettor fitted to the Corvette, depending on year and engine size. By definition, carburettors are complicated items and so it's important to proceed with some care so as not to lose any small important parts in the depths of the engine. Moreover, it is important that dirt be kept out of the carb, particularly during replacement. Even the smallest piece of grit could cause untold aggravation at a later date. On the same note, you also need to keep dirt out of the engine. It's easy to fail to do this if you don't clean the area around the car before you remove it. Degreaser and a wash off followed by compressed air is the easiest way, wearing goggles to protect your eyes.

Removal, whichever type you have, starts with the long process of disconnecting the various springs, rods and tubes. Make sure you know what goes where in order to make life easier when it comes to reassembly. Writing notes and diagrams or taking photos makes certain you make no mistakes.

Most models are equipped with a dry paper element air filter. However, some earlier models have oiled paper, with a polyurethane band or wire mesh with a polyurethane band.

SAFETY NOTE: When removing your carburettor(s), always do so when the engine is stone cold. Gasoline is extremely flammable, so take extra precautions when you work on any part of the fuel system. Don't smoke or allow open flames or bare light bulbs near the work area, and don't work in a garage where a heating appliance or any equipment that can cause sparks – such as from an electric motor – or flames, are present. If you spill any fuel on your skin, rinse it off immediately with soap and water. When you perform any kind of work on the fuel tank, wear safety glasses and have a Class B type fire extinguisher on hand. See Appendix 1.

▲ C1. Access to the carburettor is gained by first removing the air filter. In this case it is held by a single screw at its centre. The carb will now be plainly visible. It's not a bad idea to cover the open top of the carb at this point, especially if you're working outdoors, to prevent the ingress of foreign bodies. A sheet of cling film should do the trick.

▲ C2. The later model 'Vette you have, the more extraneous pipes and connections you will find, because of increasingly demanding emission regulations. Here, the crankcase breather pipe is being removed, to be followed by the vacuum advance pipe, just below it.

▲ C3. Great care should be taken when disconnecting the fuel line at the carburettor to prevent accidental fuel spillage onto the engine (just one important reason for the engine to be cold during this removal process), the paintwork or your hands. Have a clean rag to hand ready to soak up any weeping that may occur.

▲ C4. Hook off the throttle return spring ...

▲ C5. ... and then the bracket can be removed. Note the use of a back-up wrench.

▲ C6. With the bracket freed, the throttle cable can be pulled away from the carburettor. Make sure that you do not lose any of the nuts, bolts and washers and that you know exactly where they go – and in which order.

▲ C7. The choke operation lever has to be unclipped from its position on the side of the carburettor.

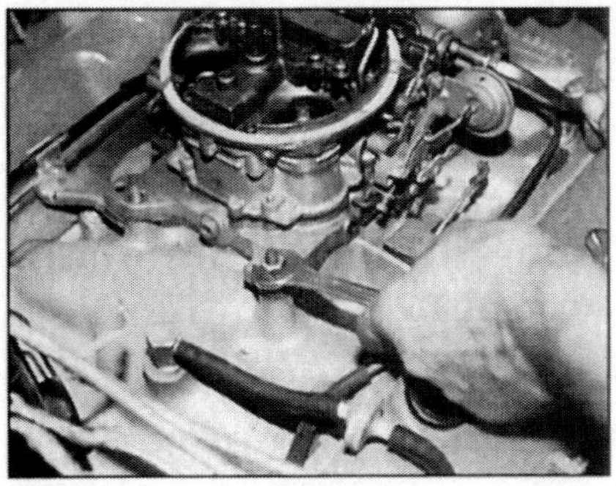

▲ C8. At this point you should have a very clear view of the carb in its entirety. In this case it sits on four studs and is secured by four flange nuts and washers. You may need to use a socket on them if they have been in place for any length of time.

▲ C9. However, you'll probably find that it's impossible to get a socket to at least one of the nuts nearest the front of the engine. If you have difficulty, patience and the copious application of penetrating oil are the keys to success. Don't get too forceful with the wrench, as you risk rounding off the flats of the nuts, and then there's some serious trouble ahead. Equally, it would be awkward to have the nut stick on the stud and bring the whole lot out, as there simply isn't room.

▲ C10. Having made a final check that all the various connections have been removed, the carburettor can be lifted off. Make sure that the area around the base of the carb is very clean at this point, so that you don't get dirt and grit accidentally dropping down into the engine. It's often the case that carburettors have been in place for many years and so will be a little reluctant to leave home. Once more, let patience be your watchword. Waggle the carb gently to break it loose of the studs. In really difficult cases, apply penetrating oil down the studs. Jarring the carb free is a last resort and you must be careful not to damage the carb itself, the studs or their threads. Only apply force downwards, not upward or from side to side. Use only a wooden mallet or a piece of wood with a hammer.

▼ C11. Ultimately, it will come free and lift off completely. Remove also the gasket (and shims if applicable) and throw it in the bin – reusing this gasket is a recipe for trouble in the future, and they cost little enough.

▲ C12. Once you have the carb in your hand (make sure you don't drop it – they're expensive!) you can run through a few basic observations. Check the correct operation of all the linkages and gas pump which squirts neat fuel into the intake bore. You may find that this last item has been disconnected by a previous owner, a common occurrence which, although it tends to give a slight flat-spot, is a way to improve fuel consumption.

▲ C13. Check the linkages and butterflies. You can see whether the butterfly spindle is worn by moving it when open. Dismantling the carburettor further should be undertaken in clinically clean surroundings. Make sure that you have plenty of small containers in which to store the jets and various connectors. Clean all components in fuel or other suitable solvent and examine for wear or damage, replacing any parts which are not up to par, including gaskets – carb repair kits are available from GM dealers and a number of outlets. Once the carburettor has been completely stripped, clean it using a proprietary cleaner. Let it air dry and then blow it through using compressed air in order to remove any small bits of debris.

ENGINE REMOVAL

While many engine operations can be carried out with the engine in the car, restoration demands that it be removed altogether. Getting the engine out is relatively simple – it is the preparatory work that takes the time! It's vital to ensure that the all pipes, wires and hoses linking the engine to the frame or body are disconnected. This, in turn, means following a strict safety code; where fuel is involved, there is always a danger of fire; brake fluid, too, is flammable and corrosive to paintwork.

IMPORTANT NOTE – AIR CONDITIONING: If your car is fitted with air conditioning, it is necessary to remove various hoses and fittings. However, this should NOT be completed until your dealer service department or air conditioning specialist has discharged the system. As it is under high pressure, it should NOT be attempted on a DIY basis – it is extremely dangerous.

In our case this particular option had not been specified. The TBI fuel injection system has to be relieved (where fitted), otherwise, the first job is to remove the battery ground terminal.

To aid access, we removed the hood altogether – it's a two-person job to be certain of not damaging the car in the process. For details of hood removal, see the relevant section elsewhere in this book.

With the engine cold, detach all the coolant hoses from the engine and drain the radiator. Labelling the various hoses and electrical/pneumatic connections by using masking tape and a felt tip pen is time-consuming but will pay dividends when it comes to reassembly. Make sure that your labels are unambiguous, and if there's any danger of confusion, take the trouble to sketch out an exact position.

▲ *ER1. Before removing the cooling fan it is essential that the engine is cold and that the battery negative terminal is disconnected. Note that many of the coolant hoses have been disconnected.*

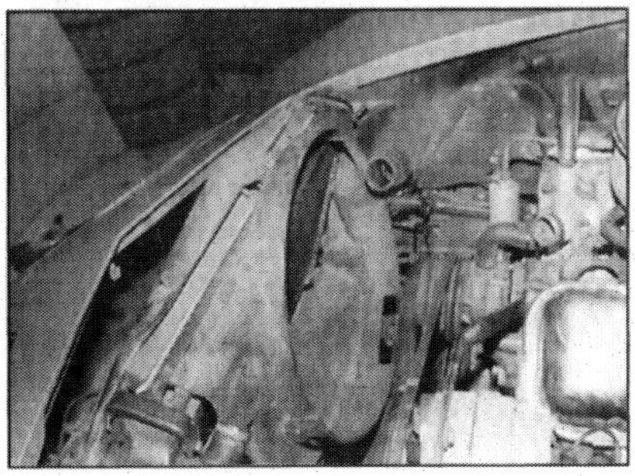

▲ *ER2. Access to the fan shroud and radiator is good. On later models fitted with an electric fan, this will have to be removed before the radiator can be released.*

▲ *ER3. The radiator and its support can be lifted out as one unit. Take great care not to scratch the bodywork as you lift it out. Take care, too, not to drip any coolant on your paintwork – if you do, wipe it off immediately, as the antifreeze solution can damage the finish. Bugs and dirt can be cleaned from the radiator by using a compressed air line and a soft brush – make sure you don't damage the cooling fins, though. Once it's clean, inspect the radiator carefully for splits or other damage – a specialist repair shop will be required to sort out any leaks. Even if it doesn't leak, you should have it reverse flushed, tested for flow rate, and pressure-tested. If there are signs of clogging, have the shop unsolder the tanks and rod the core. Make certain the radiator is fully pressure tested again for leaks.*

Alternatively ...

▲ *ER4. ... you could install a brand new unit. Make sure you choose one which meets the specifications required for your car, and make known any modifications you intend to make which could affect the performance and thus demand more from the rad. Inadequate cooling is about the easiest method there is of ruining your freshly rebuilt engine!* (Courtesy Auto Accessories of America)

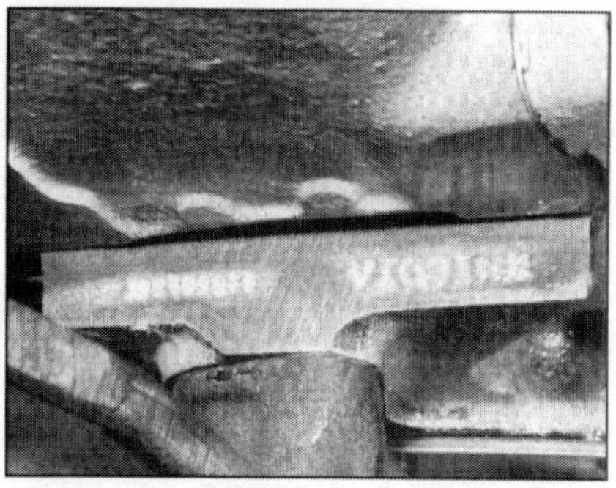

▲ *ER5. While you're in the area, check out your engine number. It should be easily read and match up with your official documents.*

▲ *ER6. The starter motor and alternator have to be removed and …*

▲ *ER7. … the various ancillaries detached, including the vacuum lines to items such as the wiper actuator …*

▲ *ER8. … the header tank, emissions system hoses, wiring connectors and ground straps. Drain the engine oil and unbolt the exhaust system from the engine.*

▲ *ER9. We removed the carburettor (see foregoing section for carburettor removal) and various linkages, and secured a piece of strong card over the inlet to prevent the ingress of foreign bodies. (Notice that the oil filler holes have similar protection.) We disconnected the fuel lines running from the engine to the chassis and ensured no dirt got into the lines by plugging and capping.*

CAUTION: Gasoline is extremely flammable, so extra precautions must be taken when working on any part of the fuel system. DO NOT smoke or allow open flames or bare light bulbs near the vehicle. Also, don't work in a garage if a heating appliance or central heating unit is present. Preferably, work on the fuel system out of doors.

POWER STEERING

Where power steering is fitted, the pump has to be unbolted but the lines and hoses should be left attached. The pump must be kept in an upright position – use wire/rope and your ingenuity to strap it thus in the engine bay.

AIR CONDITIONING

Remove the compressor and set it aside but, again, all hoses should be left attached.

▲ *ER10. Support the transmission with a jack, and position a block of wood between them to prevent damage to the transmission. It is possible to hire special transmission jacks complete with a safety chain – it's the safest and easiest way to do the job.*

On the vehicles with automatic transmission, the torque converter-to-driveplate fasteners have to be removed. Attach an engine sling or length of chain to the lifting brackets on either end of the engine, diagonally across from each other. At this point you can position the hoist and take up the slack on the chains, but do not start to lift.

WARNING: DO NOT place any part of your body under the engine when it's supported only by a hoist or other lifting device.

Remove the transmission-to-engine block bolts and the engine mount-to-frame bolts and then check carefully to ensure that there is nothing still connecting the engine to either the transmission or the vehicle.

If you already have the body off, as we had, it is possible to remove the engine and transmission as one unit – but make sure that there is someone there to prevent the latter from swinging around and that it can be supported properly.

▲ *ER11. Having removed the engine from the chassis, it's best to secure it to a purpose-made engine stand. This enables it to be worked on easily and from all directions. In this case, although the oil had been drained, the oil pan was left on the floor under the engine in order to catch the occasional drips which can make such a mess of your nice clean garage floor, especially as you're likely to get the odd surge as the engine is turned over by hand.*

TIP! REMEMBER – examine suspension A-arm bushings now, when the engine is out. On most Corvettes, suspension overhaul is far easier with the engine out of the car, so if work is needed, do it now!

ENGINE TEAR-DOWN AND REBUILD

CYLINDER HEAD REMOVAL

Cylinder head removal involves dismantling of the intake and exhaust valves and related components.

SAFETY NOTE: Before starting work on your Corvette engine, please read the Safety Notes in Appendix 1, and also:

- Keep your work area scrupulously clean and tidy. Not only will you have fewer accidents that way, you'll also have a better rebuild with less harmful dirt inside the engine.
- Use barrier cream on your hands whenever working with grease and oil, and in particular, never handle old engine oil since it is said to be potentially medically harmful.
- Have a helper to work with you since you'll work quicker, safer – and more enjoyably! The Corvette engine is a big and heavy lump of metal to work on and a helper will help avoid accidents or back strains.
- Ban smoking and all naked flames in the work area and have a suitable workshop fire extinguisher handy – not just one of the small auto-accessory extinguishers.
- Mop up all oil and gas spillages as they occur. Have plenty of clean rags available for mopping up.
- Always use the correct sizes of wrench, hoist or any other tools you may need. Overloading tools, working with dirty oily hands, having loose clothing or hair; all can lead to unnecessary accidents in the workshop.

▲ *EN1. Before starting work on the cylinder head itself, it's important to check the camshaft lobe lift, using a dial indicator, the plunger of which must be positioned directly above the pushrod. The lobe lift should be recorded and compared with the manufacturer's specifications. If any of the lobes exceed the limits, then wear has occurred and a new or reprofiled camshaft is required.*

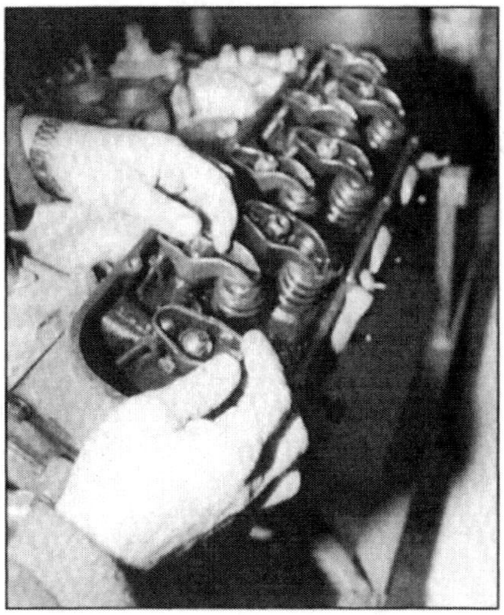

▲ *EN2. With the rocker arm cover removed, detach the rocker arm nuts, pivot balls and rocker arms from the cylinder head studs. For general overhaul work, they should be labelled and stored separately in plastic bags so they can be reinstalled in their original locations.*

Check each rocker arm for wear, cracks and other damage, especially where the pushrods and valve stems contact the rocker arm faces. Make sure that the hole at the pushrod end of each rocker arm is open, and check each rocker arm pivot area for wear, cracks and galling. If the rocker arms are worn or damaged, replace them with new ones and use new pivot balls as well.

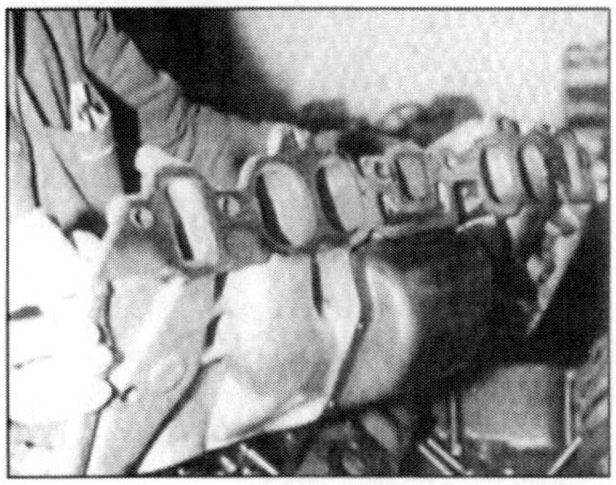

▲ *EN3. The intake manifold is removed by unscrewing the mounting bolts in quarter-turn increments until they can be removed by hand. At this stage the manifold will probably be stuck to the cylinder heads. Position a large bar under the cast-in lug near the thermostat housing and prize up the front of the manifold.*

CAUTION: Do not prize between the block and manifold or the heads and manifold or you could damage the gasket sealing surfaces, which would result in vacuum leaks.

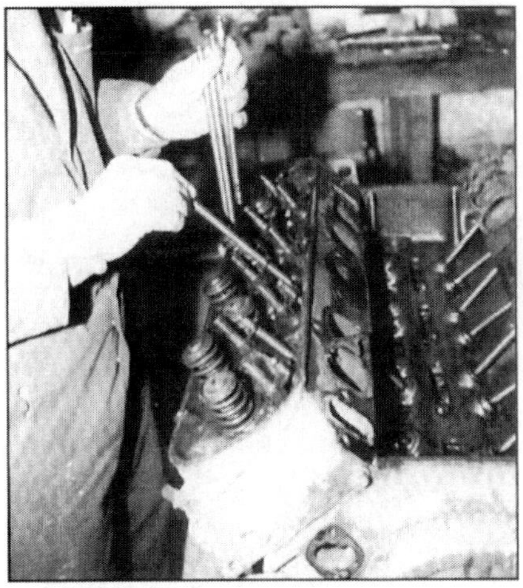

▲ *EN4. Remove the pushrods; again, they too would normally be labelled, but as new rods were to be fitted, we didn't need to keep track of which valve went in which valve guide. However, if you do label them, you will be able to link the various parts to a particular cylinder if you find trouble later on during the tear-down.*

Inspect the pushrod ends for cracks and excessive wear at the ends. Roll each pushrod on a perfectly flat surface (e.g. a sheet of glass) to determine if it's bent.

▶ *EN5. To remove the cylinder heads, loosen the head bolts in quarter-turn increments in a reversal of the tightening sequence starting at the outer ends and working in, in a criss-cross pattern to avoid warping the heads. Don't forget the row of bolts on the lower edge of each head, near the spark plug holes.*

REMOVING TIMING COVER, CHAIN AND CAMSHAFT

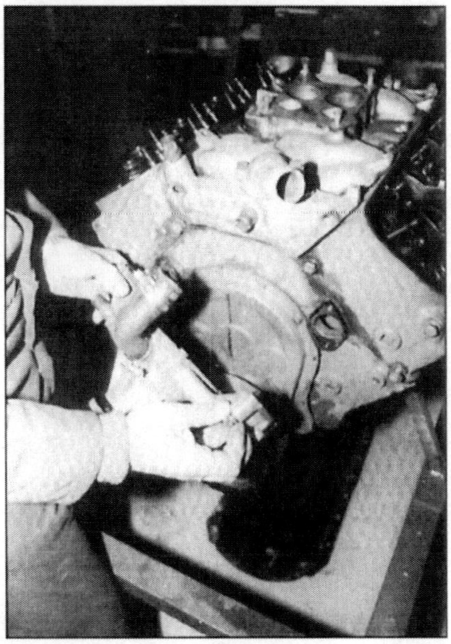

▲ *EN6. 'Vette water pumps are fitted with 'weep' or vent holes, and in the event of a seal failure, water will leak from the hole. It is usually visible from underneath the car by using a flashlight. If you're doing this job with the engine in situ, it means prior draining of the coolant system and removal of the various hoses and brackets. Any hoses which look even vaguely suspect should be replaced – their cost is peanuts compared to that of a wrecked engine caused by overheating. In this case, with the engine already out of the car and part-dismantled, life was made much easier. The pump is secured by bolts of differing lengths, so it's essential to make a note of which one goes where. A long bolt forced into a shallow threaded hole could cause terminal damage! Clean the bolt threads and the threaded holes in the engine to remove corrosion and sealant.*

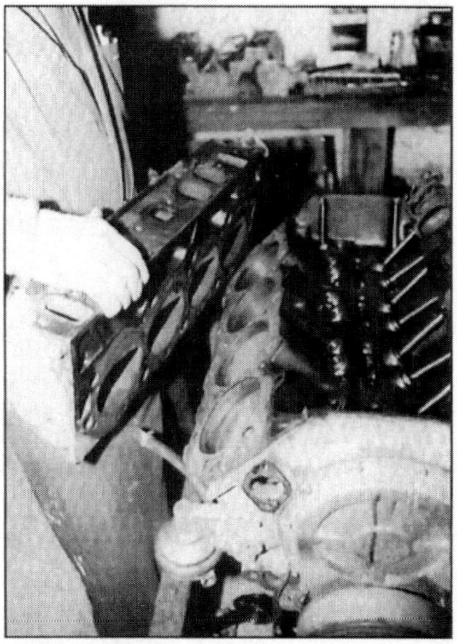

▲ *EN7. Remove the bolts and separate the crankshaft drive belt pulley from the vibration damper. Position the No. 1 piston at Top Dead Centre on the compression stroke. NOTE: If you are not removing the crankshaft, once this has been done, do not turn the crankshaft until the timing chain and sprockets have been reinstalled.*

Most, but not all, V8 engines have a large bolt threaded into the nose of the crankshaft to secure the vibration damper in position. If your engine has a bolt, remove it from the front of the crankshaft, then use a puller to detach the vibration damper. CAUTION: Do not use a puller with jaws that grip the outer edge of the damper. The puller must be the type that utilizes bolts to apply force to the damper hub only.

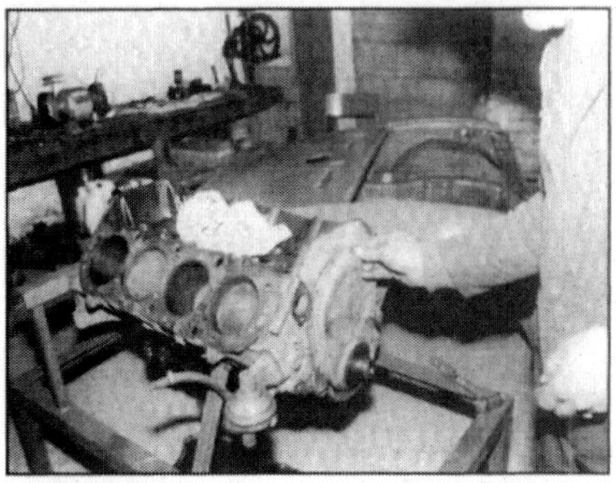

▲ EN8. On most small block V8 engines the timing cover cannot be removed with the oil pan in place – the pan bolts have to be loosened and the pan lowered by around ¼ in. If the pan has been in place for any length of time it's likely that the gasket will break as it is removed.

▲ EN9. Remove the bolts and separate the timing chain cover from the block. It's usually necessary to use a putty knife to break through the gasket seal. It's important not to damage the cover, as it is easily distorted – take your time and don't attempt to prize it off.

Use a gasket scraper to remove all traces of old gasket material and sealant from the cover and engine block. Stuff a rag into the opening at the front of the oil pan to keep debris out of the engine. Clean the cover and block sealing surfaces with lacquer thinner or acetone. Check the cover flange for distortion, particularly around the bolt holes. If necessary, place the cover on a block of wood and use a hammer to flatten and restore the gasket surface.

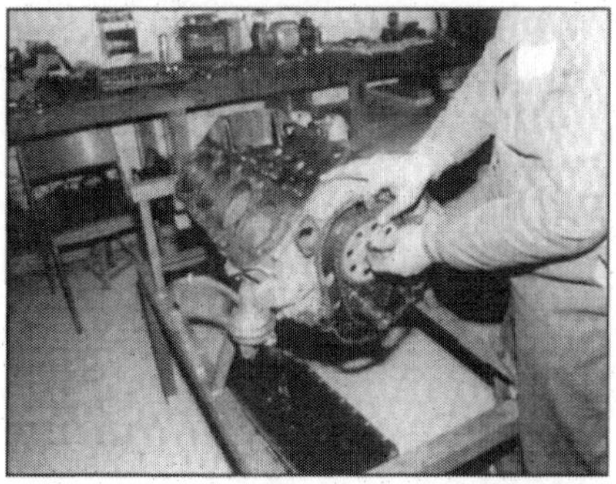

▲ EN10. Remove the three bolts from the end of the camshaft and detach the camshaft sprocket and chain as a complete assembly. The sprocket can be removed by using a two- or three-jaw puller, though care has to be taken not to damage the threads in the end of the crankshaft.

THE CAMSHAFT

▲ EN11. Before the camshaft itself can be removed, the lifters have to be extracted from the bores. There is a special tool available, though this is usually only required on higher mileage engines with a lot of varnish build-up. On newer engines the lifters can be removed using a small magnet, or even with your fingers, like a thimble. However, it is important that pliers are NOT used unless you intend to replace all the lifters and the camshaft as well. It's surprisingly easy to damage the precision machined and hardened surfaces, rendering them useless. Once out of the block, clean the lifters with solvent and dry them thoroughly without mixing them up. The lifters should be stored in a clean place and in the order in which they were removed, so that they can be replaced from whence they came.

Note that the camshaft and its lifters are 'an item'; so if you're fitting new lifters you'll need a new camshaft and vice-versa. Check each lifter wall, pushrod seat and foot for scuffing, score marks and uneven wear. Each lifter foot (the surface that rides on the cam lobe) must be slightly convex, although this can be difficult to determine by eye. If the base of the lifter is concave, the lifters and camshaft must be replaced. If the lifter walls are damaged or worn (which is not very likely) inspect the lifter bores in the engine block. If the pushrod seats are worn, check the pushrod ends.

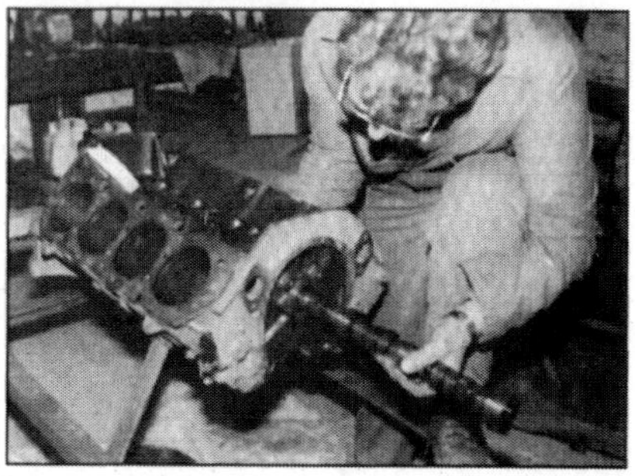

▲ EN12. A tip for removing the camshaft is to thread a 6 in.-long, 5/16 in. bolt into one of the camshaft sprocket holes and use it as a handle to ease it out. It's important to pull slowly and support the camshaft near the block so that the lobes don't nick or gouge the bearings as the shaft is withdrawn.

▲ EN13. It's worth investing in or hiring an auto body-type dent puller to remove the core plugs, for if they are driven into the block, they may prove impossible to remove. Drill a small hole in each plug and insert the puller. A large socket on an extension can be used to drive the new core plugs into their bores.

▲ EN14. Invert the block to remove the oil pump, noting that the nylon sleeve is in place. This is used to align the oil pump and the oil pump drive shaft. If you find it missing, check around in the oil pan for pieces of the sleeve and make a note that a new sleeve is required.

▲ EN15. The importance of matching rods and caps can't be overstated. Mix them up and you'll never reassemble the engine properly again! The rods and caps are built as a pair and must go back together as a pair. Use a number punch to mark each rod and cap and always mark them in the same place, such as stamping on the outer side so you can identify which way round the caps go at a glance.

▲ EN16. Alternatively, you could make small dots with a prick punch. One to eight dots, depending on rod and cap, will tell you what you need to know.

▲ EN17. Before the connecting rods are removed, check the end play with feeler gauges. Where the play is more than the service limit, new rods will be required.

▲ EN18. The connecting rod cap nuts have to be loosened half a turn each until they can be undone by hand – but take care not to drop the bearing insert out of the cap. Here, aluminium tube is being used to cover the rod bolt threads (you could use rubber hose). The reason for covering the rod bolt is to ensure that the hardened rod threads don't nick the journal surface as the crank is removed and replaced.

▲ EN19. The wear ridge at the top of each cylinder should be removed using a special ridge reaming tool – if the ridge is bad enough, failure to do this prior to removing the pistons could conceivably result in piston breakage. The piston/connecting rod assembly can be pushed out using a wooden hammer handle if necessary. Don't force it, whatever happens. If there is resistance, check that the ridges have been totally removed, but also make sure that you don't 'shoot' the piston out onto the floor!

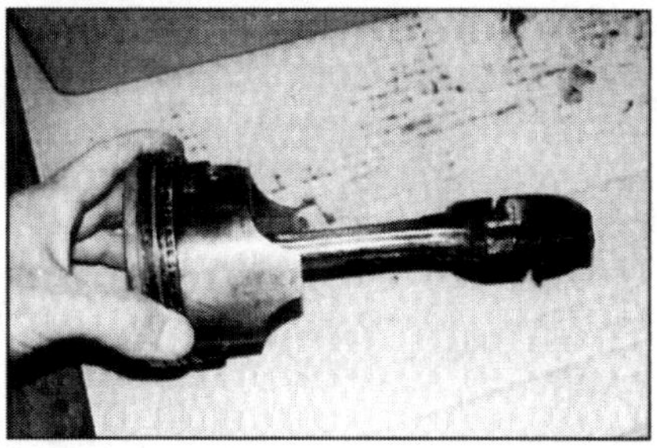

▲ EN20. One of the jobs your machine shop should be asked to do is to check the con rods for true – to make sure they are not bent – and the bearing housings for concentricity. They should also be checked for cracks. Smoothing out the seams on the sides of the con rods can marginally strengthen the con rods and also helps small cracks become visible during magnafluxing. This job is by no means essential, but if you've time to spare, the polishing process is easy and cheap to do with a fiber abrasive wheel.

CRANKSHAFT – REMOVAL AND INSTALLATION

▲ EN21. Here's another specialist job. A dial indicator is used to check the end play on the crankshaft before removing it. By prizing the crankshaft all the way in one direction, zeroing the gauge them prizing it back in the opposite direction, the end play can be accurately measured. Once again, your specialist will advise on whether the thrust washer surfaces need machining and, if so, on the sizes of oversized thrust washers that are available.

Just as with the connecting rod caps, the main bearing caps should be marked before being separated from the block, so you know exactly where they came from and where they have to go back. Ratchet out the threaded oil gallery plugs from the rear of the block and replace them with new ones.

▲ EN22. A thorough going over is required before any crank can be proclaimed fit for replacement in a rebuilt engine. Clean the crank with solvent, brushing it well into the oil holes and flushing through afterwards. Check the main and connecting rod journals for uneven wear, scoring, pits and cracks. Once more, this is a job best carried out by your specialist because not only will he have the correct size of micrometer to enable him to measure the journals correctly, he'll also have the experience to know how much wear is acceptable and when it's too much. Even so, here's a tip you could try for yourself.

A simple method to check whether the journals need to be reground is to rub a penny several times lengthwise along each one. If the copper rubs off and 'sticks' to the crank, the surface is too rough and regrinding is necessary. Remove all burrs from the crankshaft oil holes with a stone, file or scraper.

▲ EN23. Measure the diameter of the main and connecting rod journals using a micrometer – by comparing the diameter at a number of points around the journals, you'll be able to see if it is out of round. Compare your findings to the service recommendations.

▲ EN24. Sometimes, cracks in a crankshaft are obvious, but sometimes they're internal, which is the reason for taking your crank to a machine shop and having it magnafluxed.

Check the oil seal journals at each end of the crank for wear and damage. If the seal has worn a groove in the journal or if it's nicked or scratched, the new seal may leak when the engine is reassembled. It may be possible for a machine shop to repair the journal by pressing on a thin sleeve, but if they can't, then it's a case for a new crank.

▲ EN25. Although we would automatically replace the main and connecting rod bearing shells during the course of an engine rebuild, close examination of the originals can reveal important information about the condition of the engine.

Bearing failure occurs because of a lack of lubrication, the presence of water, dirt or other foreign bodies, overloading the engine, corrosion or a mixture of all those factors! Regardless of the cause of any bearing failure, it is important to correct it before the engine is rebuilt. When you are examining bearings, make sure that you do not mix them up – knowing where a specific bearing was fitted will help isolate any cause of failure.

The soft-white babbitt metal on the bearing face has worn down here to expose the copper layer beneath. It is usual for a worn crankshaft journal to go with a bearing this badly worn but don't assume the worst without having those crank journals checked for wear and for scoring.

ENGINE BLOCK CLEANING

▲ *EN26. Part of the cylinder block overhaul will have been to have the unit solvent washed and blown out with compressed air. A good machine shop will be able to soak or even 'boil' your block in solvent, so loosening even the stubbornest stains (but don't risk your laundry in there!). Even so, if you're rebuilding the block yourself and it comes back from the machine shop apparently sparkling clean, use your own supply of compressed air if you have it or take the block to an air line at your local fuel station. Don a pair of goggles and blast compressed air into every galley, channel, nook and cranny. A single metal particle can cause so much damage to an assembled, running engine, and you'll need another tear-down and rebuild if it gets into the wrong place.*

ENGINE BLOCK INSPECTION

Assuming now that the block hasn't been machined, you'll have to decide whether any machining will be necessary. Your machine shop will be able to measure the cylinder bores at top, centre and bottom, both across the line of the bores and in the same direction as the line of the bores. The measurements will be compared with the manufacturer's specifications and these measurements, taken with the evidence or lack of scuffs and scratching will help your specialist decide whether a re-bore is necessary. You'll also have to take a look at the faces on the top of the block to determine whether any refacing will be necessary. If a cylinder head gasket has been blowing for quite some time, erosion of the block surface could have taken place or, under the most severe overheating stresses, it could have distorted. Using engineer's blue and a flat plate, your specialist will be able to tell you how good your block faces are. He'll doubtless check your heads at the same time.

As we said earlier, new oil gallery plugs must be fitted. Use either Teflon tape or a non-hardening sealant such as Permatex No. 2 to ensure that the plugs stay in place and seal properly. Also fit new block core plugs with RTV sealant evenly spread around the seating face before the plug is fitted. Fit each core plug by using a large socket which fits just inside the plug and a hammer to spread each plug and hold it in place.

If you intend storing your clean, possibly machined and bare block for reassembly later, ensure that all surfaces are coated in oil, that you store it somewhere dry and that you completely cover it up so that no dirt gets into it.

▲ *EN27. Whether you tackle the honing operation yourself depends on the equipment you have to hand and your level of (justifiable) confidence. Using the traditional surfacing hone with spring-loaded stones is not usually recommended for the inexperienced home mechanic, who will find using the flex hone or bottle brush type much simpler. After oiling the cylinder bore, the brush, connected to an electric drill, is moved up and down the bore – it is important not to remove the brush whilst the drill is rotating. 'Follow the instructions supplied and take your time,' is the best advice – it's easy to make a serious and expensive mistake. When the honing is complete, the top edges of the cylinder bores should be chamfered with a small file so that the rings won't catch when the pistons are installed – make sure you don't damage the cylinder walls with the file.*

At this point the engine cleaning operation begins again, to make sure that all traces of dirt and swarf created by the honing operation are removed. All machined surfaces should be treated to a light coating of oil to prevent the onset of corrosion.

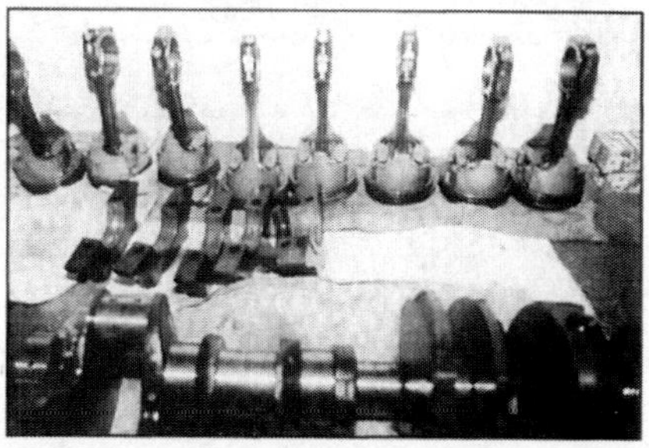

▲ *EN28. Between removing and tearing down your engine and being ready for a rebuild, there's a lot to do; you may have parts to take to a machine shop and, if you're planning a total rebuild, you'll have plenty of new parts to buy. It pays to lay out the component parts in order so that you can see exactly what you have (and what you possibly haven't got!), so that once you've started the rebuilding process you won't have to start wandering off to find some missing parts.*

CRANKSHAFT INSTALLATION

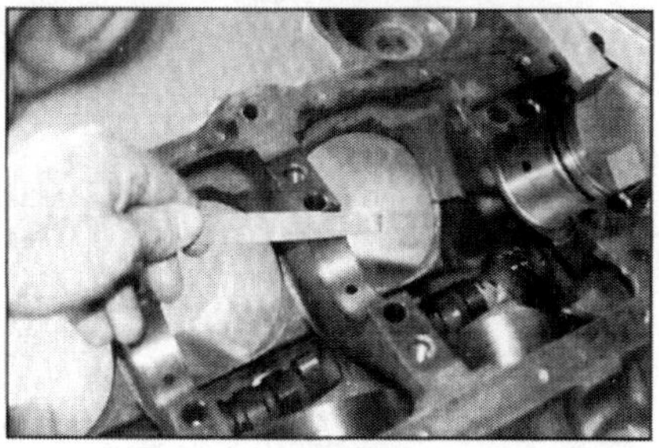

▲ *EN29. The insides of the cap and rod, where the bearing shells are fitted, must be carefully checked to ensure that there is not the slightest speck of dirt which will make the bearings run tight. Also check for carbon build-up which can take place on the outer edges of the bores. Never file the insides of the bores; a scraper will be enough to remove any built-up carbon.*

▲ *EN30. Place the engine on the stand facing upwards to prepare for the installation of the crankshaft. The back sides of the new main bearing inserts should be clean and one bearing half laid in each main bearing saddle in the block. The other halves should be fitted to the corresponding main bearing caps (as above).*

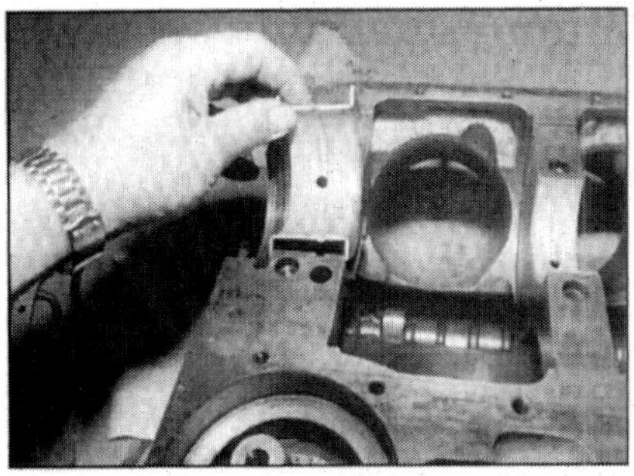

▲ *EN31. It's important to ensure that the tab on the bearing shell insert fits into the recess in the block or cap. The oil holes in the block must line-up with the oil holes in the bearing insert. Do not hammer the bearing into place and do not nick or gouge the bearing faces. Also, don't oil between cap and bearing shell.*

The flanged thrust bearing must be installed in the rear cap and saddle. Clean the faces of the bearings in the block and the crankshaft main bearing journals with a clean, lint-free cloth. Check or clean the oil holes in the crankshaft – any dirt in this area can go just one way – straight through the new bearings!

▶ *EN32. Install one seal section in the block with the lip facing the front of the engine. (If the seal has two lips, the one with the helix should face the front.)*

▲ *EN33. Make sure the seal is firmly seated and leave one end protruding about ¼ in. from the block. Install the remaining seal half in the rear main bearing cap, in the same way. The only slight difference will be that the end of the seal that you leave protruding should be the opposite end to the one that you left protruding in the block. On the back of the seal, use a very thin film of sealer, on the parts of the seal where they contact the cap and the block. MOST IMPORTANT! Oil the contact areas of the seal and the crankshaft before you refit the crank and bearing caps. If you don't, the seal's life will be very limited – perhaps to only a few minutes!*

▲ *EN34. The bearing surfaces in the block should be spotlessly clean, and then lots of engine oil or a thin layer of moly-grease should be applied to each bearing surface. Coat the thrust faces and the journal face of the rear bearing.*

▲ *EN35. Once it's certain the crankshaft journals are clean, the crank can be placed carefully in position in the main bearings.*

CHECKING THE MAIN BEARING OIL CLEARANCE / LASH

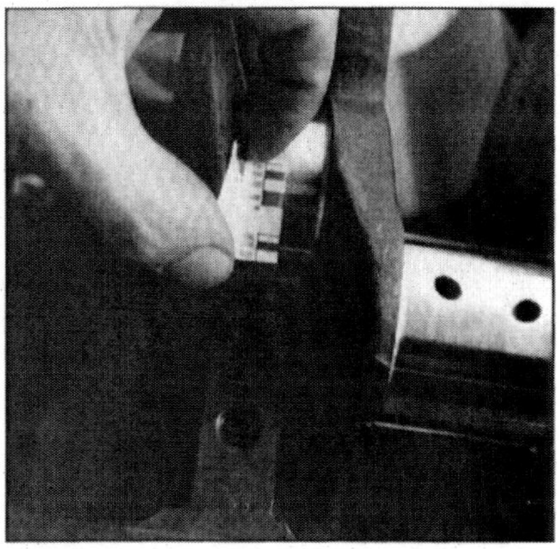

▲ *EN36. Before you properly install the crankshaft, you must check that you have the correct lash at each bearing. First, fit the crankshaft carefully in place and fit each bearing cap, tightening the caps down one at a time and rotating the crank after pulling each one down. If you find that the crank won't rotate after tightening down one of the caps, check that you haven't made a mistake and that you're using the right cap in the right place. Check also that you have been supplied with the correct bearing shells and that you've fitted them correctly. Finally suspect a build-up of dirt or carbon under one or more of the bearing shells, which leaves a reground crank done to the incorrect tolerances as your prime suspect. Take it up with your machine shop!*

What this won't have told you is whether any of the bearing clearances are too large. The only way you can check this out is to use a clever substance called Plastigage. Undo each bearing cap and, once again, tighten down, but this time with a strip of Plastigage between the bearing and crank journal. Don't rotate the crank but remove the cap yet again and compare the thickness of the crushed Plastigage to the scale printed on the wrapper to determine whether you have the correct lash or not. If the lash is too great, you could have very big problems! Once again, check that you have the correct size of bearing shells and if you have, go see what you machine shop have to say about it. You may have to use a soft-faced hammer to encourage the bearing caps free as they tend to stick to the Plastigage. You'll also have to scrape off any sticking Plastigage using your fingernail, or an old credit card but nothing hard enough to scratch the bearing or journal.

▲ *EN37. Now you're going to have to take the crank out again and reassemble it, but this time with oodles of oil on and around each journal. During the final installation of the crankshaft, apply a thin coat of gasket sealant only to the small areas of the rearmost cap or block where the ends of the two seals butt up against each other when the end cap is fitted. Don't get any sealant on the bearing face, crankshaft journal, seal ends or seal lips. Also, lubricate the seal lips with grease or lots of clean engine oil.*

With the exception of the rear cap bolts, tighten them working from the centre out. It is essential that you use a torque wrench to give the correct figure when tightening down the caps.

▲ *EN38. The rear cap bolts are tightened to 10–12 ft lb. You did remember to include the crankshaft thrust bearings didn't you? The end play in the crankshaft must be checked with a feeler gauge or dial indicator to ensure once again that the correct components and correct machining has been applied.*

On manual transmission cars there's a pilot bearing inside the rear of the crankshaft in which the front transmission shaft runs – the old one should be removed and a new one fitted. These also need plenty of lubrication.

* Note that some models have four-bolt main bearing caps. On these cars, the inner bolts should be tightened before the outers, and there are different torque figures applicable to these bolts. See your manual for details.

CAMSHAFT INSPECTION

Clean the camshaft and inspect the bearing journals for pitting, uneven wear and signs of seizure. If journal damage is found, then the bearing inserts in the block will almost certainly be damaged, too; both camshaft and bearings will need to be replaced – a job for a specialized machine shop.

The bearing journals must be measured with a micrometer to check whether they are excessively worn or out of round, and the camshaft lobes should be inspected for heat discolouration, score marks, chipped areas, pitting and uneven wear. If it passes these tests, and the lobe lift test described earlier, the camshaft can be reused.

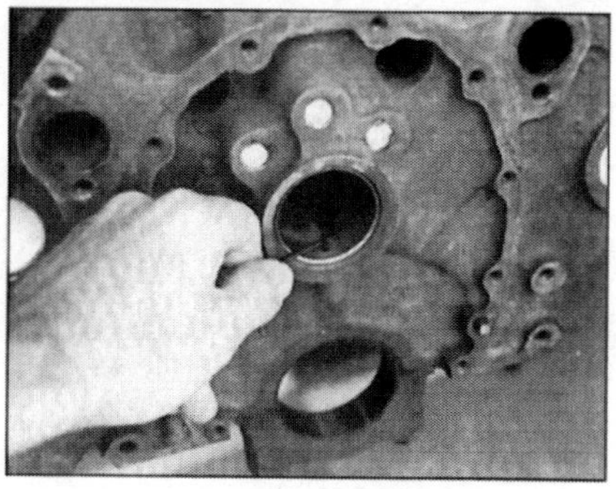

▲ *EN39. If you don't ensure that the cam bearing oil holes are properly aligned with the oil feed holes in the block, the camshaft and bearings won't be lubricated. Use a small Allen key, or something similar, to line them up. Oil for the rear bearing is fed via an annular groove, so don't expect the Allen key, or whatever, to go in much further than the thickness of the bearing.*

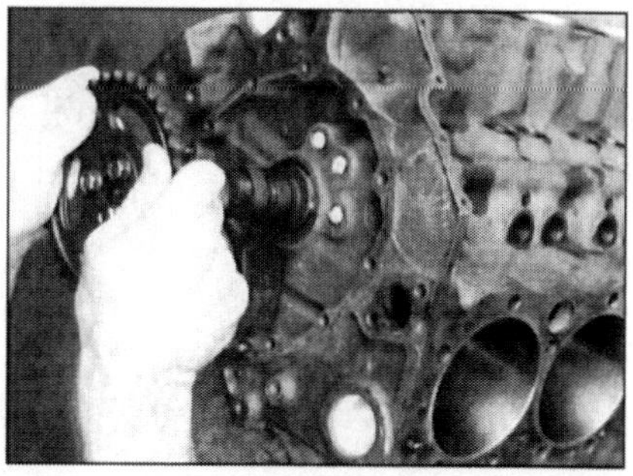

▲ *EN40. Just as with the crankshaft, the cam lobes and bearings must be coated with lubricant or with a special cam lube. As you slide the cam into the block, take great care to guide it through the centre bushings, so that the cam lobes don't damage the internal surfaces on the bushings. They will gouge the cam bearings unless a great deal of care is used. If you fit the sprocket to the end of the camshaft, that will give you something to support the camshaft with as you feed it through. The camshaft must be turned until the dowel pin is in the 9 o'clock position.*

Oil, oil, oil! Now the cam lifters must be lubricated with plenty of clean engine oil before they are fitted into the block. Original (i.e. not renewed) cam and lifters require that the lifters are replaced in exactly the positions they came from.

FITTING A NEW WATER PUMP

If you're replacing the water pump at this juncture, always compare old with new to ensure you've got EXACTLY the right model – it's easy to go wrong here. Make sure all traces of old gasket material have been removed, and apply a thin coat of RTV sealant to the engine side of the new gaskets and to the gasket mating surfaces of the new pump, then carefully mate the gaskets and the pump. Slip the bolts through the pump mounting holes to hold the gaskets in place.

Carefully attach the pump and gaskets to the engine and thread the bolts into the holes finger-tight. Install any remaining bolts (if they also hold an accessory bracket in place, be sure to reposition the bracket at this point). Tighten them to the required torque in quarter-turn increments.

FITTING TIMING CHAIN/COVER

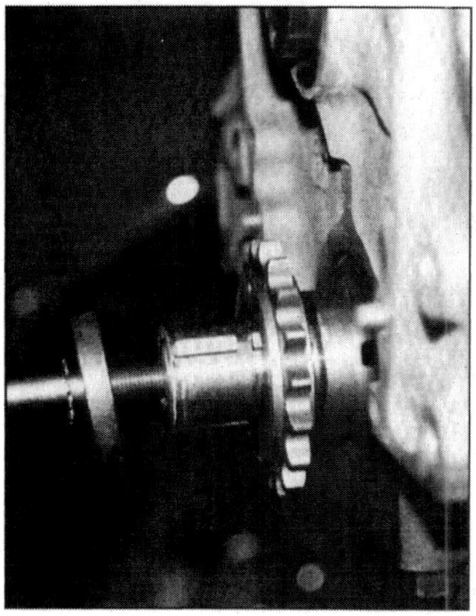

▲ *EN41. The timing chain and sprockets must always be replaced as a set – never put a new chain on old sprockets. Align the sprockets with the Woodruff key and press the sprocket onto the crankshaft with the vibration damper bolt, a large socket and some washers (or tap it gently into place) until it's completely seated. CAUTION: If resistance is encountered, do not hammer the sprocket onto the crankshaft. It may eventually move onto the shaft, but it may be cracked in the process and fail later, causing extensive (and expensive!) engine damage.*

While installing the timing gear chain, it was noticed that this particular crank got dinged at some point in handling. If it hadn't been spotted and then the raised area filed down, the lower sprocket would have wobbled. The sprocket was pulled on by using the balancer bolt and a few spacers. To make sure the sprocket seats squarely, check its runout – less than 0.002 in. is typical and easily acceptable for a sprocket. Thread lock was used on the bolts which attach the sprocket to the camshaft. The crankshaft was then spun to check for binding or tight spots in the chain.

▲ *EN42. Loop the chain over the camshaft sprocket, then turn the sprocket until the timing mark is in the 12 o'clock position. Mesh the chain with the crankshaft sprocket and position the camshaft sprocket on the end of the cam. If necessary, turn the camshaft so the dowel pin fits into the sprocket hole with the timing mark at the 12 o'clock position.*

NOTE: The No. 1 piston must be at TDC (top dead centre) on the compression stroke as the chain and sprockets are installed.

Apply a non-hardening sealant such as Permatex No. 2 or Teflon tape to the camshaft sprocket bolt threads, then install and tighten them to the specified torque. Lubricate the chain with clean engine oil.

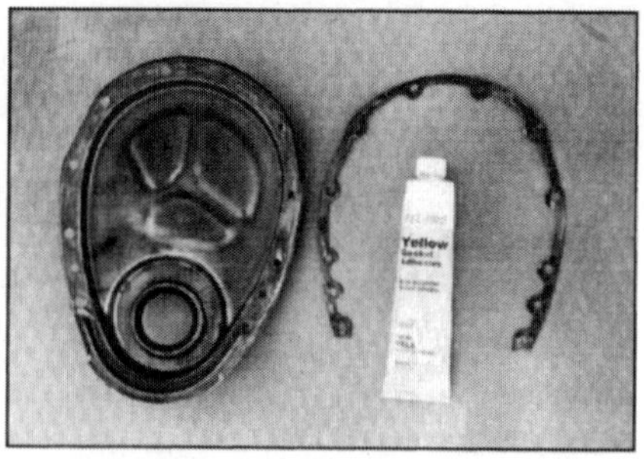

▲ *EN43. On small block engines, apply a small amount of RTV sealant to the U-shaped channel on the bottom of the cover, then position a new rubber oil pan seal in the channel. The sealant should hold it in place as the cover is installed. Tighten the cover bolts to the specified torque.*

▲ *EN44. After the timing cover was installed, the matching diameters of the harmonic balancer and crankshaft were inspected and lightly greased. If you don't have a special tool to help pull the balancer on, use a threaded rod or a 3 in. long 7/16 in. fine thread bolt. If you do any tapping use only a small hammer and wood block on the balancer.*

PISTON AND CON-ROD CLEANING AND INSPECTION

Before inspection can take place, the piston and con rod assemblies have to be cleaned and the piston rings removed using a proprietary installation tool. Care has to be taken not to damage the piston at this stage for it is easily gouged or nicked. The piston rings should always be replaced as a matter of course. Clean the carbon deposits from the top (crown) of the piston using a piece of fine emery cloth or a hand-held wire brush. (DO NOT use a wire brush in an electric drill – it will surely damage the pistons.)

Use a piston ring groove cleaning tool to remove carbon deposits from the ring grooves or, in the absence of such a tool, use a piece of an old piston ring. Again, care is required to ensure it is only carbon and not metal that is removed. With these deposits removed, a solvent cleaner can be used.

Once dry, make sure that the oil return holes on the back sides of the ring groove are clear and that the pistons are not damaged or worn excessively. Normal piston wear appears as even vertical wear on the piston thrust surfaces and slight looseness of the top ring in its groove. If the engine is not being rebored and the pistons are OK, then there is no need to replace anything other than the piston rings.

If you do have your block rebored, you'll need new (oversize) pistons as a matter of course. These, along with suitable piston ring sets, can be obtained from specialist suppliers such as Auto Accessories of America, which is where our new ring set came from. Don't be tempted to reuse rings, even if you're reusing the pistons – these relatively inexpensive items can cause untold grief and expense.

CYLINDER HEAD DISMANTLING AND INSPECTION

▲ *EN45. Check the appearance of the valves and head gasket 'foot print' before the parts are cleaned. Now's the time to spot any problems. Inspect the head carefully for cracks and evidence of coolant leakage. If you do find any cracks, a new cylinder head is required, although surface damage or warping (a machine shop check) may possibly be overcome by having the heads skimmed.*

▲ *EN46. The heads were taken to the work bench where a spring compressor was used to remove the valves.*

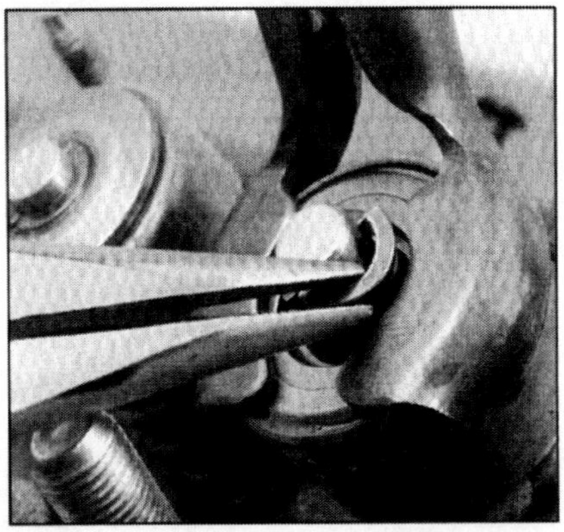

▲ *EN47. The keepers were removed using a pair of long-nose pliers. Then the retainer, shield, springs and spring seat were also removed along with the oil seal from the valve stem. The valves should be labelled so that they can be returned to their own valve guides if suitable.*

The valve faces have to be checked for signs of uneven wear, deformation, cracks, pits and burned spots. Check the valve stems for scuffing and galling, and the necks for cracks.

Rotate the valve and check for any obvious indication that it's bent. Look for pits and excessive wear on the end of the stem. Should you find any of these problems, it means another visit to your friendly machine shop.

▲ *EN48. Cleaning and examination of the cylinder head will tell you which way to go now. Examine the valve seats in each of the combustion chambers, looking for pits, cracks or severe burning. Any signs of these problems puts the head in need of a valve service beyond the scope of the home mechanic.*

Use a dial indicator attached securely to the head to check the valve-to-guide lash by measuring the lateral movement of the valve stem. (Normally a machine shop job).

With the valve in the guide and approx 1/16 in. off the seat, the total stem movement should be divided by two to calculate the actual lash. Any doubt as to their suitability should lead you to the machine shop to have them professionally checked.

The coolant passages tend to clog with scale, which has to be removed. You can screw a tap through all threaded holes to remove any traces of corrosion and thread sealant. If you have access to a compressed air line, it can be used to blow out from the holes any debris created by the cleaning process (ALWAYS wear goggles when using compressed air in this way). Items such as the rocker arms and pivot balls, valve springs etc., can be cleaned with a special solvent and allowed to dry.

The cylinder heads were cleaned using a decarbonizing chemical which makes the process simpler but which also demands great care in its use; by its nature, it's very caustic.

▲ *EN49. Use a straight edge and a feeler gauge to check the head gasket mating surface for any signs of warpage. If there is found to be warpage which exceeds the specified limits, it will need to be resurfaced at a machine shop. Note that the intake manifold flanges may also need to be machined. Scrape away all traces of old gasket material and sealing compound from the head gasket, intake manifold and exhaust manifold sealing surfaces. Be extremely careful not to damage the soft aluminium cylinder head – if required, you can use a special gasket removal solvent which chemically softens the gaskets and makes removal much easier.*

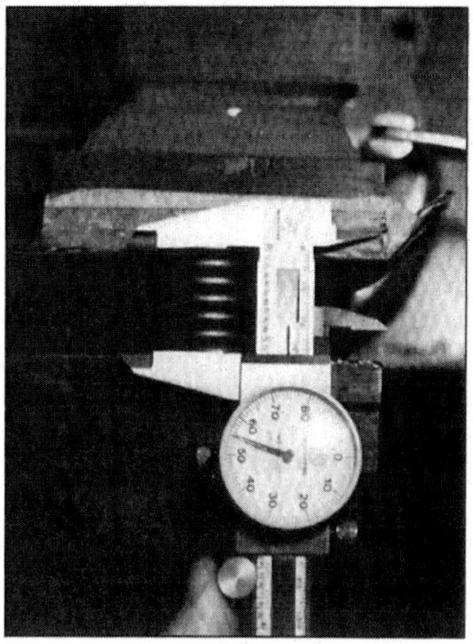

▲ *EN50. Look for obvious signs of wear or damage to the valve springs and use a dial gauge or vernier caliper to measure the length of the spring; compare it with the specifications and note that springs that have sagged cannot be reused.*

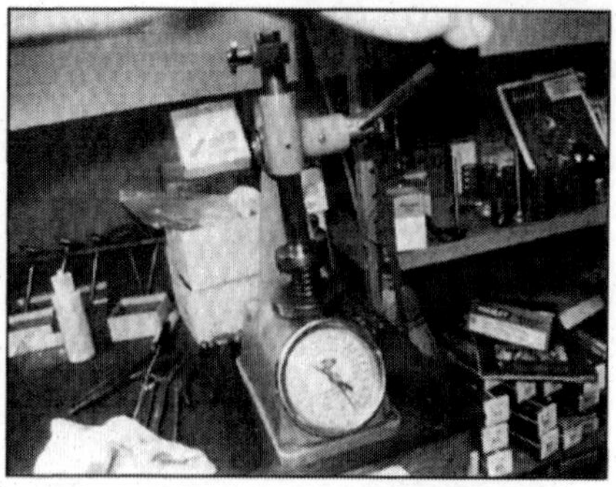

▲ *EN51. A machine shop will be able to tell you if the springs have sufficient tension to be used in a rebuild. Stand each spring on a flat surface and check it for squareness and check the spring retainers (or rotators) for wear and cracks. Again, any suspect parts should be replaced – the damage they could cause far outweighs their minimal cost.*

CHECKING PISTON RING END GAPS

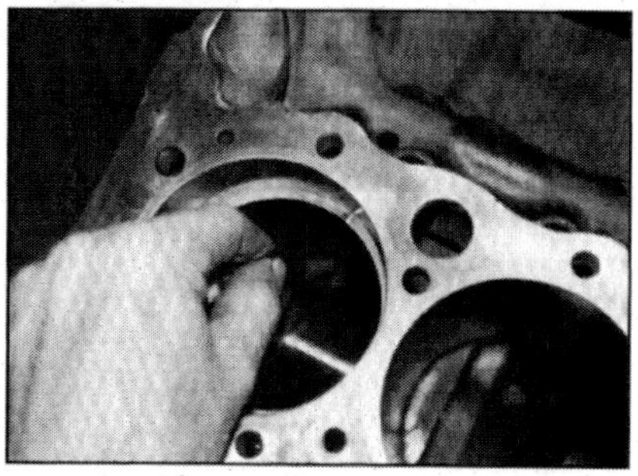

▲ *EN52. Prior to fitting to the pistons, the ring side lash and end gaps must be checked. Lay out the piston/connecting rod assemblies and the new ring sets so that they can be matched with the same piston and cylinder during the end gap measurement and engine reassembly. This requires a clean workbench and some planning ahead – most of us don't usually have large clean areas of this size available!*

Insert the top (No. 1) ring into the first cylinder and square it up with the cylinder walls by pushing it in gently with the top of the piston. The ring should be near the bottom of the cylinder, at the lower limit of ring travel.

To measure the end gap, slip your feeler gauges between the ends of the ring until a matching gauge is found.

The feeler gauge should slide between the ring ends with just a small amount of drag.

Compare the measurement to the specifications for your engine. If the gap is too small, it will have to be enlarged otherwise the ring ends may come into contact with each other when the engine is running, and cause severe damage. The gap can be enlarged quite easily by filing the ring ends with a fine file. To save wear and tear on your finger ends, mount the file in a vise (with soft jaws to prevent damaging the tool), slip the ring over the file with the ends contacting the file face and slowly move the ring to remove the excess metal from the ends. Don't leave any burrs!

A gap slightly larger than that specified isn't a real problem unless it is more than 0.040 in. adrift. If this is the case, you probably haven't got the right rings for your engine, so change them.

INSTALLING THE PISTON RINGS

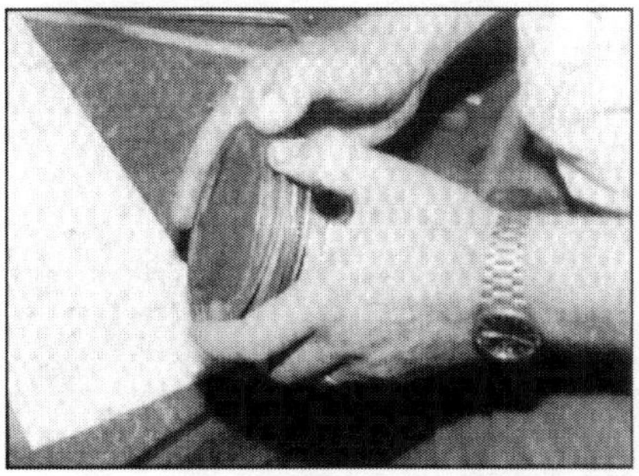

▲ *EN53. The oil control ring (the lower, No. 3 ring) comprises three components, and should be fitted first. Slip the spacer/expander into the groove, making sure that the anti-rotation tang (where applicable) is inserted into the drilled hole in the ring groove. Install the lower side rail. Do not use a piston ring installation tool on the oil ring side rails, as they are easily damaged. Place one end of the side rail into the groove between the spacer/expander and the ring land, hold it firmly in place and slide a ringer around the piston while pushing the rail into the groove. Then install the upper side rail in the same manner. Once the oil ring components have been fitted, check and make sure that the upper and lower side rails can be turned smoothly in the ring groove.*

Next, install the No. 2 (middle) ring. It's important to note that it is stamped with a mark which must face up, that is toward the top of the piston. Equally, do not mix the top and middle rings. Use a piston ring installation tool to install the ring into the middle groove but don't expand the ring any more than is necessary.

The top No. 1 ring, shown in this photo, is fitted in the same manner, making sure that the mark is facing up. That's one piston – seven more to go!

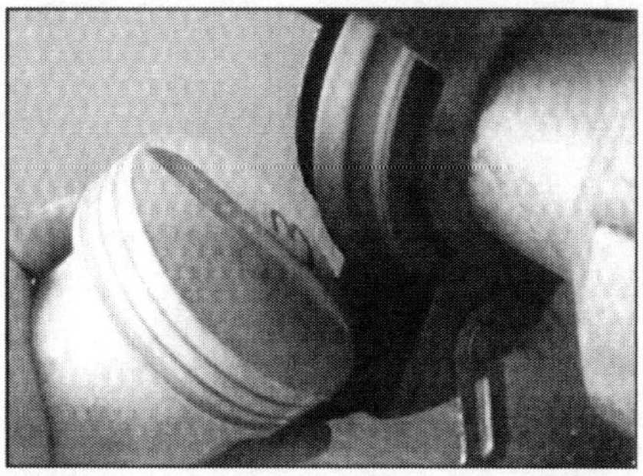

▲ *EN54. A fiber wheel can be used to smooth the sharp edges on the piston relief. This helps reduce the likelihood of hot spots and detonation.*

▲ *EN55. As shown here, you should only remove an absolutely minimal amount of material. The edges can also be smoothed with very fine grade sandpaper, but in either case do it before installation of the rod and rings so the piston can be thoroughly cleaned.*

PISTON AND CON-ROD INSTALLATION

Before installing the piston and con rod assemblies, the cylinder walls must be perfectly clean, the top edge of each cylinder must be chamfered (check that your machine shop has done it) and the crankshaft must be in place.

▲ *EN56. Lay the bearing half in place in the bottom of the connecting rod, after having cleaned it first. Ensure that the tab on the bearing fits into the recess in the rod. Take your time – excessive force should be avoided as it's very easy to gouge or nick the bearing surfaces. The bearing should not be lubricated yet.*

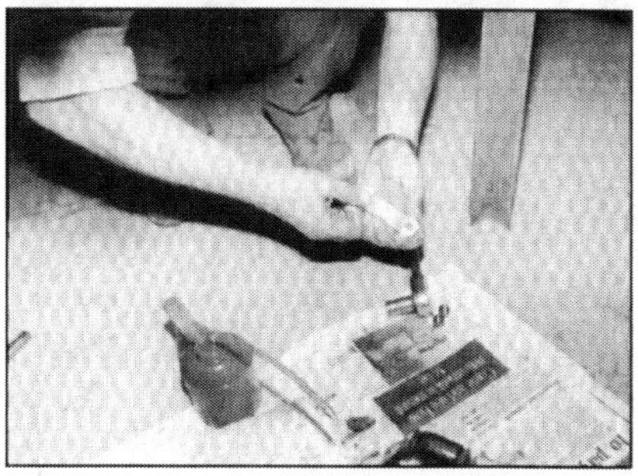

▲ *EN57. Take the other half of the bearing shell and install it in the rod cap. Once again, all parts should be clean and the tab in the insert should match appropriately. It's a good idea to slip a section of plastic or rubber hose over each connecting rod cap bolt.*

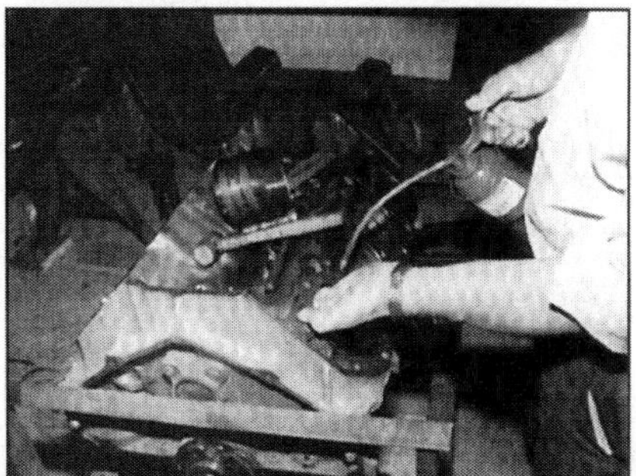

▲ *EN58. Rotate the crankshaft until the No. 1 connecting rod journal is at BDC (bottom dead centre) and apply a light coating of engine oil to the cylinder walls.*

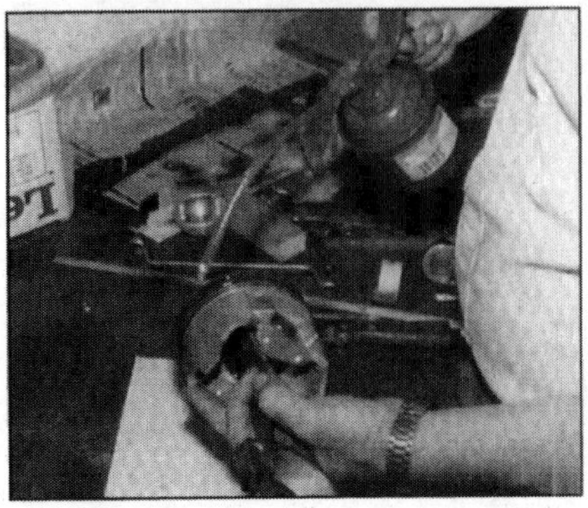

▲ EN59. Position the piston ring gaps at 120 degree intervals around the piston. Lubricate the piston and rings with clean engine oil ...

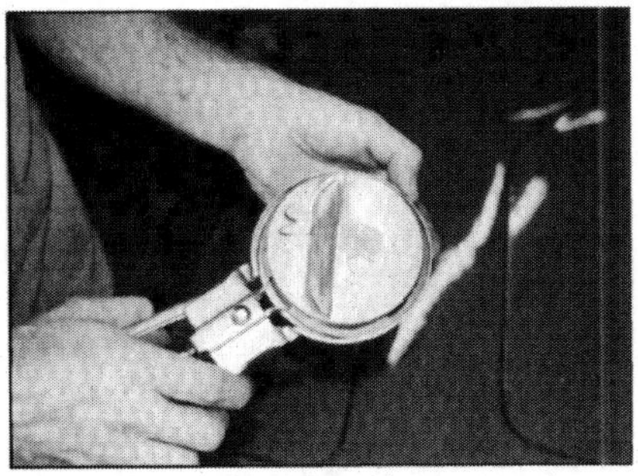

▲ EN60. ... and attach a piston ring compressor to the piston. Leave the skirt protruding around ¼ in. to guide the piston into the cylinder. The rings have to be compressed until they're flush with the piston sides.

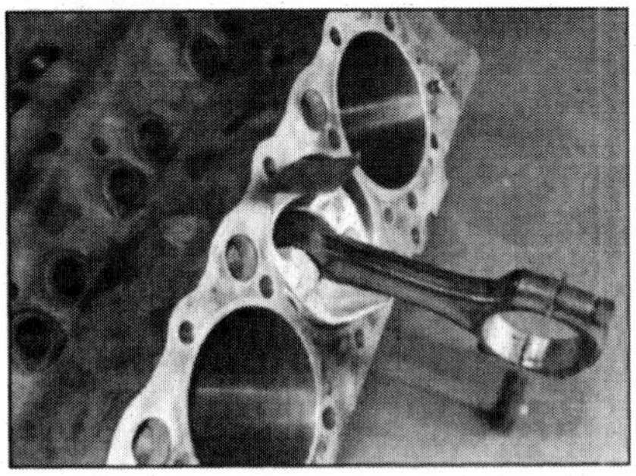

▲ EN61. Piston skirt lash can be checked by temporarily inserting the piston upside-down and testing the fit with .001 and .002 brass shim stock. This lash is normally measured more accurately by a machinist using an expensive dial bore gauge during the final honing and while the torque plate is installed. However, this final check gives you confidence that no subsequent rubbing of parts occurred.

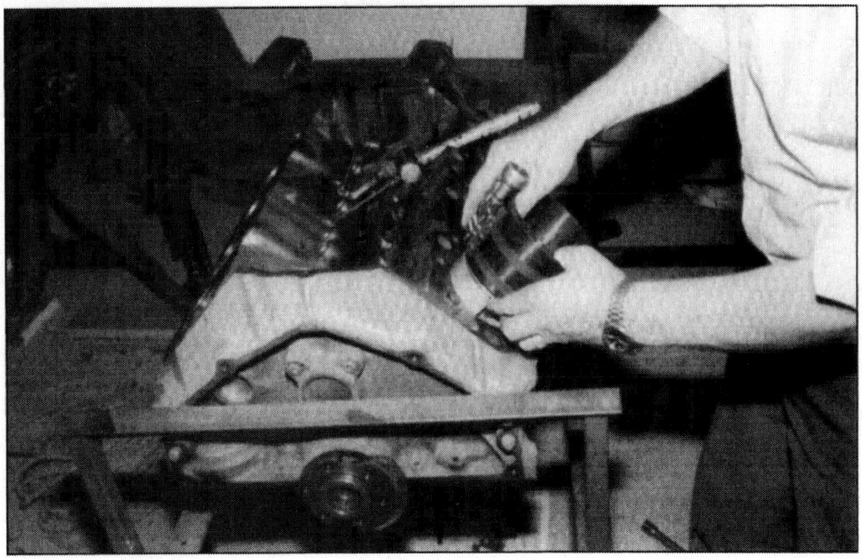

▲ EN62. With the notch on top of the piston facing the front of the engine, insert the piston/con rod assembly into the No. 1 cylinder bore and rest the bottom edge of the ring compressor on the engine block. Tap the top edge of the ring compressor to make sure it's contacting the block around its entire circumference.

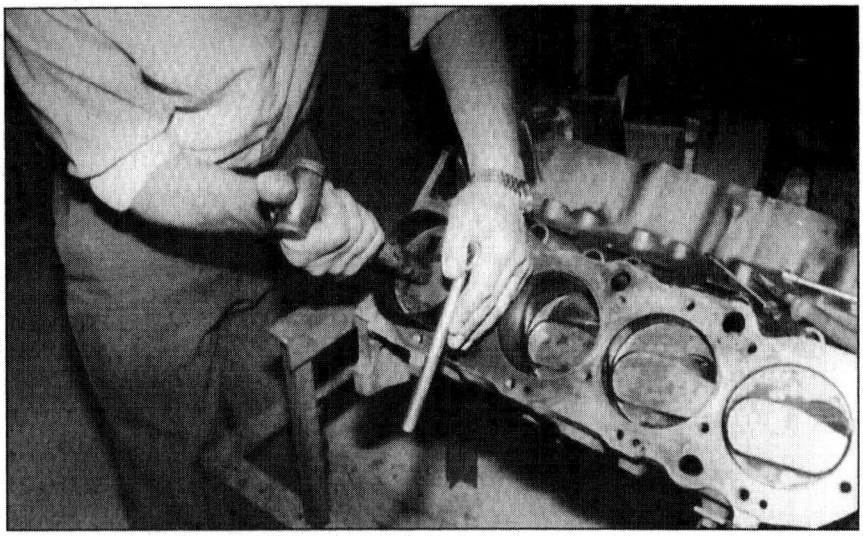

▲ EN63. Gently tap around the top of the piston with the end of a wooden hammer handle while guiding the end of the connecting rod into place on the crankshaft journal. It is natural for the piston rings to try to pop out of the ring compressor just before entering the bore, so you'll need to maintain some downward pressure at this point. Work slowly and, if any resistance is felt as the piston enters the cylinder, stop straight away! If it binds, find out why. Forcing at this point could damage the rings and/or piston and/or cylinder bore, all of which would be expensive and wasteful in terms of time and money. It helps to have an assistant guide the rod bolts over the crank journal. With the piston/con-rod assembly successfully in place, the con-rod bearing oil lash should be checked before the rod cap is permanently bolted into place.

Using the same technique as for the crankshaft (mentioned earlier) use Plastigage and tighten up the con-rod caps to the required torque before undoing the nuts and removing the caps in order to check the width of the Plastigage.

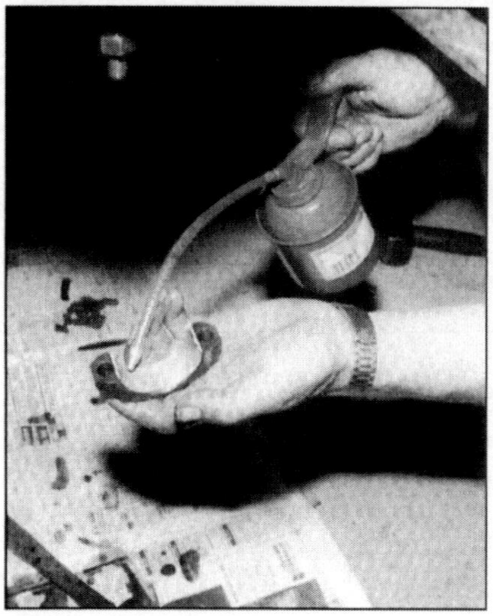

▲ *EN64. When you're happy that the clearances are correct, carefully remove all traces of the Plastigage (use a credit card or fingernail) and apply a copious amount of engine oil or a thin smear of moly-grease to both bearing surfaces. The piston has to be pushed down into the cylinder to expose the face of the bearing insert in the connecting rod – be sure to slip the protective hoses over the rod bolts first.*

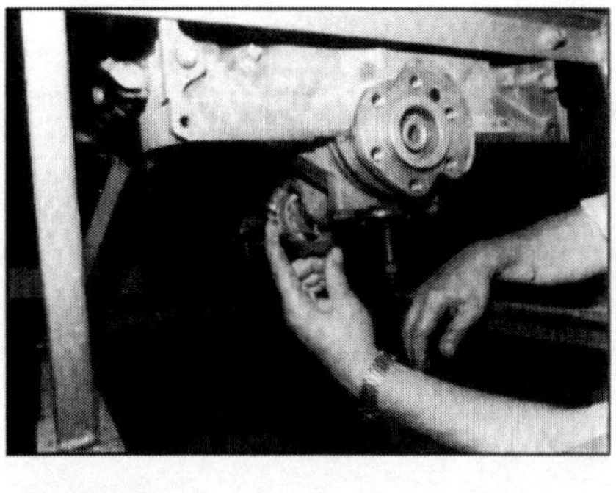

▲ *EN65. With the con-rod in place on the journal, remove the protective hoses, install the rod cap ...*

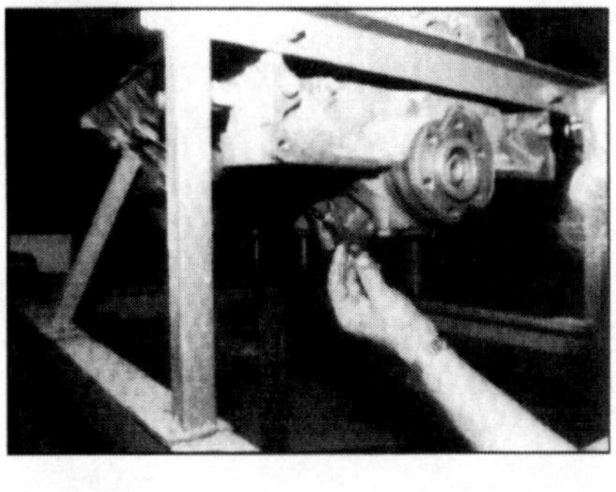

▲ *EN66. ... hand tighten the securing nuts ...*

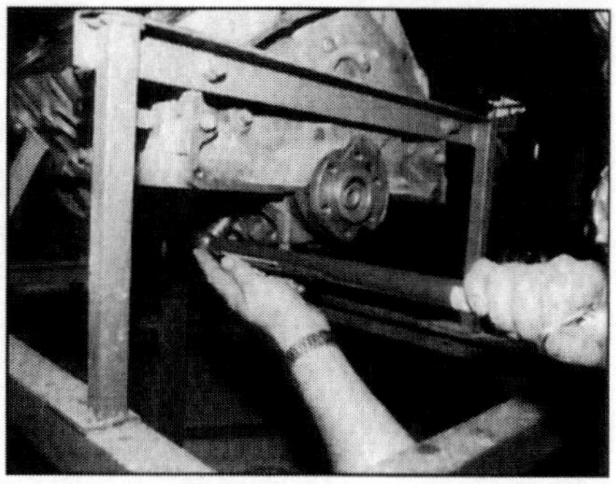

▲ *EN67. ... and tighten the nuts to the specified torque.*
Repeat this operation for the other seven piston/con-rod assemblies.

CYLINDER HEAD REASSEMBLY

Regardless of whether or not the heads were sent to a machine shop for servicing, they must be absolutely spotlessly clean before you start reassembly. If the heads were sent out for servicing, the valves and related components will already be in place.

Three different types of valve stem oil seals are used on these engines, depending on year, engine size and horsepower rating. The most common is a small O-ring which simply fits around the valve stem just above the guide boss. A second type is a flat O-ring which fits into a groove in the valve stem just below the valve stem keeper groove.

On most high horsepower applications, an umbrella type seal which extends down over the valve guide boss is used over the valve stem. In most cases, the umbrella type seal is used in conjunction with the flat O-ring type seal.

Drop the spring seat or shim(s) over the valve guide and set the valve springs, shield and retainer/rotator in place.

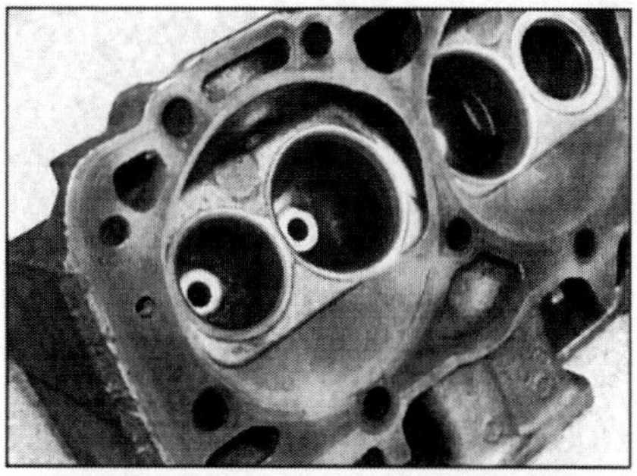

◀ *EN68. In this particular case, new valve seat inserts were not needed, but the valve seats were recut with a three-angle valve seat cutter, and new bronze valve guides were installed.*

◀ *EN69. Start at one end of the head, lubricate and install the first valve. Apply moly-base grease or clean engine oil to the valve stem. These replacement exhaust valves from TRW are stainless steel, while the intakes are a high silicone alloy.*

◀ *EN70. Compress the springs with a valve compressor, not forgetting to install the O-ring seal in the lower groove of the valve stem. Make sure the seal is not twisted. It must lie perfectly flat in the groove. Position the keepers in the upper groove, then slowly release the compressor and make sure the keepers seat properly. Apply a small dab of grease to each keeper to hold it in place if necessary.*

Repeat the procedure for the remaining valves. Once all the valves are in place in both heads, the valve stem O-ring seals must be checked to make sure they don't leak.

This procedure requires a vacuum pump and special adapter. As this is a special GM tool (Reference J-23994) you'll have to check your budget to see if you can run to its purchase or whether it's better to have your machine shop run the tests for you.

The adapter is positioned on each valve retainer or rotator and vacuum applied with the hand pump. If the vacuum cannot be maintained, then the seal is leaking and should be checked and replaced if necessary. Measure the distance from the head to the valve spring retainer to check the height of the spring when it is installed. The measurement is taken from the top of each spring seat or shim(s) to the top of the oil shield (or bottom of the retainer/rotator – the two points are the same). Use a steel rule (graduated in 1/32 in. increments) or a dial caliper. Don't forget to subtract the thickness of the oil shield if it is used. If the height is greater than that specified, shims can be added under the springs to correct it. NEVER add shims to the springs to the point where the installed height is less than specified. If the heads were sent out for service, then the installed height should be correct, but no-one's perfect, and it definitely pays to take the small amount of effort to run this check.

Apply moly-grease to the rocker arm faces and the pivot balls, then install the rocker arms and pivots on the cylinder head studs. Thread the nuts on three or four turns only. Note the new GM valve guide seals on the intake valve.

▲ EN71. Valve rotators can be used on the exhaust valves to reduce the risk of damage from hot spots. Although the stock O-ring seals were already installed on the exhaust valves, they were added to the intake, too. They may help a few years down the road.

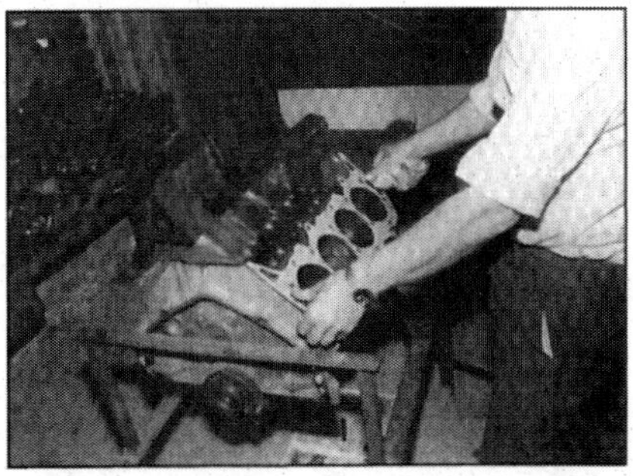

▲ EN72. Before fitting the cylinder heads, you must ensure that the gasket surfaces on both head and block are spotlessly clean. The gasket should be placed in position, with the bead up, over the dowel pins. It's also the time to check that the bolt threads are free from dirt and other foreign matter, as this can detrimentally affect those all-important torque readings.

If your engine needs a steel gasket, coat both sides of it with a quality sealer. Spread it thinly and evenly across the gasket – a good way to achieve this is to use a domestic paint roller. If, however, your engine requires a steel/asbestos composite gasket, do NOT use any sealer at all.

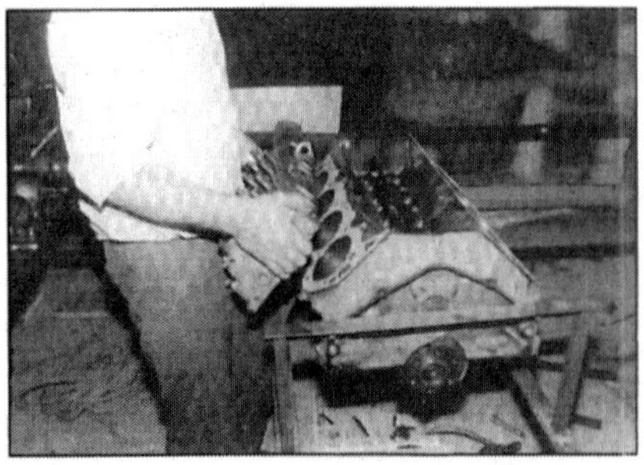

◀ EN73. Then, the cylinder head can be eased gently over the pins and into position on the gasket. The threads of the cylinder head bolts should be coated with a sealing compound and installed just finger tight.

▲ EN74. Next the head bolts should be tightened up a little at a time according to the sequence shown here. Take your time at this stage, for too much pressure could easily cause the head to distort – an expensive business!

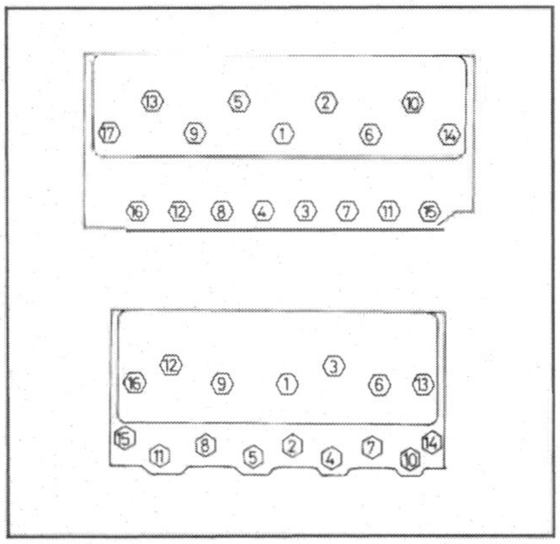

▲ EN75. Use a quality torque wrench to ensure that you do not exceed the recommended torque figures. Note the differences between the small block (below) and big block heads.

▲ EN76. With the camshaft, timing chain and sprockets in place, the lifters should be lubricated with engine oil and replaced in their original positions.

If you're reusing the original pushrods, make sure that they are going back where they came from. Not a problem here, as they are brand new.

▲ *EN77. Lubricate the lower end of each pushrod with clean engine oil or moly-base grease and install them in their original locations. Make sure each pushrod seats completely in the lifter socket. Apply moly-base grease to the ends of the valve stems and the upper ends of the pushrods before positioning the rocker arms over the studs.*

▲ *EN78. Set the rocker arms in place, then install the pivot balls and nuts. Apply moly-base grease to the pivot balls to prevent damage to the mating surfaces before engine oil pressure builds up. Be sure to install each nut with the flat side against the pivot ball.*

▲ *EN79. Before installing the intake manifold, clean the gasket and seal surfaces of the cylinder heads, the block and the manifold itself. You should have used a gasket scraper to remove all traces of carbon and old gasket material while the heads were off but you can give a final clean-up, with lacquer thinner or acetone – if there's old sealant or oil on the mating surfaces when the manifold is installed, oil or vacuum leaks may occur.*

Use a tap of the correct size to chase the threads in the bolt holes, then use compressed air (if available) to remove the debris from the holes.

WARNING: Wear safety glasses or a face shield to protect your eyes when using compressed air.

TIP: When working on the heads and the block, cover the lifter valley with rags to keep debris out of the engine. Use a vacuum cleaner to finally check that the intake ports in the heads are clear.

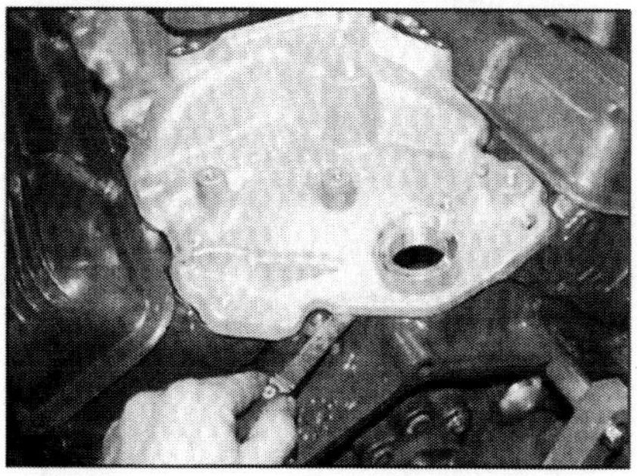

▲ *EN80. Apply a thin coat of RTV sealant around the coolant passage holes on the cylinder head side of the new intake manifold gaskets – there's normally one hole at each end.*

Position the intake gaskets on the heads and set the intake manifold on top of them, taking care not to disturb the gaskets. Make sure that all intake port openings, coolant passage holes and bolt holes are aligned correctly and the THIS SIDE UP mark is visible.

Check the clearance between the manifold and the block. If the clearance measures substantially less than the thickness of the rubber end gaskets, they will eventually be forced out when the manifold is tightened down. In that case, degrease the surfaces carefully and apply a thick bead of silicone sealer just before final installation of the manifold. This intake manifold is the new GM performance model which produces the same horsepower as the previous high-rise designs but provides increased hood clearance.

Install the front and rear end seals on the block. Note that most seals have either rubber spikes which fit into matching holes in the block, or rubber tabs which fit over the edge of the block to locate the seals. On some early models there's a tab on the rear seal which may have to be cut off for a proper fit. Refer to the instructions with the gasket set for further information. Apply a dab of RTV sealant to the seal-to-gasket joints. NOTE: Later models do not have rubber seals; a bead of RTV sealant is used instead. Make sure it extends up onto the gaskets at each end.

Fit the manifold end seals first and then the side gaskets on the cylinder heads. It's wise to use sealing compound around the water passages.

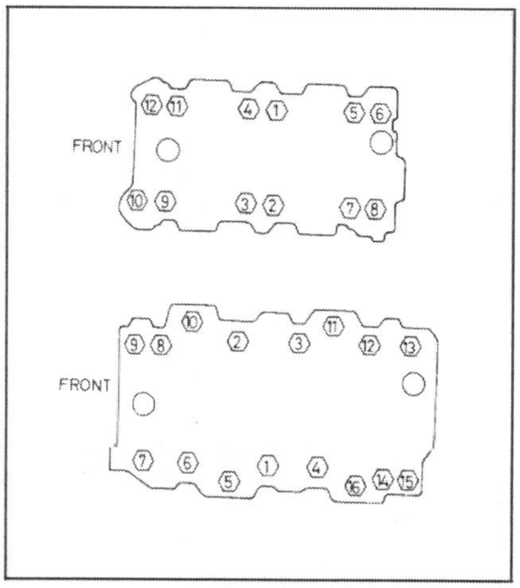

▲ *EN81. Apply a non-hardening sealant (such as Permatex No. 2) to the manifold bolt threads, then install the bolts. While the sealant is still wet, tighten the bolts to the specified torque in the sequence shown here. Tighten in a criss-cross pattern in quarter-turn increments until they are at the required torque. Work up to the final torque in three steps. The diagrams here show the bolt tightening sequence for all except 1982 models. Top is small block, bottom is big block.*

▲ *EN82. The result of all those patient hours of hard work is worth it, don't you think? This is a freshly rebuilt small block engine which, with a little care and consideration will do its owner proud for many, many years to come.*

Thanks are gladly given to Claremont Corvette and Auto Accessories of America for supplying the excellent quality components used in John Fletcher's project car rebuild.

FINAL HINTS

We strongly recommend that having gone to the great trouble of rebuilding your wonderful Corvette's engine to the high standards it deserves, you don't spoil the ship by fitting the wrong sort of clutch. Non-GM clutches can sometimes give problems, so we recommend that you pay extra and use only a genuine GM part.

Generally, resist the temptation to over-tighten fixings. Use the recommended torque figure, especially for fixings on thin steel parts such as oil pans, early valve covers or timing covers. You won't make oil leaks less likely; you'll make them more likely if you over-tighten! Don't glue every gasket you use. For instance, if you glue the head gaskets, how are you going to get the heads off next time you need to? Where you need to use a gasket sealer that has to dry off before you fit the gasket, make sure you apply the sealer in plenty of time before assembly so it can fully dry.

John Pfanstiehl recommends that you glue the water pump gaskets to the pump with adhesive and then the mating surfaces of the gasket and the block can be coated with silicone (*not* with gasket glue, or you'll never remove the pump again!).

Before fitting the balancer to the front of the crank, check the surface on the balancer where it runs inside the seal to make sure it hasn't been worn or damaged by the previous seal.

When fitting the intake manifold, always test the fitting bolts to ensure they won't bottom out in the head. Use flat washers on aluminium manifolds and in all cases, check installed

TORQUE SPECIFICATIONS

	Ft lb (unless otherwise indicated)
Rocker arm cover	
Small block	45 in. lb
Big block	50
Intake manifold bolts	30
Exhaust manifold bolts	
1968 to 1977 models	20
1978 onwards	
Outer bolts	20
Inner bolts	30
Cylinder head bolts	
Small block	
Step 1	30
Step 2	50
Step 3	65
Big block	
Step 1	40
Step 2	60
Step 3	80*
Timing cover bolts	80 in. lb
Camshaft sprocket bolts	20
Torsional (vibration) damper bolt	
Small block	60
Big block	85
Oil pan	
Small block	
1/4-20	80 in. lb
5/16-18	165 in. lb
Big block	
Oil pan to front cover bolts	55 in. lb
Oil pan to block bolts	145 in. lb
Oil pump bolt	65
Rear main bearing cap bolts	
Small block	
1968 to 1976 models	75
1977 onward	80
Big block	110

*On engines with aluminium heads, tighten the short bolts to 65 ft lb and the long bolts to 75 ft lb

clearances before assembly.

Be sure to replace coolant system components; hoses, belts, water pump thermostat, and flush out the radiator and also check it for leaks. Can you imagine going to all this trouble to rebuild your engine only to wreck it within weeks because the coolant system fails.

PRE-OILING THE ENGINE PRIOR TO STARTING UP

There are two very good reasons for pre-oiling an engine prior to starting it up. First, it will provide a coating of oil via the normal lubrication process but without the stresses and strains of combustion and, second, it enables checks to be made on the lubrication system in a way that is not likely to wreck a new engine. The simplest way requires that you have a fully charged battery and that you take out the spark plugs. On coil-in-cap models, disconnect the wire from the BAT terminal on the distributor cap. On separate coil models, remove the coil wire from the distributor cap and effectively ground the wire on the engine block.

SAFETY NOTE: Ensure that there is no fuel, fuel vapour or anything else flammable in the vicinity, since there will inevitably be some sparking between the coil wire and the engine block. Don't touch the high tension (spark plug or coil) cabling. This high voltage could cause injury – or worse.

If you now spin the engine for several minutes on the starter, you will see the oil pressure light go out or the pressure gauge will rise and oil will be pumped all around the engine. The advantage of this system is that oil is distributed over all of the moving parts but without any stresses or strains put on them. The disadvantage is that there won't have been any oil there for the first few seconds, except for the oil that you will have squirted onto every bearing surface, bushing and contact surface while you were building up the engine, although the stresses will be very low because – obviously – the engine won't be running.

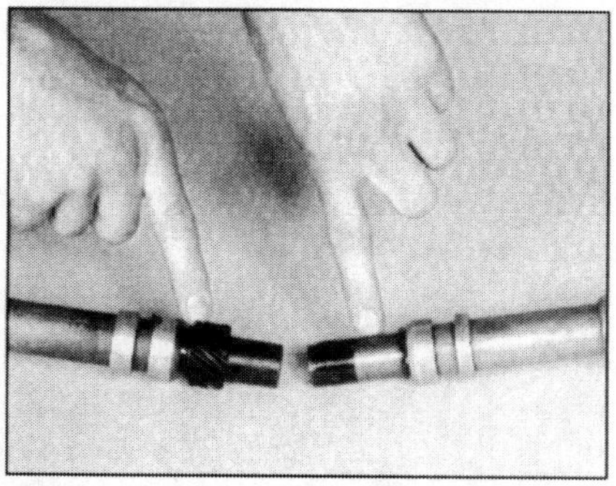

▲ *PR1. If you want to go even further, there is another way of checking the oil flow first. It demands a little effort and ingenuity, but is far simpler than stripping and rebuilding your engine (again!) because of, say, a blocked oilway. You can do this with the engine on the bench.*

The engine should be assembled except that the distributor should not be fitted and the rocker covers should be left off.

In order to operate the oil pump, it is necessary to fit a modified distributor. The gear on the lower end of the shaft has to be ground off and, (where applicable) the advance weights on the upper end of the shaft removed, so you'll need to find a Chevy distributor you can adapt, from a junkyard. The 'pre-oil' distributor should be fitted in place of the 'real' one in such a way that the lower end of the shaft mates with the upper end of the oil pump driveshaft. Once they're aligned and the distributor seats in the block, fasten the hold-down clamps and mount the upper end of the shaft in the chuck of your electric drill. By operating the drill in a clockwise direction, it will mimic the function of the engine turning. After three or four minutes, oil should flow from the rocker arm holes.

When you're satisfied that all is well, remove the distributor and plug the hole with a clean cloth, as the correct distributor will not be installed until the engine is back in the car.

- Before start-up, oil prime the engine until oil flows onto all the rockers.
- Use a good quality multigrade oil for start-up. You should not have used synthetic oil or a Slick 50 type of additive for pre-oiling the bearings or for the first few thousand miles of the engine's life.
- Change the oil and filter after running the engine for its first 20 to 40 minutes. Also, change the oil again after 100 miles or so. Don't switch to synthetic oil or use additives until the rings are broken in.
- Break the engine in gently, which means no high revving and no heavy loads at low speed in a high gear. You can gradually add speed and load after the engine has covered 3000 miles (preferably 5000 miles). By then it will be broken in and all the bearing surfaces will be lapped into one another.
- Keep the engine relatively cool for its first few hours. Leave the AC off – and avoid any summer traffic jams. In any case, keep an eye glued to the temperature gauge for the first 15 minutes to half an hour.
- Invite a friend over for start-up. As soon as the engine fires, have him help you check for fuel and oil leaks – one checking in the engine bay, the other looking underneath.
- After a complete engine warm-up and cool down, re-torque the head bolts to be safe and then do so again after around 1000 miles of running.

EXHAUST SYSTEM

When working on your exhaust system, follow these basic rules:

- Protect your eyes: wear goggles to stop dirt and bits of flaking rust going into them.
- Protect your hands with thick gloves, against sharp or hot metal.
- Only work on your exhaust when the system and engine is stone cold.
- Remove the keys from the ignition and take the usual safety precautions if you are lifting the car in any way (see Safety notes in Appendix 1). Make sure that the car is safely supported and that the wheels on the ground are chocked and blocked so it can't roll or move.
- When dismantling the system, work from the front to the back. Apply copious amounts of penetrating oil,

ideally two or three days before removing the exhaust.

- Always use new gaskets, hangers and clamps – they're cheap enough, and reusing old ones could cause major problems, especially if they fail on the road at speed.
- Smear all threaded components with an anti-freeze compound during reassembly to aid any future pull down work.
- Allow plenty of clearance between new exhaust parts and the underbody to avoid overheating the floor pan. In particular, catalytic converters get extremely hot – make sure that they and their respective heat shields are correctly fitted.

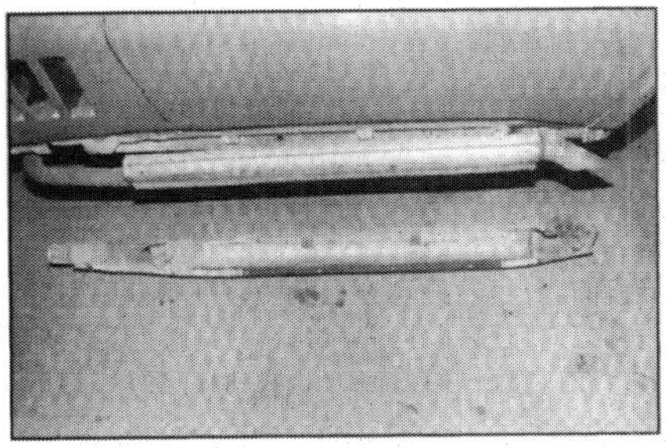

▲ *EX1. The side exhaust system was a factory option, but only in the 1969 model year. The '69 car featured on many of these pages was so-fitted, but owner John Fletcher much preferred the dual pipe system fitted as stock. As such, this Corvette's restoration included a change of exhaust system.*

Removing the side-exiting exhaust is not difficult. The heat shield (seen here on the floor alongside the muffler) is secured by six screws to the main assembly. The muffler itself is a single piece item which attaches directly to the manifold. The screws, nuts and bolts fixing the exhaust assembly are more than likely to be well-rusted and likely to give you the most hassle. Apply penetrating oil and plenty of it, ideally on each of several days beforehand, so it can do its stuff.

NOTE: Where this type of exhaust is fitted, the body rocker moulding is omitted from the lower side bodywork. In our case, it meant that John had to buy new mouldings and fit them as part of the body replacement section.

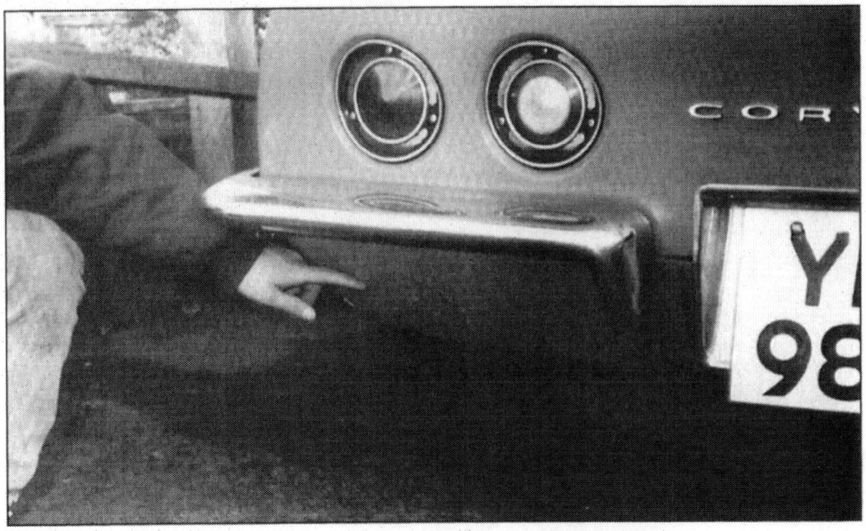

▲ *EX2. One obvious problem in changing to a rear-exiting exhaust is that there is nowhere for the muffler pipes to go when they get to the rear of the car. Anyway, that's how it looks from the outside …*

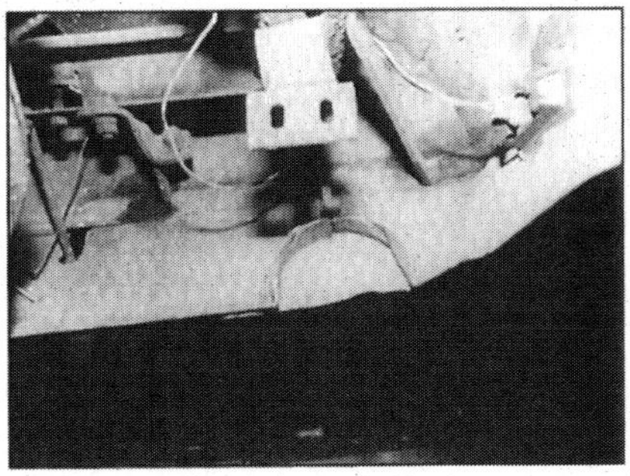

▲ *EX3. … but looking at the same view from the underside, we see that the factory effectively just blanked off the original exhaust cut-outs when the side exhaust was fitted. In fact, there are two options at this point.*

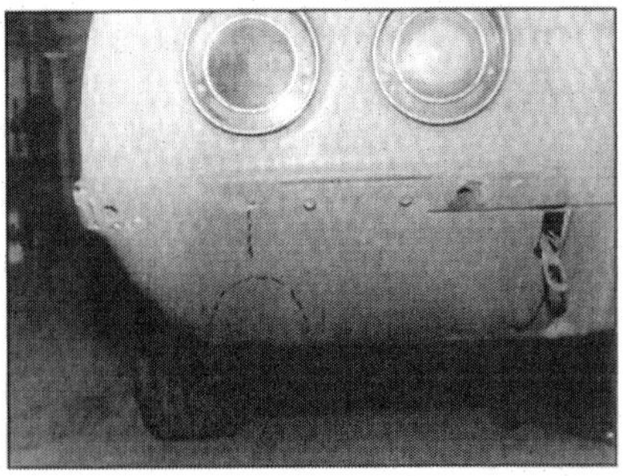

▲ *EX4. The first is to mark the exact position of the cut-out on the rear of the valance …*

▲ *EX5. … and cut out the area of fiberglass not required. Em found that the simplest way was to carefully use tin snips, making a single cut up the middle, followed by a number of smaller cuts to either side. Once most of the unwanted fiberglass had been removed, he was able to file and then sand the area to a smooth finish. Remember the usual safety precautions when working with fiberglass and when filing or cutting it: ALWAYS, wear gloves and a mask. NB: We recommend cutting fiberglass with a saw rather than snips. With the latter, there's a risk of damaging and breaking the surrounding area.*

The other option is to replace the whole of the rear exhaust panel. It is also possible to change the other way and buy panels without the exhaust holes to make changing to a side exiting system easier.

▲ *EX6. The dual pipe system comprises two sets of the following; rear muffler/pipe, centre pipe section and front pipe. Em fitted a full stainless steel system from Double S (and purchased through Claremont Corvette). It's a good investment if you intend to keep your car for a number of years. Even if you sell the car, a stainless exhaust is a good selling point, as it's one replacement item the new owner doesn't have to worry about, although not so if pure originality is your aim, of course. It's always wise to lay out any new system on the floor so that you can make sure that everything is present and correct and that you know what goes where.*

▲ *EX7. Better yet, you should put the whole system together in a 'dry run' – it's far easier than testing it whilst hanging around under the car! In this case, Em found that the mufflers couldn't be tightened onto the centre pipes. A little work with a hacksaw soon solved that problem, putting slots where they were needed.*

▶ *EX8. As you can see from these photos, Em replaced the exhaust system with the body off the car, which made life much easier. However, it's still not difficult to make the obvious error of trapping the wire for the parking brake equalizer and the seat lap belt retractor assembly. Watch out for them!*

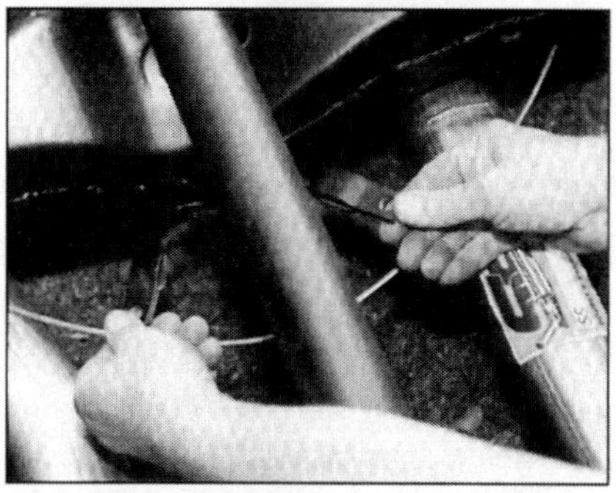

▶ *EX9. The manifold studs/nuts are sometimes problematical, with the nuts being reluctant to leave the threads. Too little patience and penetrating oil can result in a sheared stud. Take your time and, when you replace the nuts, use a copper-grease type of compound to help prevent a recurrence in the future.*

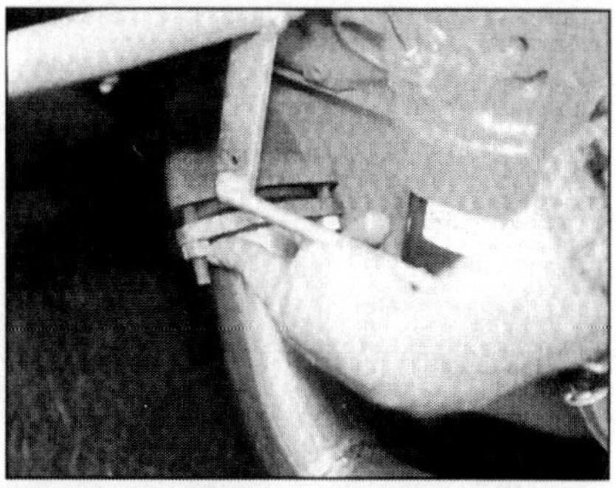

▲ *EX10. The view most other road users will see! The large bore rear pipes create a superb image of power and style and John was well pleased with them. That said, many owners prefer to have the side pipes with cleaner lines at the rear. That's the beauty of the Corvette, you can select what you want. But note how the passenger-side exhaust is much too low. The pipes had to be removed and extra bends put in them by an exhaust fitting centre before they would hang right. Be warned that stainless steel exhaust systems can be a whole lot more trouble to fit – 'cos they often don't, it seems.*

You should replace the rubber hangers as a matter of course – rubber becomes hard and brittle over a period of time and even if the hangers don't actually fail, they will not be doing the job right. Of prime importance is to make sure that, at no point, is the system too close to the bodywork or the chassis. If it is, you'll have problems with the knocking and clunking as you drive and, in some cases, the heat from the exhaust could cause the fiberglass body to overheat. Note the single leaf spring in situ *– that's covered in the suspension section.*

PART 2 TRANSMISSION

The 'restoration' car featured in many of the sections in this book was fitted with a four-speed manual 'Muncie' transmission (named after the GM gearbox plant at Muncie, Indiana), a particularly sweet-shifting unit which was phased out in the mid-seventies.

This particular 'box is unlikely to give trouble until high mileages are reached, at which point bearing wear and synchromesh trouble is likely. This usually manifests itself as jumping out of second gear while the car is on over-run.

The Turbo-Hydramatic 305 and 400 transmissions are very rugged units and, given that their fluid is changed every 50,000 miles and that only TQ Dexron is used, they will almost certainly outlast all other parts on the car.

REMOVING THE TRANSMISSION

Inside the car, remove the gear (shift) lever and the console trim plate. Room is required to enable the transmission to be lowered and pulled out. This means that the car should be raised slightly – remember the safety rules: **NEVER work beneath a car supported only by a jack, even a trolley jack.** In our case, our cause was helped by the fact that the body had been removed, so the car automatically sat much higher than normal. The exhaust system has to be removed, and on big block cars the front stud on each manifold also has to be removed. Disconnect the driveshaft at the transmission slip yoke and lower the front of the transmission and remove the yoke.

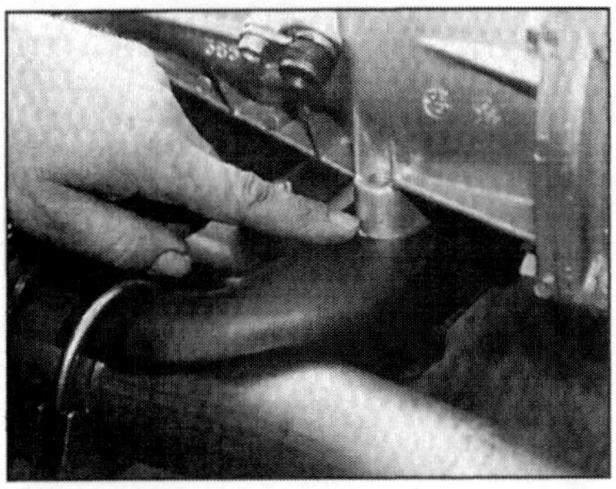

▲ *TR1. Remove the bolts which attach the transmission rear mounting to the bracket. Raise the engine slightly, using a jack and wooden block under the oil pan, until the transmission is lifted off the mounting bracket.*

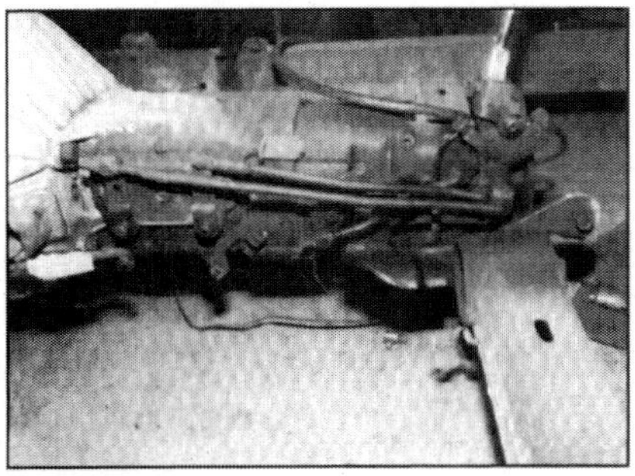

▲ *TR2. Disconnect the shift linkage bracket from the frame and remove the shift linkage mechanism complete with rods. Disconnect the shifter levers at the transmission, and the speedometer cable from the transmission (and the wiring for the TCS where fitted).*

The centre (longer) of the three rods operates 1st and 2nd gears, the upper one 3rd and 4th, and the very short rod at the bottom operates backup.

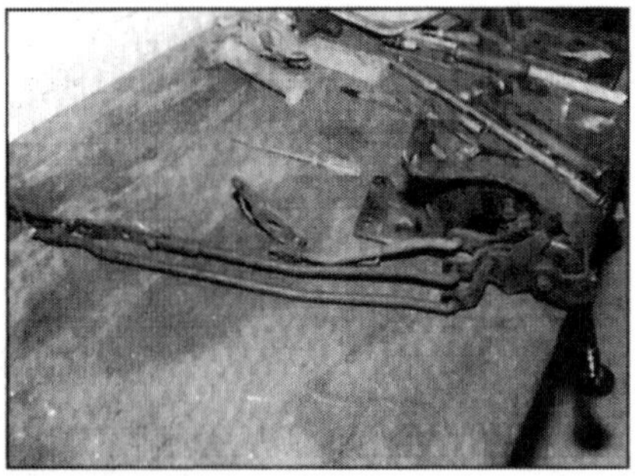

▲ *TR3. The four-speed shifter lever/rod assembly is a rugged piece of equipment, seen here off the car complete with 1st/2nd, 3rd/4th and backup rods which, when removed ...*

▲ *TR4. ... leaves us with the gear shifter and lever assembly. With the engine still supported, remove the rear mounting cross-member mount bracket, the rear mount cushion and the exhaust pipe yoke.*

▲ *TR5. Unscrew and remove the bolts which hold the transmission to the clutch bellhousing, and also the extension bottom bolt. Pull the transmission to the rear whilst turning it clockwise until it is clear of the clutch housing. Your assistant will have to lower the engine until the tachometer drive cable at the distributor just clears the ledge on the firewall. The transmission can now be moved further to the rear, until it can be tipped downwards at its front end and removed completely from the car.*

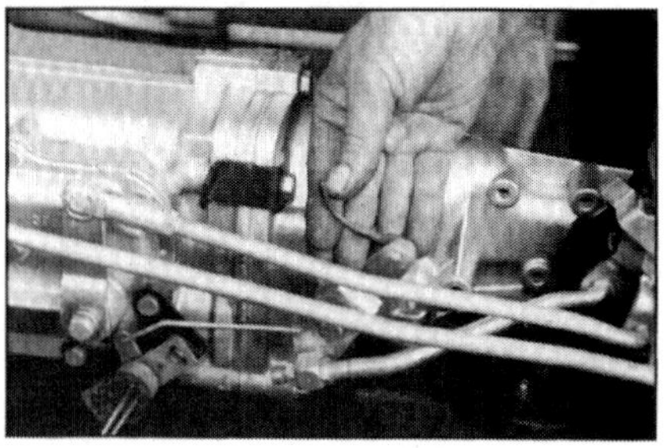

▲ *TR6. This photo shows a close up of the interlock feature which prevented the car being started in backup. A cable is attached to a lever on the side of the transmission and a rod links that in turn to an electrical switch which is wired to the ignition switch.*

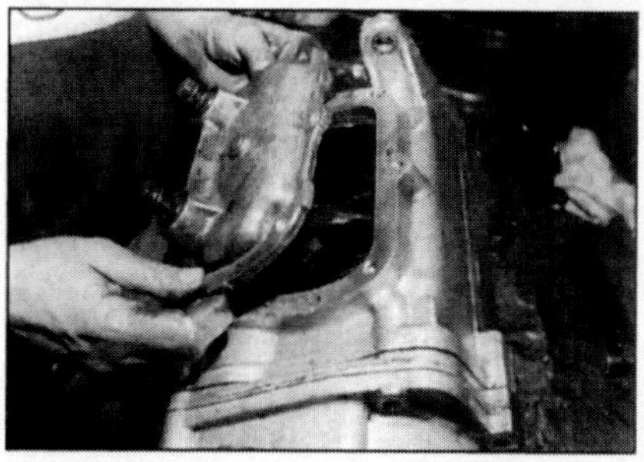

▲ *TR7. Removing the side cover offers access to the shift forks. Remove the two bolts which secure the side cover and allow the oil to drain.*

▲ *TR8. The shift fork assembly can clearly be seen here. At left is the fork for 3rd and 4th gears, and at right 1 st and 2nd gear fork. The backup shifter shaft assembly is hidden from view. At this stage, we put out the task of rebuilding the transmission to a specialist – the innards of any transmission are immensely complex, and while you may think you're saving money by refurbishing your own box, there are many cases where the problems caused by a well-meaning owner have cost him many times the standard charge of a reputable shop. Our advice is simple – unless you are well-schooled in transmission technology, leave well alone!*

▲ *TR9. With the transmission out of the car, we took the opportunity to have it thoroughly cleaned before being sprayed with a high-heat paint. Lovely!*

GEAR LINKAGE ADJUSTMENT

If your Corvette takes some rowing through the gears, there's no need for immediate panic – it may not be a full transmission overhaul on the horizon; maladjustment of the shifter linkage is a common problem. A little effort could have your gear selection as smooth as silk.

▲ *SA1. With the body on the frame (as it will be in most cases where gear shifter adjustment is required) it's difficult to visualize what goes where. This photo of the project car's part-restored '69 model shows exactly how the transmission sits in relation to the frame, engine etc.*

▲ *SA2. Start by unscrewing the gear shift knob and pulling out the spring and backup lockout. Springs have a mind of their own so be careful that it doesn't end up under the seat or somewhere equally inaccessible. Removing the lockout makes it easier to feel the position of the levers. Don't panic if your gear shifter has been rattling. Some heavy grease put on the shaft of the backup lockout before reassembly should do the trick.*

TIGHTENING UP

SAFETY NOTE: Read the safety precautions shown in Appendix 1 before going under the car for the next step.

Using a $\frac{5}{8}$ in. socket, check the transmission bolts for tightness, and also check the bolts on the frame bracket and the shifter bracket using a $\frac{9}{16}$ in. wrench. Two of the bolts are on the rear of the cross-member. The bolts which hold the shifter assembly to the bracket are commonly loose, so check all of them and tighten them up.

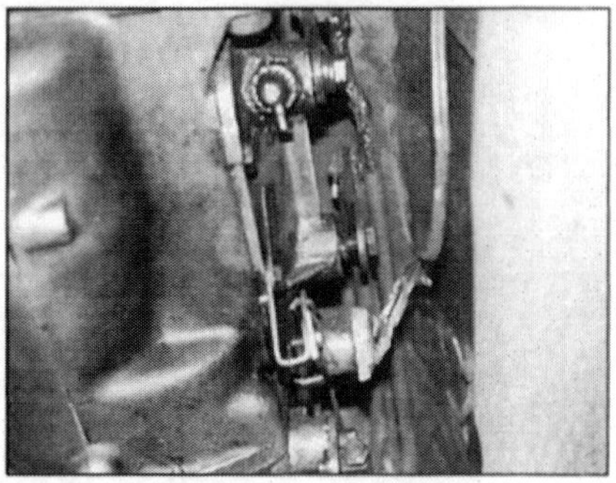

▲ *SA3. The middle lever on the transmission is the 1-2 gear lever. The bolt holding the 1-2 lever in place should have both a lock washer and flat washer beneath its head. You will need to go underneath the car to unscrew this bolt, but note that it does not have to be completely removed. Pull the lever forward at the top until it stops.*

Back inside the car, at the lower end of the gear shift lever, there is a slot or gate, in a metal plate. This gate can't be moved, so you must adjust the shift lever linkages so their slots line up with the gate. Work out where the gate is located by feel, and compare it to the position of the gear shift lever when it is all the way to the left.

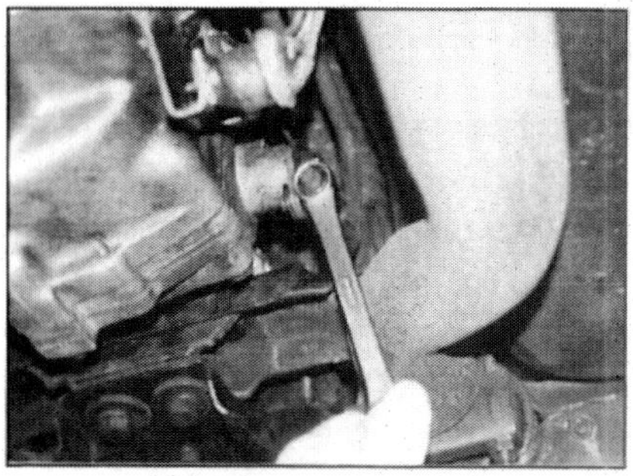

▲ *SA4. You can easily adjust the backup lever by means of two stop nuts on the linkage rod. This is the rearmost lever. Do remember in all this talk of left and right, that when you move the shift lever to the right at the top, it will be at the left underneath the car. If this is hard to figure out, confirm matters by loosening one locknut by one turn, then tighten the other lock nut and check the gear shift lever to see if you have moved it in the right direction.*

Now move the 1-2 lever back to its centre position and make sure the slot in the lever is engaging properly with the shaft, then tighten the bolt. From inside the car again, move the gear shift lever from side to side to ensure that the slot in the first and second gear position lines up with the gate. As we said before, do remember that if you want the gear shift knob to come rearward, you will have to adjust the linkage rod forward at the transmission.

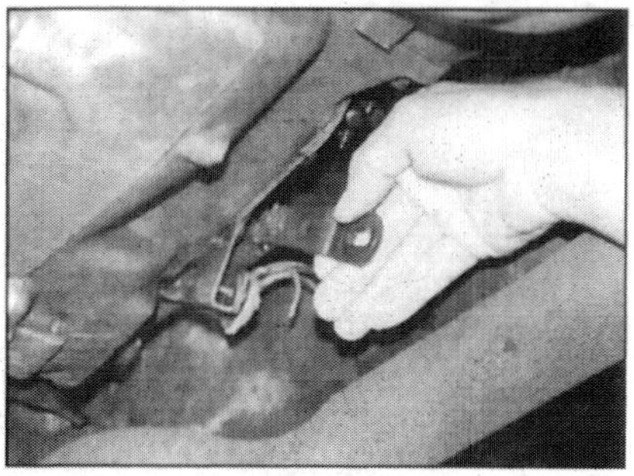

▲ *SA5. To adjust the 1-2 rod, you must unbolt the lever from the transmission and pull it down. Use a wrench to break the locknut loose and then you can turn the lever to screw its clevis on or off the rod. You have to make the adjustment a complete turn at a time so that the lever's offset is always positioned correctly. Try a turn or two, then bolt the lever back to the transmission to see if that was enough – it's trial and error, so expect to have to do this several times. When it comes to making a finer adjustment, you will have to remove the pin and separate the clevis from the lever.*

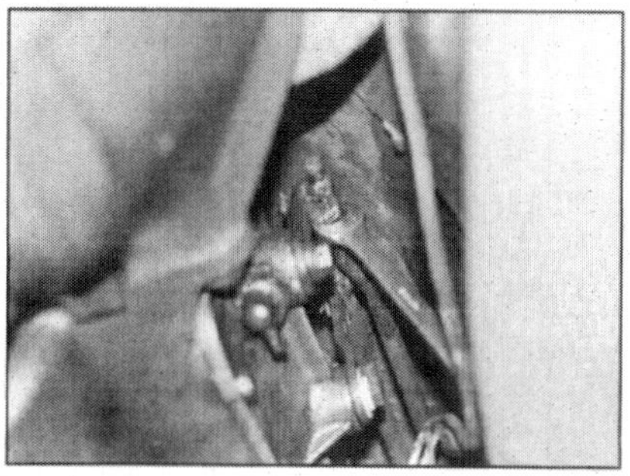

▲ *SA6. You will now have the 1-2 linkage adjusted to your liking. Now move the shift lever all the way to the right, into the 3-4 position. Keeping the gear shifter in the neutral position, i.e. not pushing it forwards or backwards into any gear, move the gear shift lever all the way to the right, in between the 3-4 positions and then shift back to the left, in between the 1-2 position. If the linkage for the 3-4 gears needs to be adjusted to align it with the position of 1-2 slot, this is carried out by turning the two locknuts on the front lever.*

The linkage rods are connected to the levers on both the transmission and the shift lever assembly by special wire spring clips. A cotter pin is used only on the clevis pin on the 1-2 lever. If your transmission has cotter pins holding the levers on, replace them with the clips. If you don't do this, there is a risk of your Corvette getting stuck in gear because a cotter pin has worn through and then fallen out of the shift linkage.

ALTERNATIVE GEAR SHIFTERS

More precise aftermarket gear shifters are available from specialists such as Hurst and others. These operate more precisely because the shifters bolt straight to the transmission. But they do transmit the vibration that the factory gear shifts leave behind, because factory shifters are bolted to the frame and are isolated from the transmission. You trade precise gear changing (shifting) for a gear shifter that moves with the torque of the engine. You may find that the shifter hits the Corvette console, in which case you may have to shim the aftermarket shifter to give it some clearance.

CLUTCH AND FLYWHEEL REMOVAL

WARNING: Dust produced by clutch wear and deposited on clutch components may contain asbestos, which is hazardous to your health. DO NOT blow it out with compressed air and DO NOT inhale it. DO NOT use petroleum-based solvents to remove the dust. Brake system cleaner should be used to flush the dust into a drain pan. After the clutch components are wiped clean with a rag, dispose of the contaminated rags and cleaner in a covered, marked container.

Corvette clutch life is usually something between 15/30,000 miles, depending largely on how sensitive the driver!

Access to clutch components is normally accomplished by removing the transmission, leaving the engine in the vehicle. If the engine is being removed for major overhaul, then check the clutch for wear and replace worn components as necessary. However, the relatively low cost of the clutch components compared to the time and trouble spent gaining access to them warrants their replacement any time either the engine or transmission is removed (unless they are new or in near perfect condition).

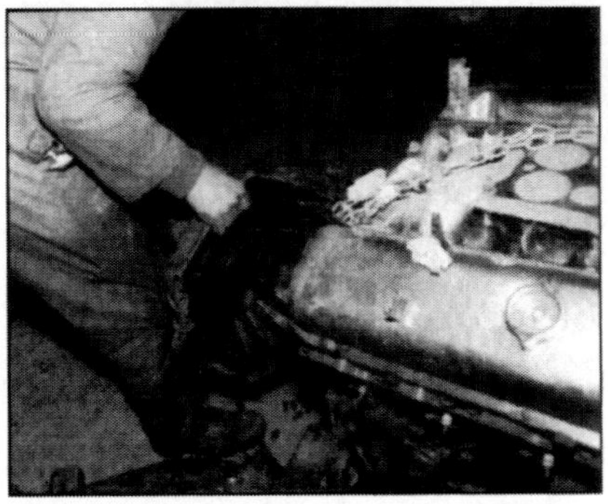

▲ *CR1. Disconnect the clutch fork pushrod and spring and remove the bellhousing-to-engine bolts to detach the housing. It may have to be gently prized off the alignment dowels with a screwdriver. As you can see here, having already removed the body made life much easier from the access point of view.*

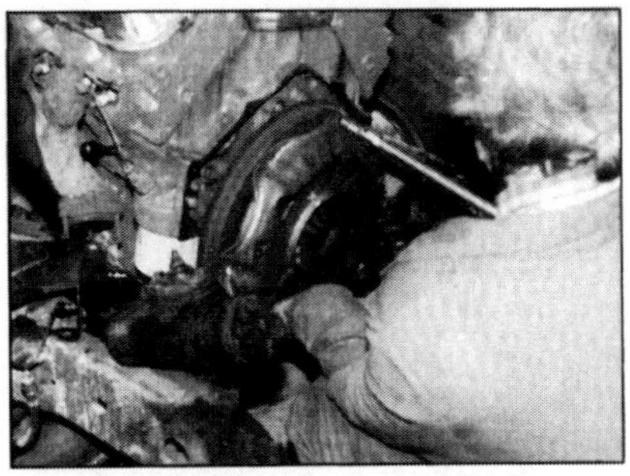

▲ *CR2. Remove the clutch fork and release bearing and support the clutch disc during removal by installing a clutch alignment tool through the clutch disc hub. Carefully inspect the flywheel and pressure plate for indexing marks. The marks are usually an 'X', and 'O' or a white letter. If they cannot be found, create marks yourself so the pressure plate and the flywheel will be in the same alignment when you reassemble.*

Turning each bolt only a quarter-turn at a time, loosen the pressure plate-to-flywheel bolts. Work in a criss-cross pattern until all spring pressure is relieved. Then, hold the pressure plate securely and completely remove the bolts followed by ...

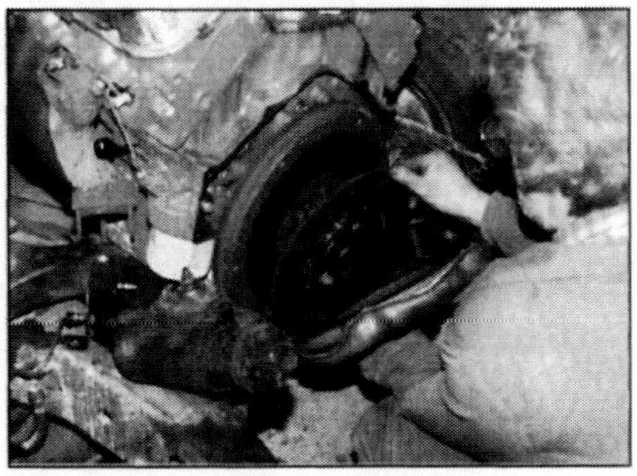

▲ *CR3. ... the pressure plate and the clutch disc. The lining on the clutch disc should be at least 1/16 in. above the rivet heads and there should be no sign of loose rivets, broken springs and other obvious damage. However, as already mentioned, these clutch components (including the release bearing) are sufficiently cheap to make their replacement a matter of course unless they are verging on the perfect.*

▲ *CR4. The flywheel bolts are, of necessity, extremely tight and it takes no small amount of effort to remove them. You'll need a suitably sized socket and a ½ in. drive breaker bar to remove them. If you have access to a pneumatic wrench, you can save some sweat, as Em is doing in this photo.*

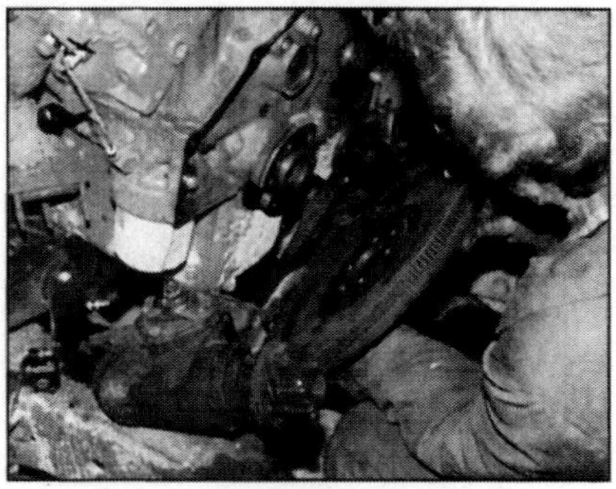

▲ *CR5. The flywheel is a very heavy piece of steel, so it is important to support it adequately, particularly when removing the final bolts.*

▲ *CR6. Having got the flywheel off, clean it of dirt and grease and make a thorough inspection – you're looking for cracks, rivet grooves, burned areas and score marks. Any light scoring can be removed using emery cloth; deeper imperfections can usually be sorted by a good machine shop, where the surface can be machined smooth and flat. It's a good idea to do this anyhow, having got this far. The ring gear teeth around its edge should be checked to ensure that none is cracked or broken.*

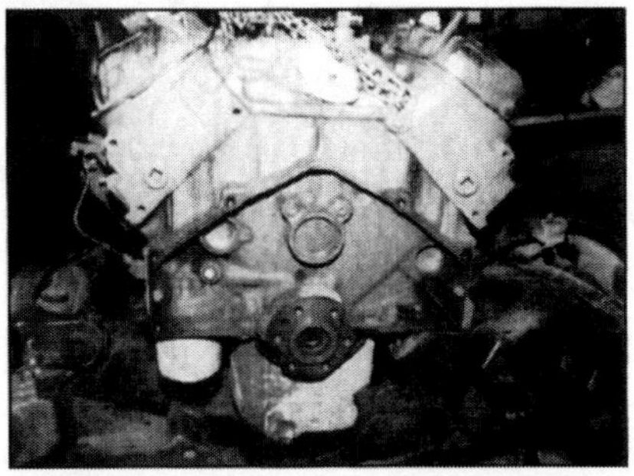

▲ *CR7. And that leaves us with just the engine, ready and waiting for whatever work is required. In this case, Em completed a complete tear-down and rebuild.*

▲ *CR8. Installation is a reversal of the dismantling procedure, but you should note that the all-important rule is that no oil or grease should come into contact with the lining of the clutch disc. Most clutch plates will be marked 'flywheel side' or similar – if you have one that isn't, it should be installed with the damper springs towards the transmission.*

At first, the pressure-plate-to-flywheel bolts should be fitted only finger tight, prior to the use of the alignment tool. Then the bolts can be tightened further, a little at a time, working in a criss-cross pattern to avoid distorting the cover. When the bolts are snug, they should be tightened to the specified torque.

DIFFERENTIAL AND AXLE SHAFTS

▲ *GD1. The driveshaft is a tube running between the transmission and the rear differential. Universal joints are located at either end of the driveshaft and permit power to be transmitted to the rear wheels. These allow for slight amounts of engine and transmission movement during normal operation.*

Normally it's hard work to actually get at these parts of your Corvette. However, this car was minus its body, so life was much easier!

▲ *GD2. The driveshaft features a splined yoke at the front, which slips into the extension housing of the transmission. This arrangement allows the driveshaft to slide back and forth within the transmission as the vehicle is in operation.*

An oil seal is used to prevent leakage of fluid at this point and to keep dirt and contaminants from entering the transmission. If leakage is evident at the front of the driveshaft, replace the seal according to details shown in your Haynes manual.

The driveshaft assembly requires very little in the way of service – basically, it either works or it doesn't! The universal joints are lubricated for life, and if problems occur they must be replaced (a procedure for which the driveshaft must be removed from the car).

The needle roller bearing universal joints in the rear driveshafts transmit all the power from the engine to the wheels and also form an essential link in the rear suspension. As such, they are more prone to wear and cause problems.

The differential is mounted directly to the chassis via a rubber isolated cross-member. Note the rust-prone gusset reinforcements to the chassis which were introduced in 1969.

THE DIFFERENTIAL

▲ *GD3. The 1968/79 models featured independent rear suspension with the differential mounted directly to the chassis via a rubber isolated cross-member. Driveshafts form the upper link of the suspension parallelogram, rubber bushes in lower link lead to poor location during hard cornering.*

POSSIBLE DIFFERENTIAL PROBLEMS

Differential problems are few, most occurring with the Positraction limited-slip type. These were optional until 1970. When the differential is hot after a good long run, a violent banging noise may be felt throughout the car when travelling at low speeds (2–5 mph) or when taking slow speed corners. It is usually caused by the Positraction plates binding rather than sliding smoothly over each other. It can often be solved by carrying out two or three oil changes (carefully, as it has to be done when the oil is hot), adding at the same time the GM additive, Part No. 1052358.

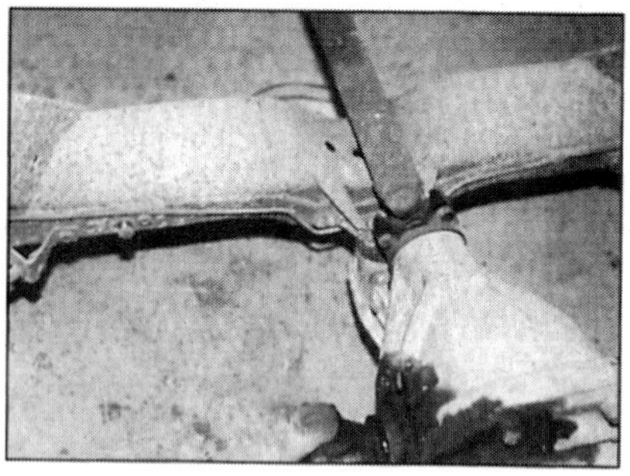

▲ *GD4. Before you start dismantling, it's advisable to use a scriber (or white paint) to mark where the driveshaft and differential flanges line up. That way, you can reassemble it the same way and prevent any possible problems caused by*

imbalance. The same applies to the two axle shafts. In this case, Em was replacing all the shafts and universal joints, so the exercise was unnecessary.

He removed the universal joint bolts and straps and taped the bearing caps to the spider during removal to prevent them dropping off. He lowered the rear of the driveshaft and then slid the front out of the transmission. To prevent fluid loss and the ingress of foreign bodies, Em placed a plastic bag over the differential and sealed it with an elastic band. Hardly hi-tech, but it works!

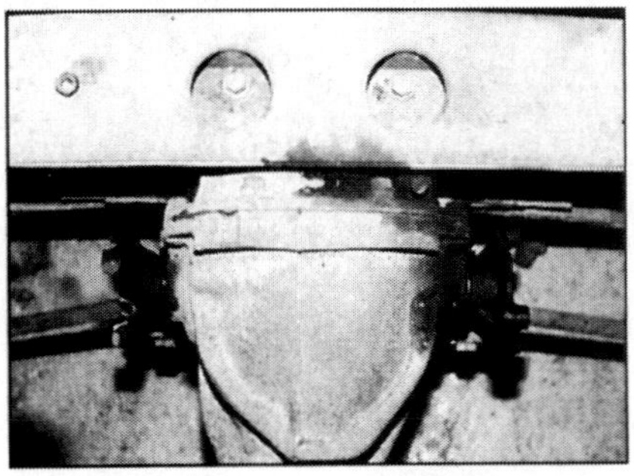

▲ GD5. The inner ends of the two axle shafts were removed in the same manner. You'll need to turn the axle shafts or wheels in order to get to the bolts – don't forget the safety rules when working under your car (see Safety notes in Appendix 1). It is necessary to bend down the four locking tabs to remove the four bolts securing the outer shaft flange to the spindle drive flange. Loosening the camber adjusting cam and rotating it so that the high point of the cam points in, the control arm will be pushed out to give some clearance between the wheel spindle and the driveshaft. Ease the axle shaft out of the outer drive flange and then remove it by withdrawing the outer end first.

REMOVING THE DIFFERENTIAL CARRIER AND COVER

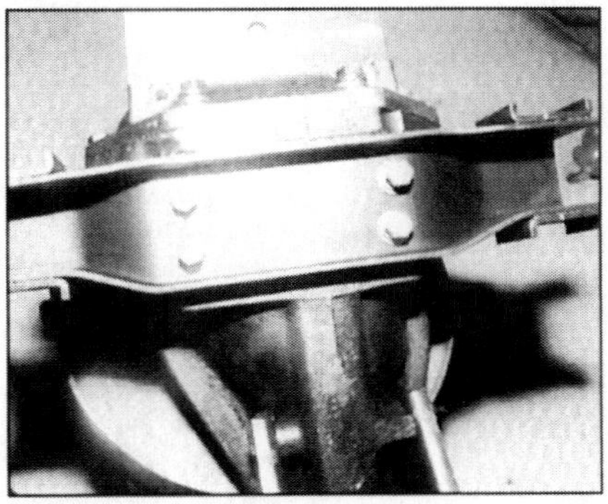

▲ GD6. The strut rod bracket has to be removed from the underside of the differential carrier. This can be left to 'hang' from the strut rods. Then the four bolts securing the rear spring to the differential carrier have to be removed. At this point you can remove either just the carrier, or the cover as well. If it's the former, splitting them will obviously allow the oil to drain out, so you'll have to be ready with an old can to prevent it messing up your garage floor.

▲ GD7. In this case Em was removing both items as an assembly. The bolt from the differential carrier front mounting bracket was removed along with its rubber bushing. The latter should be replaced as a matter of course.

◀ GD8. As with just about every other rubber bushing on the Corvette, there is a polyurethane replacement alternative. (Courtesy Vette Brakes and Products)

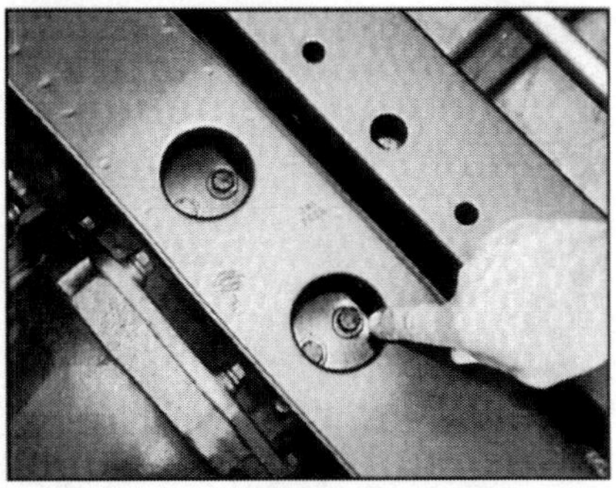

▲ GD9. The assembly has to be well supported. As well as Em's faithful helper, there's a trolley jack under there somewhere! The top of the diff cover is secured by two bolts accessed through holes in the suspension cross-member.

▲ GD10. This is no lightweight assembly and so it needs supporting correctly at this point. An extra pair of hands and trolley jack are in evidence.

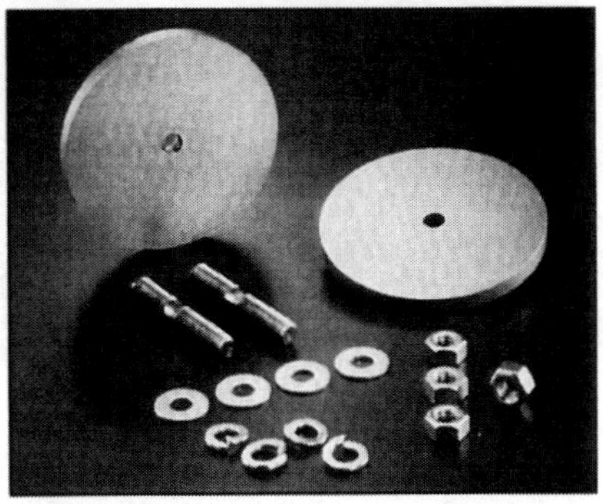

▲ GD11. When replacing the carrier, it's worth considering uprating by fitting a reinforcing kit. This kit reinforces the rubber isolators and securely retains the cross-member in the car. (Courtesy Vette Brakes and Products)

▲ *GD12. This differential was sent out to a specialist shop to be rebuilt. Here it is ready to be reassembled on the bench. Note the new diff carrier/cover gasket waiting for its new home.*

▲ *GD13. It's important that the two pieces of the differential casing are torqued up to the correct figure.*

▲ *GD14. The pristine differential carrier with the strut rod bracket and front support already fitted; it's easier on the bench than upside down on the car, even if the body has been removed!*

SOLVING DRIVESHAFT PROBLEMS

The driveshaft is a balanced unit and so, if you're replacing an original shaft, it should be thoroughly cleaned of mud and other dirt. Driveshaft problems usually reveal themselves in the form of noise or vibration while driving the car. Make a road test to verify the source of the noise.

Choose a quiet road and note the engine speed at which the problem occurs. Then run the car at the same engine speed in the other gears in the transmission. If the noise or vibration occurs at the same engine speed, regardless of the gear selected, then the driveshaft is not at fault, as the speed of the driveshaft varies in each gear.

However, if the noise of vibration decreased or disappeared altogether, it would indicate a possible driveshaft problem, such as material on the shaft which would affect the balance, missing balance weights or damaged universal joints. To check for the latter, drive again on a clear stretch of road in a high gear and then allow the vehicle to coast. Then apply the gas pedal sharply – a knocking or clunking noise points to a UJ problem.

DRIVESHAFT UNIVERSAL JOINT REPLACEMENT

▲ *GD15. Using a pair of long nose pliers (or better still, purpose-made snap ring pliers), remove the snap rings from the spider. Support the driveshaft and place in an arbor press or on a workbench equipped with a large and capable vise. Place a piece of pipe or a large socket with the same inside diameter over one of the bearing caps. Position a socket which is of slightly smaller diameter than the cap on the opposite bearing cap and use the vise or press to force the cap out (inside the pipe or large socket), stopping just before it comes completely out of the yoke. Use the vise or large pliers to work the cap the rest of the way out.*

Transfer the sockets to the other side and press the opposite bearing out in the same manner. Pack the new UJ bearings with grease, then position the spider in the yoke and partially install one bearing cap in the yoke. If the replacement spider is equipped with a grease fitting, be sure it's offset in the proper direction (i.e. toward the driveshaft).

▲ *GD16. Start the spider into the bearing cap and then partially install the other cap. Align the spider and press the bearing caps into position, being careful not to damage the dust seals, using the jaws of the vise.*

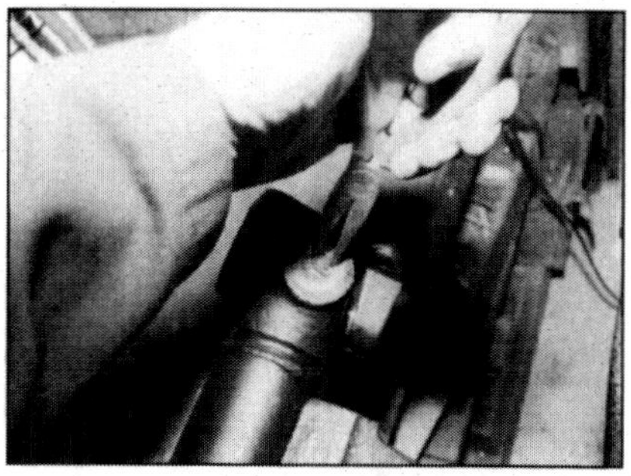

▲ *GD17. Install the snap rings. If difficulty is encountered in seating the snap rings, strike the driveshaft yoke sharply with a hammer. This will spring the yoke ears slightly and allow the snap rings to seat in the groove.*

The replacement procedure for the drive axle U-joints is very similar. The bearings should be packed with high melting point grease, and the lubricant reservoir at the end of each trunnion should be completely filled before assembly. Make sure that new dust seals are fitted to the trunnion with the seal cavity towards the end of the trunnion. Press the seal into its recess with a piece of tubing.

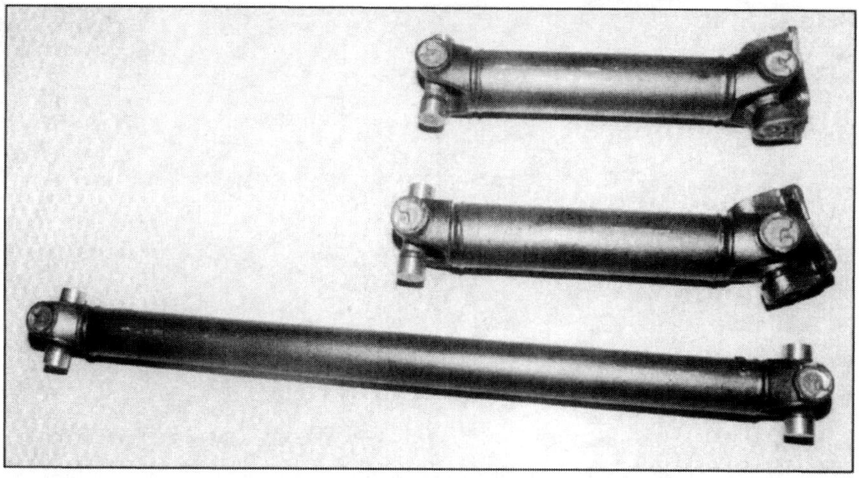

▶ *GD18. Here you can see the results of Em's labours – axle shafts and driveshaft, all fully restored with new parts, ready for reassembly on the car.*

PART 3 STEERING

SAFETY: Remember the safety rules when working on the suspension of your Corvette. NEVER go under the car when it is supported only by a jack – even if it's a heavy duty trolley jack.

Use jackstands and chock and block the wheels remaining on the ground. Wherever possible, leave the car in gear with the park brake on. Take care to protect your hands and your eyes – there's bound to be plenty of loose bits of dirt and rust etc.

▲ *SD2. The Pitman arm is connected to one end of the relay rod.*

▲ *SD3. The idler arm is fastened to the frame side rail, opposite the steering gear, and is connected to the outer end of the relay rod.*

▲ *SD1. This photo shows the basic layout of steering gear on all 'Vettes up to the 1984 model year. It is of the recirculating ball type with worm and sector gear.*

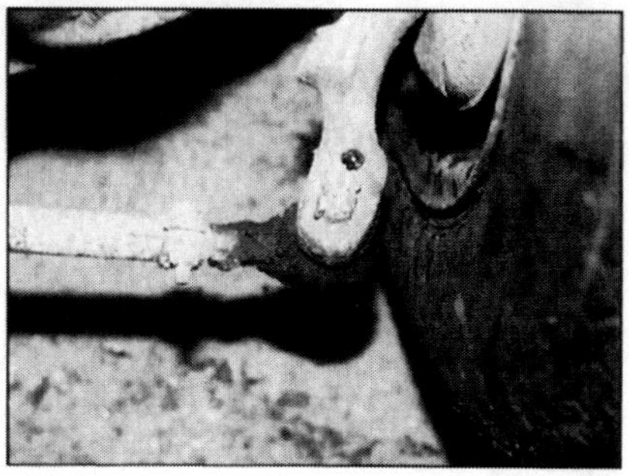

▲ *SD4. The relay rod is connected to the steering arms via two adjustable tie rods. As you can see here, there's plenty of work to be done even to get to the point where you can see the actual fasteners! Take your time and apply plenty of penetrating fluid.*

Where power steering is fitted you'll find a pump and a power piston operating through the relay rod of the steering linkage.

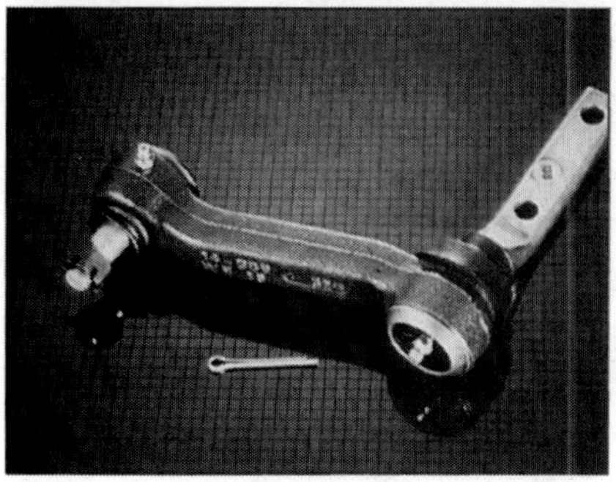

▲ *SD5. Having taken the trouble to remove worn steering components, it's not a bad idea to look at fitting an uprated replacement. A better sway bar and polyurethane bushings are common, and these are dealt with in other sections in this chapter.*

This heavy-duty idler arm helps prevent tire wear, absorbs road shock and reduces road wandering. It's designed to last longer by incorporating spring-loaded bearings and grease fittings on both bearings. (Courtesy Vette Brakes and Products)

▲ *SD6. Having got the original parts free, you'll need to really clean them up to see what their condition is under all that accumulated road dirt and grease. Any sign of serious deterioration should lead to them being thrown straight in the bin – like brakes, steering is something that needs to be totally right.*

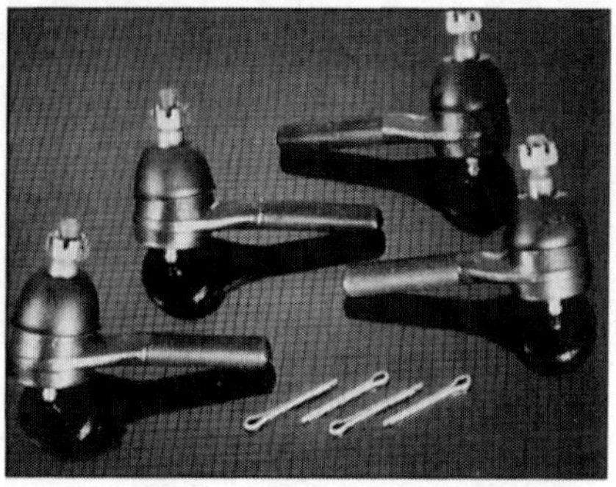

▲ *SD7. Excess up and down motion in a tie rod means one thing – replacement. Non-standard variants, as here, offer a unique dual bearing design which slows down overall wear rate.* (Courtesy Vette Brakes and Products)

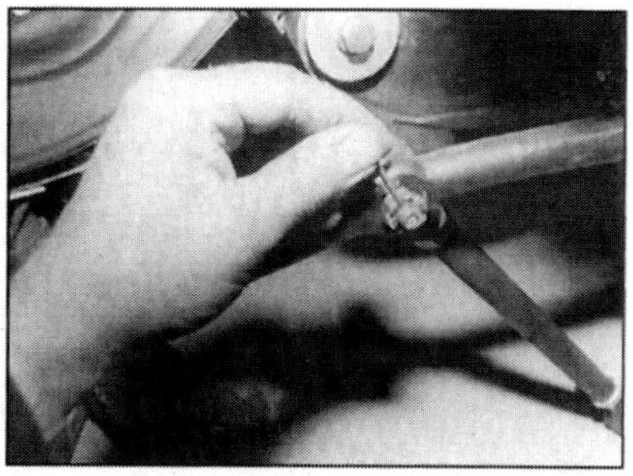

▲ *SD8. When you're working on your steering set-up, you can't fail to notice the number of castellated nuts that are used. You should also know that they are used wherever there is a need for a cotter pin. The moral is simple – if a cotter pin should be fitted, fit one; and always fit a new one. They cost little and could literally save your life.*

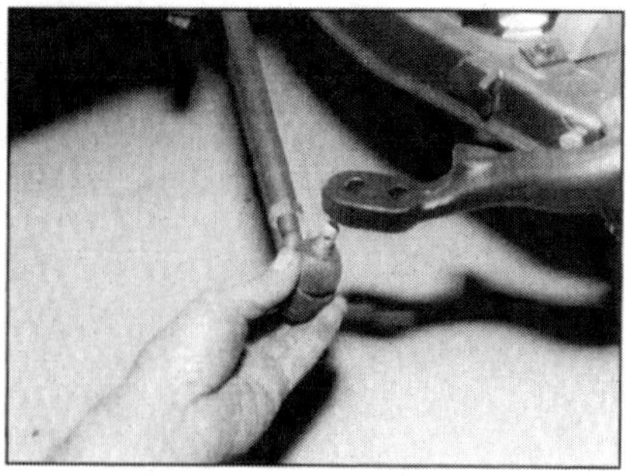

▲ *SD9. Fitting new tie rod end ball joints is easy – they're nice and clean for a start! However, removing them is likely to be more fraught, and a ball joint splitter is usually considered essential.*

HINTS AND TIPS FOR WORKING ON STEERING AND SUSPENSION

Frequently, when working on the suspension or steering system components, you may come across fasteners which seem impossible to loosen. These fasteners on the underside of the vehicle are continually subjected to water, road grime and mud and can become rusted and frozen, making them extremely difficult to remove.

To unscrew these stubborn fasteners without damaging them (or other components), be sure to use lots of penetrating oil and allow it to soak in for a while. Using a wire brush to clean exposed threads will also ease removal of the nut or bolt and prevent damage to the threads. Sometimes a sharp blow with a copper mallet and punch will break the bond between a nut and bolt threads, but care must be taken to prevent the punch from slipping off the fastener and ruining the threads. Heating the stuck fastener and surrounding area with a torch sometimes helps, too, but isn't recommended because of the obvious dangers associated with fire. Long breaker bars and extension or 'cheater' pipes will increase leverage, but never use an extension pipe on a ratchet – the ratchet mechanism may be damaged. Sometimes, tightening the nut or bolt first will help to break it loose. Fasteners that require drastic measures to remove should *always* be replaced with new ones.

REPLACING STEERING FLEXIBLE COUPLING (RAG JOINT)

PLAY IN STEERING

A Corvette will lose its razor-sharp steering over a period of years, and the reason why can come from a number of sources. For one thing, there are many ball joints in the system, and a little wear in each adds up to plenty at the steering wheel. Another reason will be soft suspension bushings, while steering box wear, power steering valve deterioration and even a softening rag joint will all add play into the steering. Replacement of many of these items is a regular service job, so ball joints, for instance, are not covered here. But before you go tearing into the steering box or assuming that you need to replace every power steering component (highly unlikely!), make a big effort to determine exactly where steering play is coming from. Unfortunately, this can require some experience of the car and unless you are fully experienced yourself, this is where the advice of a good specialist comes in. Either you will have the experience to assess where the wear has taken place for yourself, or you will be able to call upon the advice of an experienced mechanic to come to you and look over your car with you – perhaps the ideal way because then you will learn for yourself what is going on – or you will have to take you car to a specialist or GM dealership to have the system checked over for you.

The flexible coupling joining the steering column to the steering box is commonly known as the rag joint. Over a period of time the rag joint becomes mushy and introduces play into the steering where you least expect it. Some folk never think of replacing the steering rag joint because it doesn't look as if it's a component that can wear. But if you have still got sloppy steering, even though the ball joints, the A-frame bushings and the steering box are all unworn or correctly adjusted, look hard at the rag joint. A visual inspection can reveal most problems. If the material is swollen and spongy, it is time to replace the joint, and if you can see where bolts have been sliding back and forth on the rag joint, there is too much play.

SAFETY: Rag joints are reinforced with wire, and old joints can have exposed pieces of sharp wire sticking out from them. Take care when removing.

If you have to disconnect the steering column from the dash panel, do not allow it to hang down, or distortion to the column will result. Be sure the column is supported when the instrument panel mounting bracket is removed.

Removal of the rag joint itself is straightforward enough. If you are lucky and the rag joint is not riveted to one clamp, the new rag joint material can be slipped into place. But if the material is riveted to the clamp, the whole rag joint must be changed as an assembly, as described in the GM manual. In the latter case you have to drop the steering column and pull it back an inch or two.

▲ *RJ1. Remove the clamp bolt from each end of the coupling using a $\frac{7}{16}$ in. socket. You will need two different wrenches to remove the two through bolts – usually one requires $\frac{1}{2}$ in. and the other $\frac{9}{16}$ in. Sometimes the lock washers prevent the bolt from tapping out, in which case you will have to prize them off with a screwdriver.*

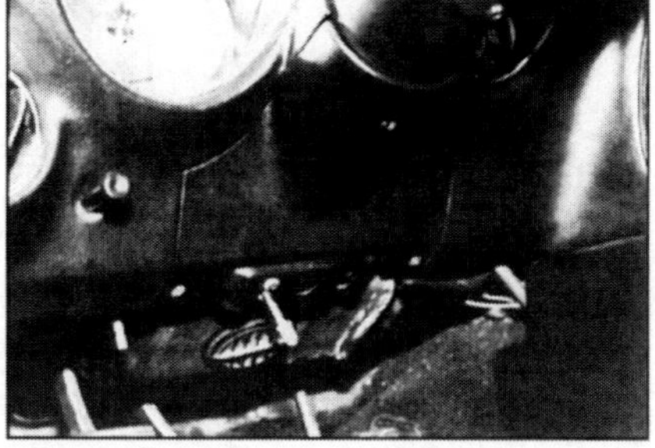

▲ *RJ2. In order to get at the steering column mounting bolts, you must first remove the three Phillips head screws holding the lower trim panel in place.*

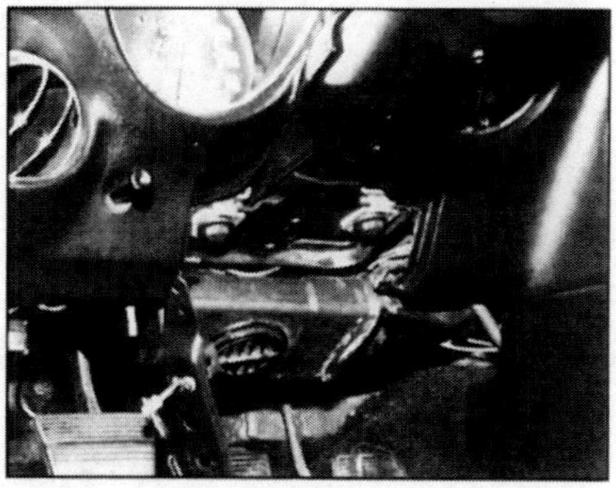

▲ *RJ3. You can now see the two column mounting bolts behind where the trim cover used to be. Do not disconnect any of the electrical connections, but do note what was said above about supporting the weight of the column, otherwise the steering column will be distorted. With the two bolts removed, the column will be free to be moved back as you disconnect the rag joint.*

Depending on whether both the clamps are going to be removed or not, will determine which components you have to mark up to carry out the following job. It is essential that the relationship between the steering box shaft and the steering shaft is maintained, otherwise the bolt won't go through when fitting the replacement rag joint, and the steering wheel may be off centre.

Tap an old screwdriver or chisel blade into the slot of the clamp so that you spread it a little. This will help greatly when it comes to sliding it up or off the shaft. If the coupling has to come off the spline shaft, use a soft hammer to tap it along and off.

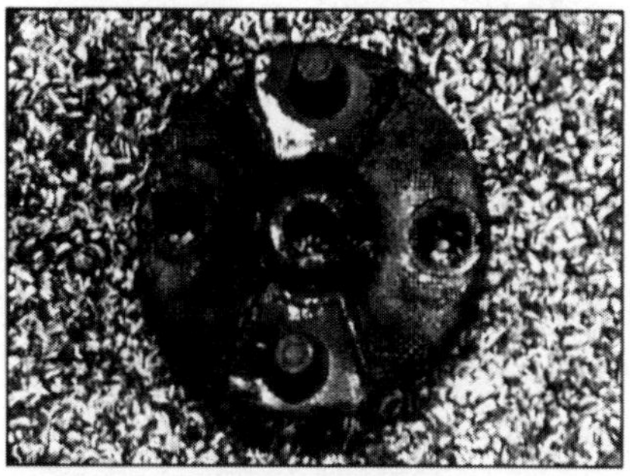

▲ *RJ4. If the old rag joint has elongated holes, as in this shot, it indicates that the bolts were moving back and forth, producing a lot of play at the steering wheel.*

▲ *RJ5. The tabs shown here fitted to this replacement rag joint were used by GM for packing purposes and should be prized off. Check carefully that your replacement joint is correct and matches the old one before going any further!*

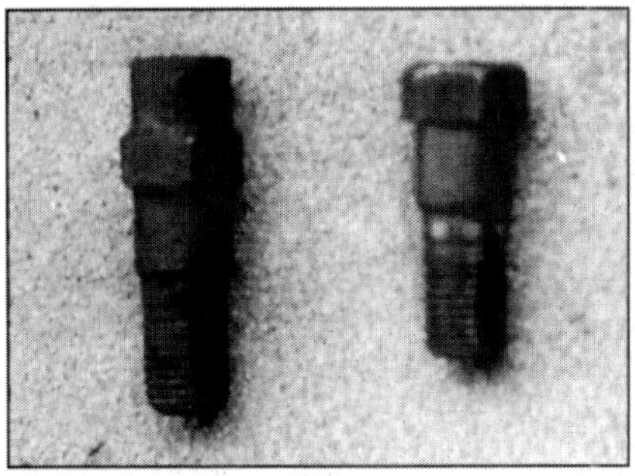

▲ *RJ6. Sometimes the bolt that comes with the replacement coupling and the original bolt are not the same. If all is well, the thread, shank and the shoulder will all be identical but the appearance will be different. If you are a stickler for originality, and the original bolt is not damaged in any way, you may want to clean and reuse it.*

▲ *RJ7. In order to get the new rag joint on, slide the new coupling onto the spline shaft using the alignment marks you made earlier. Be sure that the bolt goes smoothly through the clamp (along the shaft) and that it screws into the threads on the other side of the clamp. If any of the threads, on the bolt or in the clamp, were rusted, clean them with an appropriate die and tap or, if you don't possess them, use a sacrificial bolt cut down the length of its thread with a hacksaw – this should clean out any internal threads if you screw it in and out a few times with penetrating fluid – and ensure that the threads on your bolt are clear by running a correct size nut up and down it, again with the help of releasing fluid to clean out the threads. This won't work with threads that are damaged and you will need to resort to a tap and die.*

If any copper electrical connectors or tin head shields were fitted to the coupling originally, reconnect them. Make absolutely certain that you use the correct, original size bolts with the correct length of shank and with the washers in place, otherwise the flexible may not be properly compressed.

SAFETY: Ensure that the coupling is installed far enough for the clamp bolt to pass through the undercut on the steering box shaft.

Make certain when refitting the steering column to the dash that no wires are crimped between the steering column and the column mounting. If they are, components could stop working or, worse still, an immediate fire inside your car could be the result. On the other side of the fire wall, the backup lockout plate can be adjusted up or down if you need to carry out any fine tuning.

All of the coupling bolts must be properly fitted, and properly torqued up. If you have any queries as to the correct torque used, take the car to your GM dealer and have them torque the bolts up correctly. Similarly, if you have any queries about the correct alignment of your steering column or steering shafts – these should go back together properly, but who's to say that a problem hasn't been induced in the past, perhaps by a former owner – once again consult your GM dealership and have them check the alignments for you.

REFURBISHING POWER STEERING

This section shows you how to replace the power steering hoses and cylinder and also the power steering valve – this is because play in the steering is often caused by the valve, not by wear in the steering box.

SPECIAL NOTE: Many GM manuals call for gear oil instead of grease. The advantage of grease is that it won't leak out, and it provides better lubrication. Unless you drive your Corvette in cold weather, you're better off using grease.

▼ *PS1. It pays to orientate yourself with the job in hand, and diagrams are often clearer, especially when most of the components are covered in years of oil and road dirt! This is a typical power steering set-up.* (Courtesy Vette Brakes & Products Inc.)

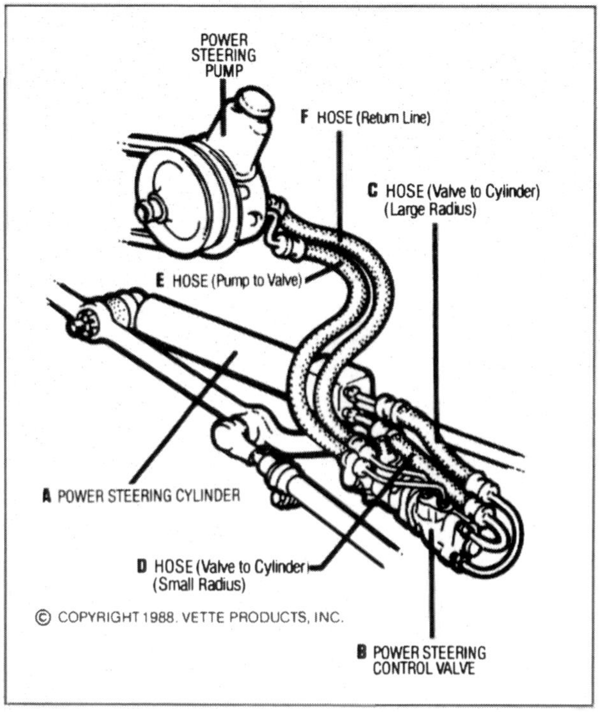

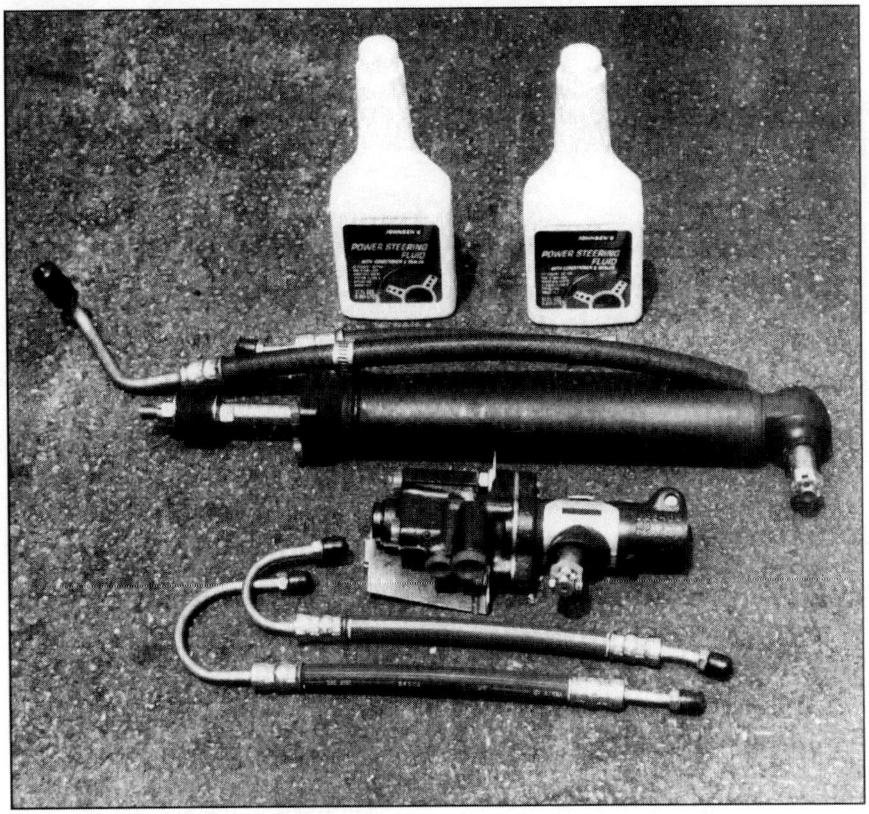

▲ *PS2. You are strongly recommended to replace all of the hoses in the power steering system, the valve and the cylinder at the same time. The rubber hoses deteriorate first on the inside (where you can't see them!) and the debris from the insides of the walls then causes problems with the valve and seals.*

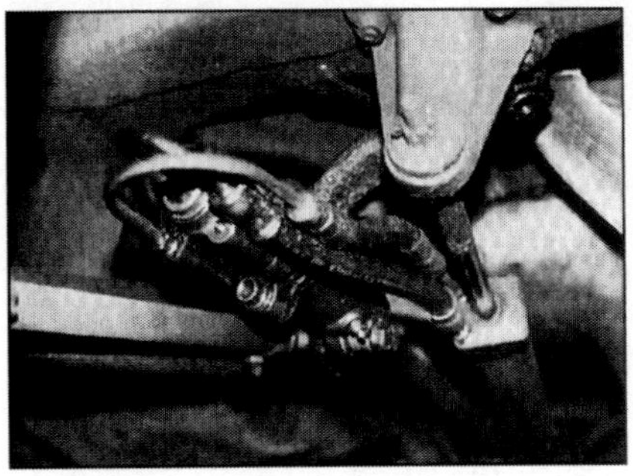

▲ *PS3. Check for fluid leaks by (obviously!) looking for fluid, but also see if the valve looks wetter and cleaner than the surrounding components. If so, fluid has been washing away the dirt. You'll also have noticed your power steering fluid level dropping. Carefully note the routing of the hoses before removing them – take a photograph beforehand to be sure. Their positioning won't be obvious when you come to install the new ones, especially since the cylinder hoses cross over each other at the valve.*

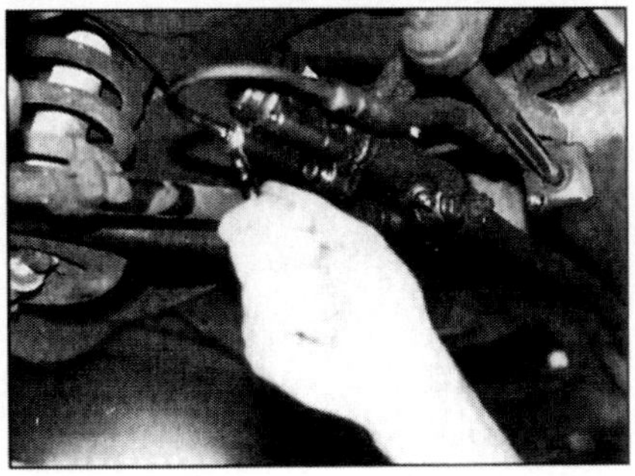

▲ *PS4. The hoses are best broken loose using flare nut wrenches so as to bear on the maximum number of flats. You'll need a $\frac{7}{16}$ in. wrench for both ends of the cylinder hoses and a $\frac{5}{8}$ in. wrench for the pump return line. Don't forget that if the fittings are stubborn and if you intend replacing the hoses in any case, you can saw through the hoses with a hacksaw and use a regular ring or socket wrench to break the hoses free.*

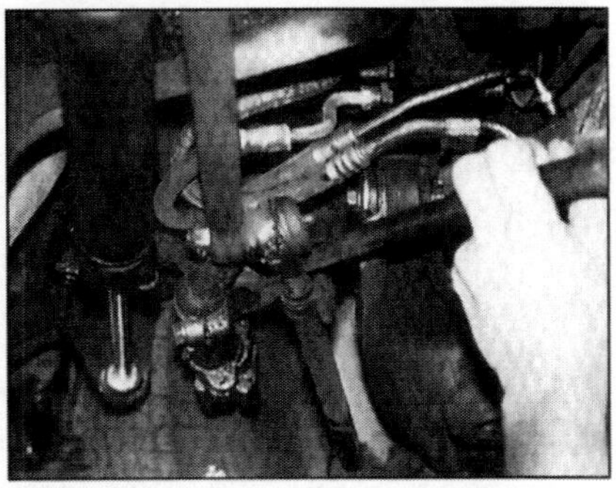

▲ *PS5. After removing the cotter pin you can unscrew the nut using a $\frac{3}{4}$ in. wrench. Screw it nearly all the way off the stud on top of the steering valve but not all the way. This is so that as the valve is popped off the Pitman arm, the valve and drag link won't fall on you. Use a tie-rod end splitter – one of the narrow-jawed type – to pop the valve loose.*

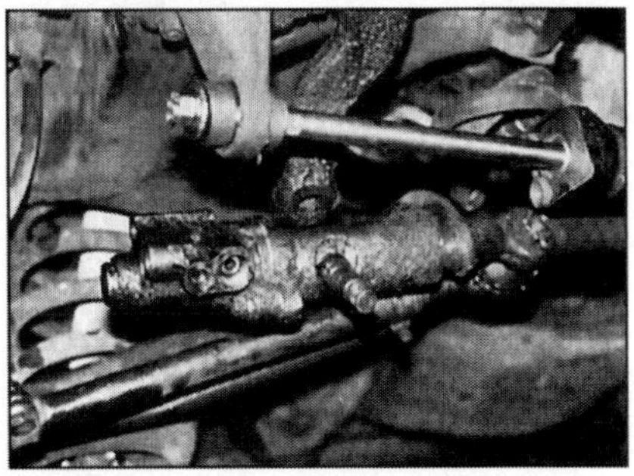

▲ *PS6. Using a $\frac{9}{16}$ in. wrench, remove the locking bolt and unscrew the valve from the drag link. You'll need to ensure that the valve stud will clear the surrounding suspension parts as it is unscrewed: pull it down and back.*

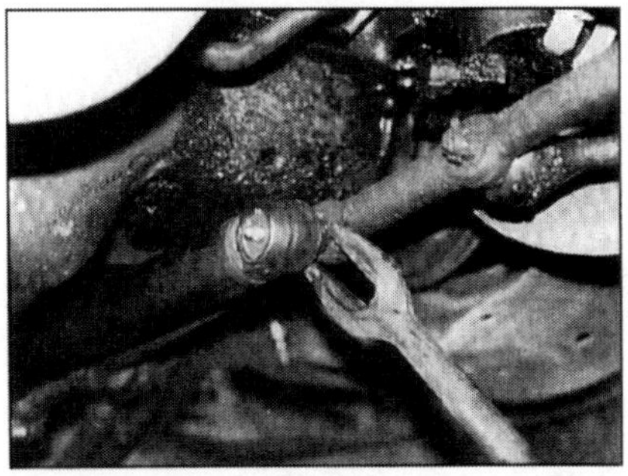

▲ *PS7. After removing the cotter pin, you can unscrew the nut part of the way and split the cylinder from the drag link.*

SAFETY: The cylinder will squirt fluid out as it is moved. Take care to keep the fluid out of your eyes, and wear rubber gloves or barrier cream to keep it off your hands.

Drain the fluid slowly from the cylinder, pumping the shaft all the way in and out. The other end of the cylinder is easily disconnected by removing the cotter pin and nut. Look carefully at the shaft and you'll see two flats, so you can grip it with a 7/16 in. open-ended wrench or adjustable wrench – but not vise grips!

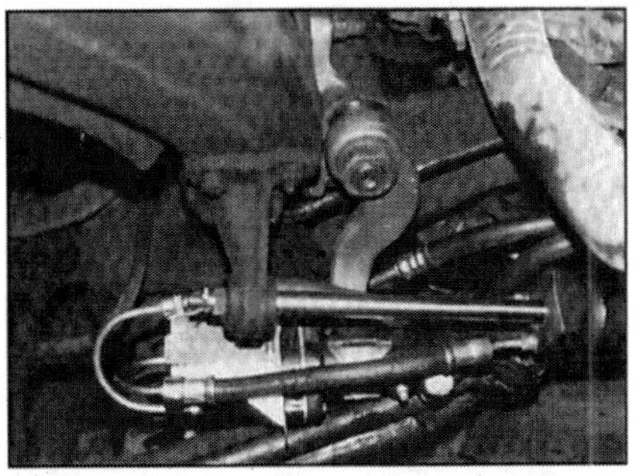

▲ *PS8. After you have bolted on the new parts, have an assistant carefully turn the wheels all the way to the left and then all the way to the right, and watch the movement of all four hoses. It is important that you ensure that hoses move freely without pulling or snagging, especially at the ends of the steering travel. You must also ensure that the hoses don't rub on any components and, in particular, the frame bracket.*

Remember to grease the fittings on the valve and on the cylinder.

To top-up the power steering fluid you'll need to have the car out of doors and the front of the car raised off the ground on axle stands so that the steering can be rotated fully in both directions without snagging. The car must have its park brake on and be in neutral (manual transmission) or park (automatic transmission). Fill the pump and start your engine, running it for no more than 20 seconds. Now shut off and top up the steering pump again. After doing this another couple of times, start the engine and slowly turn the steering wheel all the way to the right and then slowly all the way to the left, slowly working the steering wheel back and forth in both directions. Every minute or two, stop and top-up the fluid once again. When the system is full, it will contain two pints.

STEERING BOX OVERHAUL

As we said earlier, before overhauling the steering and, in particular, any of the major components, you must check that the easier-to-replace items are not worn first. The steering box is a good case in point. Your specialist should be able to adjust the steering play (lash) and, in most cases, this will be all that is needed unless the steering box is severely worn. (See *SB9* for where the play (lash) adjuster screw and locknut are situated. You'll need a specialist service job here, or carry out the advice in the GM manual.)

Also note that after having stripped off the steering box and reassembled it, you must take the car to a recognized specialist or GM dealer to have the steering assembly properly realigned.

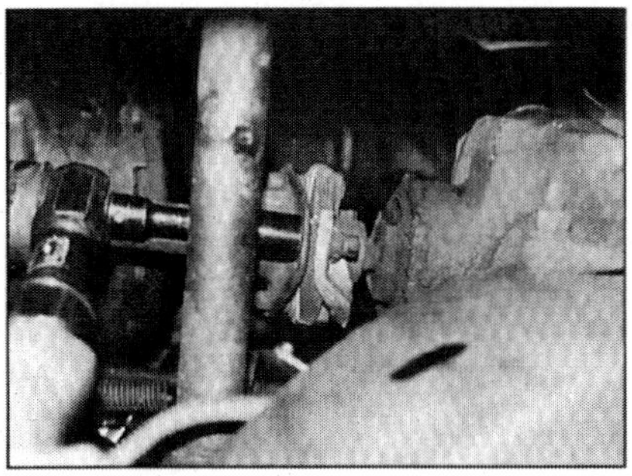

▲ *SB1. The steering box column is fitted to the main steering shaft at the rag joint. See earlier section for information on how to remove.*

▲ *SB2. The steering box is held to the frame with three nuts. You'll need to lock the mounting bolts at the rear to prevent them from turning. On automatic transmission cars removal is easier because they don't have the clutch cross-shaft getting in the way. Disconnecting the exhaust flange is all the extra dismantling required on small-block Corvettes: there's less room and therefore more dismantling on big block cars.*

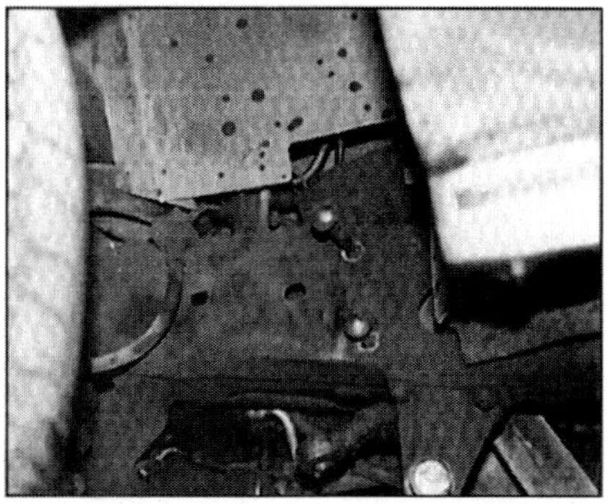

▲ *SB3. Here you can see the stove bolts – or at least two of them – with their rounded heads easily visible in the wheel well. The third is hidden behind the inner skirt. Take this opportunity to pull out the bolts, clean up the threads and run the mounting nuts onto the bolts – they are fine threaded items – to keep them safe. Lubricate the threads and paint the heads, as required.*

◀ *SB4. From the early 1970s on, the shaft had a flat side so the flange could only be installed in one position. Make sure you order the right parts for your car. Also note that different diameter threads are sometimes found on the flange studs.*

▲ SB5. *You'll need a strong puller to remove the Pitman arm, after taking off the nut and lock washer with a 1 5/16 in. wrench or socket. You may be able to beg borrow or rent a suitable puller if you don't have one in your tool box.*

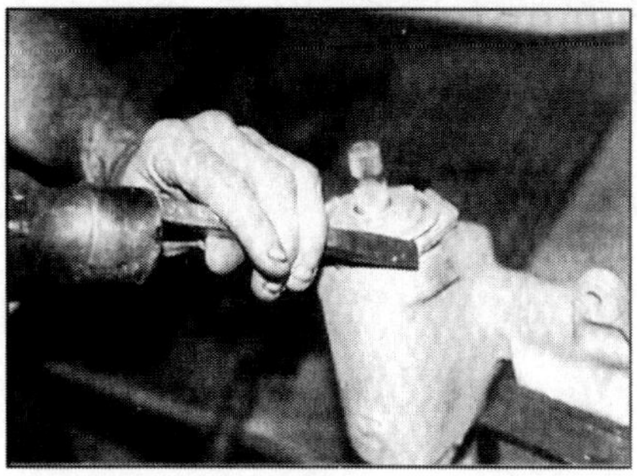

▲ SB6. *If you have to resort to tapping the lock ring loose with a drift, then so be it, but a very large adjustable wrench would be better, if you haven't got the correct size of wrench available.*

▲ SB7. *Here a pipe wrench is being used to remove the thrust bearing adjuster. This shouldn't cause a problem, as it would with a larger hexagon since it will unscrew easily.*

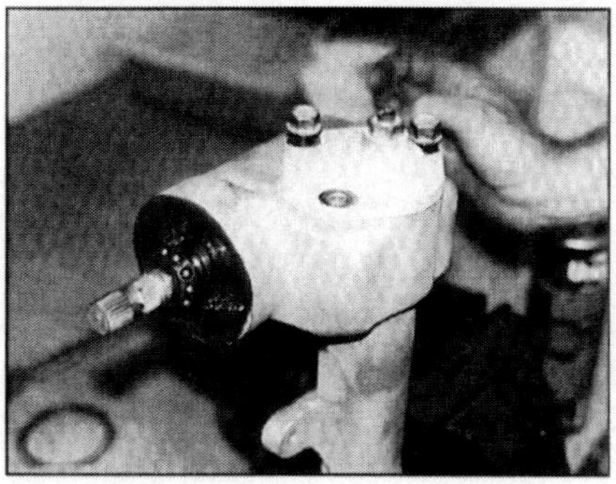

▲ SB8. *You'll need a ⅝ in. wrench to break the lock nut loose on the adjusting stud. The three cover bolts, for which you'll need a 9/16 in. socket or wrench, can then be removed.*

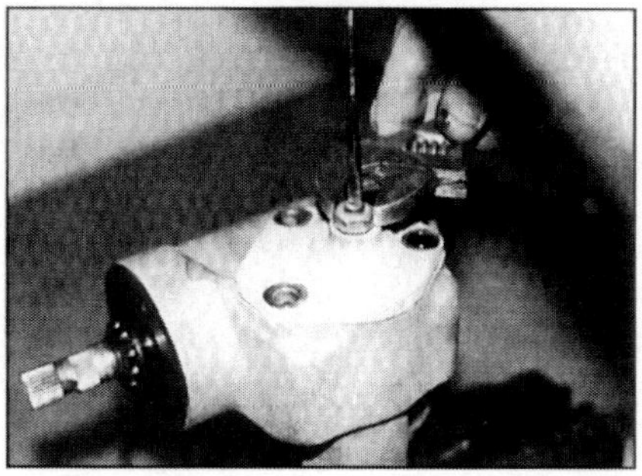

▲ SB9. *Remembering that you've previously broken loose the lock nut, hold the lock nut and use a narrow blade screwdriver to screw the stud right down inside the cover.*

▲ SB10. *You can now lift off the cover. Turn the Pitman arm shaft to its centre point and slide the shaft up and out of the housing. You'll see the gear assembly inside, and this can now be removed.*

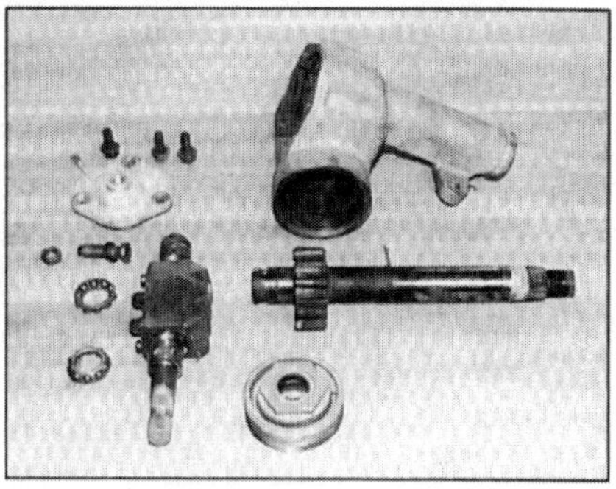

▲ SB11. *You'll need to use something like a screwdriver to pop the seals out of the housing and the thrust bearing adjuster. Lay all of the parts out on clean paper or newspaper after you have degreased and cleaned them.*

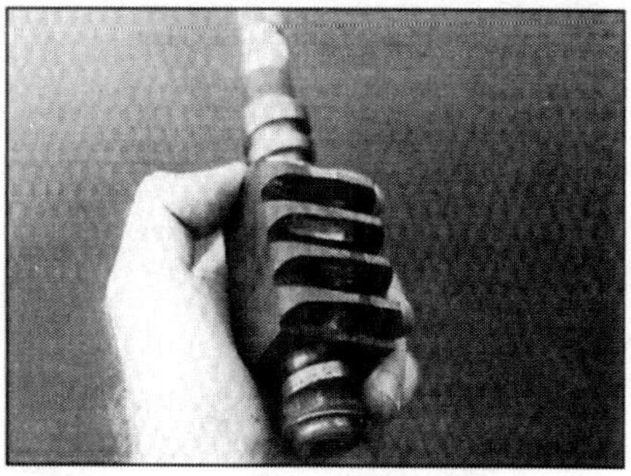

▲ SB12. *You're looking now at the teeth on the worm gear assembly and the Pitman arm shaft. Indentations will tell you that fresh parts are needed but it may take a specialist who's seen it all before to give you best advice. Worm gear replacements include the shaft and balls, and can only be purchased as an assembly – no bad thing.*

▲ SB13. *Inspect the worm gear assembly and shaft carefully and separate the balls into two groups. There should be 27 balls in each. Once disassembled, you can clean up and inspect further – remove the clamp, ball guides and then the balls – and in many cases the parts won't need replacement.*

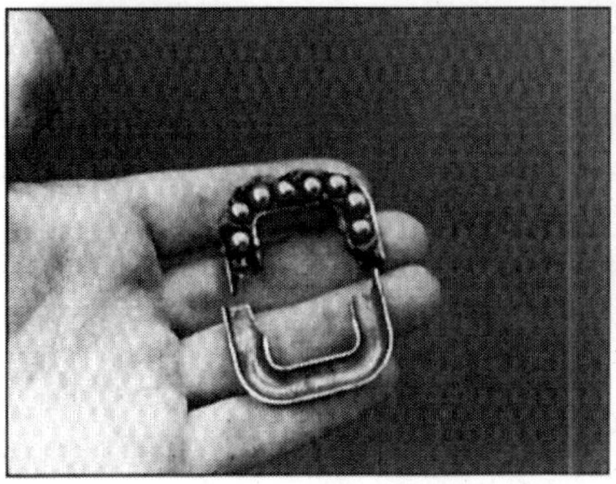

▲ SB14. *When it's reassembly time, centre the shaft and fit in as many balls as you can. Install all the balls from one group on each side and pack the guide with grease so that the remaining balls are held in by the grease, then install them. Take care not to rotate the shaft while the balls are being installed.*

▲ SB15. *It's important that the seal is refitted evenly, even though it looks like this one will need truing up before it's tapped fully home. An ideal tool is an appropriate size of socket, but take great care not to damage the seal or its housing.*

▲ SB16. *These bearing races are not highly stressed items and can last almost indefinitely. However, inspect them and replace if necessary.*

▲ SB17. *Even though GM manuals call for gear oil, John Pfanstiehl recommends using grease, which has the advantage, he says, that it won't leak out and provides better lubrication. However, you should stick with oil if your car is used in very cold conditions. Here, the housing and bearings are being packed with grease and the worm gear assembly is being fitted. Take careful note of the alignment of the shaft with the teeth. The teeth should be thicker at the bottom when the assembly is in the same position as it will be in the box. Make sure that the ball bearing is correctly engaged in its bearing race in the housing. Grease the bearing race on the thrust bearing adjuster and screw it into the case.*

▲ SB18. *Use a feeler gauge to measure the lash of the stud in the Pitman shaft. There should be .002 in. or less; if there is more you'll have to use a thicker shim washer on the stud. The Pitman arm shaft and the cover plate can then be refitted, and you'll have to use that thin bladed screwdriver to turn the stud up and into the cover.*

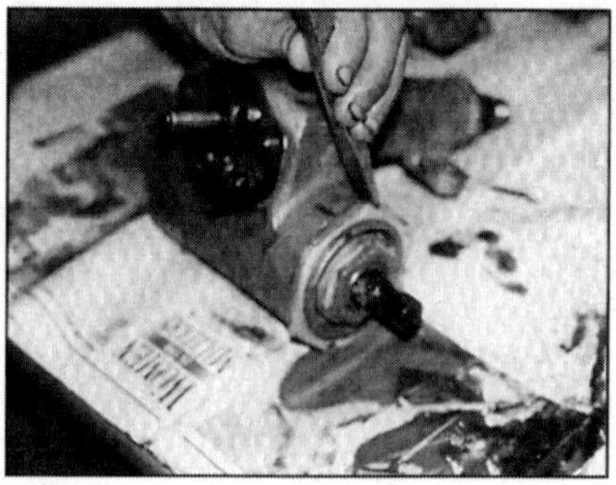

▲ SB19. *If you are not an experienced mechanic, you are strongly recommended to have the lash adjustment set by a specialist or GM dealer. For now, you can set it up just so you can drive the vehicle to your specialist. As soon as the adjuster contacts the rollers, tighten it a little further. This is because the thrust bearing needs to operate under a little pre-load. Fit the lock nut to the adjuster screw, but hold the adjuster screw with your screwdriver so that the nut doesn't get turned by the lock nut.*

Once the steering box is in the car, you can work some more on the lash adjuster screw. The one crucial thing to avoid is steering that is tight either in the straight ahead position or when turned as far as it will go, left or right. It's better to have slightly too much play in the steering and steering that doesn't tighten up on you while you drive the car to your specialist.

You can now install the Pitman arm, which can theoretically go in four different positions, but it should be obvious that there is only one that will do the job. All you have to do now, is reconnect the drag link (non power steering cars) or power steering valve and the job's complete. John Pfanstiehl recommends readjusting the steering box after driving the car for about a week – another specialist job.

PART 4 SUSPENSION

FRONT SUSPENSION OVERHAUL

Many owners take the opportunity to uprate the front suspension on their Corvette when it's overhaul time. Whether you stay 'original', go 'competition' or aim for somewhere in-between, you'll find a suspension rebuild will transform your 'Vette's drive appeal. All the options are available through Vette Brakes & Products Inc. and it's their suspension parts we feature here.

The SLA (short-long arm) suspension features independent coil springs which sit in the lower control arms, and spherical joints which connect the steering knuckle to both upper and lower control arms.

Shims in the upper control arm support shaft are used to adjust the caster and camber angle. All models feature an sway bar and taper roller wheel bearings.

The aim of the suspension is to keep the tire tread parallel to the ground only over a very limited range of the full suspension travel. So, though wide wheels and tires will make your 'Vette look 'meaner', and will assist in putting a bigger 'footprint' on the tarmac, it will also detrimentally affect the handling in other areas.

HANDLING CHARACTERISTICS

The basic handling characteristics of the car are that, when driven through a medium speed bend, some vagueness and initial understeer (where the car wants to go straight on, rather than turn in) is followed by roll oversteer, where the back end is trying to step out of line.

When the suspension has taken up the roll, the car begins to understeer as the corner continues. Lifting off the gas pedal will result in a strong oversteer.

SAFETY: Before carrying out any work on your Corvette's suspension components, read the Safety notes in Appendix 1, particularly with regard to working under the car.

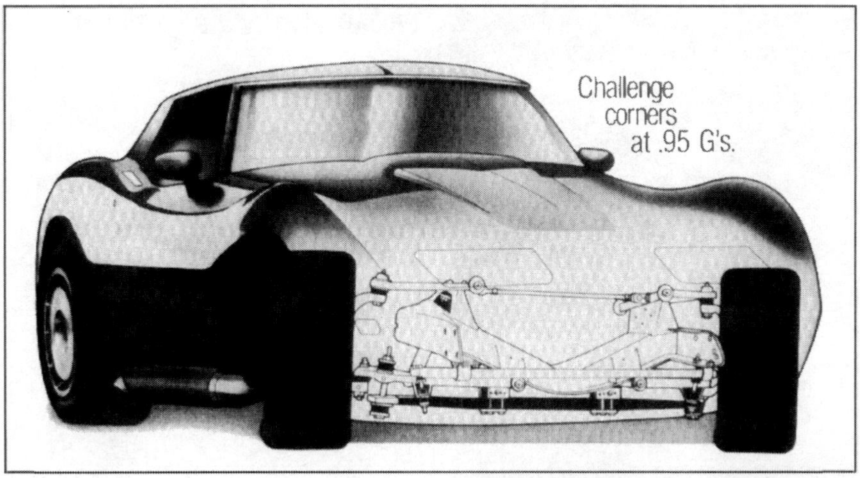

▲ *FS1. In this section we take a trip around the Corvette front end, showing how it all fits together, and reviewing various modifications and uprates.*

In general, work on suspension components requires nothing more than your standard workshop tools, though there are a few items which will make life much simpler, easier and safer – these are shown here. How far you want to go will depend on your performance requirements, your liking for originality – and the depth of your pocket! (Courtesy Vette Brakes & Products Inc.)

▲ *FS2. This photo (taken with the body off) clearly shows the workings of the suspension/brakes – it was basically the same up to 1982. It's vital that you understand exactly what goes where and the various inter-relationships before you even lift a wrench.*

One of the greatest things about Corvettes is the availability of quality spares, covering models right the way back, from specialists such as Vette Brakes & Products who supplied the components used here.

▲ *FS3. All suspension components tend to take a hammering from the elements, and damaged and rusted threads are common. This tap and die suspension set is ideal for saving time (and money!) by enabling you to clean up components and reuse them safely.* (Courtesy Vette Brakes & Products Inc.)

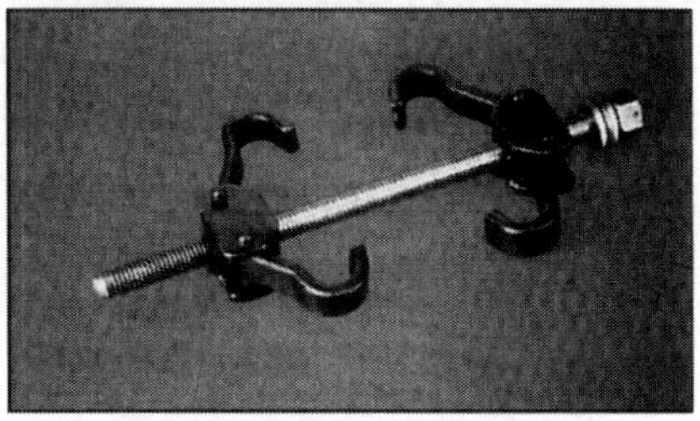

▲ *FS4. Another vital piece of equipment if you're thinking of getting serious with your Corvette's suspension is a coil spring compressor. This particular compressor is the internal type which does not interfere with coil spring towers or other suspension and frame components. Always remember that coil springs, when in place on the car, are under considerable tension and need treating with extreme care.* (Courtesy Vette Brakes & Products Inc.)

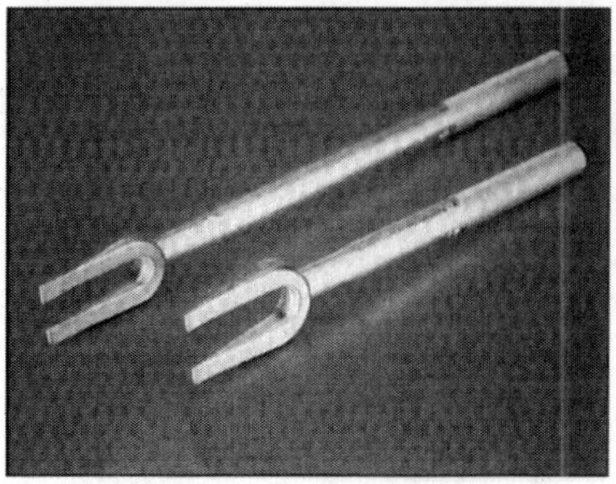

▲ *FS5. When it comes to the steering components, few can be as tricky to sort out as the tie-rod ends and the ball joints. The tie-rod end puller (at top) separates the tie rod ends from their attachment points quickly, safely and easily. The 'pickle fork' provides a similarly easy method for safe removal of ball joint from spindle.* (Courtesy Vette Brakes & Products Inc.)

SAFETY: Please read the Safety section in Appendix 1 before starting work. Remember the safety rules when working on the suspension of your Corvette. NEVER go under the car when it is supported only by a jack – even if it's a heavy duty trolley jack.

Use jackstands, and chock and block the wheels remaining on the ground. Wherever possible, leave the car in gear with the park brake on. Take care to protect your hands and your eyes – there's bound to be plenty of loose dirt, debris and rust.

FRONT SHOCK ABSORBERS

▲ *FS6. To remove a front shock absorber, raise the front end of the car and place it securely on jackstands. Use an open-ended wrench to prevent the upper (squared) end from turning, then remove the upper stem retaining nut, retainer and rubber grommet. Remove the two shock absorber lower mounting bolts and pull the assembly out from the bottom. On this particular car, the well-worn O/E shock absorbers were replaced with KYB gas-adjust dampers. At the time of writing, they were the best-selling monotube shocks on the market, which must mean something! Being charged with 280 psi nitrogen gas, they offer more positive control and complete elimination of shock fade. Better still, they carry a limited lifetime warranty. Installation is a reversal of the removal procedure.*

FRONT COIL SPRINGS

◀ *FS7. With the damper removed as described earlier, support the car securely on jackstands placed under the frame rails. Disconnect the sway bar from the lower control arm and jack up the lower control arm to relieve the tension on the control arm pivot bolts. Here you can see the front lower control arm mounting bolts and one end of each pivot shaft.*

▲ *FS8. WARNING: The coil spring is under great tension. It must be secured so it can't fly out and cause injury. Always handle a compressed spring very carefully: it can blind, maim or even kill if it releases suddenly. Wrap a chain around the spring and through the control arm, and bolt the ends of the chain together. The coil spring must be held compressed at this stage, the safest way to do this being to use a purpose-made automotive spring compressor of the type shown here.*

Remove the pivot cross-shaft nut and bolts from the inner end of the control arm and lower the jack slowly until the compressed spring can be removed. If the car isn't complete there won't be enough weight to compress the spring. This one was weighted down with sandbags!

THE POLY BUSHINGS QUESTION

John Fletcher had polyurethane bushings fitted to the A-arms on the project Corvette featured in this book but then, after some use, decided to remove them again. The reason he gave was that they squeaked, and while he wanted a Corvette that roared, he didn't want one that squeaked! This is an expensive route to follow so it clearly pays to make the right decision the first time round. If you're not sure which way to go, here are some of the pros and cons:

APPEARANCE: While rubber bushings are flat and black, poly bushings are shiny and sometimes come in strange colours.

INSTALLATION: On occasions, all bushings can be a pig to fit, but poly bushings take longer than rubber, on balance.

ON THE ROAD: Polyurethane bushings give crisper handling – but then many of the best handling cars in the world, including Z51s and ZR-1s use rubber bushings. You are likely to get a little more vibration and noise with polyurethane bushings – and that's in addition to the squeaking – although aficionados say that they *don't* squeak if they're fitted right. Read on!

LONGEVITY: Polyurethane bushings last and last, whereas rubber bushings can deteriorate in just a few short years. Could be a very major reason for switching.

▲ *FS9. The most obvious option when it comes to replacing your control arm bushings is to use the stock rubber bushings. Installation of rubber bushings is certainly the least time-consuming, although they have to be pressed out and in.* (Courtesy Vette Brakes & Products Inc.)

▲ *FS10. Alternatively, you could use polyurethane bushes in one of two ways: (a) without the outer sleeve (save and reuse your outer sleeve) or (b) with the outer sleeve. The polyurethane bushings without outer sleeves enable replacement without any pressing. The shiny bushing material and large flat chrome washers make it obvious the control arm bushings are polyurethane. These Vette Brakes & Products Inc. bushings are also graphite-impregnated to further ensure against squeaking, and were 'fitted' to produce the axial lash recommended in the directions that came with the kit.* (Courtesy Vette Brakes & Products Inc.)

The stock bushings have rubber tightly sandwiched between the inner and outer sleeves. The outer sleeve is pressed into the control arm and moves with it. The inner sleeve is held tight to the shaft by the washer and bushing bolt. During movement of the suspension, the rubber does not slide on either of the sleeves, it just flexes or twists. The amount of 'give' in the rubber allows the bushings to be installed without any 'fitting' or modification of parts.

A polyurethane bushing doesn't have the flex of rubber, so it must slide on one sleeve as the control arm moves. The bushing is held tight inside the outer sleeve and moves with it and the control arm. The bushing slides against the inner sleeve and washer which are bolted tightly to the shaft, all three of which are stationary. Polyurethane bushings have to be properly fitted. The inner sleeve must be trimmed to a length which keeps the stationary washer from putting pressure on the turning bushing.

If the sleeve is trimmed too much, the washer will squeeze the polyurethane bushing, causing it to bind and squeak during movement. If the sleeve is too long, the control arm will slide forward and backward, constantly changing front end alignment. The outer shoulder of the shaft should also be checked and the bushing trimmed if it touches.

During hard braking, rubber bushings will compress somewhat, causing the ball joints (and front wheels) to move rearward relative to the frame and other suspension components.

The same compressions can also occur during hard cornering, and this is the reason some Corvette owners install polyurethane bushings on the cars they race.

UPPER CONTROL ARM

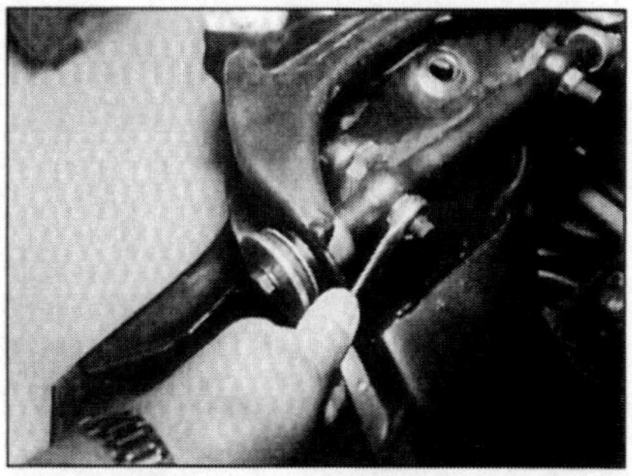

▲ *FS11. Place a jack on a firm, hard surface so it will jack as close to the outside of the lower control arm as possible. Use a solid piece of timber between the jack head and the control arm and raise the car until the tire is just off the ground. Remove the cotter pin and use a ¾ in. box wrench to break loose the upper ball joint nut. Leave the nut on by a few threads. Use a ball joint splitter to pop the ball joint out of the steering knuckle. On some Corvettes there is enough room to remove and reinstall the upper control arm without taking the wheel off. In our case, we were working not only with the wheel off, but the body as well! Two nuts attach the cross-shaft to the cross-member (see above); with these removed, the shims follow suit and the control arm can be withdrawn. A ratchet wrench helps to remove the lock nuts. Oil the threads to make the job a bit easier. Raise the ball joint end and slide the control arm off the studs.*

TIP: Use a 9/16 in. box wrench to loosen the bushing bolts to enable the control arm to pivot up easier. If the radiator shroud is in the way, either remove the shroud or tap the studs out of the frame.

▲ FS12. It is important at this stage to inspect and record the number of shims used at the cross-shaft retaining bolts. These control the caster and camber angles and must be reinstalled to their original locations. Keep the front and rear stacks separate for reinstallation later.

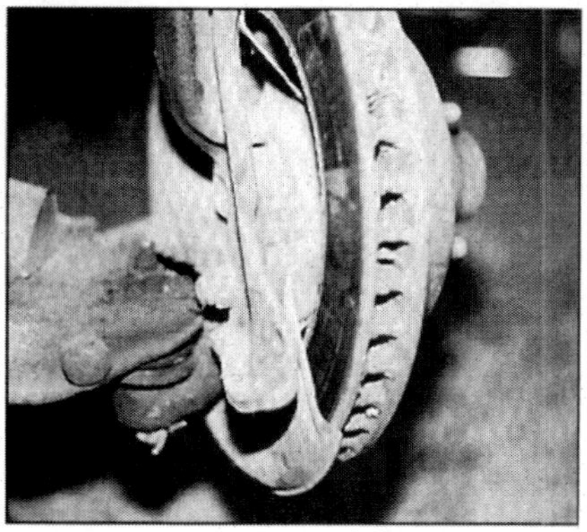

▲ FS13. After the coil spring has been removed, the lower control arm is effectively held at the lower ball joint connection. Separate this from the steering knuckle with a ball joint splitter. Note that there is (or should be!) a cotter pin through the castellated nut. When you've removed it, throw it in your trash can - these things should never be reused; despite their vital safety role they cost very little. It's a good idea to leave the nut on just a few threads while you pop the ball joint out of the steering knuckle, that way you'll keep control of what goes where.

▲ FS14. The control arm is secured at each end by two nuts, as shown here. They may need some persuasion and you'll almost certainly need to apply a drop or two of penetrating oil.

THE CONTROL ARMS OFF THE CAR

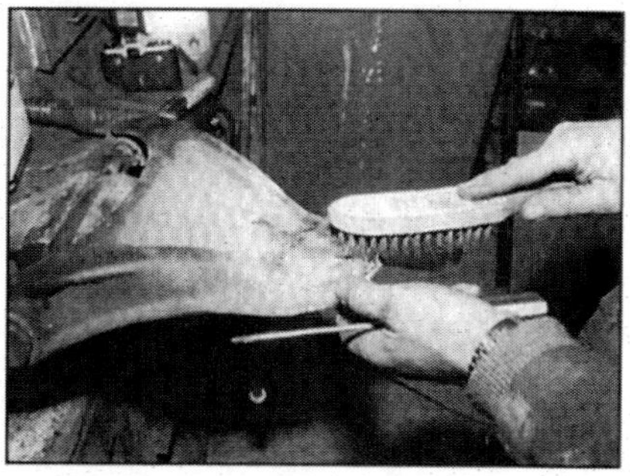

▲ FS15. It was necessary to remove the original ball joints for two reasons; Em was totally stripping the control arm in order to shot blast and paint it, and it was decided to fit a set of heavy duty ball joints. The first step in the removal process was to clean away several years worth of rust and crud ...

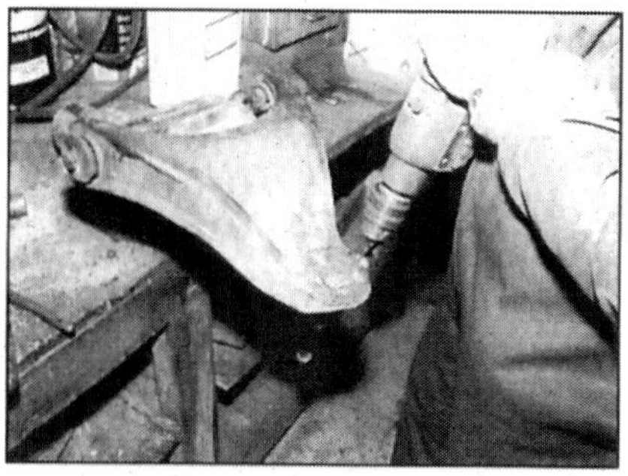

▲ FS16. ... and then the rivets securing the ball joint to the control arm were drilled out. It's important at this stage not to enlarge the holes as the drill passes through.

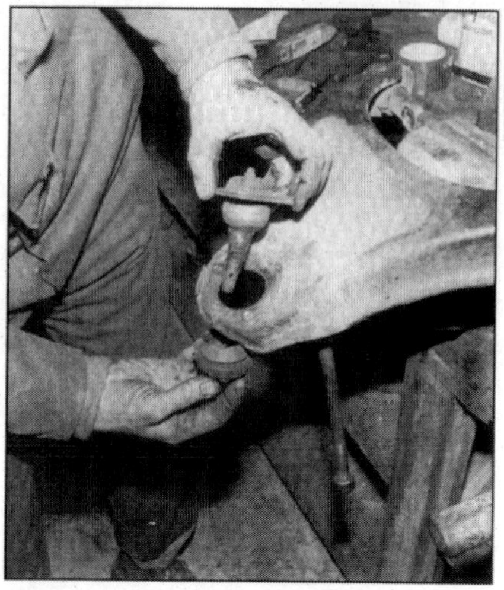

▲ FS17. The ball joint simply pulls out in two halves.

▲ FS18. Clamp the shaft in a vise and remove the bushing bolts and washers.

▲ FS19. Saw through the rubber until the blade contacts the metal sleeve. Do this all the way around and remove the rubber piece. Don't worry about scratches on the bushing's outer sleeve – the new polyurethane bushing doesn't turn against it.

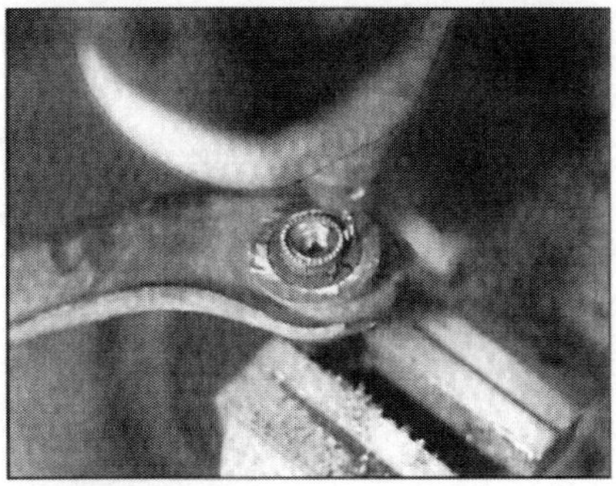

▲ *FS20. Drill down through the remaining rubber with a 5/32 in. bit. After making a number of holes, pull the remaining rubber pieces out using needle nose pliers, or tap the rubber down with a screwdriver. Again, don't worry about nicks caused by the drill. This new method is very clean, fast and easy. Alternatively you could hammer the bushings out, or have an engineering shop press them out. If the entire bushing is removed, it has to then be pressed apart so that the outer sleeves can be reinstalled. Burning the rubber is very dirty, produces tremendous amounts of toxic smoke and makes clean-up difficult, as well as the heat damaging the strength of the steel. NEVER do it! Plus, burning rubber can explode unpredictably!*

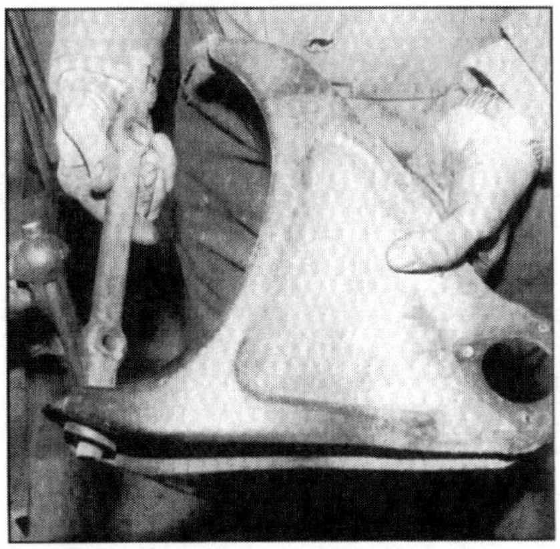

▲ *FS21. The shaft could then be pulled out.*

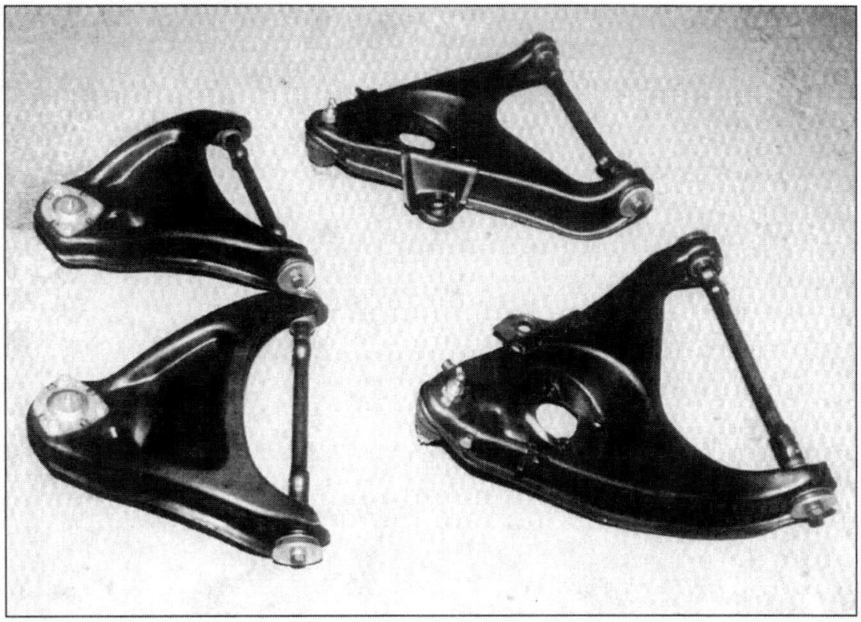

▲ *FS22. As you will have seen elsewhere, this was a total renovation job on the front suspension components. You'll probably agree that it certainly gave these control arms (upper and lower) a new lease of life!*

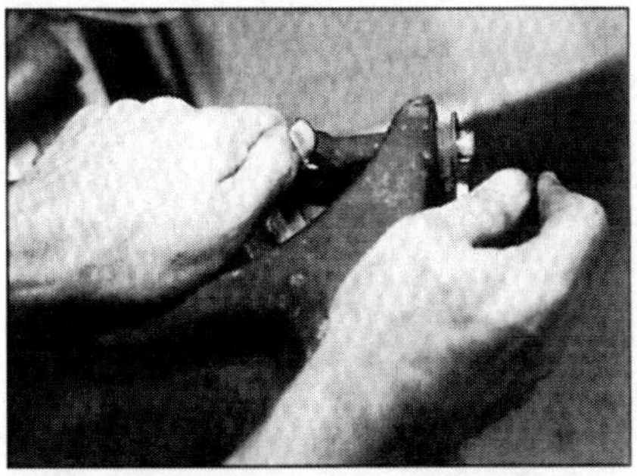

▲ *FS23. The polyurethane bushings require that you modify the parts to fit your car. The inner sleeves have to be trimmed down until the bushings have .010 in. to .020 in. axial lash when installed. Install the bushing and sleeve on one end, and then hold the remaining bushing up to the sleeve on the outer end to measure how much needs to be cut off the sleeve.*

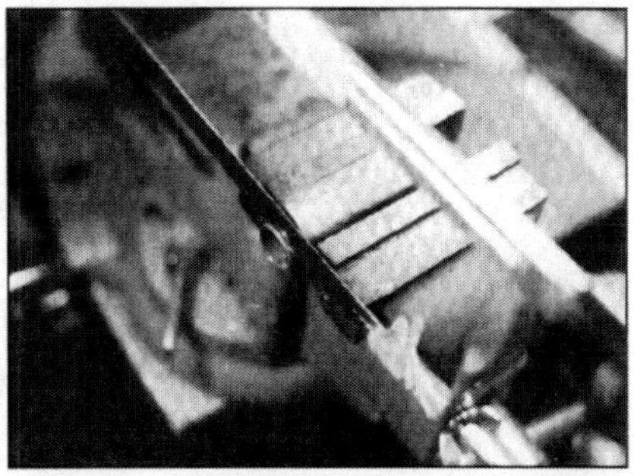

▲ *FS24. You must saw an equal amount off each sleeve, but gripping it in the vise is tricky. You could wrap cardboard around each of the vise jaws to stop the sleeve from being marked – and it's essential to do so. Also take care not to crush the sleeve in the vise. If you can find a bolt of the exactly correct size to fit inside the sleeve, so much the better. After cutting or filing the end of the sleeve, remove any burs and wash the sleeve in degreasing solution or something similar to remove any metal particles. On some polyurethane bushings the inner sleeves are almost impossible to remove; in which case you could use a mini grinder to remove the surplus metal from the ends of the sleeves.*

▶ *FS25. You could always trim off a little more from the end of the sleeve by placing it on the shaft while holding it against the grinding wheel. Don't just tear into it, though! Mark the sleeve clearly so that you can see how much you are grinding off.*

If the polyurethane bushing sleeve is too tight to remove, file or grind the sleeve while it's installed in the bushing in the control arm. Be sure to remove any burrs or other particles.

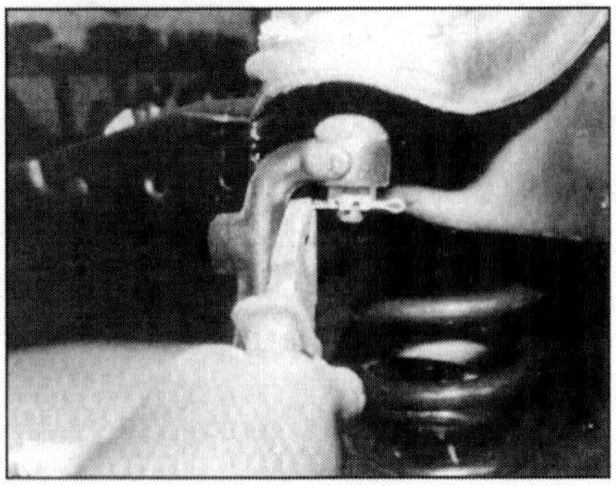

▲ *FS28. … because it is designed to secure the castellated nut holding the ball joint to the upper steering knuckle. Once in position, then the ends of the cotter pin should be spread.*

▲ *FS26. Having got everything apart, the opportunity was taken to fit a set of heavy duty ball joints. These have a self-adjusting, spring-loaded design which keeps parts in their normal operating position, whilst nylon bushings reduce internal wear. The lower ball joints take a terrific pounding, as they carry the weight of the car, and this makes them one of the fastest wearing parts on a Corvette. Conversely, the upper ball joints have a much easier life and are likely to need less attention, which is just as well because they're much more expensive than the lowers.* (Courtesy Vette Brakes & Products Inc.)

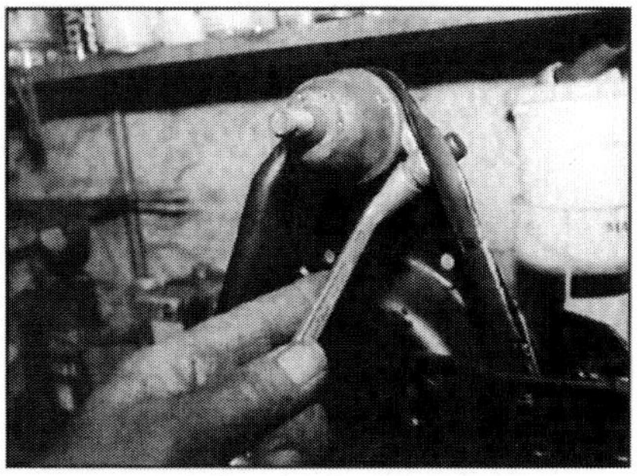

▲ *FS29. The same principle of reassembly is followed with the lower control arm. It's important that all nuts and bolts should be torqued up to the specified figure using an efficient torque wrench.*

▲ *FS27. The finished upper control arm, complete with ball joint. Note the new cotter pin in position, but not yet bent over, of course …*

▲ *FS30. Here, the control arms and ball joints are back in place, waiting for the coil spring to be refitted.*

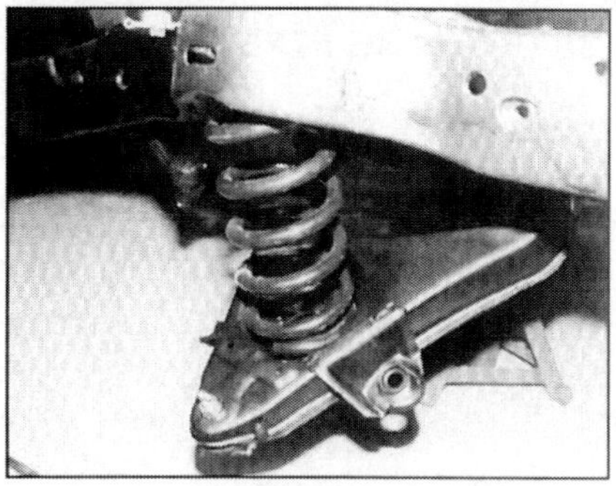

▲ FS31. Don't forget the safety rules mentioned earlier – they apply just as much when installing a spring as when removing one.

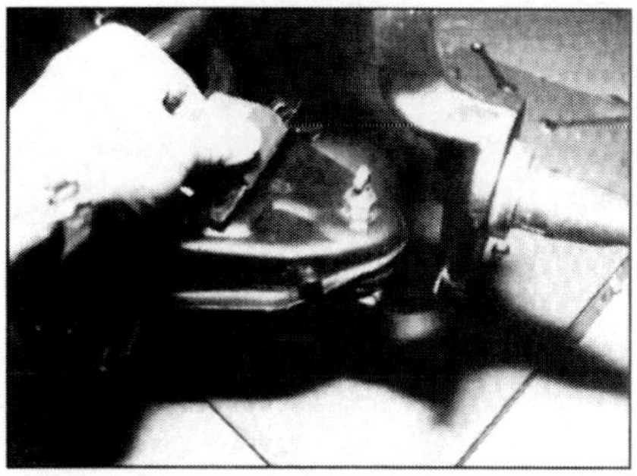

▲ FS32. With the new steering knuckle in place (you can tell it's new – there's no grease on the spindle), the lower control arm bumper can be fitted.

REFURBISHING THE SUSPENSION COMPONENTS

▲ RU1. Most suspension components can be replaced (or uprated) from specialist suppliers, but if you want to retain originality, it's yet another trip to the sandblasters. Here, we've laid out lots of horrible rusty lumps of metal that at one time were part of a Corvette's suspension – hard to believe, isn't it?

▲ RU2. Part one of the refurbishing is the process of removing all rust, paint and anything else, in this case using a sandblaster – an investment we strongly recommend.

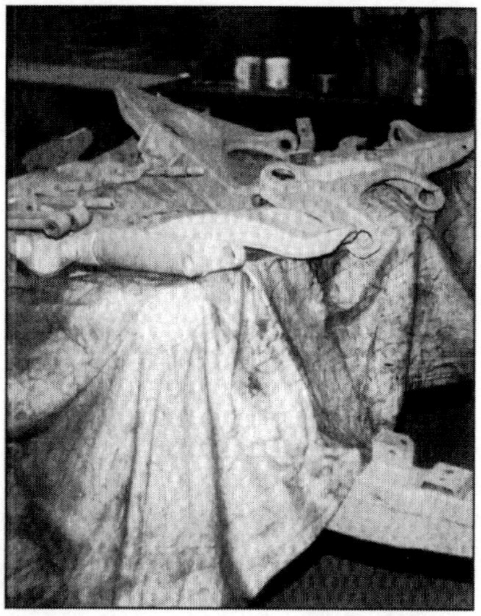

▲ RU3. As soon as the sandblasting process was complete, relevant items were sprayed with primer, as rusting starts again almost immediately. Of course, it's vitally important to make sure that every last piece of grit is removed, not only because you need a clean surface to paint, but also because grit in suspension components is a sure recipe for disaster. Take care to minutely inspect each part for damage or cracking before applying fresh paint.

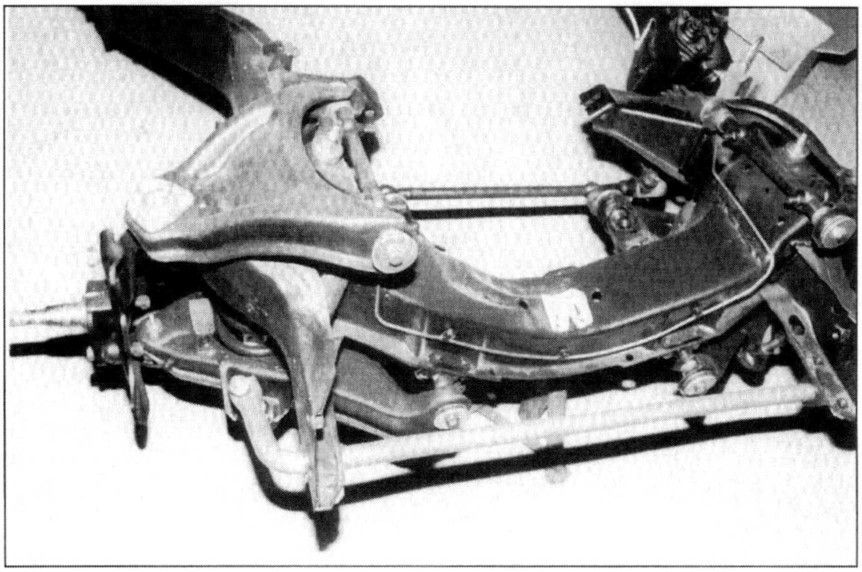

▲ FS33. This freshly restored steering and suspension assembly makes impressive viewing, doesn't it?

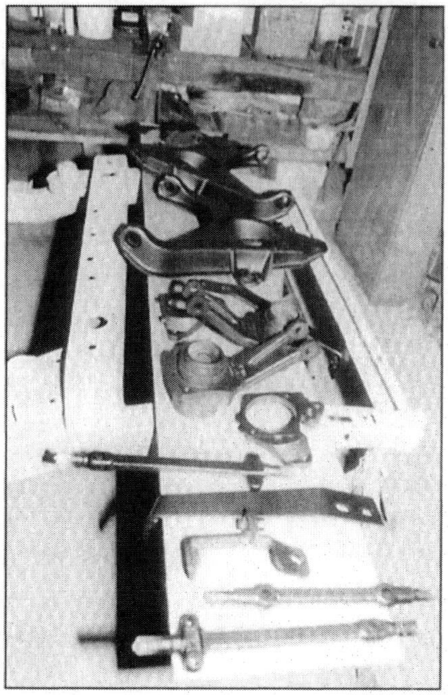

▲ *RU4. If you want to go the whole hog, you could have parts stove enamelled or plastic coated. Parts won't be stock, but they will last virtually for ever, without corroding again.*

ALIGNMENT SPECIFICATIONS

WHEELS AND TIRES

Proper wheel and tire size are very important for good handling. Twenty years ago, most cars came equipped with narrow tires with tall side walls mounted on small diameter 14 in. wheels. These gave a long, narrow footprint and the amount of tire contact with the road was relatively small. The tall side wall would flex excessively and try to 'roll' off the wheel under hard cornering. Today, low-profile tires with very short, stiff side walls mounted on 16 in. or 17 in. wheels are the norm. These have a much wider footprint and the shorter side wall will not 'roll' off the wheel while cornering. Bear in mind, though, that larger wheels and tires may effect bearing life because of the extra load put on them. Also, the wider the tire, the greater the propensity for aquaplaning – check your average annual rainfall before you buy!

ALIGNMENT

Good front-end alignment is necessary for the suspension to keep tires perpendicular to the road at all times. Toe-in or tow-out is the direction or left/right angle of the tires. It should be maintained as close to 0 degrees as possible. Camber is the vertical position of the tire. Positive camber pulls the base of the tire in while the top moves out from the car. Negative camber pushes the base of the tire out while pulling in the top. Proper camber should keep good tire contact during cornering. Caster is the position of the centre-line of the wheel relative to the ball joint. Negative caster moves the wheel forward in this relation, while positive moves it back. Positive caster is preferred as this increases steering control.

TYPE/YEAR	*CASTER*	*CAMBER*	*TOE-IN*
1963/81	2 1/4 (+/- 1)	3/4 (+/- 1/2)	1/2 (+/- 1/8)
		0 (+/- 1/2)	1/8 (+/- 1/8)
1984/85	3 (+/- 1/2)	13/16 (+/- 1/2)	5/16 (+/- 1/16)
1984 Rear		0 (+/- 1/2)	5/16 (+/- 1/16)
1985 Rear		3/8 (+/- 1/2)	5/16 (+/- 1/16)
1986	6 (+/- 1/2)	13/16 (+/- 1/2)	0 (+/- 5/16)
		3/8 (+/- 1/2)	0 (+/- 3/32)
1987/88	5 1/2 (+/- 13/16)	13/16 (+/- 1/2)	0 (+/- 3/16)
		3/8 (+/- 1/2)	0 (+/- 3/16)
1989/91	5 13/16 (+/- 13/16)	1/2 (+/- 1/2)	0 (+/- 3/16)
		3/16 (+/- 1/2)	0 (+/- 3/16)

(Reproduced courtesy of Vette Brakes & Products Inc.)

REAR SPRING REPLACEMENT

CHOOSING

▲ *RS1. Steel springs are not jobs known for their longevity. If you have a lot of miles on your Corvette, spring replacement is one of the jobs you'll eventually need to carry out. You'll also need to select the spring rate, commonly a choice between the relatively easy-riding standard springs or the firmer F-41 option springs.*

1963–82 owners have the option to purchase a fiberglass replacement. Late model cars have similar one-piece springs as original equipment. The great advantage of a fiberglass spring is its longevity – said to be up to five times greater than its steel counterpart. The other notable point about fiberglass springs is that they are much lighter than the originals. According to Vette Brakes & Products Inc., the installation of a fiberglass leaf spring kit could have your cornering capability up to 0.95g! (Courtesy Vette Brakes & Products Inc.)

DO YOU NEED TO REPLACE YOUR SPRING?

You may want to replace your steel spring for a fiberglass version, but does it really need replacing? Try this simple check: measure the distance from the ground to the body when the car is sitting on a flat level surface. If one side is lower, it can be a sign of a tired spring.

▶ *RS2. The multi-leaf spring is composed of layers of rolled steel bands separated by plastic inserts. The leaves are heat treated to maintain the proper arch, and the inserts are there to control inter-leaf friction. Though it does an adequate job of absorbing road shock, it does transfer vibration and road noise through the chassis. With age, the plastic inserts wear out, and friction between the leaves inhibit fast rebound and compression. Eventually, the leaves will fatigue and the spring is no longer able to support the car or offer a decent ride.* (Courtesy Vette Brakes & Products Inc.)

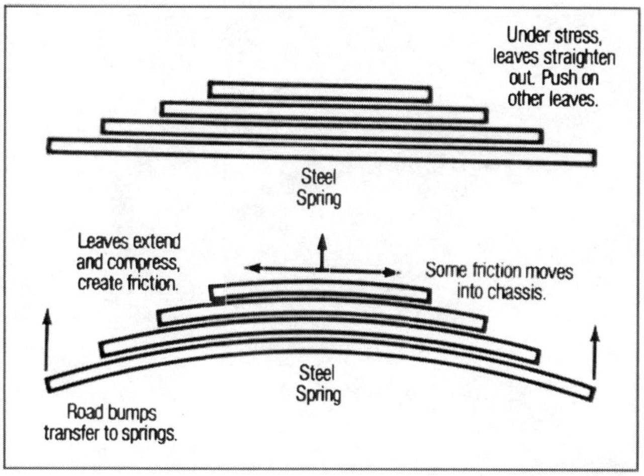

▶ *RS3. Unlike steel, the fiberglass monospring is not plagued by these problems as it is composed of long, unidirectional fibres that are bonded together in a special epoxy matrix. Because it is a monospring, there is no friction between leaves to slow its reaction to changing road surfaces. By its nature, fiberglass is more elastic than steel and will not fatigue in the way a multi-leaf spring eventually must.* (Courtesy Vette Brakes & Products Inc.)

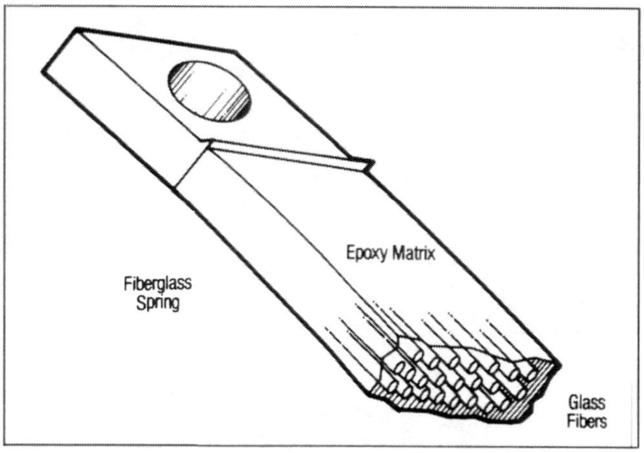

▼ *RS4. The fiberglass monospring absorbs vibrations by stretching and compressing across its surface and storing the energy. When the glass is stressed, the fibres on the bottom of the spring expand, up to 4 per cent of their original length (similar to the way a rope stretches when pulled).*

At the same time, fibres on the top of the spring resist compression the way a solid glass block resists heavy weights or sharp blows. Fibres in the middle evenly distribute the energy produced from this constant stretching and compressing. (Courtesy Vette Brakes & Products Inc.)

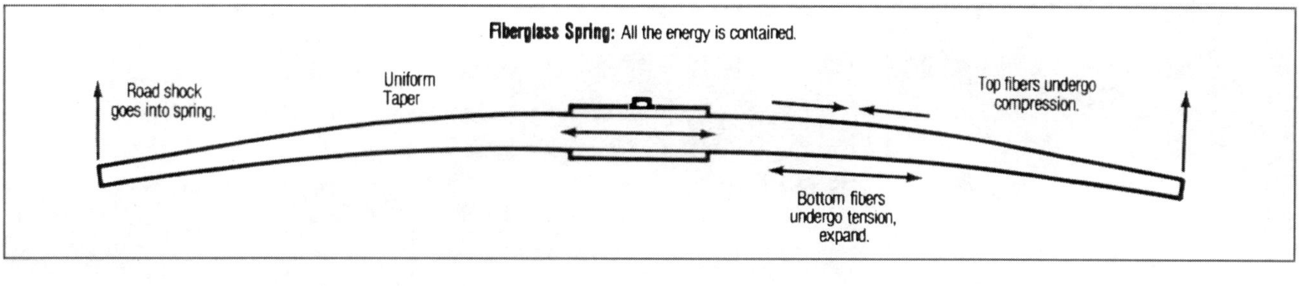

▶ *RS5. The fiberglass monospring is an impressive replacement for the ageing multi-leaf steel spring. It is designed to ride better, react faster and last longer, and Chevrolet has used this advanced technology in both the Lumina and the later model Corvettes with winning success. This diagram shows what you get and where it goes. Read on to see how to get it all there!* (Courtesy Vette Brakes & Products Inc.)

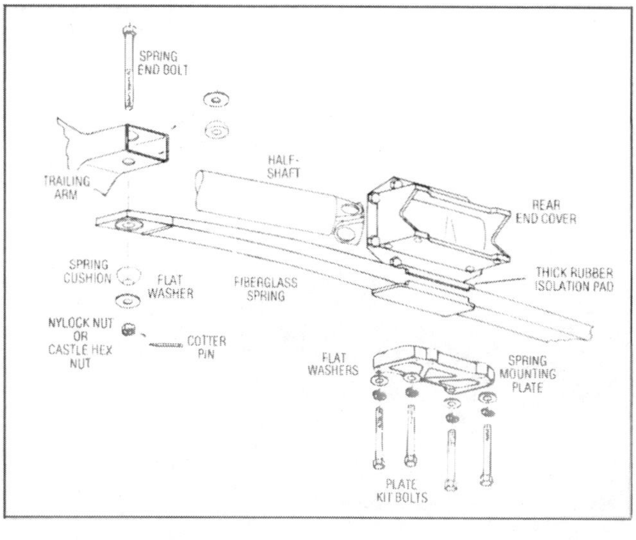

▲ RS6. After removing the cotter pin, you can break the nut loose on the spring bolt using a $\frac{7}{8}$ in. socket for the nut and a $\frac{13}{16}$ in. wrench (large) to hold the other end of the bolt. If you've thought ahead, you will have applied penetrating oil to this nut each day for several days before carrying out this work. If the nut still refuses to shift, consider using a nut splitter to break the nut open, and then replace it with a new one.

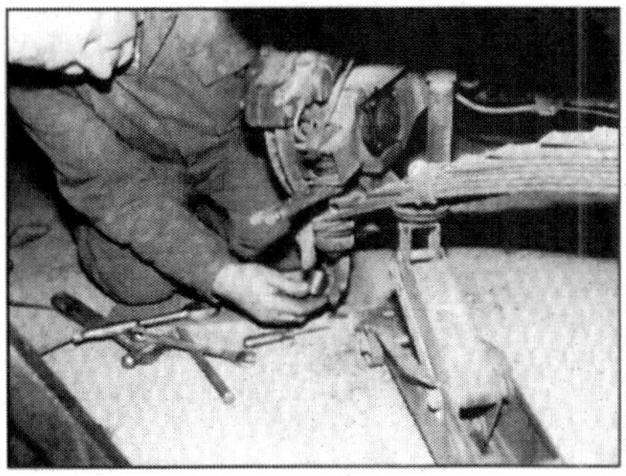

▲ RS7. Note that the car has been separately supported on jackstands so that, even if the jack slips off the spring, there's no risk of it falling on the person working on the car. It may help to use a hefty G-clamp to prevent the jack from sliding inward on the spring.

▲ RS8. If the car has a rear sway bar, its bracket must be unbolted to permit removal of the spring bolt and replacement of the upper cushion. You must be careful, as the bracket bolts often break off. Use a torch to heat the nut plate or get replacement parts to have on hand. Take care to avoid cracking or breaking the rear bolt boss on the differential cover; for one thing, replacing it is a lot of work, and for another, new exact replacements are not available for the early models.

▲ RS9. It's no surprise to learn that there are polyurethane bushings available to replace the stock rubber versions. (Courtesy Vette Brakes & Products Inc.)

◀ RS10. Reading from left to right, the new original-equipment rubber cushion appears first, followed by the same cushion after four years of hard work, and on the right a polyurethane cushion. The polyurethane cushion is likely to last longer than the rubber version, but squeaking could once again be a problem.

▲ RS11. As the rubber cushion wears, the bolts can often rub against the steel spring. You are strongly recommended to replace the whole bolt, nut, washer and cushion assembly, as shown here. (Courtesy Vette Brakes & Products Inc.)

▲ RS12. A longer bolt option is also available, the extra threads of which allow a greater range of ride height adjustment. These are available with either rubber or poly bushes. (Courtesy Vette Brakes & Products Inc.)

▲ RS13. Use caution when breaking the four bolts loose on the spring-mounting bracket. If they are frozen, use penetrating oil. Judicious application of heat is possible on the rear bolts. These two bolts have the most problems because the threads are open on the top and exposed to water.

▲ *RS14. You're likely to find plenty of evidence of corrosion all around – use a tap to clean the threads in the rear cover if required and a wire brush or die on the bolt threads. The GM replacement covers use smaller diameter bolts than the early original covers. Aftermarket replacement covers are available with the larger bolts, but their rear section is beefed up, so you should ensure that it doesn't interfere with the spare tire carrier.*

▲ *RS17. Here's how John Pfanstiehl recommends that you fit the replacement spring. Screw in the two forward bolts with the bracket hanging down towards the back. You can then slide the spring in and install the two rear bolts, but make sure the locating button on a fiberglass spring, or the spring bolt on a steel spring, slips correctly into its hole in the cover. Move the spring around a little to check that it's properly engaged. On GM replacement covers there are two spacer plates next to the bolts to help prevent the spring from twisting, which goes to show how it can pay to fit original parts. Tighten the bolts under just a little pressure at this stage, but don't torque them up fully. This will take a lot of twisting force and could cause the car to be pulled off its axle stands. Tighten the bolts to the correct torque later, when the car is standing on the ground.*

Before lowering the car to the ground, measure the thickness of the original spring and the thickness of the replacement. If the replacement is thinner, make certain the bolts don't bottom out when they are tightened. The front two are especially likely to do so. Another John Pfanstiehl tip is to remove the washers and then retighten each bolt to make sure it contacts the bracket. If not, the bolts are likely to bottom out.

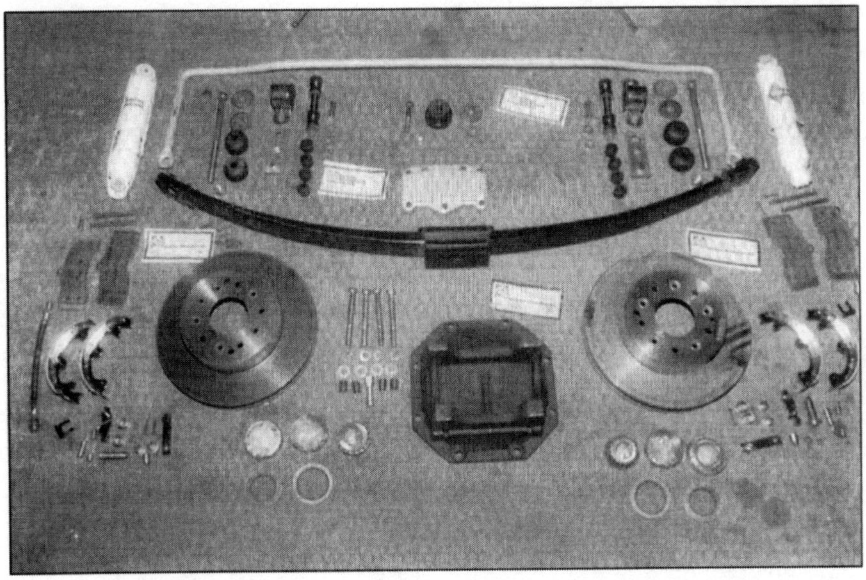

▲ *RS15. This mouthwatering array of new rear suspension (and brake!) equipment was destined for the John Fletcher restoration project car. There are certainly no excuses now for any deficiency in the stopping and handling department!*

▶ *RS16. When replacing a stock steel spring with a fiberglass monospring, a special plate is required. This is a heavy duty aluminium item which comes complete with flat washers, lock washers and bolts.*
(Courtesy Vette Brakes & Products Inc.)

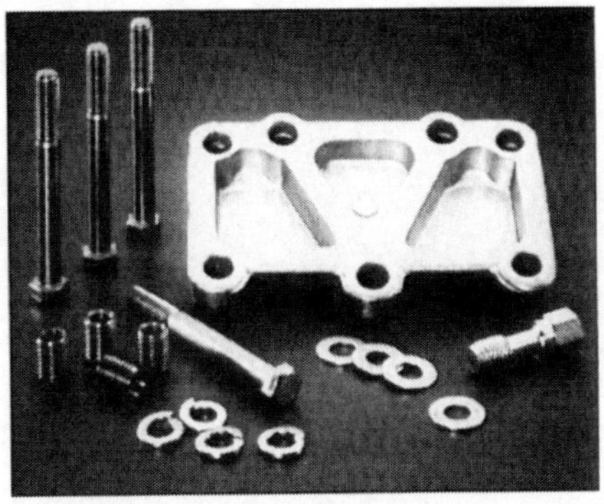

SAFETY: Ensure that you recheck the torque after about a week or 500 miles of driving. Cars with catalytic converters need a heat shield between any fiberglass spring and the exhaust pipe because of the amount of heat generated – talk to your specialist supplier to ensure they can supply you with the correct safety equipment to go with the spring – and make sure that a stock exhaust is *at least* two inches away from a fiberglass spring – it is made of plastic, after all!

REAR TRAILING ARM OVERHAUL

The Corvette independent three-link type rear suspension consists of a fixed differential carrier which is rubber mounted to the frame at three points, with the strut rods, driveshafts and torque control arms forming the three links at each wheel, and a transversely mounted leaf spring. Pivot bolts and rubber bushings are used to attach the trailing control arms to the frame side member. The trailing arm connects at the rear to the leaf spring. Four bolts are pressed into the arm for the attachment of the wheel spindles and spindle supports.

The rear wheel spindles are driven through double U-jointed, tubular drive shafts. At one end (the outboard), they are mounted to a splined spindle flange, and at the other, to the differential side gear yokes.

Tapered inner and outer roller bearings are housed in the spindle support, and bearing adjustment is achieved by using a spacer and shims of varying thicknesses.

The fork-shaped mounting brackets for the outer ends of the strut rods are also part of the spindle supports. The strut rods extend laterally from a bracket on the underside of the axle carrier to the spindle support. An eccentric cam is provided for rear wheel camber adjustment.

The shock absorbers are fixed at the top, to the frame bracket and at their lower point to a threaded stud on the strut rod mounting shaft. The multi-leaf, transverse spring is secured in the middle under the differential cover and at each end via end links to the rear of the trailing arms.

REMOVING THE REAR TORQUE CONTROL ARM

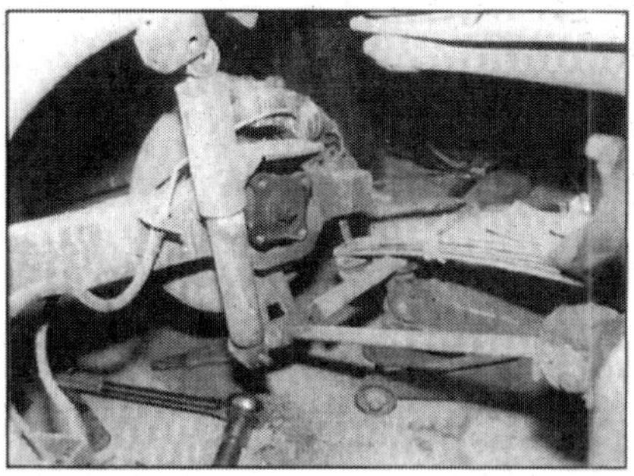

▲ *TA1. As you can see, at this stage we had taken the body off and had excellent access to everything we needed to photograph. The spring has to be disconnected at the relevant end, as well as the sway bar. If you're not completing a total rear-end overhaul, the damper can be left in place and just released at the bottom mounting. In our case, we actually removed the damper for later replacement. The four bolts that secure the outer end of the drive axle to the spindle flange had already been disconnected. Again, we had actually removed the shaft, not something you would do as a matter of course if your main aim was to service the torque control arm.*

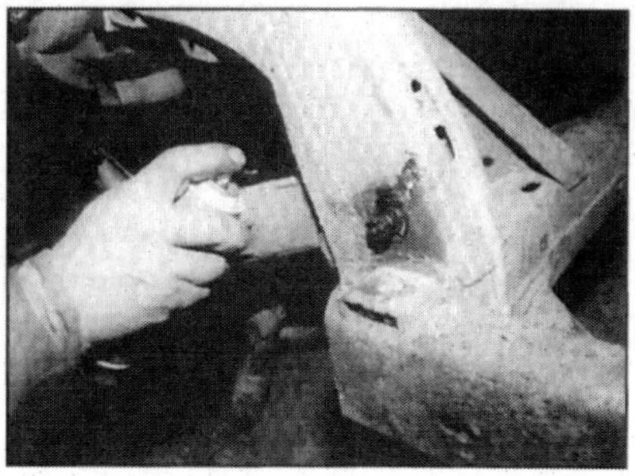

▲ *TA2. The strut rod and the parking brake cable have to be disconnected. So, too, does the brake line from the caliper and the control arm – make sure that you plug the line to prevent loss of fluid and to prevent the fluid getting onto your car and yourself! The pivot bolt is certain to be rusted in place, so spray plenty of penetrating oil in, on and around the bolt. Remove the cotter pin and throw it away – never reuse them, they're very important but incredibly cheap.*

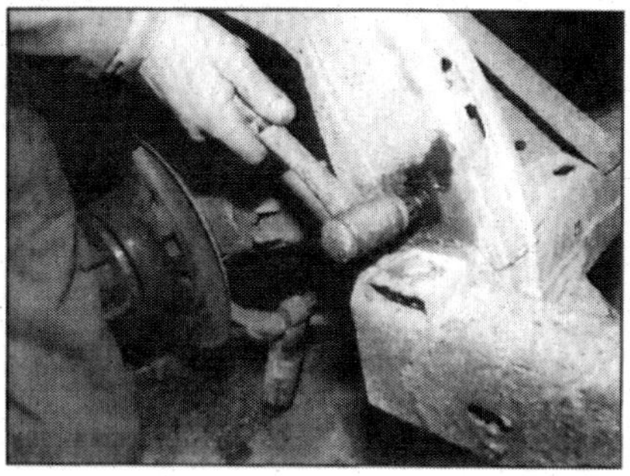

▲ *TA3. Although we could get the castellated nut off, the pivot bolt wouldn't move, despite the penetrating fluid. The nut was replaced at the end of the bolt and then tapped lightly with a copper mallet to 'encourage' the bolt to leave home.*

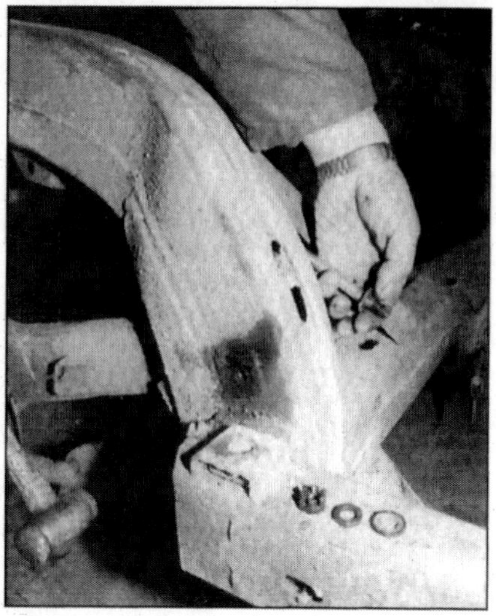

▲ *TA4. Success! When refitting the control arm, the pivot bolt should only be installed hand tight until the car is back on its wheels. Once the weight of the car is on the tires, tighten the control arm pivot bolt to the torque specified, then install the cotter pin. It then remained to pull the control arm out …*

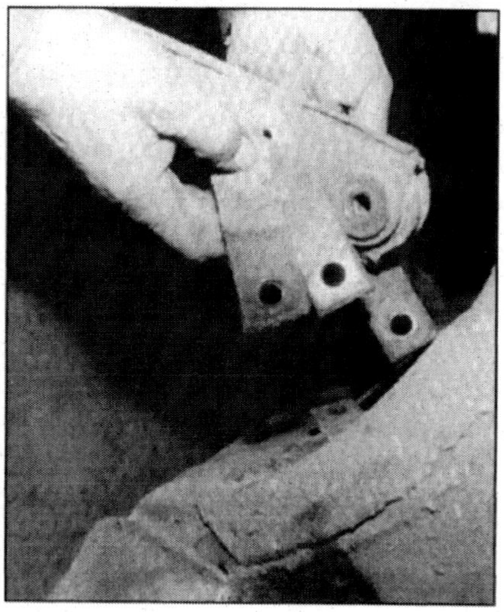

▲ *TA5. … and remove the shims. These shims (of varying thickness) are fitted at the point where the trailing arm pivots at the side member in order to adjust the rear wheel toe-in angle. Note which shim goes where – taping them together is a good idea if you're reusing the same shims.*

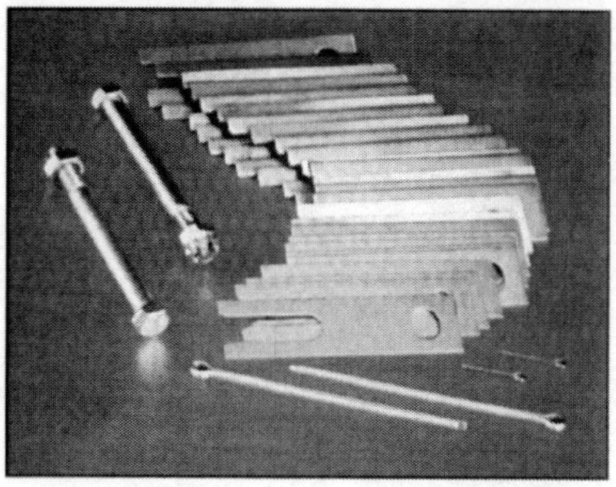

▲ *TA6. As we've already seen, rust is always a problem with suspension components, and many owners are now fitting stainless steel shim kits to make alignment much less of a chore.* (Courtesy Vette Brakes and Products Inc.)

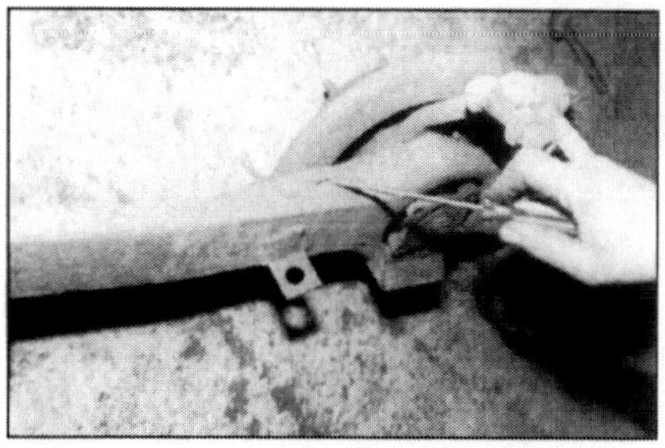

▲ *TA7. It's already been mentioned that the Corvette torque control arms are notorious for rusting. Their poor design encourages it, and this is one of the first tests you should make – if the screwdriver prizes the arm apart, you've got tin worm trouble!*

▲ *TA8. New front bushings can be installed to the control arms but the retaining flares will have to be drilled out before the old ones can be removed. After installing new bushings, the retainers will have to be flared with a special tool. Most owners find it easier – to say the least – to have them installed by a dealer or machine shop.* (Courtesy Vette Brakes and Products Inc.)

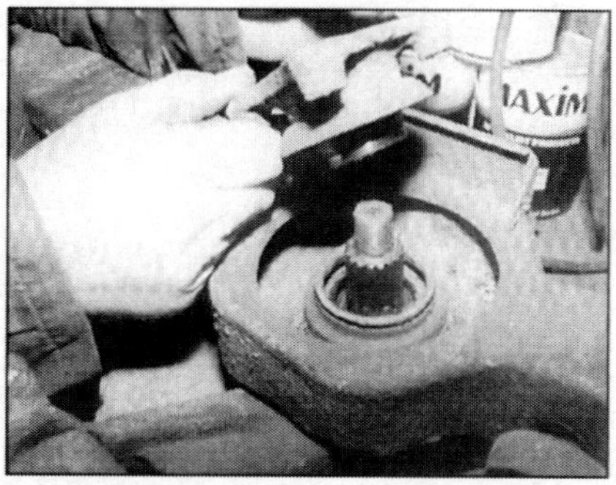

▲ *TA9. With the arm on the bench, we were able to dismantle it further in relative luxury. Once the spindle yoke was removed, the spindle itself was pressed out …*

▲ *TA10. … and this was followed by the rear brake flange bracket …*

▲ *TA11. … followed, in turn, by the control arm to spindle support. The outer bearing will have to be pressed off the spindle or removed using a special puller – take it to your local specialist or GM workshop.*

Installation of the torque control arm assembly is a reversal of the dismantling procedure, with particular emphasis on keeping dirt and grit out of the bearing surfaces and getting the right number of shims in the right places.

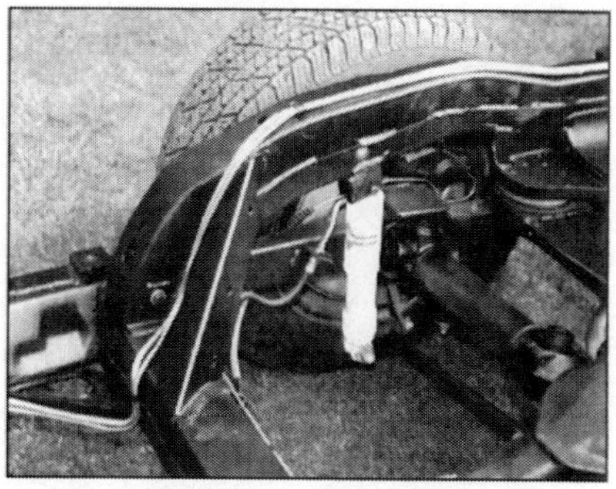

▲ *TA12. More than a little different from the first photo in this section! In this pristine condition, the rugged nature of the Corvette's suspension and allied components can clearly be seen.*

FITTING A FRONT OR REAR SWAY BAR

Rear sway bars are standard on some Corvettes and an option on others. If yours was not fitted with a rear sway bar as standard, fitting one couldn't be simpler as the mounting holes are already drilled in most Corvette's frames and rear suspension trailing arms, so it's just a matter of bolting one in place. Then, if you want to return to originality, all you have to do is unbolt it. As an aside, thicker front sway bars can be fitted in a similar way, i.e. as a bolt-off, bolt-on replacement.

The advantage of a sway bar is that it helps to keep the car level during cornering, and generally improves handling, although it also tends to give an increase in oversteer when you fit a sway bar to the rear, and an increase in understeer when you fit one to the front. Suspension modifications are always a matter of trade-offs, and another down side to fitting a sway bar is that the suspension becomes harsher. The idea of the sway bar is to transmit some of the loadings from one side of the suspension to the other, so it follows that as one wheel hits a bump, the movement in the trailing arm will be transmitted to the other side. However, if an improvement in handling is what turns you on, and stiffer suspension doesn't turn you off, here's how to fit a rear sway bar.

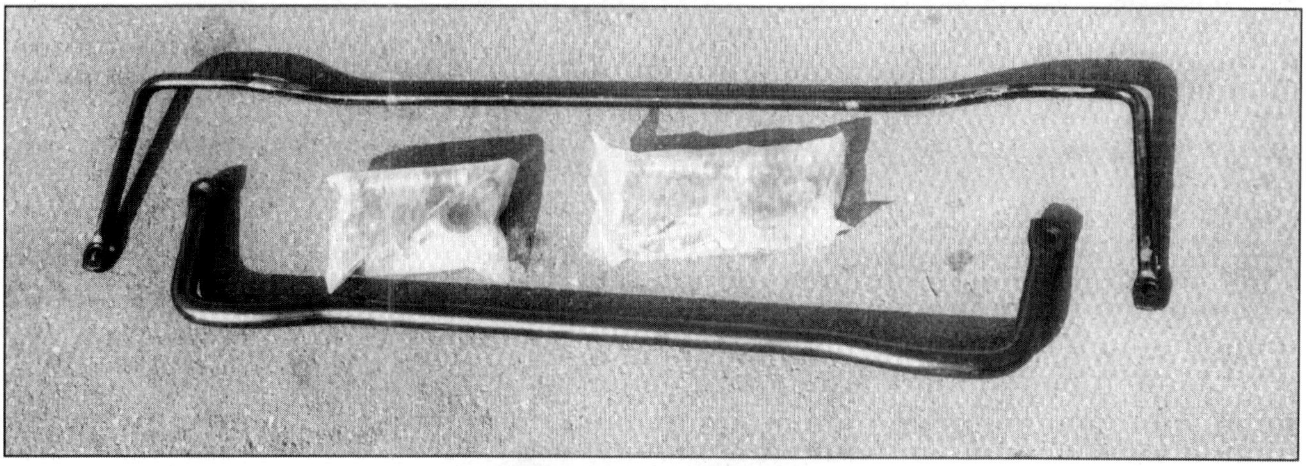

▲ *AB1. Specialists such as Vette Brakes & Products Inc. supply both front and rear sway bar kits in varying thicknesses, and each kit will include all the brackets, bushings and fittings required. The only additional stuff you'll need will be four new 5/16 in. mounting bolts of the correct type and strength, with lock washers and flat washers.*

▲ *AB2. You start by attaching the relevant bracket to the trailing arm by a threaded nut plate. Two bolts hold the nut plate in place. Fit but don't tighten the first until you've fitted the second.*

The polyurethane bushing will be difficult to get on to the sway bar, so do it while it's off the car. The part of the bar where the bushing will rub should be sanded clean and greased before the bushing is forced into place. This could be a good way of cutting out that dreaded squeaking effect from the polyurethane.

Note that the centre of the bar has an offset so that it provides clearance with the spare tire carrier. Be sure to use a tap to clean the threads so there's no risk of breaking the nuts loose inside the frame.

▲ *AB3. Two important points regarding the link nut: you must make sure that the nut goes on sufficiently far for the nylon, locking-part of the nut to be engaged, but not so far that the spring is completely compressed when the trailing arm is in the down position. The link spring is a necessary part of this set up because the bushings don't compress, unlike the original factory bars with rubber bushings.*

▼ *AB4. It's even easier to change a front sway bar, although once again remember the earlier advice about cleaning out the nut threads so they don't break loose inside the frame. You don't have a spare tire to worry about at this end of the car; instead GM gives you a fan shroud that the bar must clear.*

These sway bars work so that when the suspension is extended, the hard bushings are forced apart, and this compresses the spring. You must back the nut off to the point where the spring is not completely compressed when the suspension is fully extended.

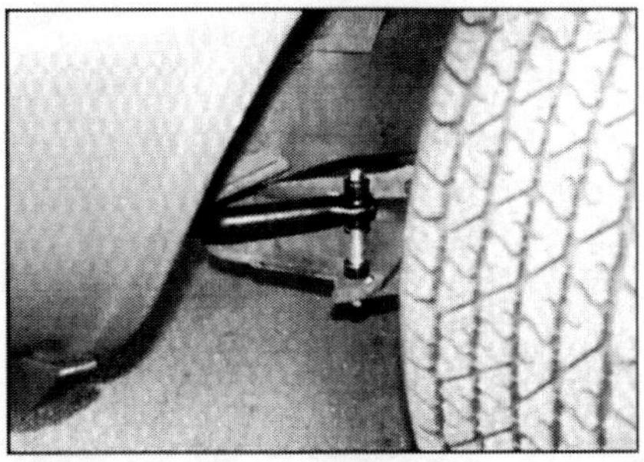

▲ *AB5. Check the angle of the link when the car is resting on the ground once more. It should be nearly straight at regular riding height.*

Take note that whenever you fit a non-GM sway bar kit, it has to be adjusted carefully, and that you may need a few flat washers (use the plated type or, better still, stainless steel) or even U-shaped front-end shims to adjust the clearance. When the bracket has been fixed to the frame, make sure the bar pivots freely before fixing it to the A-arms. If it's tight, you'll have to have recourse to the adjustments we've just mentioned.

FITTING TIPS

Before installing the polyurethane bushings to a front or rear sway bar, sand the paint off the bar where the bushing will work, and grease the contact area. The delrin bushing and spring combination which comes with this Vette Brakes & Products Inc. kit may be too noisy for some tastes. Thorough as ever, VB&P supply an alternative soft polyurethane link bushing which you can try, along with the steel spacer sleeve, if you want a slightly softer, quieter set up.

SWAY BAR BUSH AND END LINK

It's a pleasant surprise to find that one of the fastest wearing parts on a Corvette suspension is also one of the easiest to replace – you should be able to replace your sway bar links and bushings in around 30 minutes. The link rubbers undergo a lot of deflection at the sway bar and break down quickly. The photos here show polyurethane link kits and bushings, but the steps are the same for replacement of the original rubber links and bushings.

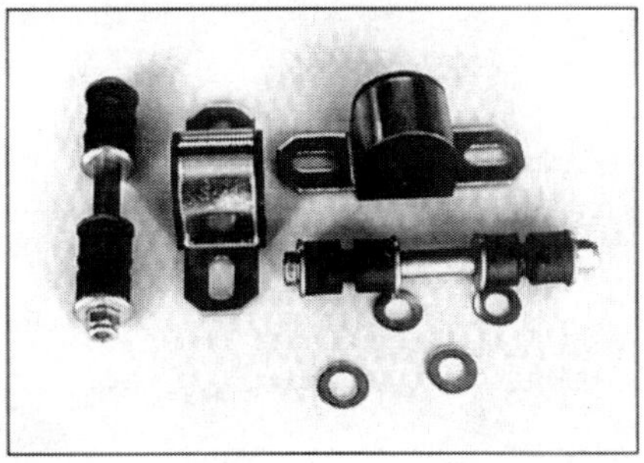

▲ *AB6. As we said earlier, sway bar bushes do wear and require replacement at regular intervals, and the polyurethane link kit shown here will extend the life of those bushings at the risk of a slightly harsher and noisier suspension and the further risk of squeaks, although to be fair the manufacturers usually put that down to faulty installation. If you just replace stock rubber bushings with polyurethane, the original brackets are installed as normal, retaining a stock appearance.* (Courtesy Vette Brakes and Products Inc.)

▲ *AB7. You can buy the end links with stock-type fittings on the end …* (Courtesy Vette Brakes and Products Inc.)

▲ *AB8. … or there are the adjustable types described earlier in this section which enables the driver to adjust the coupling of the sway bar to suit particular driving requirements.* (Courtesy Vette Brakes and Products Inc.)

INSTALLATION

Clean up and oil the exposed threads and remove the four bolts. Mark one side of the bar; it is very easy to replace it back-to-front! Typically, the bar curves down to provide more clearance for the radiator.

Open the slit on the polyurethane bushing with a screwdriver. Look closely for the seam – it can be difficult to tell on which side the bushing is slit. After the slit is opened, insert a wider screwdriver or a chisel and twist it to prize the slit further apart. Grease both the sway bar and the inside of the bushing before snapping the bushing onto the car. Use a cloth to get a good grip on the bushing, and don't worry, the bushing can only go on one way.

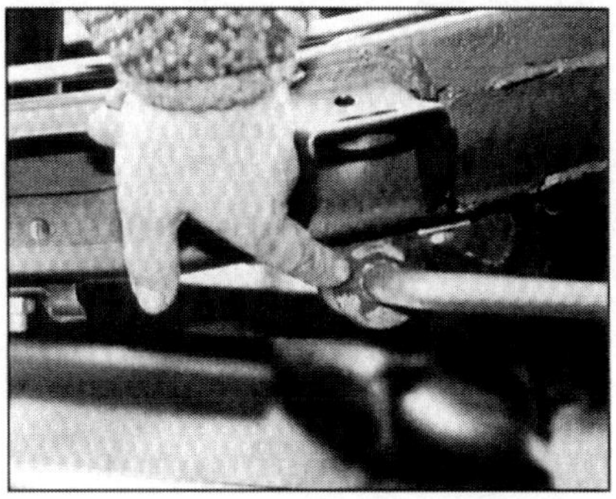

▲ *AB9. This is a stock sway bar fitted with polyurethane bushings, and this is the bushing bracket referred to earlier which must be adjusted so that the bushing is not tight on the bar.*

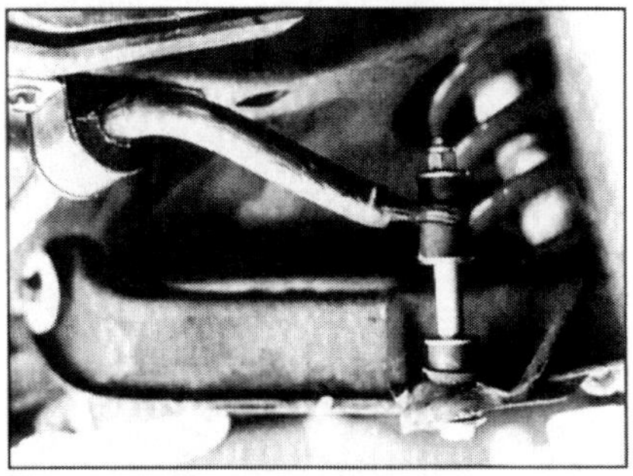

▲ *AB10. We saw the adjustable type of sway bar link earlier – this is the stock end link. It's a tip worth remembering to reassemble all suspension fixings and threads using anti-freeze compound. You never know: it might be you who also has to carry out the next overhaul!*

▲ *AB11. Note the difference between the stock end links (albeit with polyurethane bushings) and this adjustable version. The ability to tailor this aspect of the suspension to your driving needs makes this option well worth considering.*

REAR END PROBLEM SOLVING

There are a lot of expensive things to go wrong in the rear end of a Corvette, and Corvette drivers, like the rest of the human population, can be split into optimists and pessimists. The optimists drive on in the knowledge that there's nothing important going on back there – and that usually means they drive the car until real damage has been caused. The pessimists believe that every time a stone gets stuck in a tire, a major overhaul is called for – and that means unnecessary sleepless nights. The good news on Corvette rear end clunks is that two of the reasons for them are easy and cheap to sort out, so let's dip into the John Pfanstiehl fund of knowledge and see where those little clunkers can be found.

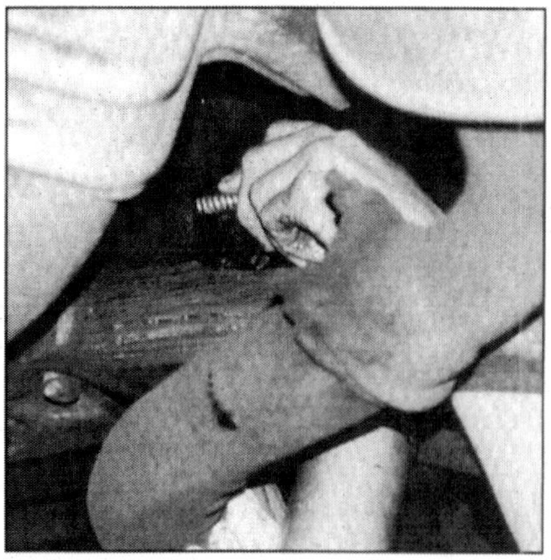

▲ *SC1. Oil seepage around the front of the differential carrier usually looks worse than it is. Don't use the presence of oil in this area as your guide as to whether the pinion seal needs replacement: only do so if the fluid level goes down noticeably on a couple of successive oil changes. As John Pfanstiehl says, it's not that seal replacement is such a big deal; it's that disturbing the pinion nut can lead to other problems such as early bearing failure in worst cases. Obviously, oil on the differential carrier isn't exactly a 'clunk' – but to a pessimist it's just as bad!*

SAFETY: Be careful to keep old oil off your hands – wear gloves – and be especially careful of hot exhausts on 1975 and later cars.

If new fluid doesn't cure the problem, try a particular John Pfanstiehl tip: refill the differential carrier with ATF and drive the car around a few times in tight circles in the parking lot. Pump out the ATF and refill with GM oil and a suitable additive. IMPORTANT: make sure that the additive is suitable for a limited slip differential and won't actually stop the LSD from working – unless that's what you want it to do.

◀ *SC2. You're taking off from stationary and making a turn and you get a clunk from the rear end along with unusual sideways shifting. The likely suspect sounds frightening enough: sticking Positraction clutches in the limited slip differential. This sounds really scary, but the solution is simply to refill with fresh GM Positraction fluid and additive. Here's how to do it.*

Drive the car until the differential fluid is warm and remove the filler plug. On most models, you'll need a 5/8 in. eight-sided socket. You'll need one of those cheap electric drill-operated pumps and a piece of rubber hose to suck the oil out, because there's no drain plug hole – thanks GM.

▲ *SC3. If you get a sharp clunk as you sharply let off the gas pedal when on the move, suspect the rubber upper bushing on the carrier front mount. The lower bushing is built to last – and usually does so. The upper one can often be mashed – compare the old bushing at right in this illustration with the new ones.*

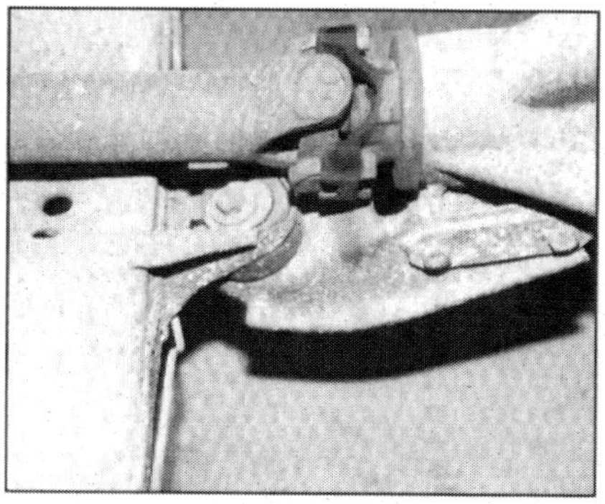

▲ *SC4. Grip the bolt head with a 5/8 in. box wrench while the nut is removed with an 11/16 in. socket, and remove the bolt. The bolt will probably have to be tapped or prized to break it loose. Taking great care not to crush the brake line, use something to take pressure off the lower bushing so it can be slid out. Once the lower bushing is out, it is easier to remove the upper bushing because there is now more clearance.*

Wear in the inner side of the carrier bracket has been a Corvette problem for some years, and GM have gradually added more and more reinforcement to the bracket over the years, first by welding a washer onto the bracket and later by welding on thicker plates. If you see signs of movement, particularly at the front bolt, consider installing a later model bracket.

When reassembling, place the upper bushing in position, prize the carrier bracket (reused or new) back down and insert the lower bushing bolt, ensuring that the bolt hole and bushing holes have been centred. Take note that the smaller flat washer goes against the bolt head and the larger flat washer and lock washer go on the bottom, against the nut. Tighten to about 65 lb/ft but take note that this kind of tightening can cause a car to be pulled off its axle stands, so take the very greatest of care if the car is supported off the ground. The compression in the rubber will be taken up until the sleeve inside the upper bushing bottoms out.

SUSPENSION UPRATES

AVAILABLE IMPROVEMENTS – CHOOSING

There are many aftermarket kits and products around to improve the Corvette's handling, some of which are shown in this section. However, as with most other aspects of customization, what you want from your car will be a personal thing and it's a wise move to try out some of the more extreme modifications before you buy and install them yourself. Often, what seems an ideal suspension set-up can become quite literally 'hell on wheels' when it comes to day-to-day driving conditions. Though those shown here are good value for money, many are far from cheap and we wouldn't want you to rush out to buy something that really didn't suit your style of driving.

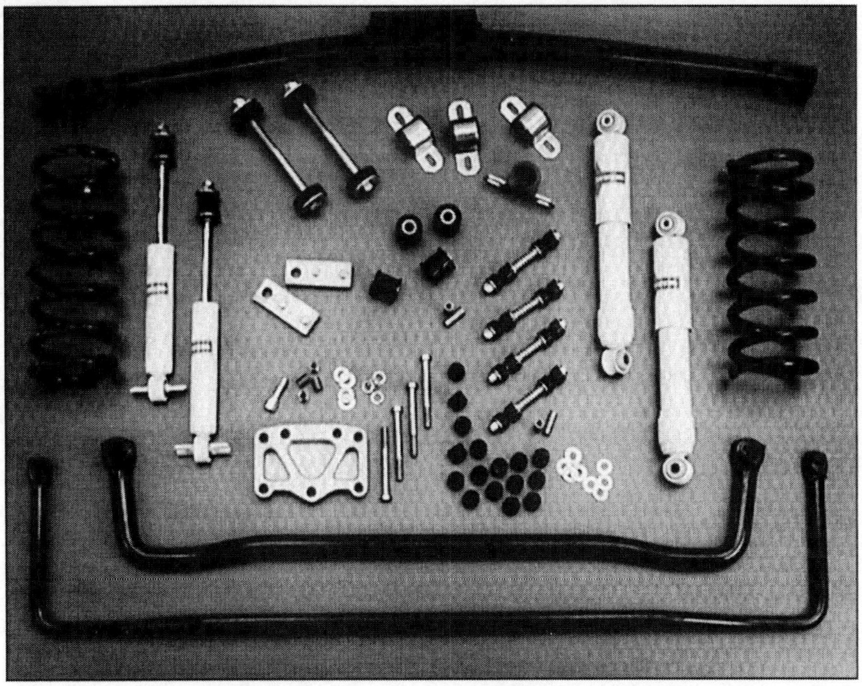

▲ *SU1. For an uprate to improve both ride and handling but without breaking the bank, the Grand Touring suspension kit is about right; it neatly bridges the gap between the rather languid stock set-up and the hard-charging nature of the more performance oriented kits.* (Courtesy Vette Brakes & Products Inc.)

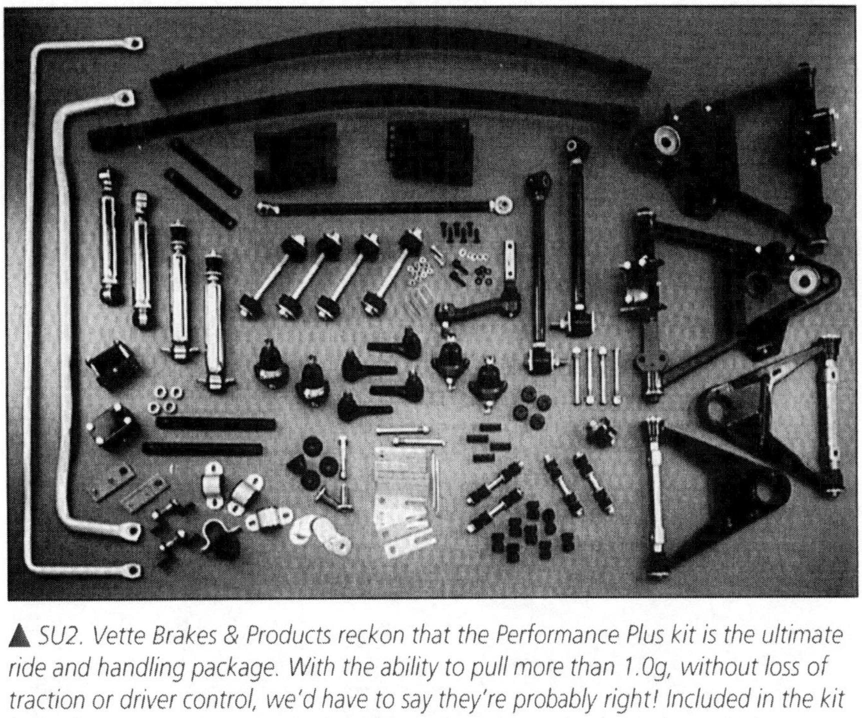

▲ *SU2. Vette Brakes & Products reckon that the Performance Plus kit is the ultimate ride and handling package. With the ability to pull more than 1.0g, without loss of traction or driver control, we'd have to say they're probably right! Included in the kit is the front transverse monospring system, Carrera gas shocks and rear dual-mount monospring.* (Courtesy Vette Brakes & Products Inc.)

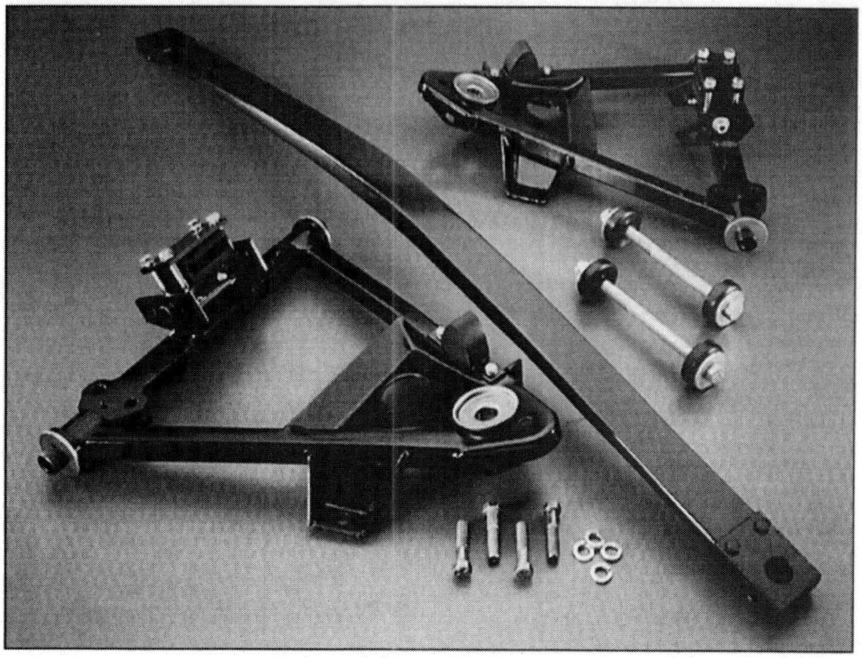

▲ *SU3. If you want to update your pre '83 'Vette to the latest specification, you can throw away your coil springs and replace them with a fiberglass monospring. According to Vette Brakes & Products, you can swap from 'old fashioned' coil springs to a front monospring in half a day. After removing the stock lower control arms and front coil springs, install the existing ball joints on the new lower control arms. Bolt the new control arms onto the frame, connect the ball joints onto the wheel spindles and torque the bolts. Bolt the spring rate adjuster blocks to the cross-shafts and reinstall the shock absorbers. Bolt the fiberglass monospring to the rate adjustment blocks and attach the spring ends to the control arms with ride height adjustment bolts. The kit includes the fiberglass spring itself, mounting plates and adjustment blocks, and lower control arm spring mount discs.* (Courtesy Vette Brakes & Products Inc.)

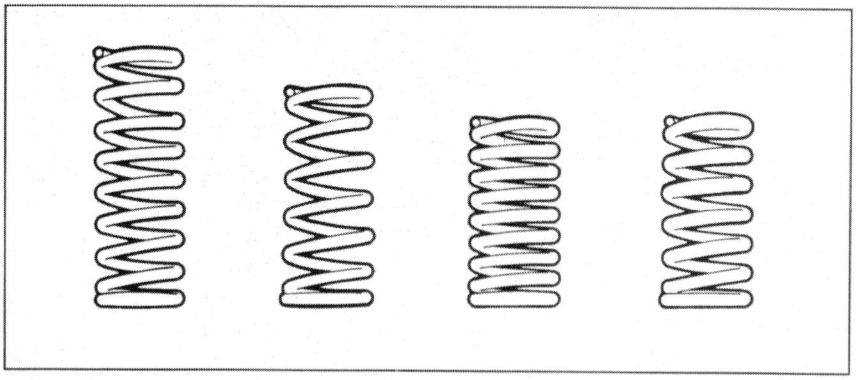

▲ *SU4. Aftermarket coils such as those shown here are shorter than the stock coils because they use larger diameter rods wound into fewer spirals – hence, they are stronger. The two springs shown to the left of this diagram represent springs under no load at normal height; those to the right are springs under load, compressed in the suspension.* (Courtesy Vette Brakes & Products Inc.)

HOW COIL SPRINGS AFFECT RIDE HEIGHT

Uprating your coil springs is one way to restore ride height and eliminate that sagging front end and mushy ride. Vette Brakes & Products Inc. offer five different spring rates – it's important to choose the right one to suit your requirements and driving style.

This guide gives approximate details for Corvettes as delivered from the factory using normal accessories and stock wheels. The use of aftermarket wheels/tires or high pressure gas shocks may change the ride height. Also note that each non-stock spring rate will provide a different ride height. The lower the spring rate, the higher the ride height (necessary to keep the car from bottoming out).

280 lb/in stock GM coils

1963/82 .... Stock height

460 lb/in Grand touring coils

1965/67 big block .... Stock height
1965/67 small block .. 1 in. above stock
1968/74 big block .... 1in. below stock
1968/82 small block .. Stock height

550 lb/in Street/slalom coils

1965/67 big block .... Stock height
1963/67 small block .. 1in. above stock
1968/74 big block ... Not recommended Insufficient ride height or use with coil spacers.
1968/82 small block .. 1in. above stock

860 lb/in

1963/82 .... For racing only – will lower ride height

(Courtesy Vette Brakes & Products Inc.)

STRUT ROD BUSHING MODS

▲ *PX1. The simplest uprate when it comes to strut rods, is to replace the stock rubber bushings with polyurethane ones. These will improve ride, road holding and hold alignment for longer.* (Courtesy Vette Brakes and Products Inc.)

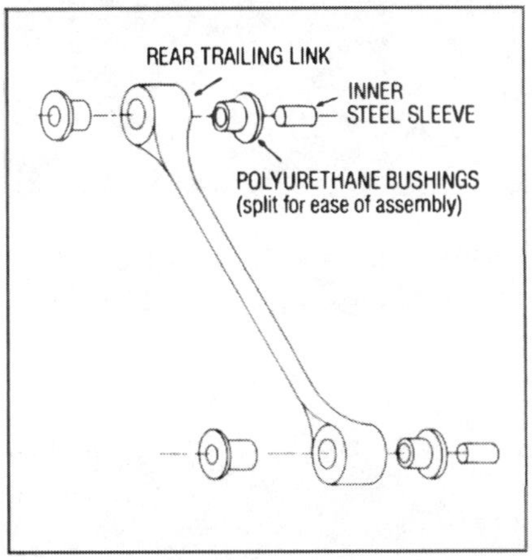

▲ *PX2. When purchased as a set of eight bushings, they are split for ease of assembly. The old rubber bushings should be removed using a press and small piece of pipe. The inside of the trailing arm should be cleaned thoroughly before installing the new poly bushings, and the outside surface of the inner metal sleeve should be greased before fitment.* (Courtesy Vette Brakes and Products Inc.)

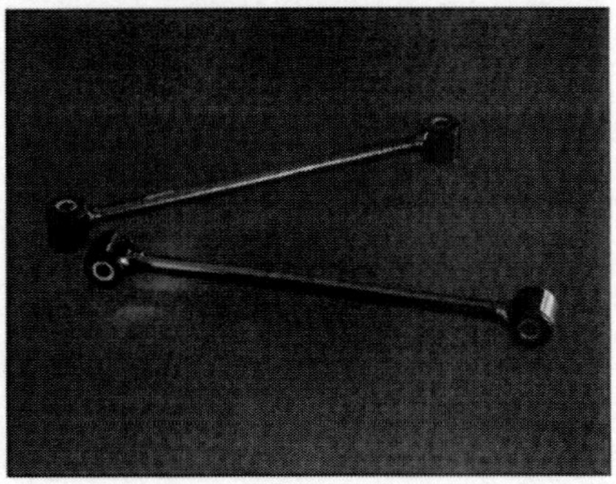

▲ *PX3. Alternatively, you could replace your original strut rods with stock items but ready-fitted with polyurethane bushings. These are for '63–'79 model Corvettes.* (Courtesy Vette Brakes and Products Inc.)

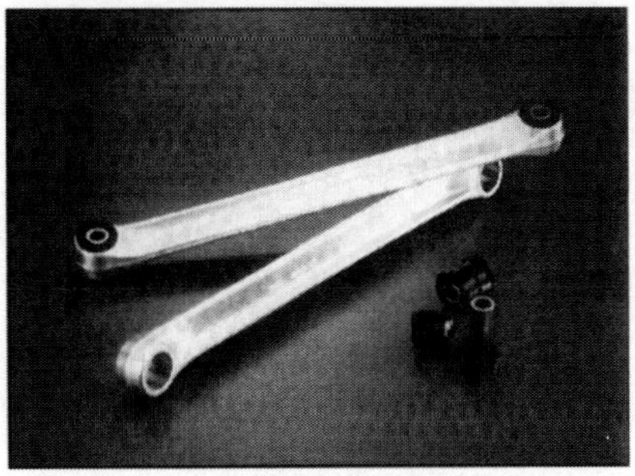

▲ *PX4. There are also versions available for '84–'91 models. The uprated bushings will last much longer than the stock rubber items.* (Courtesy Vette Brakes and Products Inc.)

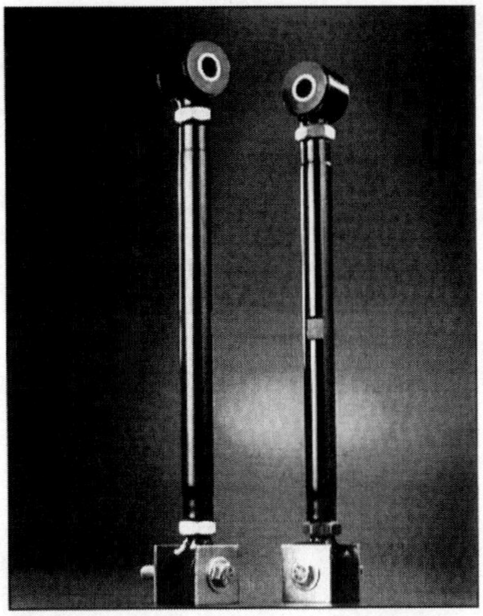

▲ *PX5. Fitting poly adjustable strut rods will allow you to get even more in the way of suspension performance from your Corvette. They enable you to keep your rear tires in vertical alignment and decrease the negative camber settings that wear down your tires and create an excessive load on your Corvette's drive line.*

These particular custom rods feature stronger DOM swedged tubing and adjustable threaded rod ends. The advantage is that the rods can be shortened or lengthened simply by turning the rods. When the required setting has been achieved, the settings are locked by turning the jam nuts. According to Vette Brakes & Products Inc., these strut rods offer some three times the adjustment range compared to stock camber adjustment cams, offering added control over rear camber adjustment. As the name suggests, the rods shown here also feature polyurethane bushings, complete with steel in sleeves in rod eyes. With the polyurethane acting as a bearing, the strut rods can pivot freely with the suspension and, of course, the nature of the material means that they last up to 10 times longer than the factory supplied rubber bushings. (Courtesy Vette Brakes and Products Inc.)

▲ *PX6. First, place the car on jack stands – remember the safety rules. Using common wrenches and a shock removal tool (as shown here) ...* (Courtesy Vette Brakes and Products Inc.)

▶ *PX7. … remove the spring bolts, lower shock mounting shafts, strut rod mounting brackets and strut rods. For added strength, weld your existing strut rod bracket tabs onto the bracket. (The tabs were only spot welded at the factory.) Reinstall the brackets and new strut rods, shock mounting shafts and spring bolts. Lower the car to the ground and roll it back and forth. Loosen the strut rod jam nuts and adjust the rear camber settings as follows.* (Courtesy Vette Brakes and Products Inc.)

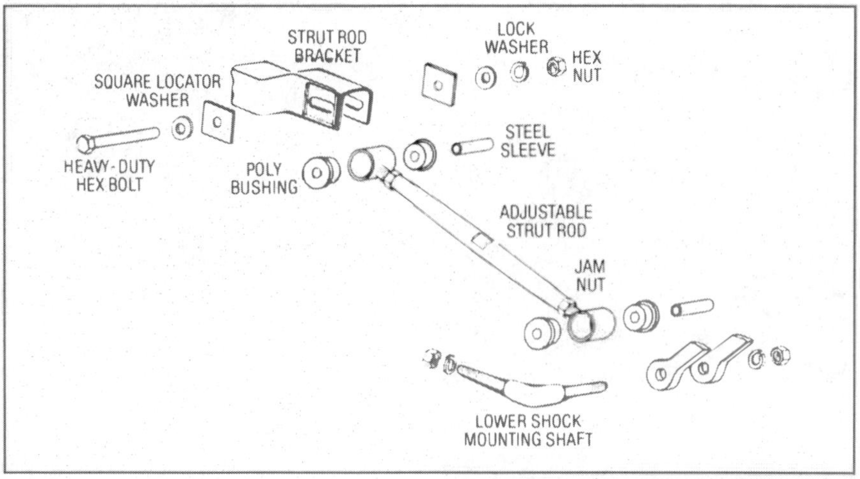

ADJUSTING REAR CAMBER

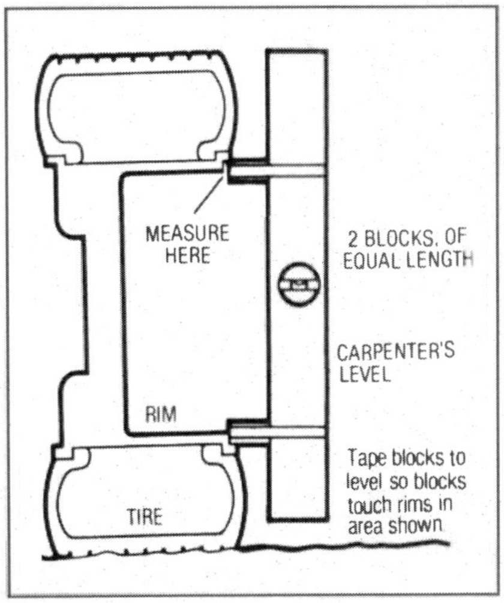

▲ *PX8. A simple way to estimate camber can be done with two identical wooden blocks and a carpenter's spirit level. Tape the two blocks onto the side of a level. Stand the level in a vertical position parallel to the tires. Blocks will be at right angles to the tire. Adjust the camber so that the blocks barely graze the edges of the tire rims. To achieve a beginning point of ½ degree negative camber, the upper block should be about ⅛ in. distance from the upper tire rim on a 15 in. wheel. Adjust the strut rods to the desired setting. Each complete turn of the strut rod adds or subtracts approximately ⅛ in. in block-to-tire measurement, or ½ degree of camber.* (Courtesy Vette Brakes and Products Inc.)

▶ *PX9. Because of their lowly position in life, camber adjustment cams can rust to the point where it's hard to get correct adjustment, or where they simply freeze altogether. Replacing with stock cams is the answer.* (Courtesy Vette Brakes and Products Inc.)

▶ *PX10. If only adjustment were really as easy as this! On a restoration project car with the body off, all is crystal clear. In real life, it means getting under the car where encrusted road dirt and rust makes it a dirty job.*

▼ *PX11. Installation of rear end links will be easier with the rear of the car jacked up – remembering the safety precautions – don't get under a car supported only by a jack. Then, with the car sitting at the normal ride height, tighten the lock nuts until there is approximately a 0.020 in. to 0.030 in. gap between the spring coils. CAUTION: Do not overtighten. If the car corners too flat or if the tire contact isn't correct, back off the adjustments to get proper roll coupling and handling.* (Courtesy Vette Brakes and Products Inc.)

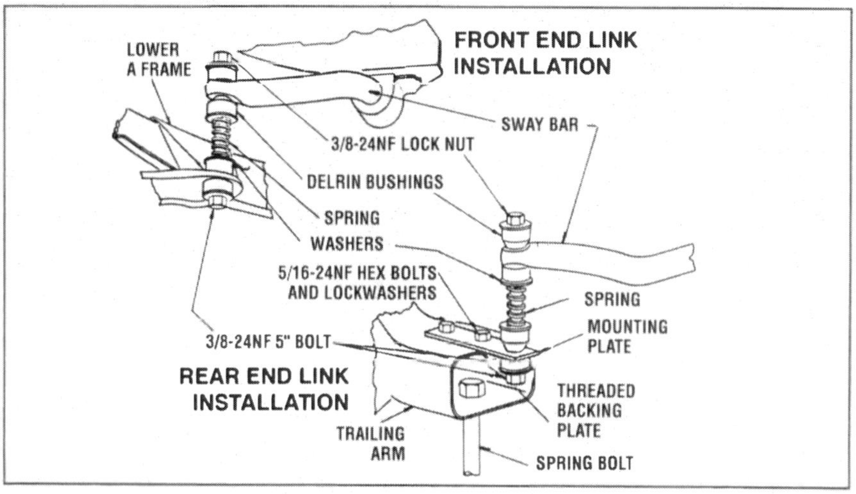

REAR WHEEL BEARINGS

The weakest parts of 1963–1982 Corvettes are its trouble-prone rear wheel bearings. However, much of the pain and suffering of this once-dreaded task has been eliminated by a new system of servicing and repair; replacement of the rear wheel bearings as an assembly. Of course, this problem is only relevant to 1963 and later Corvettes since that was the year that GM introduced independent rear suspension. The following doesn't apply to earlier models.

Some manuals and handbooks recommend regreasing the rear wheel bearings and also readjusting them for play. In practice, and according to Corvette expert John Pfanstiehl, both of these are likely to cause far more problems than they solve. Basically, there are what John calls 'design flaws' in the Corvette's rear bearings between 1963 and 1982 and the car has become progressively more prone to problems as the years go by because extra loadings are placed on them by increases in tire size and weight.

▲ *WB1. A small adapter tool for repacking rear wheel bearings is available from Chevrolet and from aftermarket suppliers, but not only is it a huge amount of work to use the tool, because of the amount of dismantling that's needed, it is just the inner bearing that gets any grease. Worse still, many mechanics don't understand the correct loading to put on the bearing when they reassemble it and wheel bearing failure has often occurred after an unskilled mechanic has over-tightened the bearing preload. In short, the grease that was put into a Corvette's wheel bearings when they were first built-up is unlikely to go anywhere, so regreasing them could be a waste of time and could well cause more problems than it solves. Consider leaving well alone! And the same applies to any workshop manual recommendations to adjust the bearings. Bearings don't wear down in the same way as bushings, so it's unlikely that extra play will be found in the bearing compared with what was there when it was new, unless the bearing is way beyond the point of no return – see later.* (Courtesy Vette Brakes & Products Inc.)

▲ *WB2. Most cars wheel bearings give the driver plenty of attention that they are about to fail, from increased rumbling and noise, especially on cornering, and the same is true of Corvette front wheel bearings. Failure is sudden and without warning, and when this happens at highway speeds, the bearings can literally friction melt, as these examples show.*

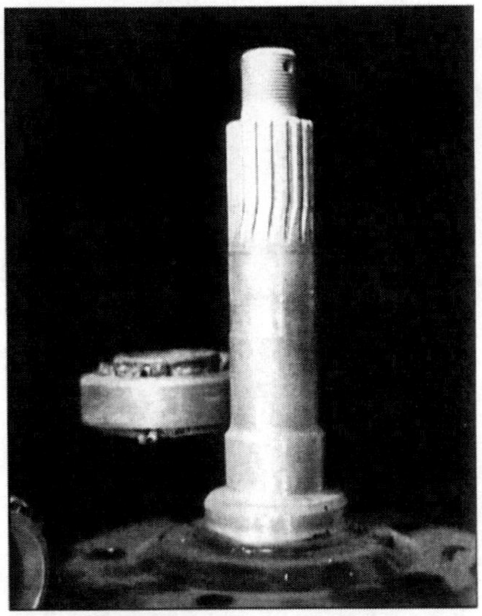

▲ *WB3. Another common occurrence is for spindles to be so affected by the heat of a bearing failure that they twist. The higher horsepower Corvettes use stronger types of spindles so this is far less likely to happen.*

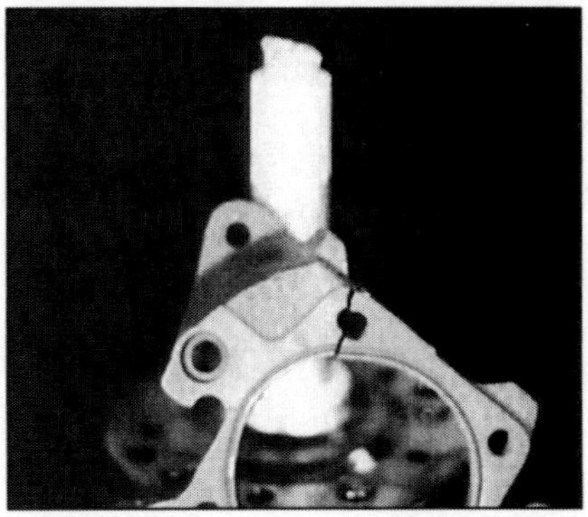

▲ *WB4. There is a special works spindle press for these bearings, but John Pfanstiehl claims that caliper mounting brackets can be cracked or bent when this press is used. In fact, he says that quite a number of expensive parts can be ruined during the repair or servicing of Corvette wheel bearings.*

PREVENTIVE STEPS

The traditional way of checking wheel bearings, by spinning the wheels and listening for noise, won't tell you anything because of the noise in the rest of the drive line on independent rear suspension cars. John Pfanstiehl's recommended solution is to remove the brake pads or caliper, disconnect the axle shaft, ensure the parking brake shoes are backed away from the rotor and then spin the rotor and listen. If you hear any roughness or feel any vibration, you should remove the spindle flange for a close inspection, and then the cage and sides of the inner bearing rollers can be seen. Any evidence of pitting or wear means that the bearings are on their way out and you should consider the solution suggested later in this section.

You can carry out a simpler check on bearing play but this won't necessarily tell you a great deal, except that if there is excessive play, there's every chance that the bearing will fail imminently, and if there's no play at all, ditto. You can lift the car and support it on jack stands by the frame. Put one hand on the bottom of the tire and the other on the top and pull out on the bottom while pushing in on the top. If you do this hard, you will very likely discover a large amount of play, but if you take a look, you'll see that this is because of the axle shaft and yoke moving in and out of the differential. If a workshop tells you you have excessive play in the rear end, check that this is not what they have got in mind. Play of this sort can be quite alarming but is common and not necessarily a problem. Before you get to that large amount of play, you should feel a smaller amount of play in the wheel bearing itself. It may help to detect the play if you have an assistant move the wheel while you place your hand on the wheel bearing assembly. You should be able to feel the play as well as see it. Remember, if there's none there, this is potentially an even bigger problem!

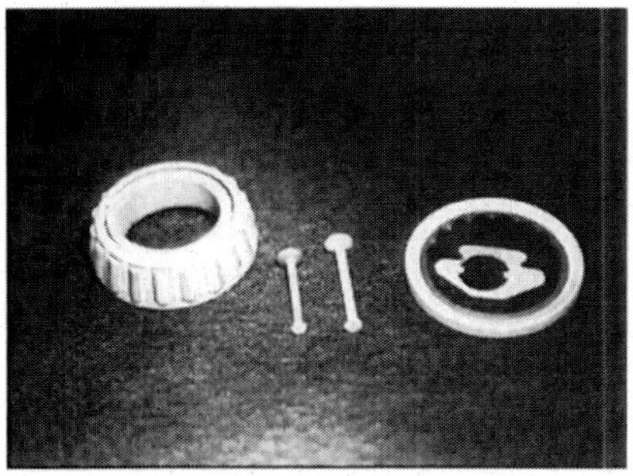

▲ *WB5. This photograph is from John Pfanstiehl's chamber of horrors, and illustrates new Corvette aftermarket rear wheel bearing replacement parts that potentially add to the problem rather than solving it. Here's a parking brake hold down pin that will allow the shoes to hit the spindle, a badly mishapen anchor tab and a seal that might be able to catch fish, but sure as hell couldn't seal.*

SOLUTION

▼ *WB6. It may appear more expensive in the short run, but it could very well work out far less expensive and far safer in the long run to replace the wheel bearing assembly with the kind of built-up completed units that are now available for Corvettes. This has been constructed by a specialist workshop who know what the problems are and have overcome them on the workshop bench. It won't necessarily make the bearings last any longer, but at least you know that the unit will be right to start off with. It can be fitted with your stock rotor or with a new one and saves plenty of time and potential trouble.* (Courtesy Vette Brakes and Products Inc.)

FOOTNOTE: So you still fancy fitting wider wheels to your Corvette? When you look at the potential problems with wheel bearings on '63–'82 Corvettes, you suddenly realize that adding extra stress to those rear wheel bearings isn't such a good idea after all.

PART 5 BRAKES

One of the worst Corvette problems concerns the cast-iron brake calipers with their four pistons. Disc brakes were introduced for the last three models of the original Sting Ray series and were modified for 1967 to remove the first headache – corrosion behind the hard plastic insulator at the rear of the disc pad and piston. This had been very sensibly incorporated to prevent heat from the pad boiling the fluid, but had been found unnecessary in normal road use. Corrosion of the aluminium piston often resulted in the insulator cracking and breaking off, allowing road dirt and water past the outside dust boot.

Unfortunately, not only does water penetrate and rust the polished bores of the cast-iron caliper from the outside, but moisture also reaches the bores from the inside. What apparently happens is that the brake fluid itself, being hygroscopic, absorbs the water from the atmosphere, rusting the bores and corroding the aluminium pistons.

The result of this is that the air is drawn past the seal which no longer has a polished seat and leads to the spongy brake pedal feel familiar to Corvette owners all over the world.

This kind of corrosion takes around four years to mature from new. Fluid does not necessarily pour down the inside face of the wheels and tires, and the caliper can look quite dry externally. Air will, however, still be drawn into the system, either when the brake pedal is released after application or as a result of the pumping action of a slightly out of true brake rotor or loose wheel bearing.

Criticism of GM is easy, but it has to be remembered that, apart from giving a slightly high unsprung weight, the Corvette brakes are superb. And they would point out (and we would wholeheartedly agree) that standard brake fluid should always be changed every two years to prevent just such problems; the propensity for the fluid to absorb water is known the world over and doesn't just affect the 'Vette.

Alternatively, by draining off the original and switching to silicone brake fluid, which is non-hygroscopic, you can remove the possibility of internal rusting.

Note: Many car manufactures now recommend renewing brake fluid *every year!*

BRAKE MASTER CYLINDER

▲ *MS1. The hydraulic system consists of two separate circuits, one operating the front brakes and one operating the rear brakes. The master cylinder has separate reservoirs for the two circuits, and in the event of a leak or failure in one, the other will remain operational. A visual warning of circuit failure or air in the system is given by a warning light activated by displacement of the piston in the pressure differential switch portion of the combination valve from its normal 'in balance' position. On models from 1968 to 1975 the valve includes a pressure switch which illuminates a warning light if hydraulic pressure drops in either the front or rear brake circuits. The valve is designed to assure full pressure to one brake system should the other fail The switch will stay in the 'warning' position once a failure has occurred, and the only way to turn the light off is to repair the cause of the failure and apply a very high brake pedal pressure. On post '75 models, the valve also includes a proportioning function, whereby under heavy braking it reduces hydraulic pressure to the rear brakes to prevent early rear wheel lock-up. Note that the fluid level should always be within ¼ in. of the top of the reservoir.*

▲ *MS2. Conventional brake fluid is particularly corrosive to paintwork, so it pays to be extremely careful when you're dismantling your braking system! It is also hygroscopic – it absorbs water. This means that, over a period of time, not only will your braking become less efficient (as the fluid and the moisture have different boiling points) but also your steel brake pipes will start to corrode from the inside. These problems can be avoided by using silicone brake fluid.* (Courtesy Vette Brakes & Products Inc.)

MASTER CYLINDER

▲ *MS3. The master cylinder is attached to the firewall or the power brake booster, depending on which Corvette model you have. In this case, it was the former. Removal is quite simple, just two nuts hold the cylinder in place, and with the hydraulic lines disconnected (careful with that brake fluid!), it can be drawn away from the car. Once clear of the car, pour out excess fluid from the cylinder into a bucket, or similar, and then press in the pushrod by hand to expel any fluid remaining.*

SAFETY: We strongly recommend that you do not consider rebuilding brake master cylinders or calipers unless you are a trained mechanic and fully competent to do so. Take advantage of one of the fully built-up units as illustrated here.

▲ *MS4. If you are rebuilding your master cylinder, take great care to prevent the ingress of dirt once the top is removed. You can see the twin cylinders quite clearly here, there being one for each circuit – the cylinder at right here (nearest the front of the car) is the one for the front wheels, the cylinder at left operates on the rear wheels.*

▲ *MS5. There are two pistons, a primary (seen here on the right) and a secondary. To remove them, grip the cylinder in a vise, depress the pushrod and prize out the locking ring from the inside of the cylinder. (Bendix versions have a stop screw on the bottom of the cylinder which has to be removed before the pistons can be released.) Having got this far it's wise to replace at least the seals, or better still ...*

▲ MS6. ... do as Em did and replace the seals and the primary piston assembly. Seal rebuild kits and piston/seal kits are available as proprietary items. Replacing the pistons into the cylinder is a reversal of the disassembly procedure, though the pistons need dipping in clean brake fluid before insertion.

▲ MS7. If the bores of the master cylinder are scored too badly to be reusable, then you can obtain replacements as a complete unit quite easily, either as factory rebuilt units or for most models new master cylinders are still available. (Courtesy Vette Brakes & Products Inc.)

▲ MS8. Replacement master cylinders are also available for power brakes, complete with the power booster unit, or power boosters are available separately. (Courtesy Vette Brakes & Products Inc.)

BRAKE HOSES AND PIPES

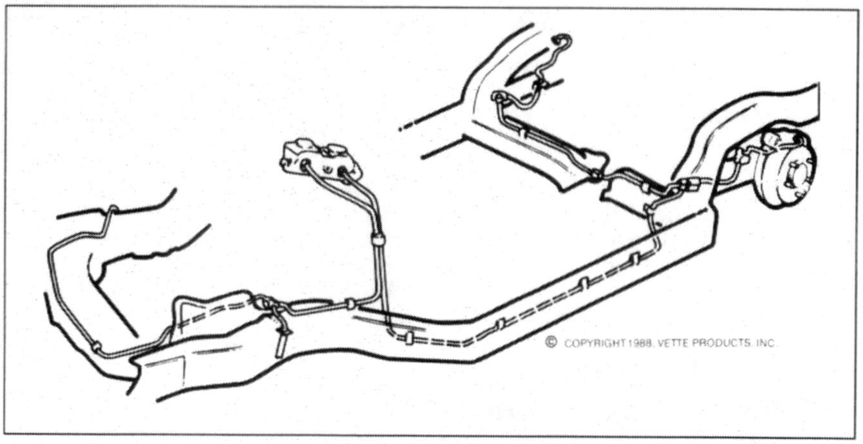

▲ PH1. The hydraulic pressure is carried from the master cylinder to the individual wheel cylinders by a system of steel brake lines and flexible hoses. It is important to note that you should only ever use brake pipes with double flare joints for a stronger seal. And it's equally important to note that when replacing metal brake lines, only special steel tubing should be used – this is designed to withstand the high pressures developed by the hydraulic braking system and to resist corrosion. Ordinary copper tubing should NOT be used. This diagram gives an indication of the basic Corvette brake line layout. (Courtesy Vette Brakes and Products Inc.)

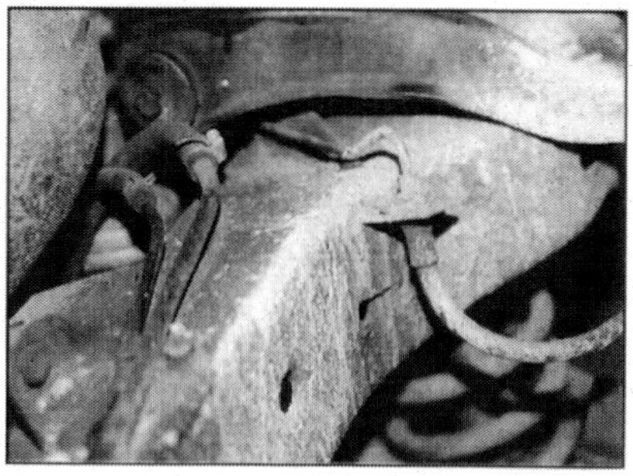

▲ PH2. Brake connections on a Corvette of any age will require a little forethought if problems are to be avoided. Years of all the elements can throw at them is going to make those threaded unions reluctant to open up. They're likely to be covered in a thick layer of encrusted road dirt, so the first job is to wire-brush the flats and threads of the flare nut. This is frequently a trouble spot and it is wise to clean here ...

▲ PH3. ... and spray plenty of penetrating oil, preferably a few days before attempting to break it free. A ⅜ in. tubing wrench is strongly recommended. Watch the tubing to make sure it is not twisting as the flare nut is turned. If it does so, try gripping carefully with pliers. If the brake line has to be replaced, the driver's side brake line is very short and therefore relatively easy to change compared to the other side.

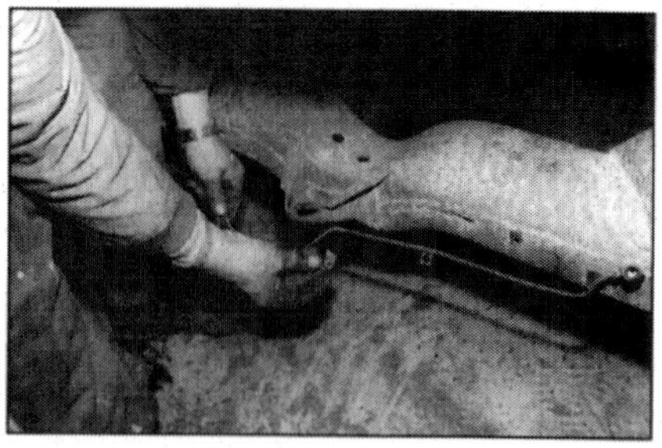

▲ PH4. The metal brake lines are secured to the chassis by brackets. Even if you're replacing the brake lines, take some care removing these brackets. As usual, you'll probably need plenty of penetrating fluid. If you're bending your own replacement pipes, don't accidentally bend the originals as you remove them – it's easily done.

▶ PH5. *Where the various lines meet, they are joined by a junction box. Take careful note of the intricate way the pipes are bent to wind an accurate path around the car's innards.*

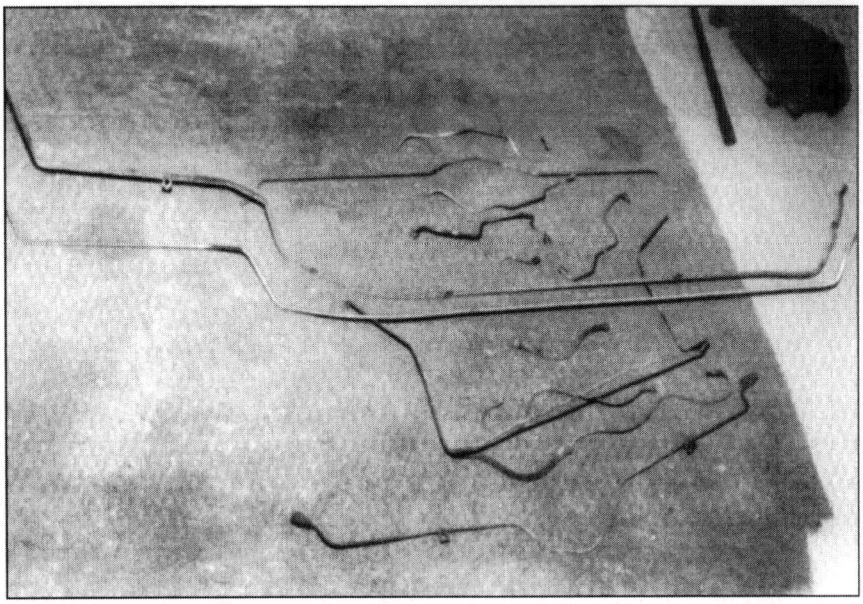

▲ PH6. *All lengths, shapes and angles; the collection of steel brake lines after they'd been removed from the 'project' '69 big block Corvette. If your car is as old as this, reusing old brake lines is not to be advised. Even if there is no obvious sign of corrosion, it could still be there working from the inside, thanks to the hygroscopic nature of standard brake fluid.*

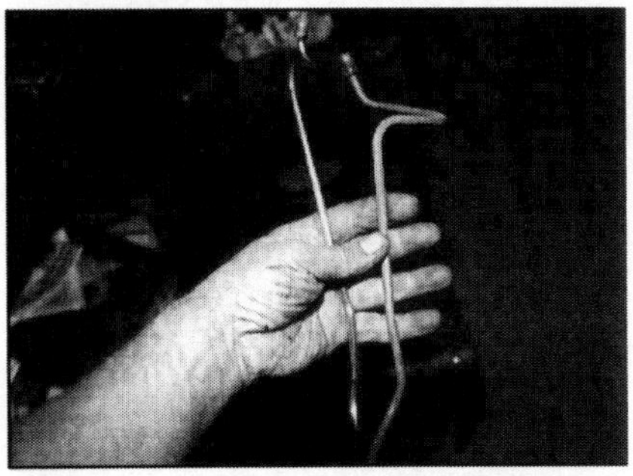

▲ PH7. *It is possible to take straight lengths of steel tubing and create new brake line sections, but you need to be very skilled: the new item must mimic exactly the original, and don't forget, you're working in 3D! Em used many years of experience to create the new one (left) to fit precisely where the old one went.*

TRAILING ARM BRAKE LINES

▼ PH8. *But for most of us, it's better to buy new pipes, ready to fit. These shown here are the trailing arm brake lines. They are spiral armoured with double flared ends and have been pressure-tested. Unless you are a wizard at creating perfect flared joins, this is the way to go – you only get one chance to get your braking system right!* (Courtesy Vette Brakes and Products Inc.)

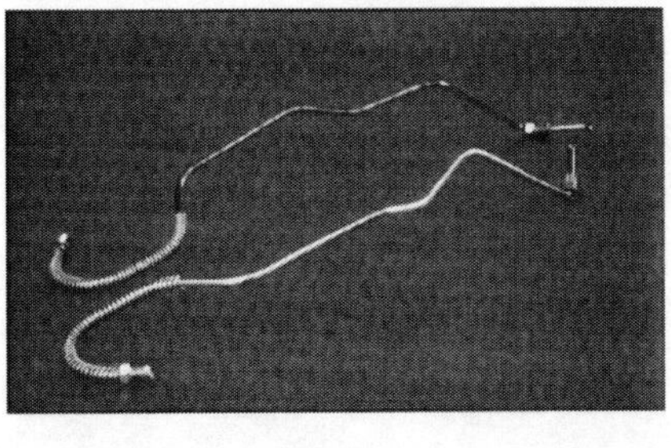

STOCK RUBBER BRAKE LINES

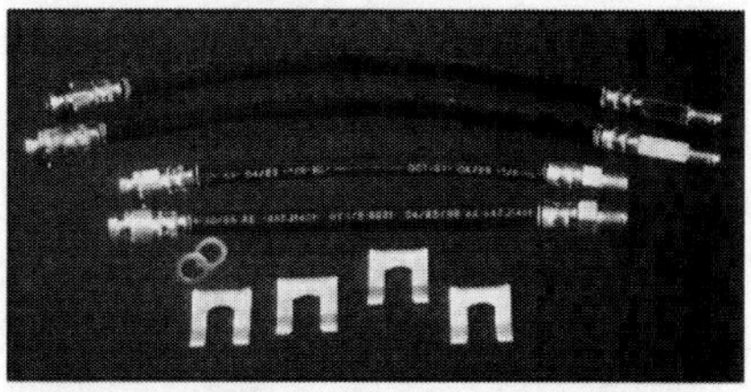

▲ PH9. *The flexible rubber hoses which carry the hydraulic pressure from the steel lines to the wheel cylinders should be of equally high quality. If they are not, they could bulge (which itself will cause a drop-off in braking performance) and ultimately split. They have to be flexible and pliable, but at the same time immensely strong in order to combat the twisting forces which pull them every which way as the car is driven. Again, buying stock lengths with the unions ready fitted is the best option.* (Courtesy Vette Brakes and Products Inc.)

BRAIDED BRAKE LINES

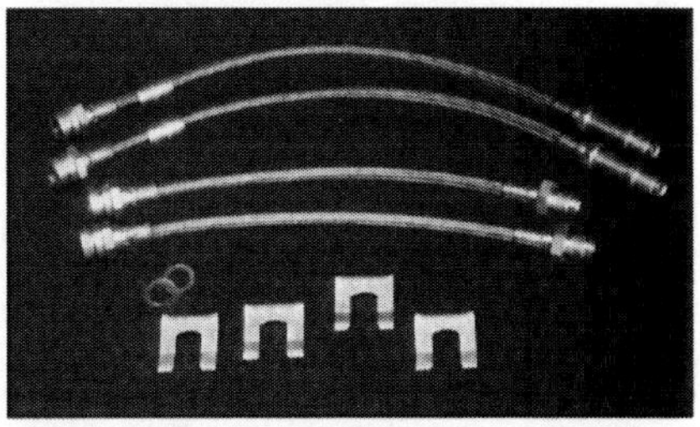

▲ PH10. *An alternative is to use braided brake lines as used in most forms of motor racing. These really are the ultimate!* (Courtesy Vette Brakes and Products Inc.)

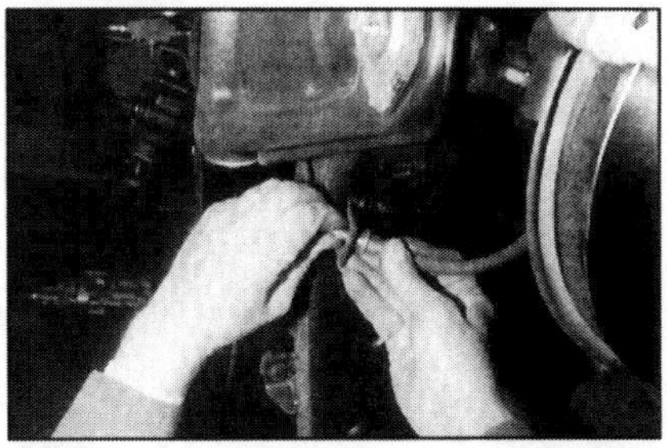

▲ PH11. *The brake line support brackets have a 12 point hole to aid location. When fitting the new hoses, it is important that they should be inserted at the angle which causes them to twist the least – they will be under enough stress without giving them some extra to cope with.*

▲ *PH12. New brackets, new metal lines and a totally restored chassis frame – now is the time to apply plenty of wax rustproofing! Make sure that the lines follow the original route and that they aren't bent or twisted as you tighten the bracket securing nuts.*

ROAD TESTING

When your brakes have received any attention (from a complete tear-down to a simple pad change) it is vital that you have a qualified workshop check them thoroughly before you use the car 'in anger'. When the brakes have been checked they should be quick and responsive with plenty of 'bite'. A spongy pedal indicates a system which has not been bled correctly. Similarly, if the car veers to one side during heavy braking, something is seriously amiss – possibly a sticking caliper or contaminated pads.

Before even taking to the highway, get the car off the ground (remembering the safety rules – see Appendix 1) and check all metal lines and flexible hoses for leaks, paying particular attention to the unions.

In addition, check carefully around each wheel to ensure that fluid is not leaking from the wheel cylinders or from around the bleeder. After pumping the pedal, make sure that the fluid level in the master cylinder has not noticeably diminished.

It's a time-consuming process, but it's better to dig out any potential problem areas sooner rather than later – when it may be too late!

FRONT BRAKE OVERHAUL

Make sure that the vehicle is secure before you start working underneath – it should never be supported only by the wheel-changing jack – use solid axle stands and chock and block the wheels where appropriate. Please read the Safety section in Appendix 1 before you start working on your car's brakes.

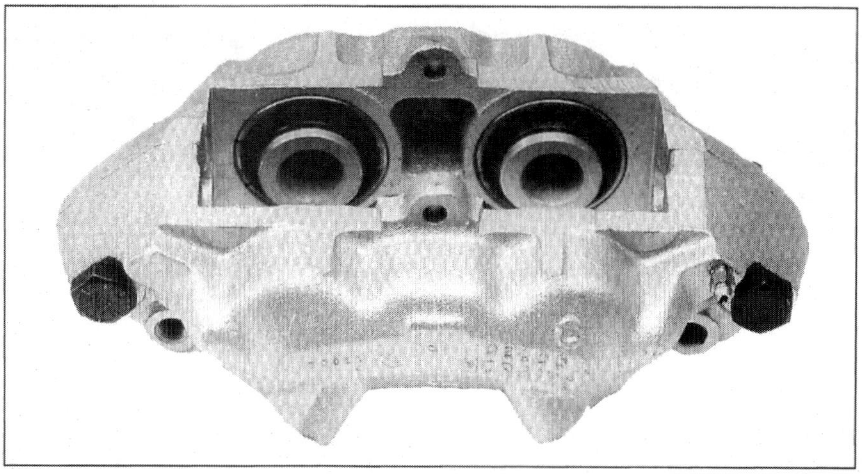

▲ *DB1. When you start work on your braking system, you may find that your car already has stainless steel (lined) calipers fitted already. If you're not sure, look for a number or other identifier stamped on the outside of the caliper, usually on the flat, milled area near the top of the pads. If you haven't got stainless calipers, they're a worthwhile investment when you come to overhaul your front brakes.* (Courtesy Vette Brakes & Products Inc.)

If you are looking to change the pads (or shoes on older cars) remove the cover from the brake master cylinder, making sure that it is placed where it cannot become dirty and that no dirt can enter the master cylinder itself. Place a rag around the top of the master cylinder to soak up any fluid which spills out as the calipers are eased apart. Don't let that fluid get on your bodywork, it dulls or even strips paintwork!

DISMANTLING

As usual, it's a good idea to work on one side at a time. That way, if you get in a fix, you can use the side you haven't touched as a reference point to see how it should be done. If you're in the process of a complete restoration, where you will not be reassembling for weeks, or even months, it's a good idea to take plenty of photos or use a video camera – either will aid greatly when it comes to put everything back together.

BRAKE PADS

Always replace brake pads in sets – never replace the pads on only one wheel. Also, the dust created by the brake system may contain asbestos which is extremely harmful to your health. NEVER blow it out with compressed air and don't inhale it. An approved filter mask should be worn when working on the brakes, and under no circumstances use gasoline-based solvents to clean brake parts. Use brake cleaner or denatured alcohol only!

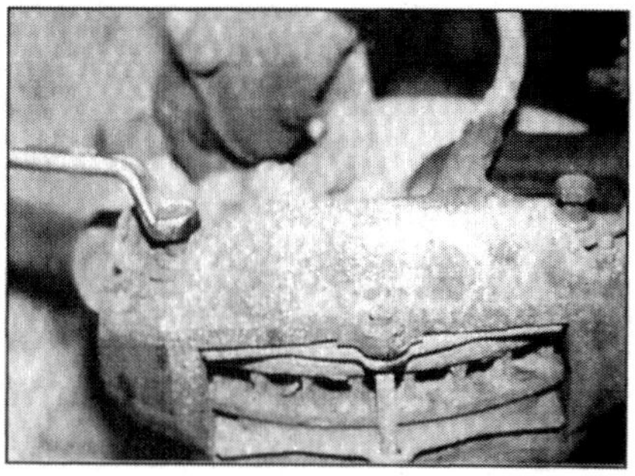

▲ *DB2. Calipers don't have to be in place long before they start to rust up like this one. They are likely to be tricky to remove, especially if they have started to corrode badly. Penetrating oil, patience and a good strong 5/8 in. wrench are required.*

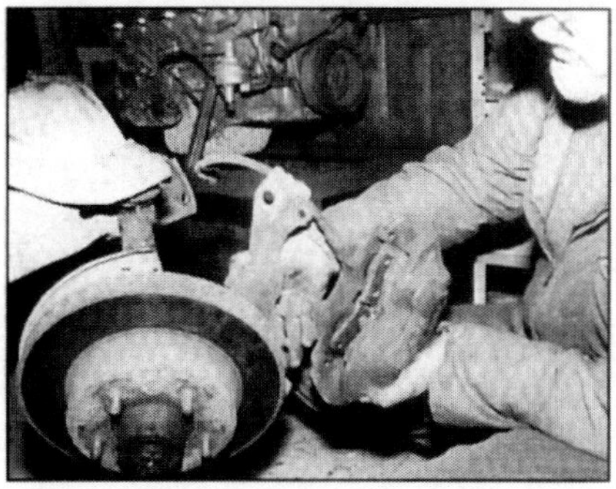

▲ *DB3. With these bolts removed, the caliper, complete with pads and pistons, will simply pull away from the rotor. Note that the brake line from the caliper to the frame is still attached. This is because we had the luxury of having the body off. In most cases, before the caliper is removed, the brake hose at the frame bracket should be detached.*

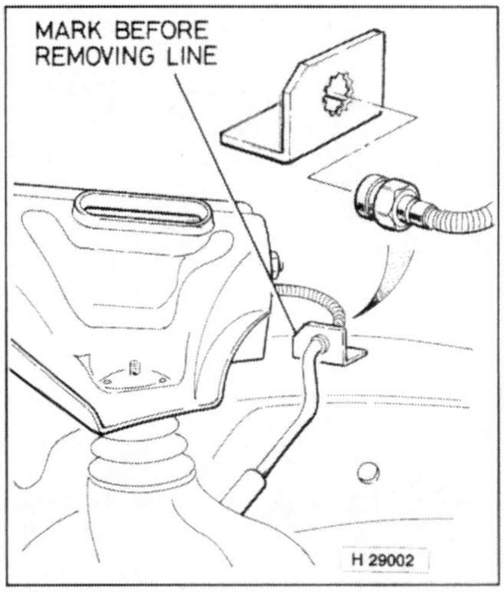

▲ *DB4. You should use a backup wrench to avoid twisting the fuel line. Use a flare nut wrench to avoid rounding the corners off the fitting and have a cloth ready to mop up any spillage as the pipe comes adrift. It's a good idea to mark the position of the brake hose to the frame bracket so that the hose isn't twisted when it's reinstalled.*

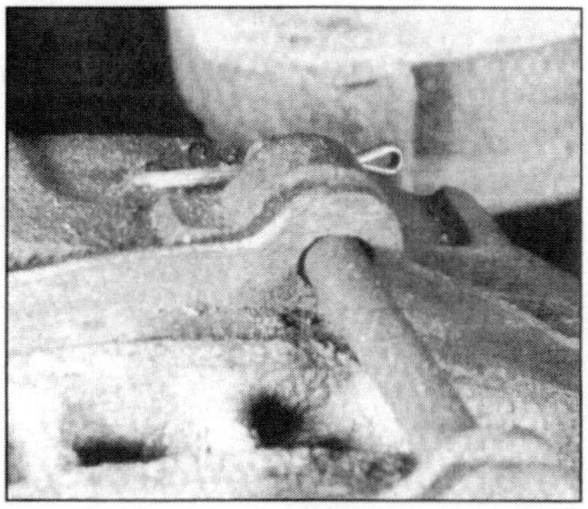

▲ *DB5. To get out the pads, the pad retaining pin has to be removed. This is held in place by a cotter pin which has to be withdrawn using needle-nose pliers …*

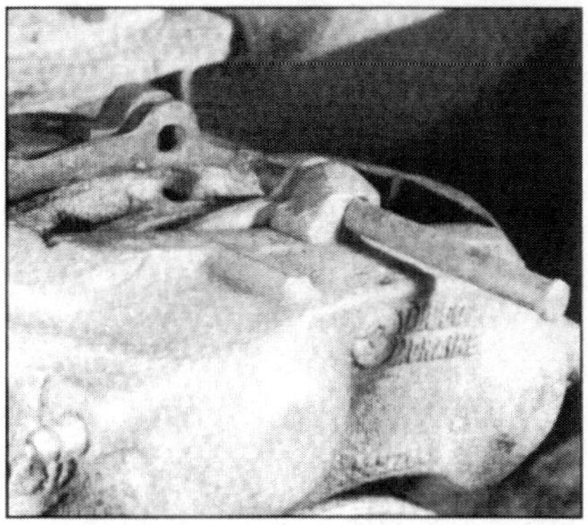

▲ *DB6. … before the pad retaining pin can be eased away.*

▲ *DB7. The pads should then pull out easily. If they are reluctant to come out, use the hole in the top of the pad backing plate to waggle them a little or for inserting a strong screwdriver to add a little leverage. With the pads removed, note that all the pistons should be approximately the same distance out from the caliper. If the caliper is not centred on the rotor, the installation of the thicker (new) pads could cause the brakes to lock up. It is not uncommon to find this situation on the rear, caused by a caliper mounting bracket being bent during wheel bearing work.*

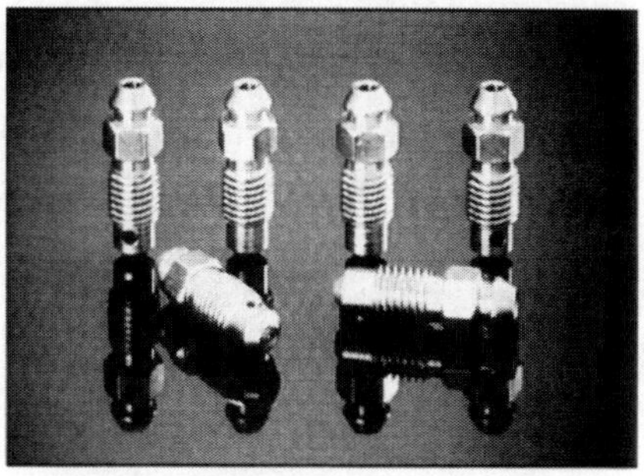

▲ *DB8. One problem you're likely to have at this stage is that the bleeder screws will have rusted solid. The upshot is often that they have to be drilled out. Replacement with stainless bleeders and plugs will save you having this hassle next time you service your brakes.* (Courtesy Vette Brakes & Products Inc.)

REMOVING THE FRONT ROTORS

▲ *DB9. Before removing the rotors it's important to make a visual check on their condition, both inside and out. Light scratches and shallow grooves are no cause for concern, but deeper score marks (0.015–0.38mm plus) are a more serious problem. It is possible to get rotors skimmed, although it may be more cost-effective in the long run to replace the rotors altogether – always as a pair, of course, even if one rotor is unmarked.* (Courtesy Vette Brakes & Products Inc.)

CHECKING ROTOR RUNOUT

It's a good idea at this point to check the runout of the rotor, especially if you've been experiencing some degree of judder as you brake. Place a dial indicator against the outer edge of the rotor, set the indicator to zero and turn the rotor slowly by hand. Compare the runout with that listed in the specifications for your particular model. Anything under 0.002 in. is excellent, but readings over 0.004 in. can cause problems such as pumping of air into the system, resulting in a loss of brake pedal. Don't take chances – if you're not sure, replace them! Even when you've fitted new rotors, it's wise to run this check once they're fitted.

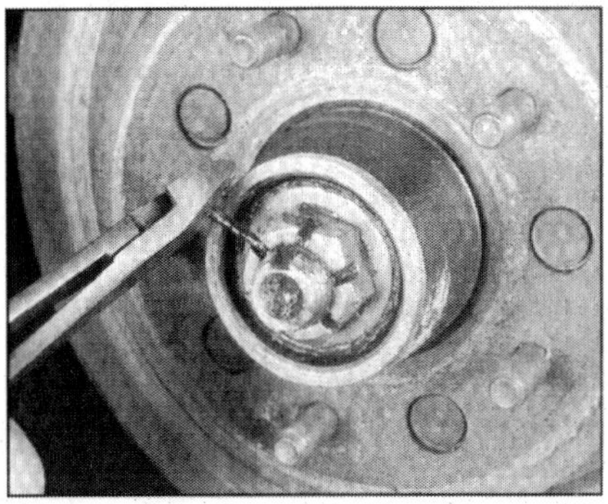

▲ *DB10. Having removed the calipers (as described earlier) the rotors can be removed. After easing out the wheel bearing dust cap the cotter pin must be removed. Whatever you do, DON'T reuse these, but replace with a new pin every single time you remove one.*

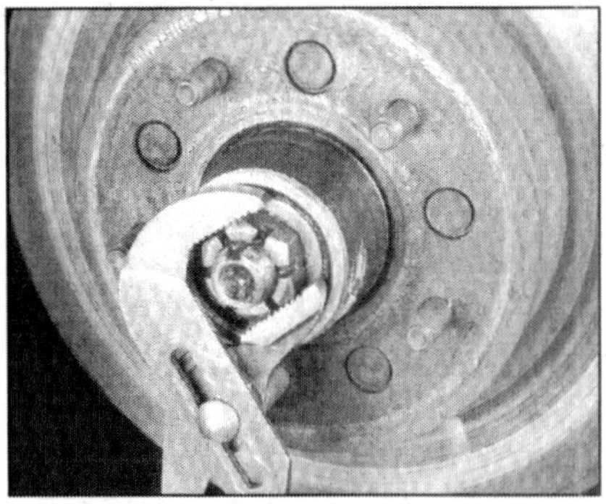

▲ *DB11. The spindle nut should be quite easy to release, as it is required only to be just more than finger tight so as not to ruin the bearing which lies beneath.*

▲ *DB12. You may have to wiggle the hub/rotor assembly slightly to encourage the outer wheel bearing from its grease-packed home. At this point, you can pull the hub assembly off the spindle.*

▲ *DB13. Once the hub assembly was off the spindle, Em needed to separate hub from rotor. So, the rivets were dot-punched ...*

▲ *DB14. ... prior to drilling out the centres. It is important to ensure that the rotor and hub cannot move while the drilling operation is being carried out. Em used a pillar drill and vise.*

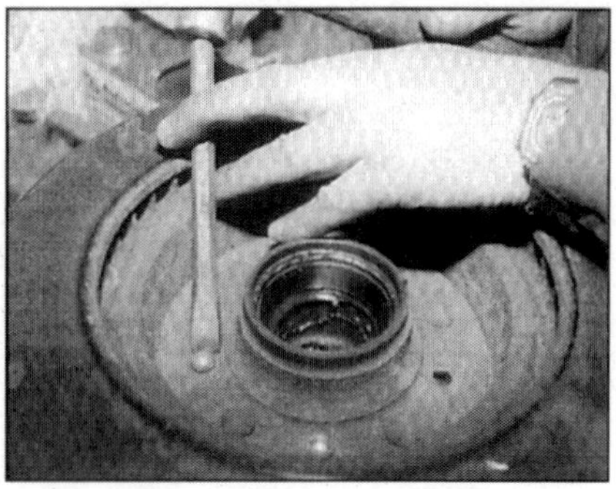

▲ *DB15. With holes drilled right through their heads, these rivets weren't in a position to put up much of a fight when attacked with a hammer and sharp cold chisel. Don't forget that eye protection is essential when performing this kind of work.*

▲ *DB16. With the rivets out, the two components can be separated.*

▲ *DB17. The seal in the rear of the hub is quite simple to remove, requiring just a screwdriver or similar to prize it out.*

▲ DB18. However, the inner seal needs a little more persuasion – that hammer and chisel again …

▲ DB19. … before it can be removed. Take extreme care not to damage the surfaces inside the hub.

STOCK OR UPGRADE?

▲ DB20. Unless your aim in life is to beat all in the Concourse displays, it's often a good idea to upgrade while components are stripped down. There are many and various ways you can do this. For example, the rotors and calipers seen here offer 25 per cent more braking power for 1985–87 Corvettes. They are OEM Girling two-piston calipers and heavy-duty 13 in. rotors designed for use with factory 17 in. wheels (aftermarket 17 in. wheels may not have the correct caliper clearance – check before fitting). (Courtesy Vette Brakes & Products Inc.)

▶ DB21. Another upgrade idea is to opt for aluminium hubs. Those shown here are precision machined and are matched to heavier-than-stock vented rotors for improved strength and heat transference. (Courtesy Vette Brakes & Products Inc.)

CLEANING AND REPLACING THE FRONT HUBS/ROTORS

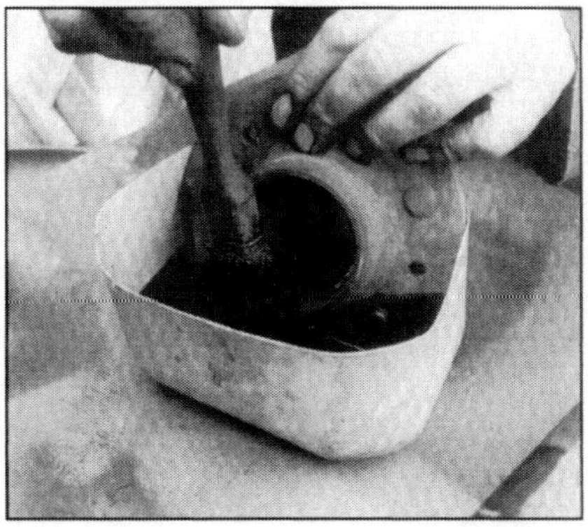

▲ DB22. Having taken out the bearings and seals, the next job is to thoroughly clean and degrease the hub, scraping off any corrosion and crud.

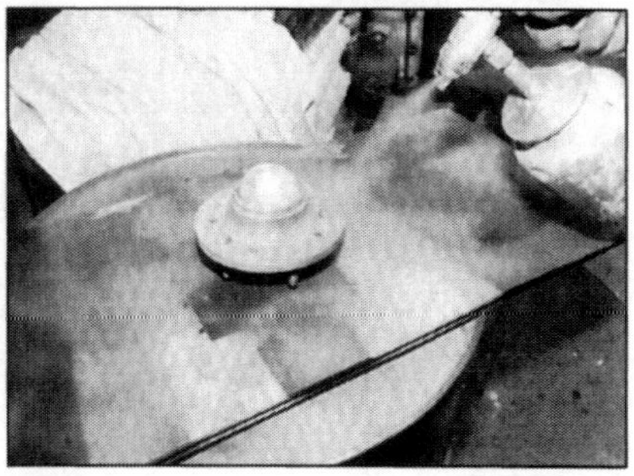

▲ DB23. Note the fitment of a very non-standard aerosol cap to keep the paint away from the bearing surfaces inside the hub. A little ingenuity goes a long way in restoration work!

▲ DB24. The new outer bearing is being offered up in position by hand …

▲ DB2 5 … but it requires a mechanical press to fully push the bearing home. If you try doing this with a bench vise, note that it is essential that the bearing goes in perfectly level and evenly.

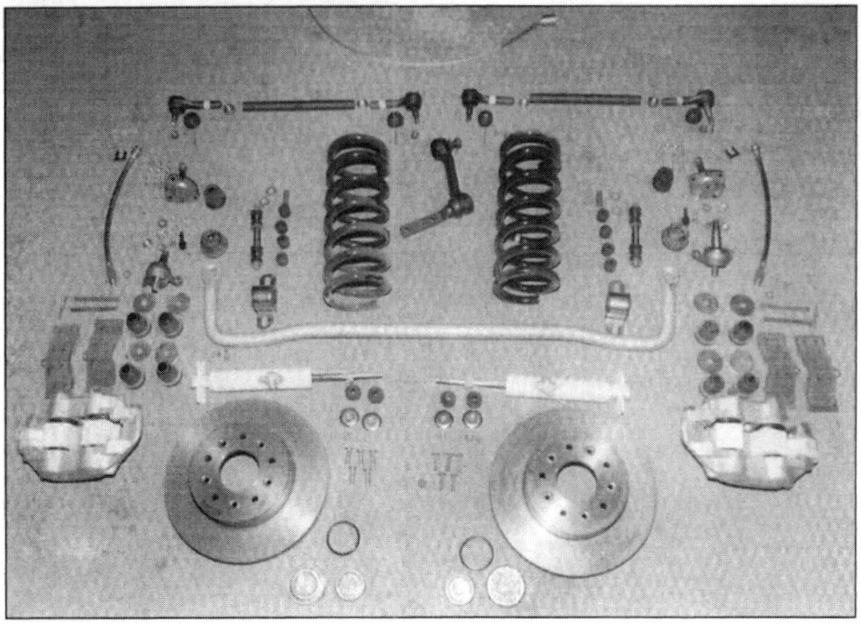

▲ *DB26. A schematic layout of the front brake/suspension parts prior to installation. Some of the parts, such as the calipers, were restored, but others, such as the rotors, were new replacements from Vette Brakes & Products Inc. It is not usually necessary to turn rotors unless they have been really badly chewed-up by a worn pad. But in our experience, turning rotors on a lathe often increases the run-out and, in any case, Corvette rotors don't have much metal to lose before they are out of spec. Check your manual for the minimum permitted thickness for your model of Corvette. If you have only a little glaze on your rotor, a little scuffing with 220-grit paper should be sufficient. The rim makes a handy holder while sanding the inner surface, but make sure that you cover the bearing area to keep out that damaging grit.*

▲ *DB27. Though these rotors were riveted to the hub, it is not strictly essential for them to be re-riveted on reassembly, as they are held in place by the calipers and the road wheel itself. (The rivets were fitted by GM to aid the vehicle assembly process on the production line.) However, in order to retain something in the way of original features, Em chose to do so. Having finished the riveting operation, the inner bearing seal was installed with a little grease around both seals to stop them running dry …*

▲ *DB28. … and the new hub assembly fitted to the spindle. It is, of course, important to keep the new rotor free from grease, which means that the hands doing the work have to be clean!*

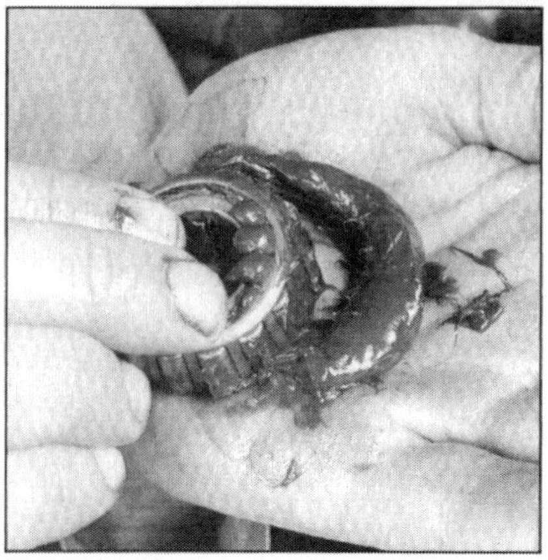

▲ *DB29. The new outer bearing should be well-greased before installation – work it well in to the bearing rollers on the palm of your hand …*

▲ *DB30. … before sliding the bearing into position on the spindle …*

▲ *DB31. … adding the washer …*

▲ *DB32. ... and the spindle nut. This should be turned to be barely more than hand tight – no more than 12 ft lb of torque. Spin the hub in a forward direction to seat the bearings and remove any grease or burrs which could cause excessive play later. Check that the tightness of the nut is still around 12 ft lb and then loosen the spindle nut until it is loose – no more. Finally, hand tighten (do NOT use a wrench of any kind) until it is snug. At this point, hub end play should be 0.001–0.005 in. If you don't have a dial indicator, which you will need to measure the end float, carry out the rest of the work as described here and then take your car to a qualified specialist or GM dealership and have them check the end float. Carrying out the work as described here will make your Corvette safe to drive carefully to your nearest dealership.*

Install a new cotter pin – always a new one! – through the hole in the spindle and spindle nut. If the nut slots don't line up with the spindle hole, loosen the nut the minimum amount until they do. Don't be tempted to tighten the nut until the slot lines up with the hole; over-tight bearings can lead to dangerous problems. You must not untighten the nut more than one-half flat to install the cotter pin. If you think you need to turn it more, you're not doing the job properly, so check the alignment of those holes once again!

Bend over the ends of the cotter pin until they're flat against the nut and, if necessary, cut off any extra length which could interfere with the dust cap. You can now install the dust cap, tapping it into place with a hammer.

CALIPER REPLACEMENT

▲ *DB33. We recommend that most Corvette restorers replace faulty calipers with new or rebuilt components. Only experienced, trained mechanics are recommended to stop and rebuild their own calipers.*

Before installing the caliper, fit both pads, their retaining pin and its cotter pin. If you are refitting a caliper to a vehicle where the hydraulic system is still connected, ensure that you have taken the cap off the master cylinder and placed large quantities of absorbent rag around the neck of the master cylinder so as to catch the brake fluid that will inevitably flow over the master cylinder neck.

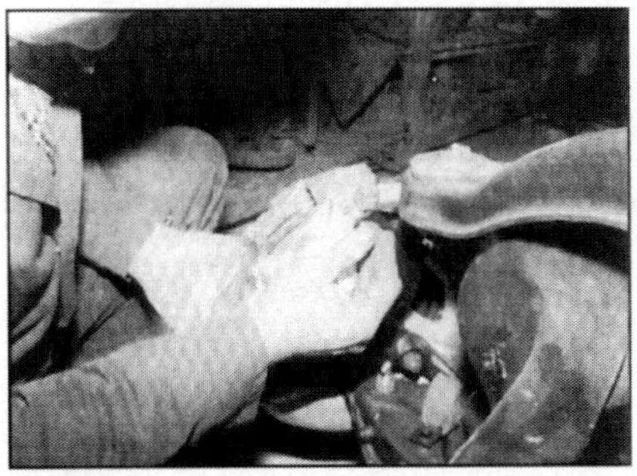

▲ *DB34. Install both pads, their retaining pin and its cotter pin. Spread the pads apart, pushing the pistons into the caliper and place a $\frac{5}{16}$ in. socket or an object of similar width between the pads. This makes it much easier to install the caliper assembly onto the rotor. Insert the pad retaining pin, making sure not to get any grease onto the pads themselves.*

TIP: File a small chamfer on the leading edges of the brake pads to reduce the chances of brake squeal. If any fingerprints get on the pad linings, use sandpaper to remove them.

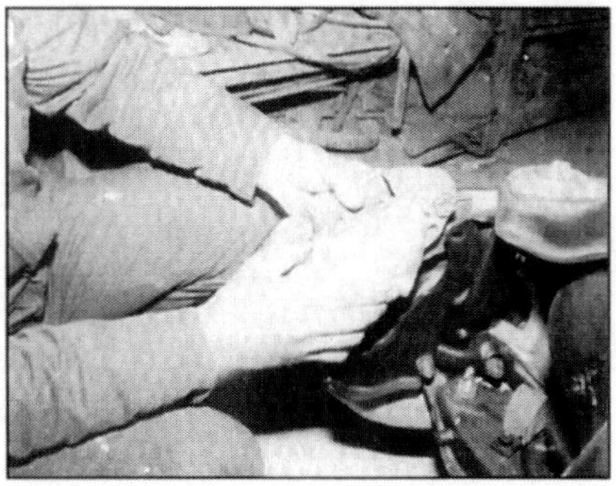

▲ *DB35. The caliper and piston assembly should then slide easily into place over the rotor allowing the mounting nuts and bolts to be fitted. After tightening the caliper to 130 ft lb (check your manual for any model variations), make sure there is clearance between the rear of the pads and the caliper and don't forget to fit a new cotter pin in the pad retaining pin.*

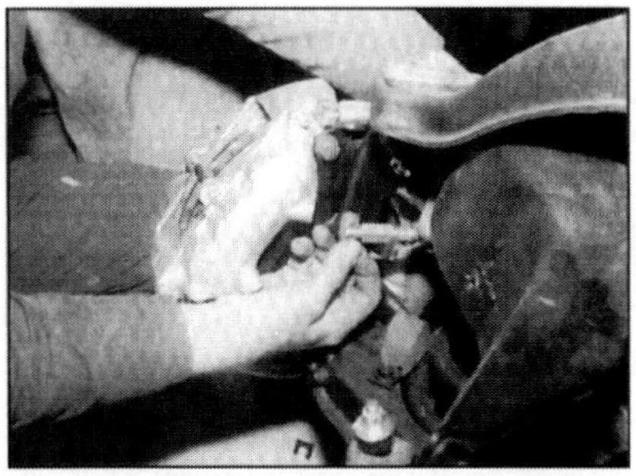

▲ *DB36. Install the brake hose and its copper gasket into the caliper and tighten it before connecting the other end to the brake line.*

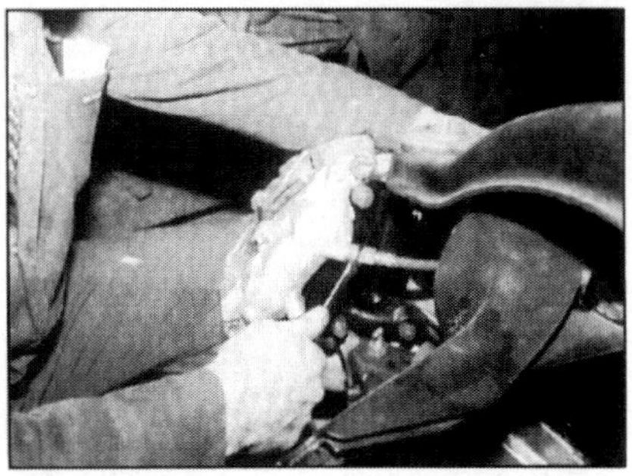

▲ *DB37. Make sure the flare nut turns freely into the brake hose and that it is not cross-threaded. Then turn the hose so it curves a bit away from the frame, and install the bracket clip. Turn the wheel all the way in both directions to check the positioning of the brake hose; it should never go under or against the*

frame, nor should it be positioned so it is pulled or rubs against any moving part. If it is OK, tighten the flare nut.

Unless the hydraulic system has not been broken into, you will have to replenish and bleed the hydraulics. We recommend that if you carry out this work yourself, you follow the instructions in your Haynes workshop manual and have a trained mechanic check your brakes before using the car on the road.

▼ *DB38. If you own an earlier Corvette, with drum brakes, there's no problem with obtaining repro. parts to keep the show on the road! Not as many specialists sell parts for these earlier cars, but your search should not be too arduous – that's what comes from owning the world's most favourite sports car!*

Illustrated here is the Vette Brakes & Products Inc. complete drum brake overhaul kit for 1963–1964 Corvettes: new drums are available, too. A custom option is the drum brake–disc brake conversion kit, but if you follow this route, keep those original parts! You, or a future owner, may wish to switch back to stock and genuine, original parts will become worth more (if they are restorable) as the years go by.

REAR BRAKE OVERHAUL

As with the front calipers, we recommend that you should only rebuild your own rear calipers if you are a trained mechanic. Otherwise, have a qualified mechanic carry out the work for you, or purchase new, rebuilt – or even uprated components – from top specialist suppliers.

▲ *RB1. The rear brake calipers are removed in almost exactly the same way as the front. The calipers are held by two bolts (use a ⅝ in. socket) …*

▲ *RB2. … and with the brake pipe removed, they can be lifted clear.*

▲ *RB3. With the calipers removed, further dismantling can take place on the bench. Note that the hose has already been broken loose from the caliper using a long ⅝ in. open-end wrench. If you aren't changing the hoses, you probably won't have to disconnect the other end – and that can save a lot of trouble if the threads of the brake line are rusty – which is almost a foregone conclusion. Just unscrew the caliper from the hose and then renew the copper washer on the hose when it is refitted. Remove the two caliper mounting bolts with a strong ⅝ in. socket and then lift the caliper off. The pistons are not difficult to remove. A screwdriver is usually all that is needed to prize off the dust seals. Take out the bleeder screws, too, prior to cleaning the calipers. A torch may be required to break them loose. Stainless steel replacements are available from Vette Brakes & Products Inc. Note: it is a good idea to routinely inspect Corvette calipers for signs of dampness, which signifies a leak.*

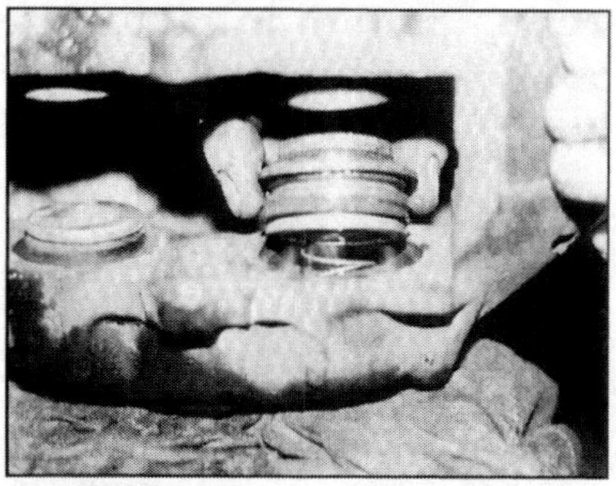

▲ *RB4. This close-up shows the piston assembly, which comprises a piston boot (on top), the piston itself, a seal and a spring. With the pistons removed …*

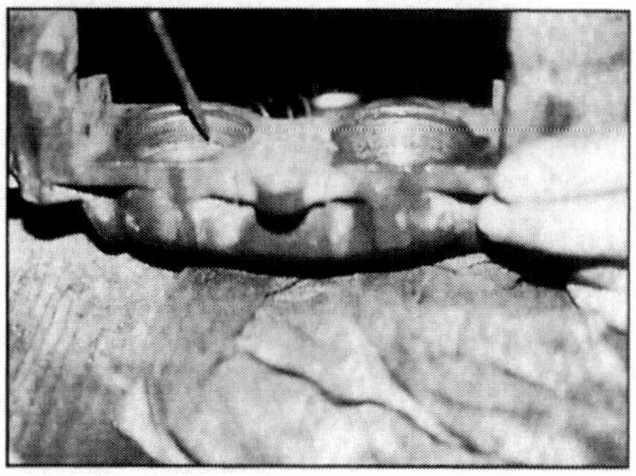

▲ *RB5. … the state of the piston bores can be judged. The lower half of the bore is the sealing area and should be free from pits or deep wear marks. As you can see, there is plenty of rust here (as there was on all caliper bores, both front and rear), which does not make for good sealing properties! This is because that non-silicone brake fluid is hygroscopic and absorbs water. Over a period of time this creates rust on the inside of the wheel cylinders and the piston bores, as here. Using silicone fluid pre-empts this, but on the downside, some claim that silicone fluid is more abrasive. If you mix it with regular fluid it loses its properties anyway.*

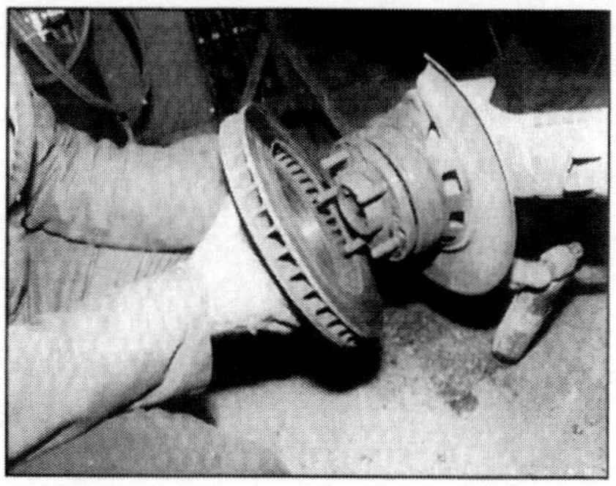

▲ *RB6. To dismantle the rear brake assembly, the rivets holding the rotor to the hub (on some models these are screwed together) have to be drilled out as shown earlier on the front hub assembly. With the rotor removed, there is access to the parking brake shoes (see following section). If you were simply replacing the shoes, then this could be effected without removing the drive spindle flange.*

PARKING BRAKE AND REAR HUB

The weak point of the braking system (pre-1988) is the parking brake. This operates by a conventional pull-up hand lever mounted in the console between the seats, rather than by the foot operated pedal found on many cars. It actuates a separate internally expanding shoe and drum system, the drum being on the inside of the back brake rotor. Its operation is through levers and adjusters which are steel stampings without proper bushings, and these freeze up and rust in service. Because the caliper and brake rotor must be removed to do all but the most simple adjustments, this part of the car is ignored by many owners. Corvettes from 1988 onward have a totally different disc-operated parking brake system.

As you will see, many manufacturers now make stainless steel components for the parking brake system, and these really can be recommended as ways to solve the corrosion problems in this area.

▲ *PB1. The cable-actuated parking brake operates on the rear wheels only and is activated by the lever mounted in the centre console. Having made sure that the lever was fully released, the spring at the equalizer must be disconnected and the rear nut from the cable removed to allow the front cable to hang down.*

▲ *PB2. Inside the car, the trim cover has to be removed from the parking brake lever. In the example here, the seats had already been removed, but if your seats are in situ, you'll need to push down the seat cushions in order to gain access to the lower mounting screw on each side of the cover.*

▲ *PB3. Don't forget to remove the park brake warning switch, which is secured by a single screw and washer.*

▲ PB4. The primary cable (from the lever to the equalizer) can then be pulled away. This cable passes around the primary cable pulley …

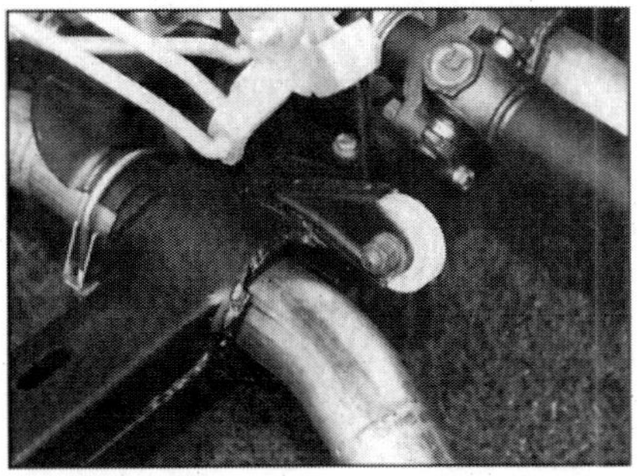

▲ PB5. … mounted on the frame in between the driver's side exhaust and the prop shaft.

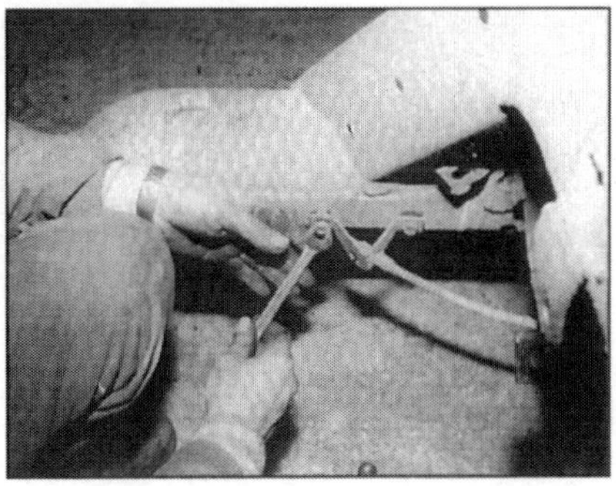

▲ PB7. The cable clip retainers have to be removed from the forward rear support bracket …

▲ PB8. … and the rear bracket. Once the clips have been removed …

▲ PB9. … the cable can be pulled free.

▲ PB10. The end of the cables can then be disconnected from the brake operating levers on the backplates. Using a screwdriver, turn the adjuster screw to expand the brake shoes and push them inwards so that the front shoe hold-down plate is visible from the side.

Turn the axle flange plate until the access hole in the plate is in alignment with the head of the shoe hold-down spring pin. Depress the spring retaining cup, insert a pair of needle nose pliers through the hole in the plate, grip the pin and turn it through 90 degrees. Remove the spring/cup and then repeat for the remaining shoe. Turn the adjuster screw until the shoes are fully retracted then prize the shoes apart at the adjuster. At this point, you can remove the adjuster and spring. Prize the shoes apart at the anchor pin, lifting them up and away, allowing the straight section of the shoe return spring to pass between the outer end of the anchor pin and the axle flange plate.

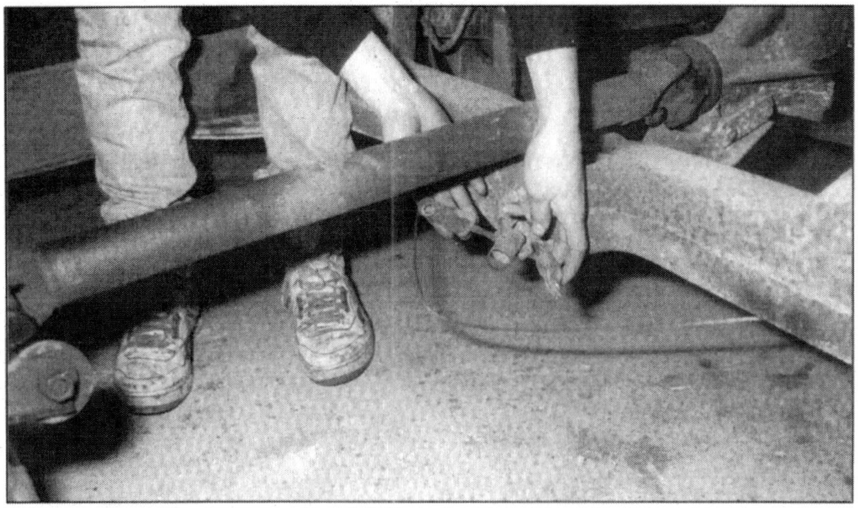

▲ PB6. This leaves the rear cable which should be removed from the equalizer, as here.

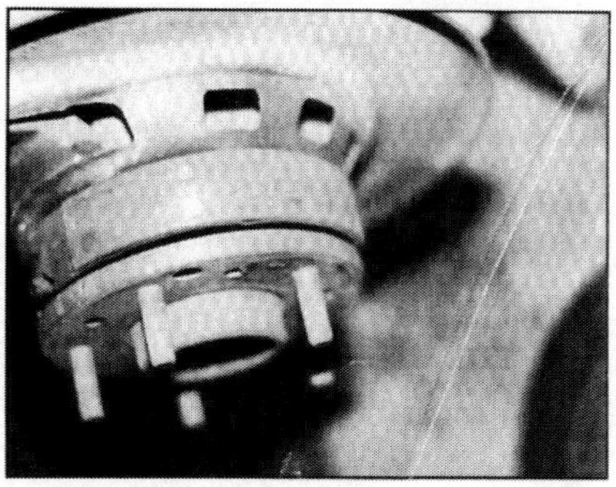

▲ *PB11. With the rear rotor removed, the parking brake shoes can clearly be seen at the back of the hub. In this case, they are more than a little worn, down to the rivets. As they actually have little work to do (by definition, the car is not moving when the park brake is applied) this implies that these particular shoes have been around for a long time and must have been badly adjusted, or frozen in the 'on' position.*

▲ *PB12. The drive spindle flange is secured in the same way as the front hub – a spindle nut and cotter pin. By removing the latter, the nut can be undone and the spindle flange removed. You can see the poor state of the grease and general condition of the hub and parking brake assembly; long overdue for some attention.*

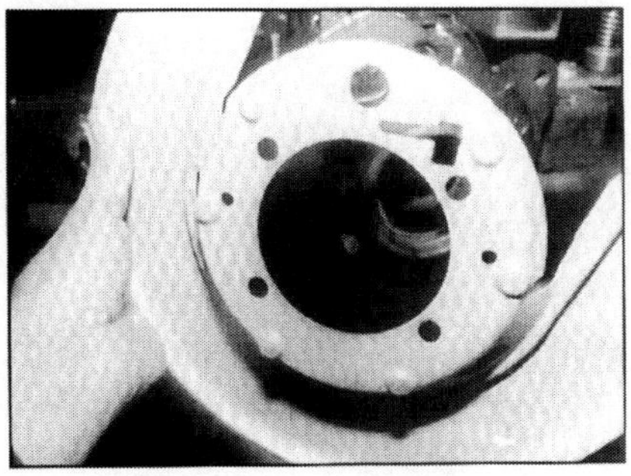

▲ *PB13. The backplate, complete with parking brake shoes etc., can be taken off by removing the four nuts which hold it in position. As this was part of a complete restoration, we removed the control arm complete with the assembly.*

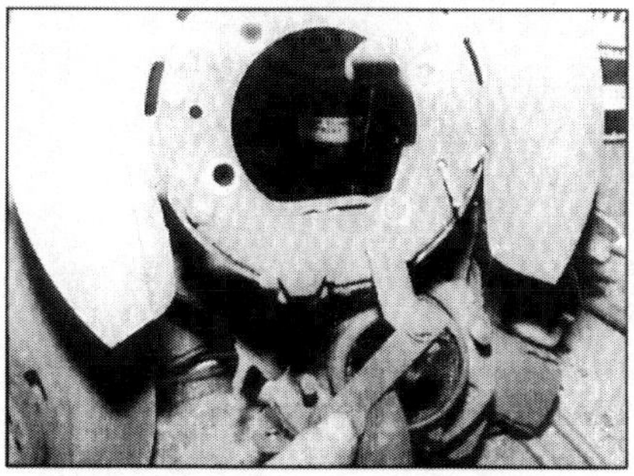

▲ *PB14. A view of the backing plate showing the parking brake actuating lever, a device which is prone to seizing solid. Note the general condition of the backing plate ...*

▲ *PB15. ... compared with the same item and its fellows after a trip to the shotblasters. At this point, these particular pieces had been cleaned up and resprayed ready for reinstallation.*

◀ *PB16. Close up, they look even better. This rear plate is good enough to pass for brand new.*

▲ *PB17. Reassembly of the parking brake assembly is a reversal of the dismantling procedure, but much cleaner and easier! Wherever possible it pays to treat the threads of any fasteners to a coating of grease, preferably of the copper-impregnated variety.*

▲ *PB18. Having gone this far you will want to replace all the parking brake components, not just the shoes. After all, every mechanical component has a limited useful life, even those springs and cup washers.*

▲ *PB19. As with the front bearings described earlier, ensure plenty of grease is worked into the bearing roller cage.*

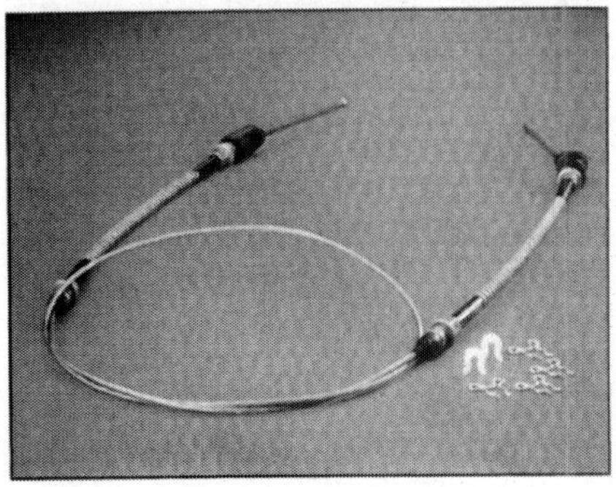

▲ *PB20. You can kill the rust bug on parking brake cables for all time by fitting stainless steel cable sets.* (Courtesy Vette Brakes & Products Inc.)

▶ *PB22. Worn or corroded parking brake shoes can be replaced with stock items or, once more, with stainless steel upgrades. Parking brakes are such a pain that the extra expense of all the stainless steel parts will be worth paying if you intend keeping your Corvette.* (Courtesy Vette Brakes & Products Inc.)

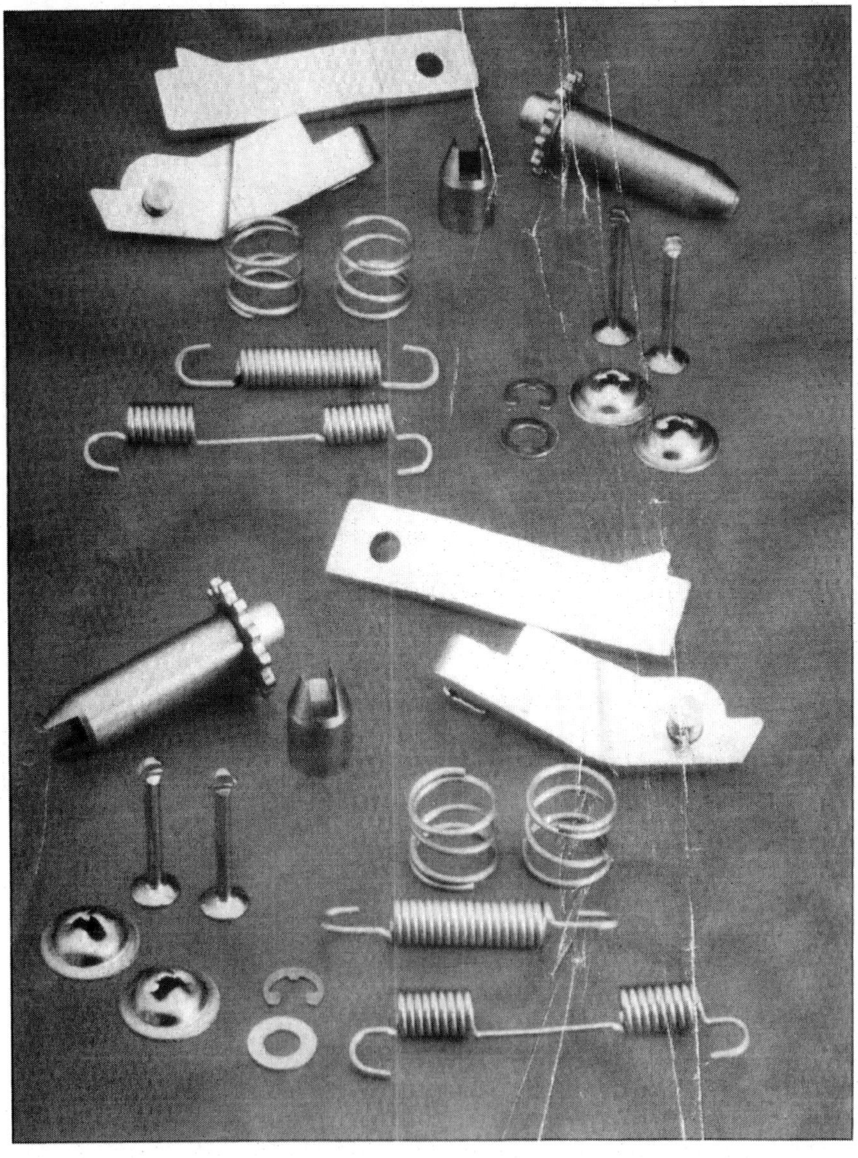

▲ *PB21. You can even replace the whole assembly with stainless steel components. This kit includes interfaces, springs, retainers, auto-adjusters and hardware.* (Courtesy Vette Brakes & Products Inc.)

PARKING BRAKE ADJUSTMENT

▲ *PB23. Adjusting the parking brake has to be carried out with the rear wheels off and with the brake cables at the equalizer loosened off, with a little slack at the cables. Rotate the brake rotor until you see the adjuster screw through the hole in the drum section of the rotor assembly. Insert a screwdriver (it needs to be a long one, as shown here) - turn the adjuster screw until the brake is locked. Then back it off six or seven notches. Adjust the nuts at the equalizer to take up any slack from the cable, though not so much that the levers at the backplates are moved from their fully off position. When the parking brake lever is applied over 14 clicks, the rear wheels should be locked. With the parking brake lever off, the rear wheels should be able to be turned by hand without any signs of brake drag.*

BRAKE TROUBLESHOOTING CHART

(Reproduced courtesy Vette Brakes & Products Inc.)

Leaking brakes	Check and/or replace rubber flex hoses seals and/or dust boots. Also change brake fluid – dirty fluid can cause leaks.
Soft pedal	Bleed brakes and check for leaks. Also measure rotors for uneven runout which can cause brakes to draw air even though there are no signs of leakage.
Low pedal	Front calipers are probably switched. This puts the bleeder screw in the wrong position and even though fluid will come out, air will not, causing the pedal to go to the floor.
Rumbling noise	Check or replace the wheel bearings.
Brakes squeal	Check for glaze build-up on rotors and worn pads. Knock off glaze with 150 grit or finer sandpaper and thoroughly clean or replace pads.
Brakes stick	If rubber hoses are twisted, they can act like a check valve, allowing fluid into the caliper, but not out. Check or replace rubber hoses. Also, check or replace the master cylinder if hoses are all right.
Brakes pull	Check for uneven pad wear, twisted flex hoses or fluid on pads and rotors. Remember to check frame alignment and condition of front-end parts, because trouble there could easily be mistaken for a brake pulling problem.

Chapter 6

Electrics and Fiber Optics

RETRACTABLE HEADLIGHTS

The Corvette's retractable headlights are vacuum operated. The vacuum circuit incorporates a vacuum tank to provide a reserve for several actuations of the headlights, even without the engine running. If the light control knob is pulled, then the headlights will be raised and switched on. If the manual control knob is pulled, then the headlights will be raised but without their being switched on.

In the event of a failure, or during overhauls and fault-finding sessions, the headlights can be raised manually by disconnecting the vacuum hoses from the actuators at the rear of the headlight and then pushing upward on the light housing from under the grille.

IMPORTANT NOTE: Never close the headlight doors by hand.

▼ *HD1. Disconnect the four hoses from the relay under the centre of the front panel – hood open. These are usually color-coded with corresponding letters moulded into the relay valve connectors. If you have any doubts at all as to their location, use masking tape and a ball pen to mark the location of the hoses before their removal.*

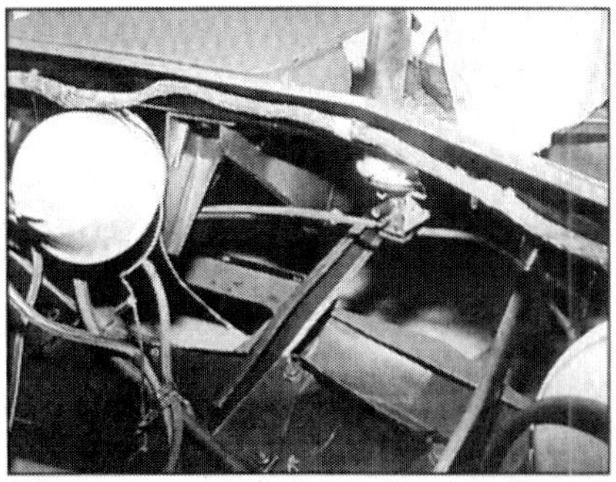

▶ *HD2. Remove the two screws that secure the relay ...*

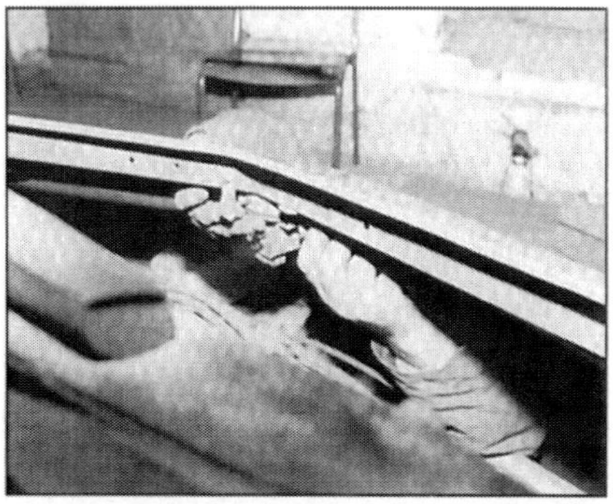

▶ *HD3. ... and remove it from the engine bay.*

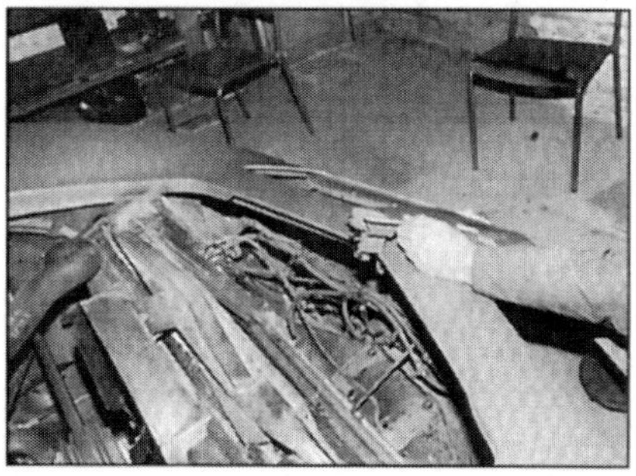

▼*HD4. The headlights, doors and frame assemblies can be removed as complete units, with vacuum actuators attached, so you can refurbish them properly.*

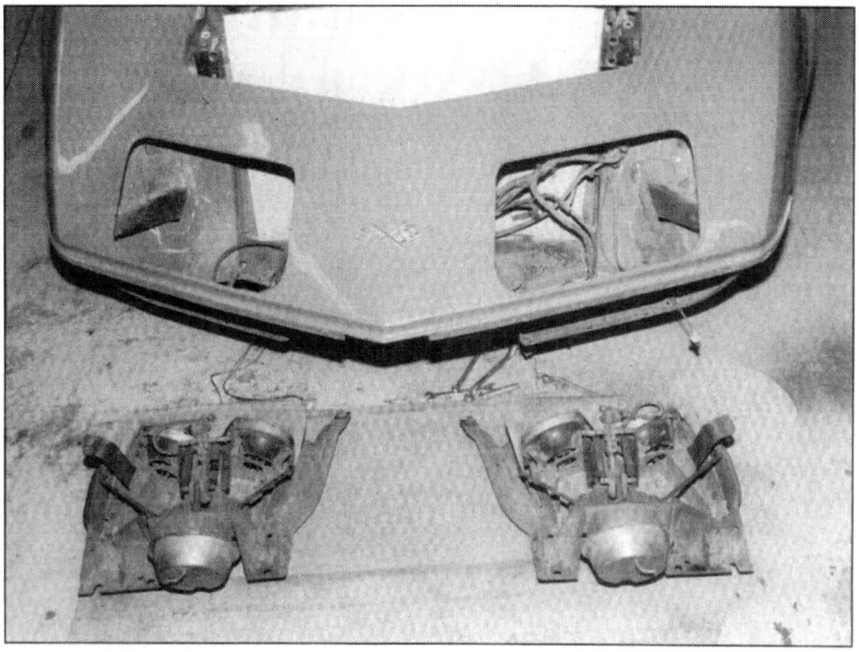

▲ HD5. Now you can dismantle them at your leisure; working upside down in a tricky part of the car is a sure way to lose nuts and washers for ever! The complexity of these assemblies makes it, once again, an excellent time to break out the camera or sketch pad to record what goes where – and how.

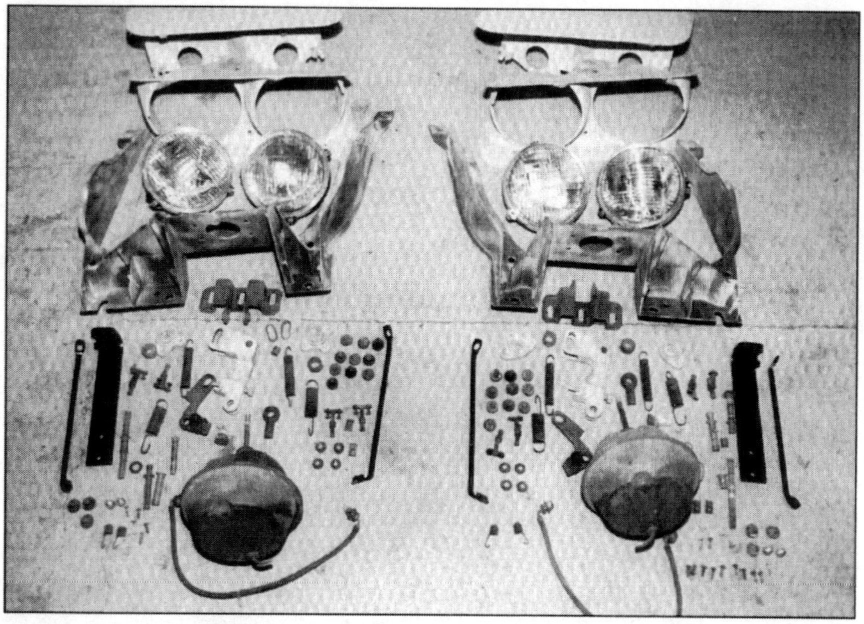

▲ HD8. When the two assemblies have been totally dismantled, it's hard to believe just how many different components there are. Now you can see why it's so vital to make notes and sketches and take photographs when dismantling these parts. The GM manual is not a complete guide to reassembly!

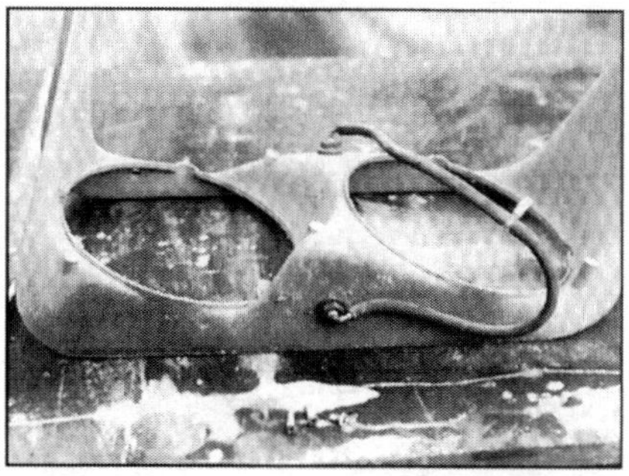

▲ HD6. A feature which many owners of '69 and '70 cars are unaware of is the headlight washing system. Water jets spray the outer (dipped) beams when the windshield washer switch is depressed and the headlights are in the raised position. In this view of the rear of the headlight bezel, you can see the pipe connecting to two nozzles.

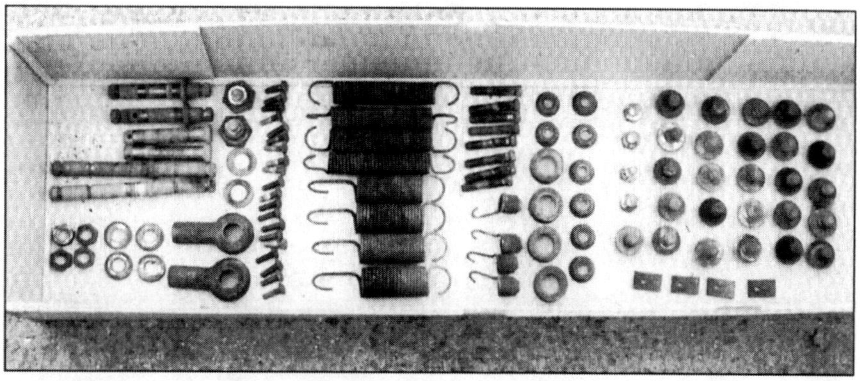

▲ HD9. All these metal parts required sandblasting before replacement. If you don't have access to sandblasting equipment, remove whatever surface rust you can (by using a wire brush and sandpaper cloth) repaint, and apply grease or other suitable lubricant where applicable.

▲ HD7. Each assembly features twin 5¾ in. sealed beam headlight units. The glass should be checked carefully for signs of cracks which can lead to condensation, which in turn can lead to a rusting of the reflector and a diminution of the lighting intensity.

RESTORATION TIPS

▼ HD10. An actuator fits on the back of each headlight unit and operates the flip-up headlight unit under vacuum pressure. Three different styles of rear tubes were used on actuators over the years, although they are all functionally interchangeable. 1968 and early 1969 actuators (top centre) had a smaller diameter tube. The tube on the currently available replacement actuators (lower left) is the same as used on late-model Corvettes and is identified by a tube bent close to the can. Like the 1968 actuator, the rear tube on 1969 to late 1970s Corvettes (lower right) was bent almost ¾ in. away from the can.

▶ HD11. Restorers should look for the pad-stamped codes which were evident on the rear surface of the actuator for a number of years. This isn't in the NCRS judging book yet, although more information is being compiled. The original hoses and their color-coding stripes clean up remarkably well. Some aftermarket hose harnesses have color stripes which are extremely bright. The actuators metal finish is cadmium plate with a gold irridite sealer. Aftermarket replacement actuators won't normally have these factory numbers.

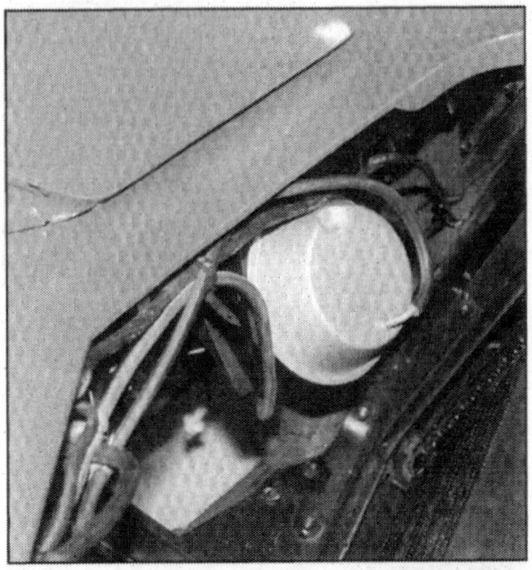

▲ HD15. … and when complete the headlight units can be added. Em fits the headlamp spring back into place.

▲ HD12. By this time, John Fletcher's paint shop had painted those parts that needed it: the bezels and cover housings – and the chrome pieces had been replated. A considerably more encouraging sight than the motley selection of bits and pieces seen earlier!

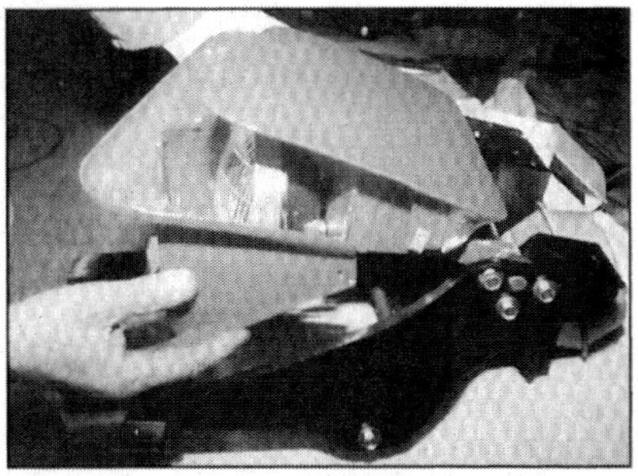

▲ HD16. Then the housings are fitted back onto the support assemblies.

▶ HD13. Em pieced together the support assemblies, complete with the springs. Note that Em's giving the thumbs up to two of the adjustment points; his right thumb is on the opening rod adjustment mechanism and his left is 'pointing' to the up-and-down movement mechanism. You can see here how the pushrod coming out of the front of each actuator acts on the sprung operating levers for the flip-up mechanism.

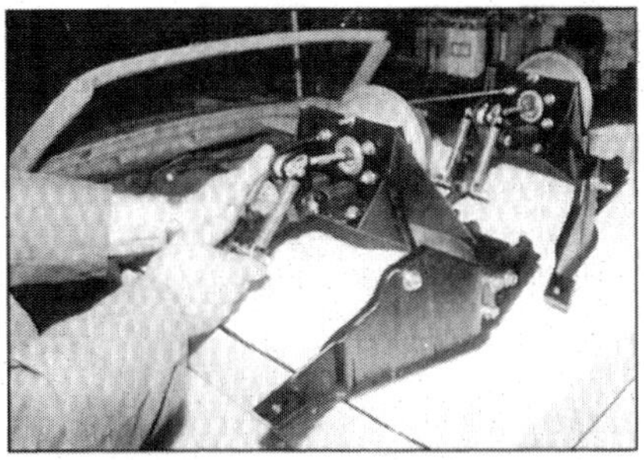

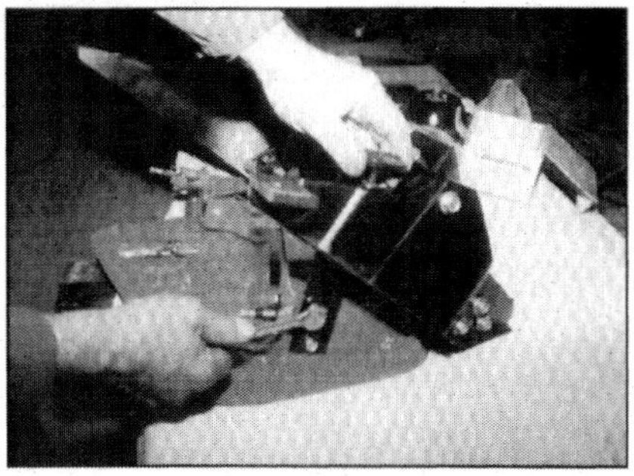

▲ HD17. It requires some care not to damage the newly painted items. New bolts screws and fixings should be used, or the originals cleaned up and reused, if you're striving for complete originality.

▶ HD14. The housing's components were built up in the same meticulous manner …

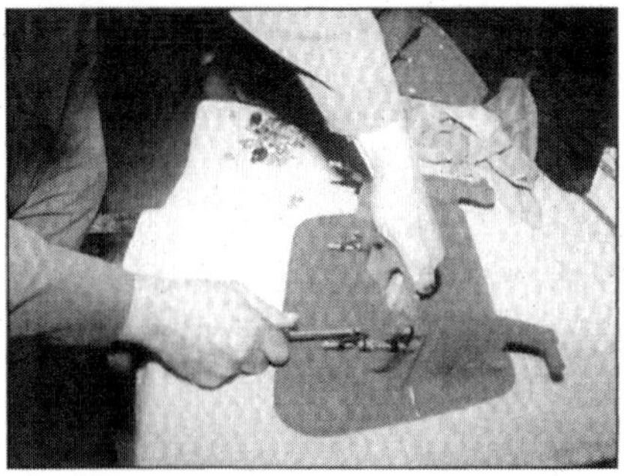

▲ HD18. Once the headlight bezels have been added …

▲ *HD19. ... the two assemblies are ready to be refitted. Note the brand new washer system rubber tubing – an essential, every time!*

▲ *HD20. Fitting the complete units was made much easier by the fact that it was a complete restoration, and many of the parts that would normally get in the way, notably the radiator and shroud, had been removed. Here, the units are in place, waiting for the addition of the relay actuator and, of course, the hoses linking all the various vacuum-operated parts together.*

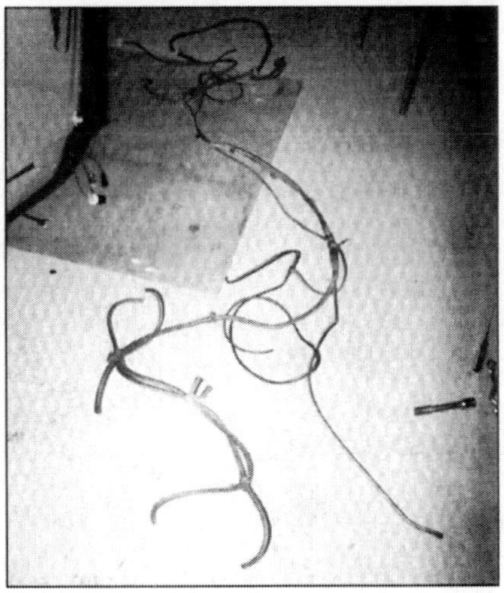

▲ *HD21. And when it comes to hoses, the Corvette has more than its fair share! At this stage in the proceedings, it's best to simply replace all hoses with new. Over a period of years the rubber deteriorates and can harden and crack, with an ensuing loss of pressure. Where you have problems removing tubing from a nipple, it is sometimes possible that an aerosol silicone spray will encourage it to slide away. If it doesn't, then it will have to be cut. The routing of the tubing around the car is vital – clearly it should not be trapped or positioned where any moving parts will foul it. If you're striving for total originality, you'll want to replace color-coded hoses, like-with-like, wherever possible.*

▼ *HD22. The lift-up headlamps are such a vital component in these model years of Corvette – both aesthetically and for practical reasons – that it's a great feeling of achievement to see them working like new again.*

ADJUSTMENT

There is the provision for adjustment up and down, in and out and of the opening rod mechanism. Always make sure that when the lights are fully extended, the micro-switch on the door opening linkage shuts off the warning light.

HEADLIGHT HANG-UPS – HOW TO CURE THEM

Of course, you may not want, or need, to dismantle the entire assemblies, just improve their efficiency. If they no longer raise and lower when required, if they shake about as you drive at night or if they close with a worrying 'crack', you might well find that the actuator assembly and its adjustments are the likely culprits. Here's how to check through your system.

FAULT FINDING

It helps to understand the basics of the vacuum system on a Corvette's headlights. The three following line drawings explain how this works but note that to some extent, these systems overlap and they've been divided into three just so you can understand the different parts of the system.

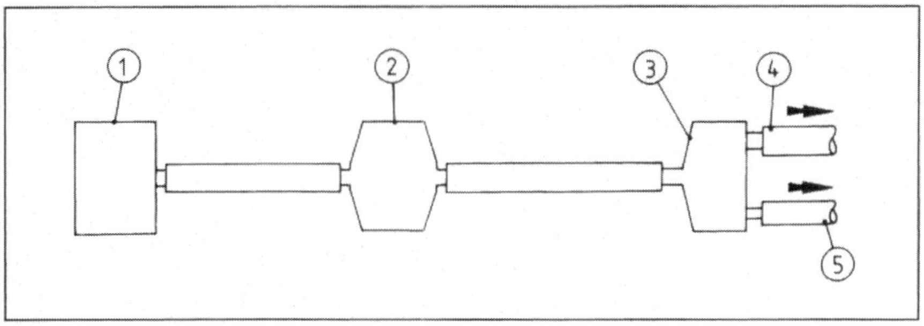

▲ HD23. The vacuum source circuit (1) delivers energy in the form of a vacuum to the power (4) and control (5) circuits, and is rather trouble prone. Rubber hoses will deteriorate and split when they are subjected to heat and oil – and that's just what they get when they are sighted near the Corvette's engine. The filter (2) doesn't often give problems but the one-way valve (3) can be a pain!

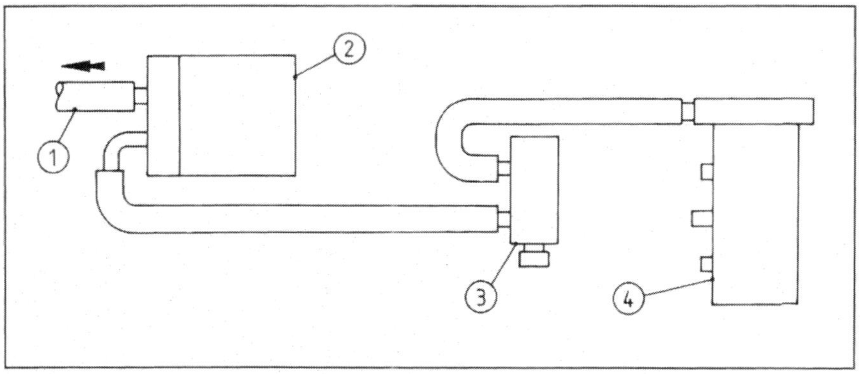

▲ HD24. The control circuit is the part of the system that you operate directly from inside the car. The override switch (3) enables you to raise the headlights during car washing without having to turn them on. This is a fail-safe system in that vacuum pressure holds the headlights down. If the vacuum fails, the headlights 'fail' in the open position. 1. Vacuum source; 2. Headlight switch; 3. Override switch; 4. Vacuum relay.

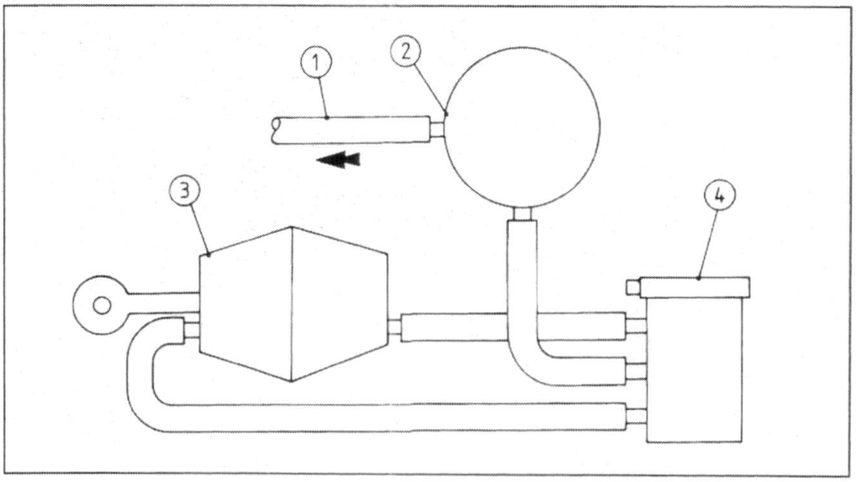

▲ HD25. Once the control circuit has operated the vacuum relay (4), the power circuit switches vacuum from one side of the actuator to the other to raise and lower the headlights. Things won't work properly if the hoses are connected up wrongly (obviously!) or if the diaphragm is leaky. 1. Vacuum source; 2. Vacuum reservoir tank; 3. Headlight door actuator; 4. Vacuum relay.

▲ HD26. The first thing to check for is leaks at the various connections. Start at the source which is where the rubber hose connects to the intake manifold, behind the carburettor.

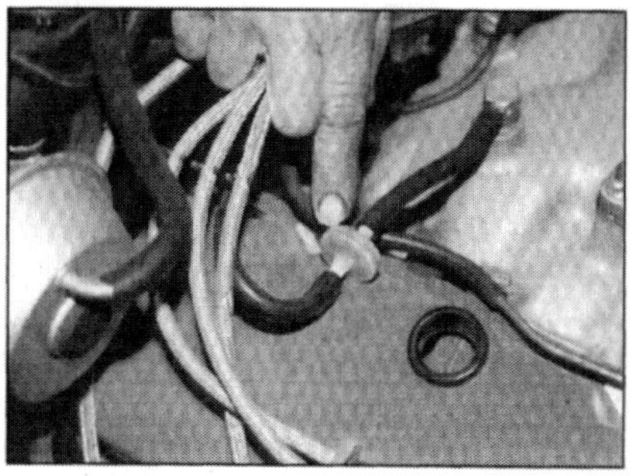

▲ HD27. The filter, shown in HD23, is next in the line. It should have a small filter screen inside, but this often gets sucked into the engine. Some specialists don't advise replacing these filters because they reckon it's better to have an occasional piece of dust go into the system rather than the entire metal screen! But the choice is yours, of course. Check the connections and also look or blow through the valve to make sure it passes air easily. Leave attached for an original appearance, but note that they do go yellow with age.

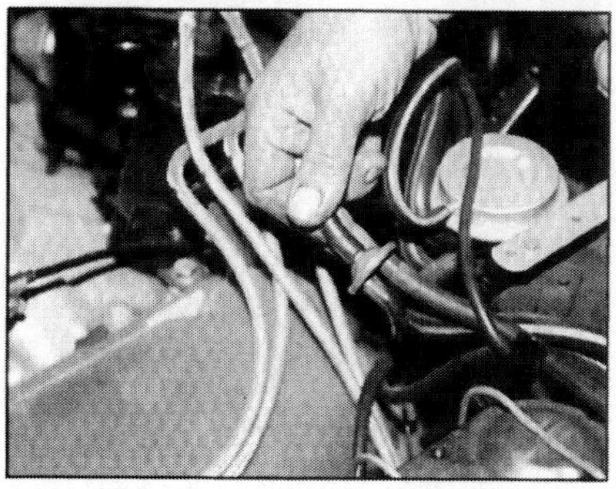

▲ HD28. Next comes the aluminium one-way valve. The job of this valve is to stop the vacuum from bleeding away when the engine is shut off. Blow air into the valve from the engine side. It should only 'suck' in this direction, so if you can blow air, the valve has failed. This is an AC-Delco part and can be purchased from many local part stores, not just GM dealerships.

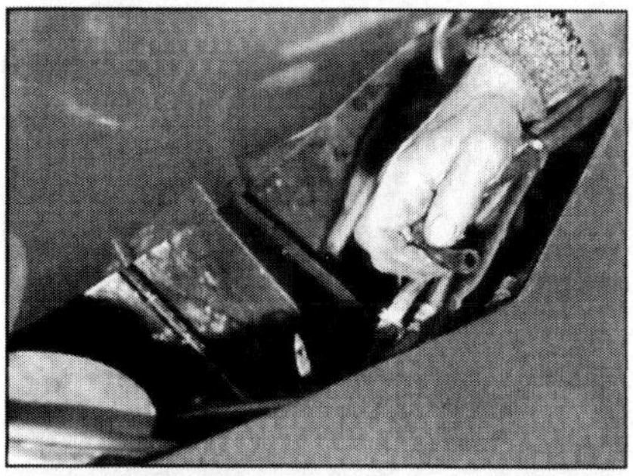

▲ HD29. Now go to the vacuum relay which you'll find under the front panel, at the centre, in front of the hood.

Pull off the small hose from the 'mushroom head' on the top of the vacuum relay and, with the engine running and the headlight switch pushed in (and the override switch up), there should be a vacuum at the end of the hose. With either the headlight switch out OR the override switch down (or both), there should not be a vacuum.

If this is all OK, replace the small top hose and remove the top and bottom large hoses (in a vertical line of three, on the lower section of the vacuum relay), also leaving the centre large hose in place. Now, when the headlight switch is in and the override switch is in the up position, there should be a vacuum at the top stub on the vacuum relay (the one from which you removed the top large hose). Move either the headlight switch or the override switch and the vacuum should go to the other, lower stub. If it doesn't, the valve is faulty and needs replacement.

Next, pull off the centre hose and close off the end with your finger. If there is a strong vacuum, all is well. If not, check for a hole in the vacuum reservoir tank. But note the following:

1 Corvettes from 1973–1982 used the cylindrical member ahead of the radiator as a vacuum tank.

2 On 1970–1982 Corvettes there's another control valve. If yours is one of these models, pull off the centre hose from the other control valve and plug it with a pen or bolt. If you now have a strong vacuum, you will probably find a strong leak on the side which has been plugged.

In all cases, if none of these does the trick, follow the hose back to the filter and intake manifold and check for splits or pinching in the hose.

Now you must establish which is the rubber vacuum hose connecting the rear (radiator-side) of each actuator (the metal cylinders behind each headlight). Pull each hose in turn from the vacuum relay and try sucking on the hose. Once you have built up some vacuum, slide your tongue over the hose opening and wait a few seconds to see if the vacuum disappears. Do this when the headlights are in the down position and you won't have to suck so much air out to get a vacuum.

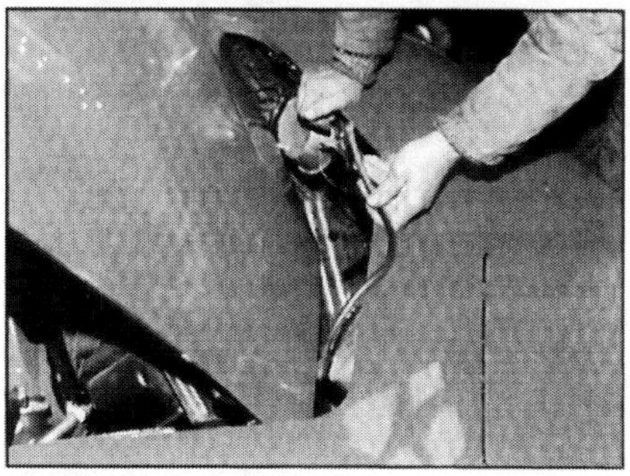

▲ HD30. Having tested the operation of one headlight regulator, now try the same on the other actuator, just behind the headlamp on the other side of the car. Now you can raise the headlights, and disconnect the hose which goes to the other side of each actuator from the vacuum relay and, once again, test it for pinching, splits or leaks.

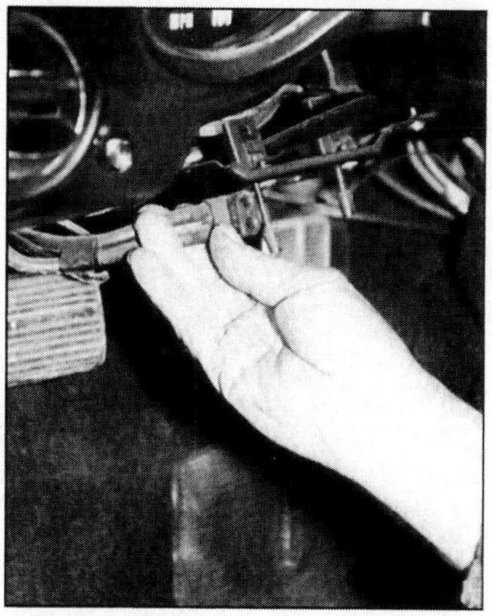

▲ HD31. If vacuum is not being properly delivered to the vacuum relay, check the override switch. There should be vacuum at the upper hose connector when the headlight switch is in and the override switch is up. Moving either switch should cause the vacuum to stop. If this top connector is OK, unplug the lower hose and this should have vacuum when the light switch is in, and no vacuum when the headlights are on.

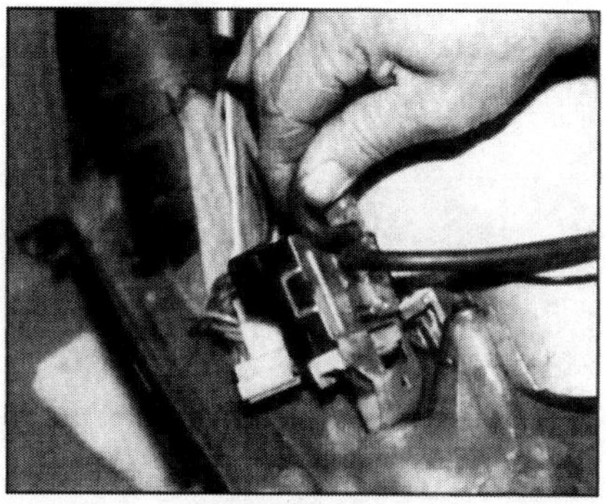

▲ HD32. You should also check connections to the headlight switch (it combines electrical and vacuum functions through a relay) because this can also develop minor leaks. If you have a hissing noise, this could also be caused by the hoses being reversed – check them out!

ACTUATOR REPLACEMENT

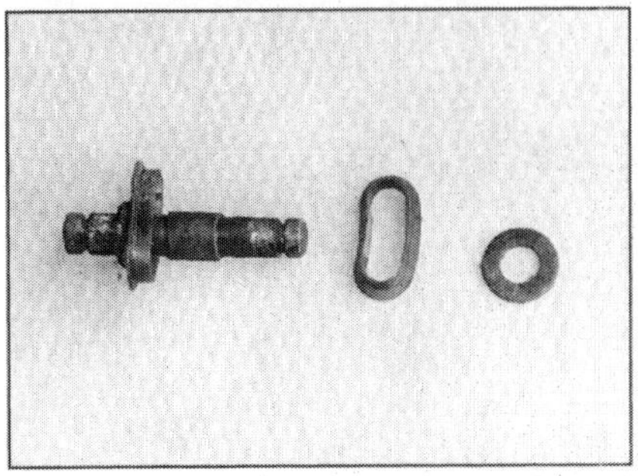

▲ *HD33. The actuators are attached at access points which can be reached through the air intakes in front of the radiator. Slide out the shaft by pushing on the end of the shaft which has the cotter pin hole. These shafts are frequently rusted in position and it will help if you apply penetrating oil first. If you still have a problem, clamp vise grips on the other end of the shaft and turn it as you pull it out. If the nylon bushing sleeves stay attached to the shaft, no problem: you don't need to remove it.*

▲ *HD34. Use a ½ in. swivel socket on a long extension to remove the four actuator mounting nuts. You can then pull the actuator towards the rear of the car and out. Installation of the new actuator is carried out in the reverse order. However, you will also have to adjust the eyelet on the actuator rod, and note that the left and right actuators are not identical; they have different rear tubes. If you have fitted the correct actuator to the correct side of the car, the rear tubes will point towards the outside of the car.*

HEADLIGHT DOOR ADJUSTMENT

Screw the eyelet into the actuator rod until about the same number of threads are exposed as were shown on the old actuator. In all probability you will need to readjust the position after fitting, so it may not be too critical. You can now reinstall the shaft shown in HD30, and its springs, but without the cotter pin. Raise the headlights all the way up and then lower them again. If they don't come up far enough – in other words, if they don't lock into place – take out the shaft, unscrew the eyelet a few turns and try again. If the headlights don't go down far enough, you'll have to screw the eyelet further in.

▲ *HD35. The headlight door can be adjusted so it is flush with the top panel when the headlights are down. The screwdriver here is pointing at the big end of the bolt which requires the use of a $\frac{9}{16}$ in. wrench. The lower end, with a smaller hexagon on it will usually round off. It's a good idea to put masking tape all around the edges of the headlight doors, both on the top panel and on the doors themselves to save any scratching as you line up the headlights.*

▲ *HD36. It's difficult to show this with the headlight in the car, but there's another adjustment which is the headlight door stop bolt which limits the height of the headlight door in the up position. It's actually a bolt with a lock nut, and if the bolt doesn't contact its stop, the headlights will bounce and flutter as you drive along. If the screw is too far in the opposite direction, the mechanism may not lock properly or will be difficult to unlock and close. The bolt should be adjusted so it contacts just before the mechanism is all the way up, so it encourages the mechanism to lock firmly into place.*

ELECTRIC HEADLAMP DOOR MOTORS

▲ *HD37. Some earlier and later Corvettes were fitted with electric headlight door motors – intrinsically less complex to trouble-shoot than the push-me-pull-you vacuum system.*

Some original equipment electric door motors are still available new, but the biggest source of problems comes from the plastic drive gears because they break easily. Replacement drive gears are also available for most models – though they're not cheap! But then, if you can't buy a motor because they're no longer made, you'll be glad of the stronger brass drive gears available for some models. (Courtesy Auto Accessories of America)

WINDSHIELD WIPERS AND WASHERS

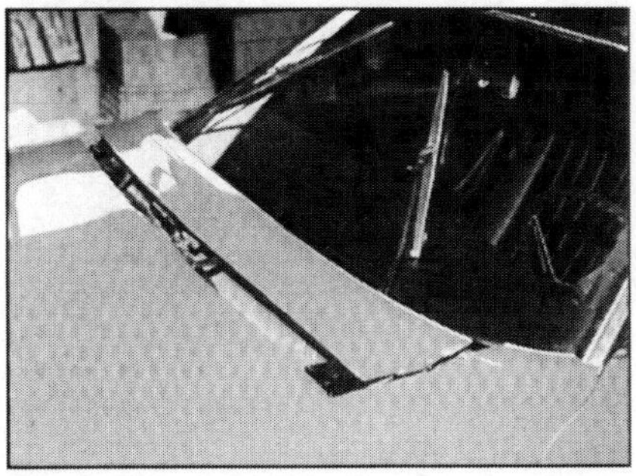

▲ *WL1. The 'hidden wipers' system was used on cars from 1968 through 1972. Left alone, the wiper flap system usually functions well for many years. Most problems occur when owners have been working on the instrument panel and stereo or when crash damage has been repaired.*

▲ *WL2. When the wipers are switched on, an automatically controlled wiper mechanism cover panel is raised by a vacuum operated actuator on the firewall in the engine bay.*

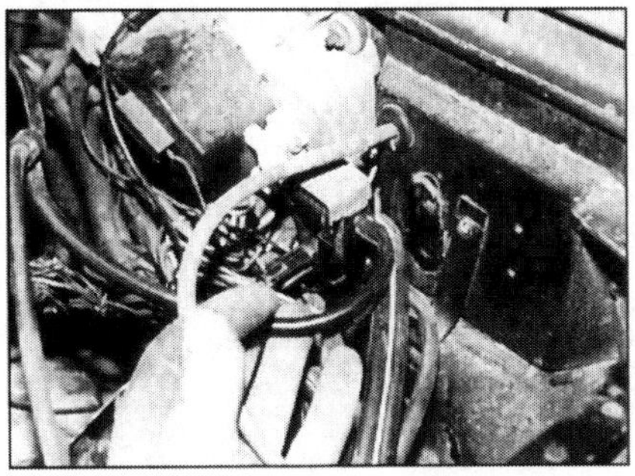

▲ *WL3. A limit switch is incorporated into the circuit to prevent the wiper motor operating until the cover panel is fully open. Also, an interlock valve (shown here) located in the right-hand side of the plenum prevents the panel closing until the wiper arms are safely parked.*

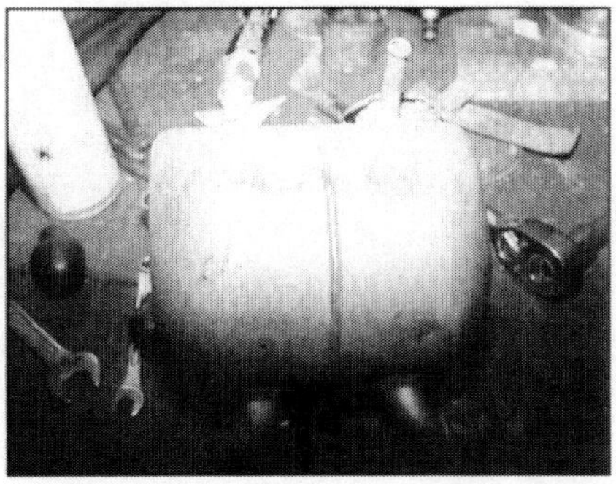

▲ *WL4. The vacuum source is the same one used to operate the headlight doors, whereby a 'stock' of air is retained in the header tank, allowing operation of either headlight doors or wiper cover panel even when the engine is not running. A relay valve, required to send the vacuum to the correct side of the door actuator for either the opening or closing cycle is situated near to the tank on the right front fender. The header tank is prone to corrosion after a few years, and problems with both systems simultaneously may be a rust hole in the tank, as seen here.*

▲ *WL5. A brazed repair is possible though new tanks are available quite cheaply, and replacement is simple.*

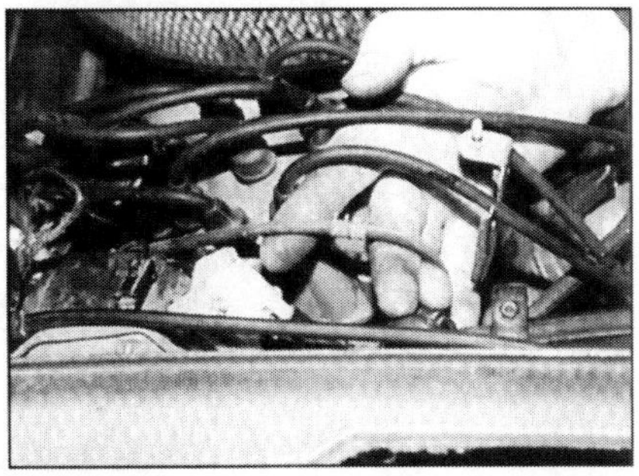

▲ *WL6. Maintenance consists of regularly inspecting the electrical and vacuum connections and replacing the cover panel bumpers when they have become compressed below $\frac{3}{16}$ in. thick.*

▲ *WL7. The wiper door can be opened manually, for remedial work or more serious dismantling of the wiper system. With the wiper motor in the 'parked' position, open the hood and disconnect the battery ground cable. Remove the rubber plug from the front of the wiper door actuator, then insert a screwdriver and push the internal piston to the rear to open the wiper door.*

▲ *WL8. This photo shows the basic layout of the wiper mechanism in situ. Note the position of the vacuum header tank in the fender.*

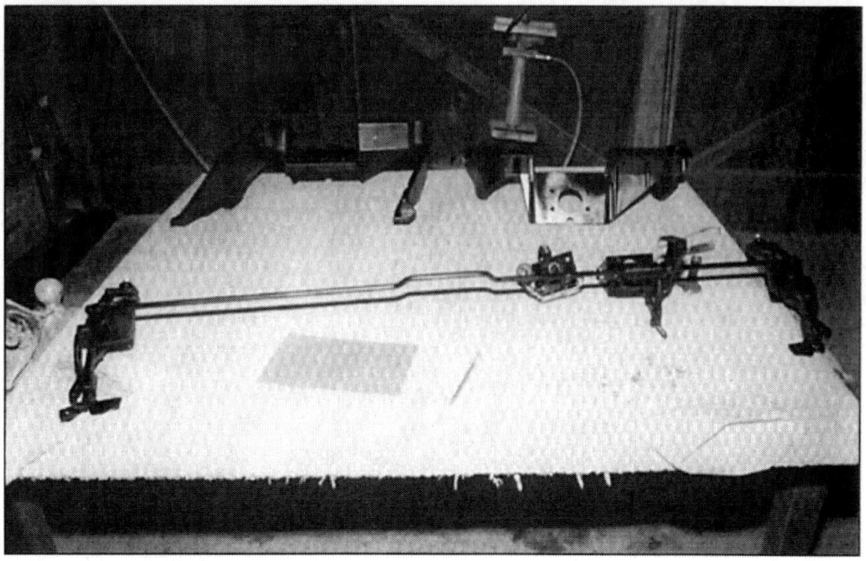

▶ *WL9. It's probably simpler to see what goes where with the complete mechanism off the car, as here. This is not the most complex collection of rods, rails and springs to be found on a Corvette (there's plenty of competition for that honour!) but it's wise to make plenty of notes and take plenty of photos when doing anything other than routine maintenance or fault-finding.*

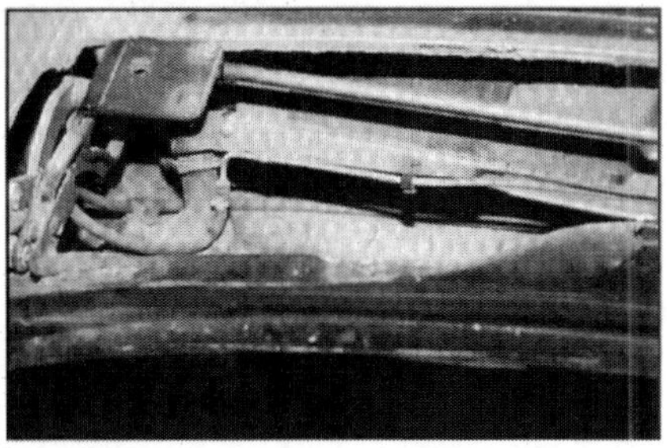

▲ *WL10. The windshield washing system was designed to put the washer fluid exactly where the wipers are by the simple expedient of running a length of tubing from the washer bottle to a metal pipe attached to the wiper arm. Both flexible and metal tubes are prone to crack and may need replacement.*

▲ *WL11. The arms can be disconnected complete with the pipes. They clip onto the arms, and care should be taken when removing them that they do not get bent out of shape or cracked.*

WINDSHIELD WIPER PROBLEMS

Corvette wipers are more complex than most, partly because of their peculiar wiper action, partly because of the introduction of a door into the equation and partly because the fiberglass body makes it even more important than normal that proper grounds are in place on the electrical circuit.

CONTROL CIRCUIT RELAY

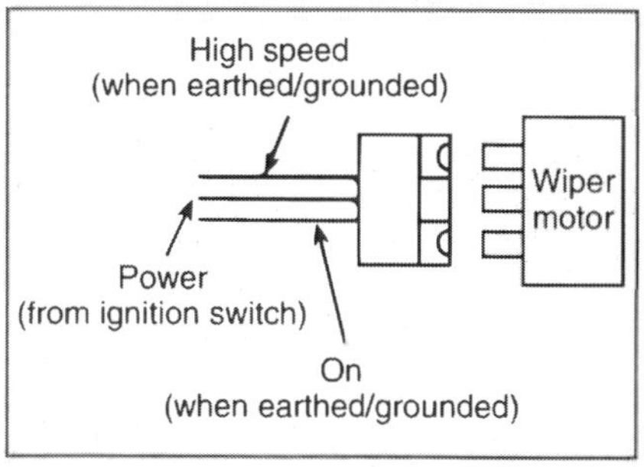

▶ *WW1. There's a relay situated inside the wiper motor for the control circuit. The centre wire of the three-wire connector is the relay power source, and the relay is activated when the top wire is grounded by the switch in the off and low-speed positions. This top wire is light blue or purple. When you switch to the high speed position and low positions, the light blue or purple wire is disconnected from ground or power. Note that models from 1969 to 1972 had another relay in the circuit that turned the wiper to low speed when the ignition was turned off.*

So that the wipers can return 'home' and park out of sight, it is necessary for the system to receive power after the switch is turned off. (Courtesy John Pfanstiehl and *Corvette Fever* magazine)

▼ *WW2. Corvettes from 1969 to 1972 have a power supply that is separate from the ignition switch, so the wipers can turn off and park automatically even when the ignition is shut off. There's an override switch fitted to these models, so you can stop the wipers at any point in their travel to make changing the blades easier. 1968 models only tell a different story again with a different motor, wiring and a unique ground interrupt circuit. See your workshop manual for these models.*

As we said earlier, the ground circuit is a particular source of problems on fiberglass bodied cars. If your wipers and heater blower refuse to work after you've had an engine or starter change, look out for the ground wire by the starter which is meant to be attached to the engine by a bellhousing or starter bolt. If it's been left off, there's your problem! (Courtesy John Pfanstiehl and *Corvette Fever* magazine)

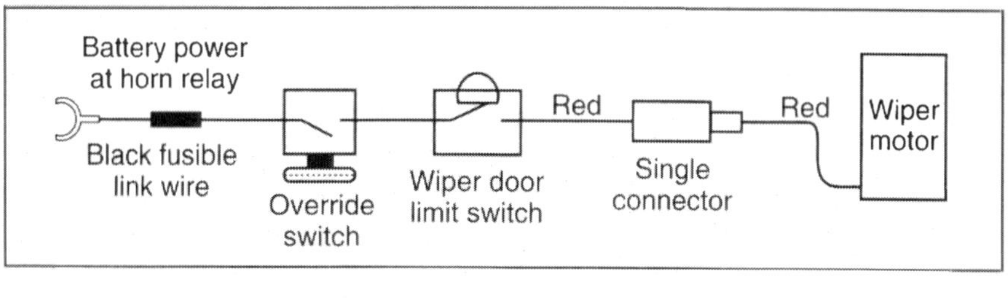

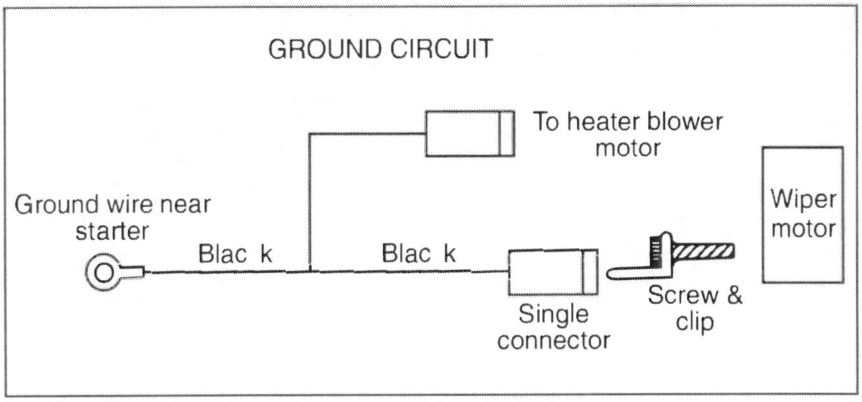

▲ *WW3. The wiper door limit switch arm should be adjusted so that the switch button is pushed when the door is open, allowing power to flow through to the wiper motor (1969–1972 only).* (Courtesy John Pfanstiehl and *Corvette Fever* magazine)

Use a test light and clip it to the power terminal on the alternator (the stud with the red wire). Touch the bare metal housing of the wiper motor to check for grounding and then unplug the single black wire and connect the test light across it and its connector. In both cases the test light should indicate and show that there's a good ground. If there's not, take a look at the black ground wire, near the starter.

If the metal gauge housing (1968–1976 cars) is broken below the wiper switch, the switch grounding will not be complete. The simplest remedy is to reconnect the upper and lower housing pieces with a strip of metal or a short jumper wire.

Touch one end of your test light to the centre wire of the three wire connector on the wiper and the other end to a ground. Turn the ignition on and the light should show. If it doesn't become live, inspect the wiper fuse. Next, unplug the three-wire connector and test the power sources. The bottom wire should only illuminate the test light when the switch is in the 'off' and 'low' speed positions. For 1969–1972 models, these tests must be carried out when the ignition is turned off. Finally, check out the wiper switch itself. Having carried out all your checks, you may have to remove the wire switch plate for further investigation if the control wires were shown not to operate properly.

▲ *WW4. Here is the connector on the back of the switch. Corvette expert John Pfanstiehl has come across quite a number of cars where the plug has been distorted so that it can be plugged in the wrong way. Naturally enough, if you find this to be the case, you've found your problem.*

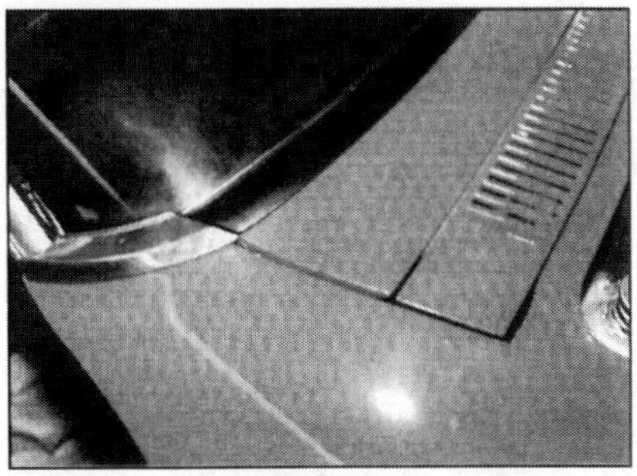

▲ *WW5. On 1968 to 1972 models, you've also got to check the alignment of the grille, and take care with the washer tubes on the wiper arms because they are liable to being torn off by the wiper door, and when they get old they will become brittle and fragile.*

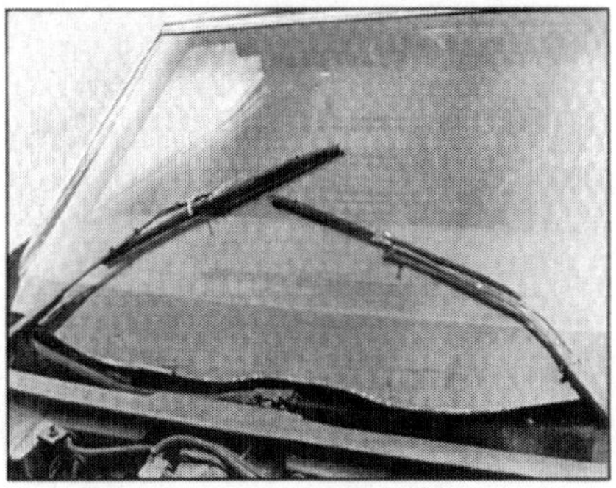

▲ *WW6. Check the wiper travel by wetting the screen and turning the wipers to 'high'. Ensure they don't hit or drag over the windshield posts. To change the adjustment, stop the wiper blades at the position where they are closest to each other, as shown here. Make certain that the blades aren't adjusted too close because then they will hit and jam. After making an adjustment, don't just turn the wipers straight back on again; use the ignition switch or the override switch (where fitted) to operate the wipers in short movements through their travel. Don't forget to carry out that final check with the screen wetted and the wipers in all speeds before assuming that all is well.*

▲ *WW7. There is a degree of adjustment available where the wiper motor crank connects to the linkage arms. The elongated slots allow you to move the arms in and out, but if you still can't stop the wiper arms from going too far, you could either increase the length of the slot or increase the bends on both of the cranks.*

FIBER OPTICS REPAIRS

The method of monitoring the lamps on 1968 through to 1971 Corvettes was so futuristic that it's still a talking point today. There's a fiber optic cord pointing at each of the car's exterior lamps and the light transmitting efficiency of the fiber optic cord is so superb that the light emitted from the other end of the cord, in the passenger compartment tells the driver if any of the lamps fail.

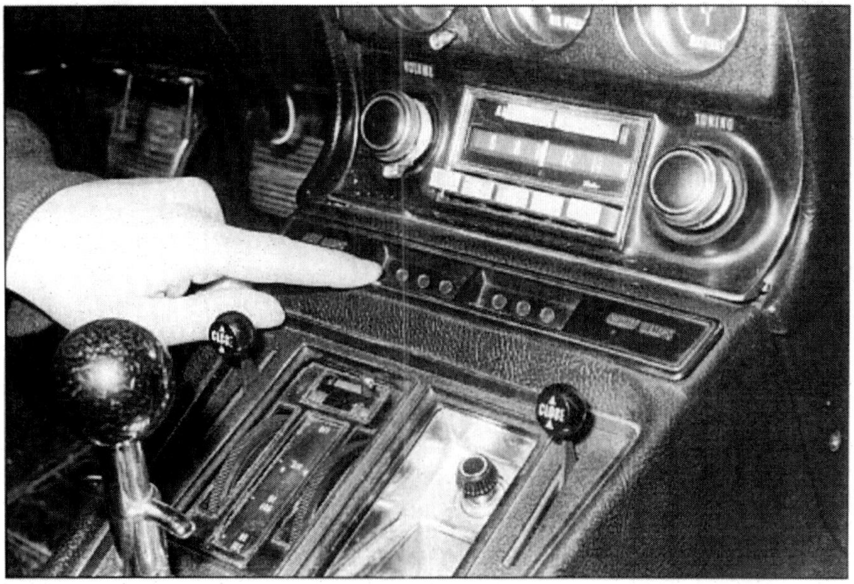

▼ *FB1. The Corvette's fiber optic system not only makes the car safer, it also reduces the likelihood of the car being stopped by an officer of the law. And with a car as high-profile as the Corvette, the fewer opportunities you give 'em, the better! The system cannot wear out since it does not depend on any moving parts or electrical connections, but unfortunately it is often damaged by hamfisted repair work or in an accident. At the time of writing, neither individual cords nor the complete harness can be ordered separately, other than as part of the entire front or rear body wiring harness. Where are you, aftermarket producers? However, fiber optic cords can be purchased from some scientific suppliers in various diameters and lengths, or from dealers in custom van parts. In the meantime, here are some brilliant ideas (pun intended!) from John Pfanstiehl on how to make the most of damaged fiber optics.*

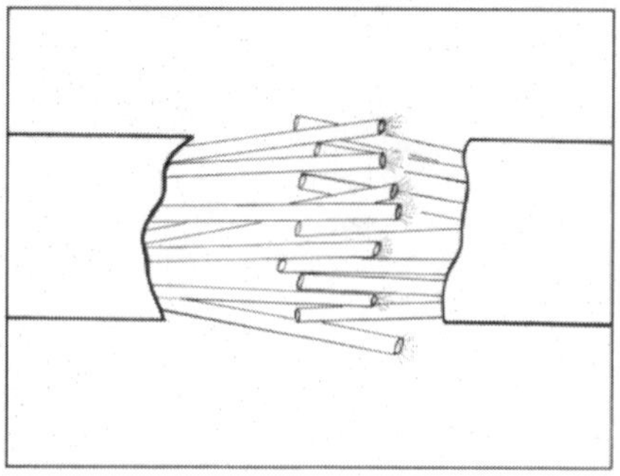

▶ *FB2. You have to remember that fiber optics transmit light like a water pipe transmits water. It's true that if you overlap the ends of the 'pipe', you won't have light dripping from under the dash – but you won't have light at the other end of the 'pipe' either.* (Courtesy John Pfanstiehl)

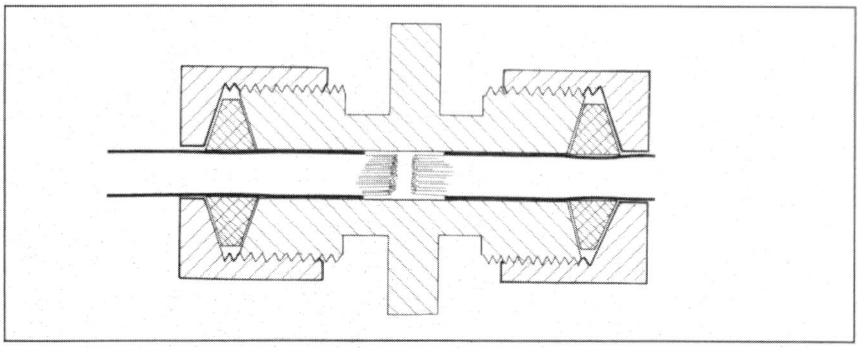

▲ *FB3. The John Pfanstiehl solution is to use a ⅛ in. compression coupling which you may have to drill to accommodate the fiber portion of the cord. Take the very greatest care to tighten the nuts to hold the jacket of the cord so that the ends of the cord are as close as possible and as precisely aligned as possible. You may be able to think up your own similar solutions, involving perhaps, a drilled nylon bushing into which the cords are pushed – but then you'd have to ensure that the cords are tightly held and not likely to pull out again.* (Courtesy John Pfanstiehl)

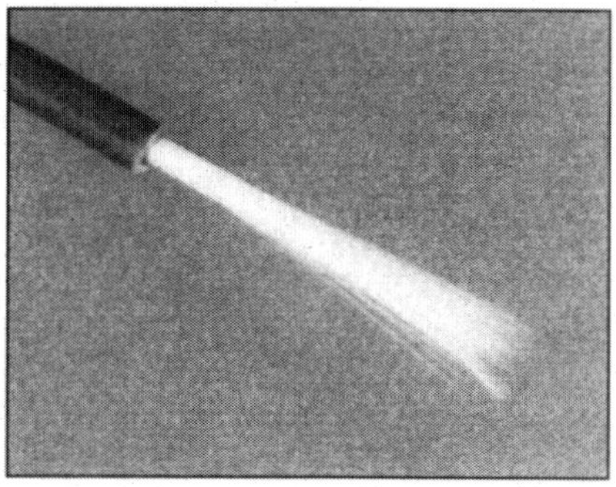

▲ *FB4. A severed fiber optic cord looks likes this and must be cut off clean before the connection can be made. If the break is very near the end of the cord, look at ways of releasing some of the fiber optic cord from the harness so that you can ease it through to a shorter route and make a direct connection to the lamp.*

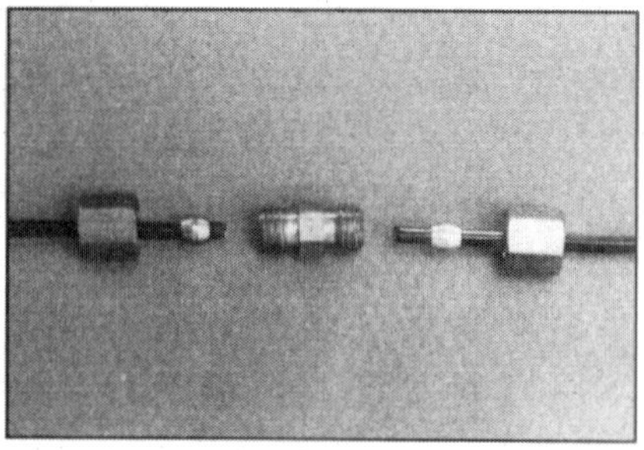

▲ *FB5. Here's the John Pfanstiehl compression union repair prior to assembly.*

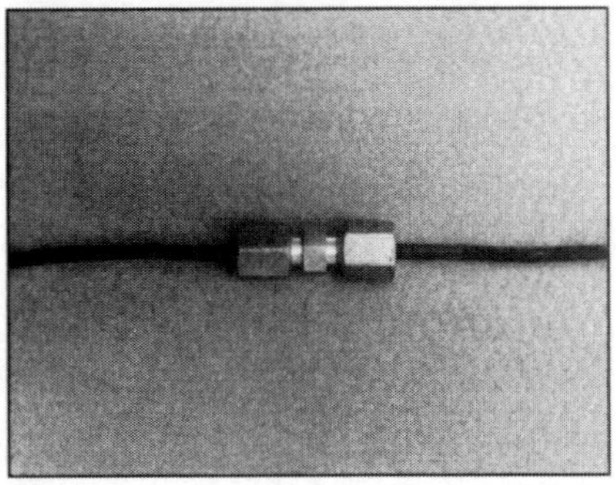

▲ FB6. The assembled repair will need to be hidden neatly out of the way. You can get away with just one repair in a line; two repairs make the light signal unacceptably dim. Do ensure that the two ends of the cord are pushed into the connector so that they are butting up against each other, transmitting the maximum amount of light between the two sets of fiber optics.

▲ FB7. Dimmed fiber optics can often be improved by cleaning road grime and dirt from the back of headlamps. On the other hand, if one fiber optic has had to be repaired leaving one turn signal light in the cockpit brighter than the other, you could try spraying a little aluminium paint over the back of the headlamp on the good side to diminish the amount of light transmitted and match it with the repaired side.

Alternatively, you could try increasing the amount of light getting into the fiber optic cord on the repaired side. There are several tricks to try here. See the following notes.

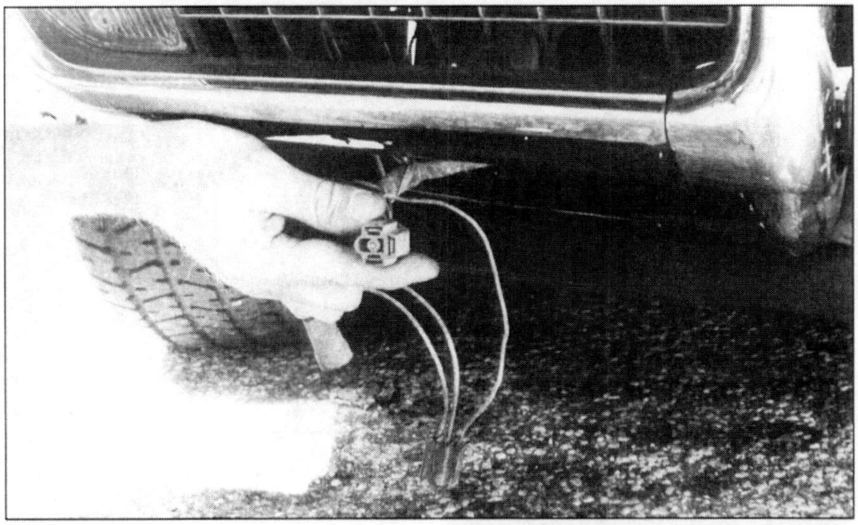

▲ FB8. You don't have to remove the headlamps to take out the headlamp terminal blocks. Get at it from under the front grille, remove the block and clean it up.

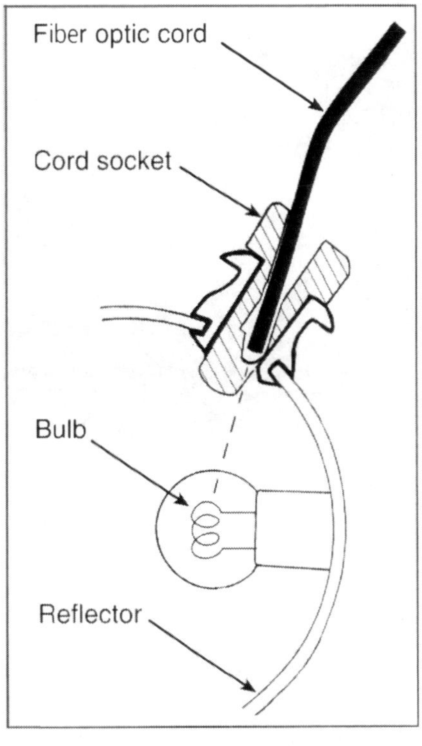

▲ FB9. The fiber optics on the turn signals can be made brighter by redrilling the cord socket at an angle so the cord is aimed more accurately at the bulbs filament. Decreasing the distance between the filament and the fiber optic by one half is said to increase the light in the passenger compartment by four times. (Courtesy John Pfanstiehl)

▲ FB 10. Access to the fiber optic ends for turn signals can be gained through openings in front of the radiator.

REPLACEMENT PARTS

As we said earlier, you may be able to purchase new lengths of fiber optic cord, but you would have to disguise them cunningly if they were not to be seen anywhere that would turn a concourse judge's head. The cost of installing an individual cord would be far less than a complete harness, however, because none of the electrical connections would have to be remade. Aftermarket harnesses are now becoming available for those occasions when you do need to replace the complete harness, such as in a total restoration or after a part fire. If you are fortunate enough to locate an original GM body wiring harness for those cars containing fiber optic cords – you can occasionally pick them up at swap meets or through enthusiast advertisements in club magazines – these are the part numbers to look out for.

CHEVROLET PART NUMBERS OF BODY WIRING HARNESSES CONTAINING fiber OPTIC CORDS

Year	*Front*	*Rear*
1968	6295258	6295248
1969	6297568	6297545
1970	8901918	8901531
1971	8901918	8901903

ALARM PROBLEM SOLVING

Of all the accolades it has received, the title of 'America's most stolen car' is one the Corvette could do without. This problem was first addressed by GM in 1968, when an alarm system was made available as an option. It was fitted as standard to all Corvettes from 1972 – the same year that the fiber optic warning light system was dropped. It's a simple system (though in later years it has become considerably more complex) whereby if the doors or hood are tampered with, a horn (mounted within the wheel arch above the left rear tire) is activated. The system is armed by inserting a key into a lock cylinder mounted on the left-hand front fender. The anti-theft relay and flasher are located within the centre console.

Alarm problems usually revolve around the switch itself, notably its neglect. It happens to many owners and it could happen to you. The door ajar and alarm switch gets bent or goes out of adjustment so that you can't use the alarm. Funny how fixing such small things gets put off. And then, one day, it's too late: the thieves have struck!

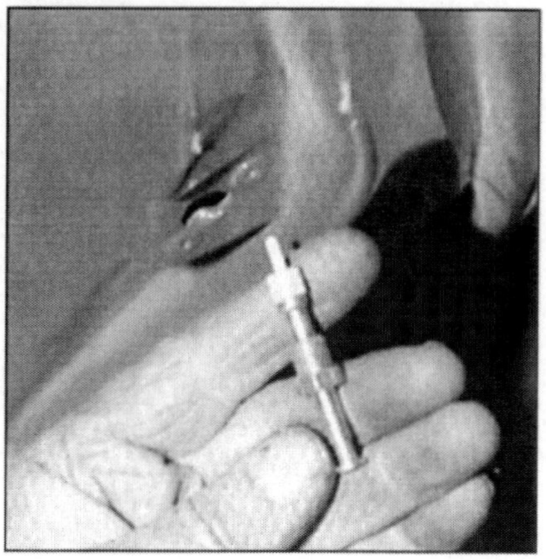

▲ *AL1. The switch at the rear of the door sticks out some way and often gets snagged by owners or passengers – you can't be careful all of the time. If the inner shaft is bent, it won't slide inside its sleeve as the door is closed. Catch it early enough, and all you'll need to do is straighten the bend with needle nosed pliers.*

▲ *AL2. Although, as you can see from the former shot, the switch is meant to tighten down into its receiving thread, vibration and movement can loosen the tight fit so the sleeve itself becomes loose. Try using a top-class adhesive to relocate the sleeve.*

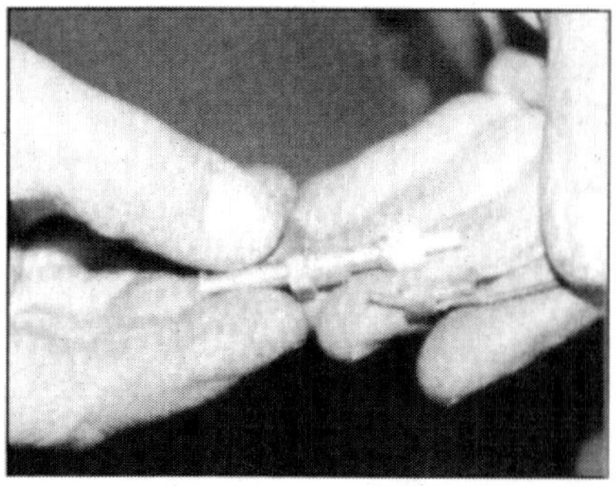

◀ *AL3. Be warned that removing the switch can be a tricky business because the threads become corroded in this position. Try applying penetrating oil each day for several days before you attempt to remove the switch and use lots of delicacy so you don't just break the sleeve free from the car's body. Once the thread is unscrewed, wiggle and turn the switch and the wires through the hole – there's an inch or more of wire to be pulled out, complete with the switch.*

Look closely at the wire ends and clean any corrosion with fine sand paper before fitting them into the new switch.

If you lose wiring inside the bodywork, don't take a hacksaw to the door jamb! Just remove the access panel inside the rear wheel well.

INTERMITTENT PROBLEM ONE

If the alarm works sometimes when the door is opened but not on other occasions, the contacts are undoubtedly at fault. Disconnect them and clean them up as described. If the contacts in the switch are too badly corroded you'll need a new switch. It's also just possible that there's a break in the end of the wire, caused by movement over the years – have recourse to that access plate once again.

INTERMITTENT PROBLEM TWO

If the alarm seems over-sensitive, and especially if the door-open light flickers on when you are driving, you'll probably find the wire ends are rubbing against the steel reinforcement panel inside the door jamb. Overcome this GM boo boo by wrapping a layer of electrical tape around the exposed metal portions of the wire ends, then re-insert the switch back into the door jamb.

▲ *AL4. The good news about the hood switch is that it is much easier to get at and less likely to corrode. The bad news is that there is more susceptibility to dirt and air-born oil particles. Clean and inspect the switch and then adjust it by bending the bracket, as shown. The alarm should go off when the hood release is operated and the hood popped up.*

Appendix 1
Workshop Procedures and Safety First

Professional motor mechanics are trained in safe working procedures, whereas the onus is on you, the home mechanic, to find them out for yourself and act upon them. However enthusiastic you may be about getting on with the job in hand, do take the time to ensure that your safety is not put at risk. A moment's lack of attention can result in an accident, as can failure to observe certain elementary precautions.

There will always be new ways of having accidents, and the following points do not pretend to be a comprehensive list of all dangers; they are intended rather to make you aware of the risks and to encourage a safety-conscious approach to all work you carry out on your vehicle.

Be sure to consult the suppliers of any materials and equipment you may use, and to obtain and read carefully operating and health and safety instructions that they may supply.

ESSENTIAL DOS AND DON'TS

DON'T rely on a single jack when working underneath the vehicle. Always use reliable additional means of support, such as axle stands, securely placed under a part of the vehicle that you know will not give way.

DON'T attempt to loosen or tighten high-torque nuts (e.g. wheel hub nuts) while the vehicle is on a jack; it may be pulled off.

DON'T start the engine without first ascertaining that the transmission is in neutral (or 'Park' where applicable) and the parking brake applied.

DON'T suddenly remove the filler cap from a hot cooling system – cover it with a cloth and release the pressure gradually first, or you may get scalded by escaping coolant.

DON'T attempt to drain oil, automatic transmission fluid, or coolant until you are sure it has cooled sufficiently to avoid scalding you.

DON'T grasp any part of the engine, exhaust or catalytic converter without first ascertaining that it is sufficiently cool to avoid burning you.

DON'T allow brake fluid or antifreeze to contact vehicle paintwork.

DON'T syphon toxic liquids such as fuel, brake fluid or antifreeze by mouth, or allow them to remain on your skin.

DON'T inhale dust – it may be injurious to health.

DON'T allow any spilt oil or grease to remain on the floor – wipe it up straight away, before someone slips on it.

DON'T use ill-fitting wrenches or other tools which may slip and cause injury.

DON'T attempt to lift a heavy component which may be beyond your capability – get assistance.

DON'T rush to finish a job, or take unverified short cuts.

DON'T allow children or animals in or around an unattended vehicle.

DON'T park vehicles with catalytic converters over combustible materials such as dry grass, oil rags, etc., if the engine has recently been run. As catalytic converters reach extremely high temperatures, any such materials in close proximity may ignite.

DON'T run vehicles equipped with catalytic converters without the exhaust system heat shields fitted.

DO wear eye protection when using power tools such as an electric drill, sander, bench grinder, etc., and when working under the vehicle.

DO use a barrier cream on your hands prior to undertaking dirty jobs – it will protect your skin from infection as well as making the dirt easier to remove afterwards; but make sure your hands aren't left slippery. Note that long-term contact with used engine oil can be a health hazard.

DO keep loose clothing (cuffs, tie, etc.) and long hair well out of the way of moving mechanical parts.

DO remove rings, wrist watch, etc., before working on the vehicle – especially the electrical system.

DO ensure that any lifting tackle used has a safe working load rating adequate for the job, and is used precisely as recommended by the manufacturer.

DO keep your work area tidy – it is only too easy to fall over articles left lying around.

DO get someone to check periodically that all is well, when working alone on the vehicle.

DO carry out work in a logical sequence and check that everything is correctly assembled and tightened afterwards.

DO remember that your vehicle's safety affects that of yourself and others. If in doubt on any point, get specialist advice.

IF, in spite of following these precautions, you are unfortunate enough to injure yourself, seek medical attention as soon as possible.

FIRE

Remember at all times that gasoline is highly flammable. Never smoke, or have any kind of naked flame around, when working on the vehicle. But the risk does not end there – a spark caused by an electrical short-circuit, by two metal surfaces contacting each other, by a central heating boiler in the garage 'firing up', or even by static electricity built up in your body under certain conditions, can ignite gasoline vapour, which in a confined space is highly explosive.

Always disconnect the battery ground terminal before working on any part of the fuel system, and never risk spilling fuel on to a hot engine or exhaust.

It is recommended that a fire extinguisher of a type suitable for fuel and electrical fires is kept handy in the garage or workplace at all times. Never try to extinguish a fuel or electrical fire with water.

FUMES

Certain fumes are highly toxic and can quickly cause unconsciousness and even death if inhaled to any extent. Gasoline vapour comes into this category, as do the vapours from certain solvents such as trichloroethylene and those from many adhesives. Any draining or pouring of such volatile fluids should be done in a well-ventilated area.

When using cleaning fluids and solvents, read the instructions carefully. Never use any materials from unmarked containers – they may give off poisonous vapours.

Never run the engine of a motor vehicle in an enclosed space such as a garage. Exhaust fumes contain carbon monoxide which is extremely poisonous. If you need to run the engine, always do so in the open air or at least have the rear of the vehicle outside the workplace.

If you are fortunate enough to have the use of an inspection pit, never drain or pour gasoline, and never run the engine, while the vehicle is standing over it; the fumes, being heavier than air, will concentrate in the pit with possibly lethal results.

THE BATTERY

Never cause a spark, or allow a naked light, near the vehicle battery. It will normally be giving off a certain amount of hydrogen gas, which is highly explosive.

Always disconnect the battery ground terminal before working on the fuel or electrical systems.

If possible, loosen the filler plugs or cover when charging the battery from an external source. Do not charge at an excessive rate or the battery may burst.

Take care when topping up and when carrying the battery. The acid electrolyte, even when diluted, is very corrosive and should not be allowed to contact the eyes or skin.

If you ever need to prepare electrolyte yourself, always add the acid slowly to the water, and never the other way round. Protect against splashes by wearing rubber gloves and goggles.

MAINS ELECTRICITY

When using an electric power tool, inspection light, etc., which works from the mains, always ensure that the appliance is correctly connected to its plug and that, where necessary, it is properly grounded. Do not use such appliances in damp conditions and, again, beware of creating a spark or applying excessive heat in the vicinity of fuel or fuel vapour.

Also, before using any mains powered electrical equipment, take one more simple precaution – use an RCD (Residual Current Device) circuit breaker. Then, if there is a short, the RCD circuit breaker minimizes the risk of electrocution by instantly cutting the power supply. Buy from any electrical store or DIY centre. RCDs fit simply into your electrical socket before plugging in your electrical equipment.

IGNITION HT VOLTAGE

A severe electric shock can result from touching certain parts of the ignition system, such as the HT leads, when the engine is running or being cranked, particularly if components are damp or the insulation is defective. Where an electronic ignition system is fitted, the HT voltage is much higher and could prove fatal. Consult your handbook or main dealer if in any doubt. Risk of injury while working on running engines, e.g. adjusting the timing, can arise if the operator touches a high voltage lead and pulls his hand away on to a projection or revolving part. On some vehicles the voltage used in the ignition system is so high that it can cause injury or death by electrocution.

WELDING AND BODYWORK REPAIRS

It is so useful to be able to weld when carrying out restoration work, and yet there is a good deal that could go dangerously wrong for the uninformed – in fact more than could be covered here. **For safety's sake** you are strongly recommended to seek tuition. In addition, all of the information and instructional material produced by the suppliers of materials and equipment you will be using must be studied carefully. You may have to ask your stockist for some of this printed material if it is not made available at the time of purchase.

In addition, it is strongly recommended that *The Car Bodywork Repair Manual*, published by Haynes, is purchased and studied before carrying out any welding or bodywork repairs. Consisting of 292 pages, around 1,000 illustrations and written by Lindsay Porter, the author of this book, *The Car Bodywork Repair Manual* picks the brains of specialists from a variety of fields, and covers arc, MIG and 'gas' welding, panel beating and accident repair, rust repair and treatment, paint spraying, fiberglass work, filler, lead loading, interiors and much more besides. Alongside a number of projects, the book describes in detail how to carry out each of the techniques involved in car bodywork repair, with safety notes where necessary. As such, it is the ideal complement to this book.

COMPRESSED GAS CYLINDERS

There are serious hazards associated with the storage and handling of gas cylinders and fittings, and standard precautions should be strictly observed in dealing with them. Ensure that cylinders are stored in safe conditions, properly maintained and always handled with special care, and make constant efforts to eliminate the possibility of leakage, fire and explosion.

The cylinder gases that are commonly used are oxygen, acetylene and liquid petroleum gas (LPG). Safety requirements for all three gases are: Cylinders must be stored in a fire resistant, dry and well-ventilated space, away from any source of heat or ignition and protected from ice, snow or direct sunlight. Valves of cylinders in store must always be kept uppermost and closed, even when the cylinder is empty. Cylinders should be handled with care and only by personnel who are reliable, adequately informed and fully aware of all associated hazards. Damaged or leaking cylinders should be promptly taken outside into the open air, and the supplier and fire authorities should be notified immediately. No one should approach a gas cylinder store with a naked light or cigarette. Care should be taken to avoid striking or dropping cylinders, or knocking them together. Cylinders should never be used as rollers. One cylinder should never be filled from another. Every care must be taken to avoid accidental damage to cylinder valves. Valves must be operated without haste, never fully opened hard back against the back stop (so that other users know the valve is open) and never wrenched shut but turned just securely enough to stop the gas. Before removing or loosening any outlet connections, caps or plugs, a check should be made that the valves are closed. When carrying cylinders, close all valves and appliance taps, and extinguish naked flames, including pilot jets, before disconnecting them. When reconnecting, ensure that all connections and washers are clean and in good condition, and do not overtighten them. Immediately a cylinder becomes empty, close its valve.

Safety requirements for acetylene: Cylinders must always be stored and used in the upright position. If a cylinder becomes heated accidentally or becomes hot because of excessive backfiring, immediately shut the valve, detach the regulator, take the cylinder out of doors well away from the building, immerse it in, or continuously spray it with, water, open the valve and allow the gas to escape until the cylinder is empty. If necessary, notify the emergency fire service without delay.

Safety requirements for oxygen are: No oil or grease should be used on valves or fittings. Cylinders with convex bases should be used in a stand or held securely to a wall.

Safety requirements for LPG are: The store must be kept free of combustible material, corrosive material and cylinders of oxygen.

Cylinders should only ever be carried upright, securely strapped down, preferably in an open vehicle or with windows open. Carry the suppliers' safety data with you. In the event of an accident, notify the Police and Fire Services and hand the safety data to them.

DANGEROUS LIQUIDS AND GASES

Because of flammable gas given off by batteries when on charge, care should be taken to avoid sparking by switching off the power supply before charger leads are connected or disconnected. Battery terminals should be shielded, since a battery contains energy and a spark can be caused by any conductor which touches its terminals or exposed connecting straps.

When internal combustion engines are operated inside buildings, the exhaust fumes must be properly discharged to the open air. Gasoline must be contained in metal cans which should be stored safely. In any area where battery charging or the testing of fuel injection systems is carried out there must be good ventilation, and no sources of ignition. Inspection pits often present serious hazards. They should be of adequate length to allow safe access and exit while a car is in position. If there is an inspection pit, gasoline may enter it. Since gasoline vapour is heavier than air it will remain there and be a hazard if there is any source of ignition. All sources of ignition must therefore be excluded.

WORK WITH PLASTICS

Work with plastic materials brings additional hazards into workshops. Many of the materials used (polymers, resins, adhesives and materials acting as catalysts and accelerators) readily produce very dangerous situations in the form of poisonous fumes, skin irritants, risk of fire and explosions. Do not allow resin or two-pack adhesive hardener, or that supplied with filler or two-pack stopper to come into contact with skin or eyes. Read carefully the safety notes supplied on the tin, tube or packaging.

JACKS AND AXLE STANDS

Special care should be taken when any type of lifting equipment is used. NEVER even consider working under your car using only a jack to support the weight of it. Lifting jacks are for raising vehicles not supporting them, and must be replaced by adequate supports before any work is begun on the vehicle. Axle stands are available from many discount stores, and all auto parts stores. These stands are absolutely essential if you plan to work under your car. Simple triangular stands (fixed or adjustable) will suit almost all of your working situations. Drive-on ramps are very limiting because of their design and size.

When jacking the car from the front, leave the gearbox in neutral and the brake off until you have placed the axle stands under the frame. Make sure that the car is on level ground first! Then put the car into gear and/or engage the handbrake and lower the jack. Obviously, DO NOT put the car in gear if you plan to turn over the engine! Leaving the brake on, or leaving the car in gear while jacking the front of the car will necessarily cause the jack to tip (unless a good quality trolley jack with wheels is being used). This is unavoidable when jacking the car on one side, and the use of the handbrake in this case is recommended. If the car is older and if it shows signs of weakening at the jack tubes while using the factory jack, it is best to purchase a good scissors jack or hydraulic jack – preferably trolley-type (depending on your budget).

WORKSHOP SAFETY – SUMMARY

- Always have a fire extinguisher at arm's length whenever welding or when working on the fuel system – under the car, or under the bonnet.
- NEVER use a naked flame near the gas tank.

- Keep your inspection lamp FAR AWAY from any source of dripping gasoline; for example, while removing the fuel pump.
- NEVER use gasoline to clean parts. Use kerosene or mineral spirit.
- NO SMOKING!

If you do have a fire, DON'T PANIC. Use the extinguisher effectively by directing it at the base of the fire.

PAINT SPRAYING

NEVER use two-pack, isocyanate-based paints in the home environment or home workshop. Ask your supplier if you are not sure which is which. If you have use of a professional booth, wear an air-fed mask. Wear a charcoal face mask when spraying other paints, and maintain ventilation to the spray area. Concentrated fumes are dangerous!

Spray fumes, thinners and paint are highly flammable. Keep away from naked flames or sparks.

Paint spraying safety is too large a subject for this book. See Lindsay Porter's *The Car Bodywork Repair Manual* (Haynes) for further information.

FLUOROELASTOMERS – MOST IMPORTANT! PLEASE READ THIS SECTION!

Many synthetic rubber-like materials used in motor cars contain substances known as fluoroelastomers (containing fluorine). These are commonly used for oil seals, wiring and cabling, bearing surfaces, gaskets, diaphragms, hoses and 'O' rings. If they are subjected to temperatures greater than 315°C, they will decompose and can be potentially hazardous. Fluoroelastomer materials will show physical signs of decomposition under such conditions in the form of charring or black sticky masses. Some decomposition may occur at temperatures above 200°C, and it is obvious that when a car has been in a fire or has been dismantled with the assistance of a cutting torch or blow torch, the fluoroelastomers can decompose in the manner indicated above.

In the presence of any water or humidity, including atmospheric moisture, the by-products caused by fluoroelastomers being heated can be extremely dangerous.

According to the Health and Safety Executive, 'Skin contact with this liquid or decomposition residues can cause painful and penetrating burns. Permanent irreversible skin and tissue damage can occur.' Damage can also be caused to eyes or by the inhalation of fumes created as fluoroelastomers are burned or heated.

If you are in the vicinity of a vehicle fire or a place where a vehicle is being cut up with cutting equipment, the Health and Safety Executive recommend the following action:

- Assume, unless you know otherwise that seals, gaskets and 'O' rings, hoses, wiring and cabling, bearing surfaces and diaphragms are fluoroelastomers.
- Inform firefighters of the presence of fluoroelastomers and toxic and corrosive fume hazards when they arrive.
- All personnel not wearing breathing apparatus must leave the immediate area of a fire.

After fires or exposure to high temperatures:

- Do not touch blackened or charred seals or equipment.
- Allow all burnt or decomposed fluoroelastomer materials to cool down before inspection, investigation, tear-down or removal.
- Preferably, don't handle parts containing decomposed fluoroelastomers, but if you must, wear goggles and PVC (polyvinyl chloride) or neoprene protective gloves whilst doing so. Never handle such parts unless they are completely cool.
- Contaminated parts, residues, materials and clothing, including protective clothing and gloves, should be disposed of by an approved contractor to landfill, or by incineration according to national or local regulations. Original seals, gaskets and 'O' rings, along with contaminated material, must not be burned locally.

Symptoms and clinical findings of exposure:

A Skin/eye contact:

Symptoms may be apparent immediately after contact, or there may be considerable delay following exposure. Do not assume that a lack of immediate symptoms means there has been no damage; delays of minutes in treatment can have severe consequences:

- Dull throbbing ache.
- Severe and persistent pain.
- Black discolouration under nails (skin contact).
- Severe, persistent and penetrating burns.
- Skin swelling and redness.
- Blistering.
- Sometimes pain without visible change.

B Inhalation (breathing) – immediate:

- Coughing.
- Choking.
- Chills lasting one or two hours after exposure.
- Irritation.

C Inhalation (breathing) – delays of one to two days or more:

- Fever.
- Cough.
- Chest tightness.
- Pulmonary oedema (congestion).
- Bronchial pneumonia.

FIRST AID

A Skin contact:

- Remove contaminated clothing immediately.
- Irrigate affected skin with copious amounts of cold water or lime water (saturated calcium hydroxide solution) for 15 to 60 minutes. Obtain medical assistance urgently.

B Inhalation:

- Remove to fresh air and obtain medical supportive treatment immediately. Treat for pulmonary oedema.

C Eye contact:

- Wash/irrigate eyes immediately with water followed by normal saline for 30 to 60 minutes. Obtain immediate medical attention.

Appendix 2

British and American Technical Terms

As this book has been written in England, it uses English phrases and spellings. Some of these differ from those used in America. Normally, these cause no difficulty, but to make sure, a glossary is printed below. (See also, the note on page 5)

British	American
Accelerator	Gas pedal
Aerial	Antenna
Anti-roll bar	Stabilizer or sway bar
Bonnet (engine cover)	Hood
Boot (luggage compartment)	Trunk
Boot lid	Trunk lid
Bottom gear	First gear or low gear
Bulkhead	Firewall
Cam follower or tappet	Valve lifter or tappet
Carburettor	Carburetor
Catch	Lock
Choke/venturi	Barrel
Clearance	Lash
Crownwheel	Ring gear (of differential)
Disc (brake)	Rotor/disk
Drop arm	Pitman arm
Drophead coupé	Convertible
Dynamo	Generator (DC)
Earth (electrical)	Ground
Estate car	Station wagon
Fast back	Hard top
Fault finding/diagnosis	Trouble shooting
Float chamber	Float bowl
Freewheel	Coast
Gudgeon pin	Piston pin or wrist pin
Gearchange	Shift
Gearbox	Transmission
Halfshaft	Axle
Handbrake	Parking brake
Hood	Soft top
Hot spot	Heat riser
Indicator	Turn signal
Interior lamp	Dome lamp
Layshaft (of gearbox)	Countershaft
Motorway	Freeway, turnpike, Interstate
Number plate	License plate
Paraffin	Kerosene
Petrol	Gasoline
Petrol tank	Gas tank
'Pinking'	'Pinging'
Propeller shaft	Driveshaft
Reverse	Back-up
Rocker cover	Valve cover
Saloon	Sedan
Seized	Frozen
Side indicator lights	Side marker lights
Side light	Parking light
Silencer	Muffler
Sill	Rocker panel
Spanner	Wrench
Split cotter (for valve spring cap)	Valve spring retainer
Split pin	Cotter pin
Steering arm	Spindle arm
Sump	Oil pan
Tappet	Valve lifter
Thrust bearing	Throw-out bearing
Top gear	High gear
Trackrod (of steering)	Tie-rod
Transmission	Drive train
Tire	Tire
Vice	Vise
Wheel nut	Lug nut
Windscreen	Windshield
Wing/mudguard	Fender

Appendix 3
Technical Specifications

Where a specification is carried forward from one year to the next we show it as 'As for previous model except for:' and then the changes that took place in that year are detailed. If your Corvette's specifications differ from those shown here, check Appendix 4 'Options' before assuming that your car is fitted with non-standard parts.

N.B. These are technical specifications: interiors and body specs are not covered here.

1953 Base Model Corvette
Engine
Ohv inline 6
150 horsepower at 4200 rpm
235.5 cu in., 3.56 in. bore x 3.96 in. stroke
Compression ratio 8.0:1
3 x Carter single-throat carburettors

Transmission
Two-speed automatic transmission
3.55:1 axle ratio

Suspension
Front: coil springs, tube-type shock absorbers, sway bar
Rear: leaf springs, tube-type shock absorbers, rigid axle

Wheels and Tires
Steel disc wheels
Tire size: front and rear 6.70 x 15 tubed type

Brakes
Drum

Dimensions
Wheelbase: 102 in.
Front track: 57.0 in.
Rear track: 58.8 in.
Curb weight: 2886 lb

1954 Base Model Corvette
As for previous model except for:
Engine
Only the 6-cylinder was used in 1954. Horsepower increased during the year.

Wheels and Tires
Tires changed at unspecified date from tubed to tubeless.

1955 Base Model Corvette
As for previous model except for:
Engine
Ohv in-line 6
155 horsepower at 4200 rpm
Ohv V8
195 horsepower at 5000 rpm

In 1955 both the six and a V-8 were considered to be 'base' engines. All but a few 1955 Corvettes had the V-8 engine.

Transmission
Manual 3-speed transmission
Single dry-plate clutch

Dimensions
Curb weight: 2805 lb

1956 Base Model Corvette
As for previous model except for:
Engine
Ohv V-8
210 horsepower at 5200 rpm
265 cu in., 3.75 in. bore x 3.00 in. stroke
Compression ratio 9.25:1
Single four-barrel carburettor

Transmission
3.70:1 axle ratio

Dimensions
Curb weight: 2875 lb

1957 Base Model Corvette
As for previous model except for:
Engine
220 horsepower at 4800 rpm
283 cu in., 3.875 in. bore x 3.00 in. stroke
Compression ratio: 9.5:1

Dimensions
Curb weight: 2849 lb

1958–1960 Base Model Corvette
As for previous model except for:
Engine
230 horsepower at 4800 rpm

Dimensions
Curb weight: 2926 lb (1958)
2975 lb (1959)
2985 lb (1960)

1961 Base Model Corvette
As for previous model except for:
Transmission
3.36:1 axle ratio

Dimensions
Curb weight: 3035 lb

1962 Base Model Corvette
As for previous model except for:
Engine
250 horsepower at 4400 rpm
327 cu in., 4.00 in. bore x 3.25 in. stroke
Compression ratio: 10.5:1

Dimensions
Curb weight: 3065 lb

1963–1964 Base Model Corvette
As for previous model except for:
Suspension
Rear suspension: single transverse leaf spring, tube-type shock absorbers, independent with lateral struts

Dimensions
Wheelbase: 98 in.

Front track: 56.25 in.
Rear track: 57.00 in.
Curb weight: 3015 lb (1963)
3125 lb (1964)

1965–1966 Base Model Corvette
As for previous model except for:
Engine
Compression ratio: 10.5:1 (1965),
10.25:1 (1966-on)

Wheels and Tires
Tire size: front and rear 7.75 x 15

Brakes
Disc, all four wheels

Dimensions
Front track: 56.8 in.
Rear track: 57.6 in.
Curb weight: 3135 (1965)
3140 (1966)

1967 Base Model Corvette
As for previous model except for:
Engine
300 horsepower at 5000 rpm

Dimensions
Front track: 57.6 in.
Rear track: 58.3 in.
Curb weight: 3155 lb

1968 Base Model Corvette
As for previous model except for:
Wheels and Tires
Tire size: front and rear F70 x 15

Dimensions
Front track: 58.3 in.
Rear track: 59.0 in.
Curb weight: 3210 lb

1969 Base Model Corvette
As for previous model except for:
Engine
300 horsepower at 4800 rpm
350 cu in., 4.00 in. bore x 3.48 in. stroke

Dimensions
Front track: 58.7 in.
Rear track: 59.4 in.
Curb weight: 3245 lb

1970 Base Model Corvette
As for previous model except for:
Transmission
Single dry-plate clutch
Four-speed manual
Axle ratio: 3.08:1 (automatic)

Dimensions
Curb weight: 3285 lb

1971 Base Model Corvette
As for previous model except for:
Engine
270 horsepower at 4800 rpm
Compression ratio: 8.5:1

Dimensions
Curb weight: 3202 lb

1972 Base Model Corvette
As for previous model except for:
Engine
200 horsepower at 4200 rpm

Dimensions
Curb weight: 3305 lb

1973–1977 Base Model Corvette
As for previous model except for:
Engine
190 horsepower at 4400 rpm (1973)
195 horsepower at 4400 rpm (1974)
165 horsepower at 3800 rpm (1975)
180 horsepower at 3800 rpm (1976–77)

Wheels
Tire size: front and rear GR70 x 15

Dimensions
Rear track: 59.5 in.
Curb weight: 3416 lb (1973)
3388 lb (1974)
3529 lb (1975)
3541 lb (1976)
3534 lb (1977)

1978 Base Model Corvette
As for 1977 except for:
Engine
185 horsepower at 4200 rpm
Compression ratio: 8.2:1

Wheels and Tires
Tire size: front and rear P255/70R x 15

Dimensions
Curb weight: 3572 lb

1979–1981 Base Model Corvette
As for previous year except for:
Engine
195 horsepower at 4200 rpm (1979)
190 horsepower at 4200 rpm (1980–81)

Transmission
3.55:1 (1979 automatic)
3.07:1 (1980 automatic)
2.72:1 (1981 manual)
2.87:1 (1981 automatic)

Dimensions
Curb weight: 3503 lb (1979)
3336 lb (1980)
3307 lb (1981)

1982 Base Model Corvette
As for previous year except for:
Engine
200 horsepower at 4200 rpm
Compression ratio: 9.0:1
Throttle body injection carburettors

Transmission
Three-speed automatic transmission
2.87:1 axle ratio

Dimensions
Curb weight: 3342 lb

1983
No official 1983 model year cars were sold to the public. Instead, cars built from March 1983 were all designated '1984 model year' cars.

1984 Base Model Corvette
In 1984 there was a complete model change, so we have detailed all the changes for that year, and this is the last year we have covered in this section.

As for previous year except for:
Engine
205 horsepower at 4000 rpm

Transmission
Four-speed automatic with overdrive and high stall torque converter, four-speed manual with computer-controlled overdrive (manual override) in 2nd, 3rd and 4th gear.

Suspension
Front: Upper and lower A-arms, sway bar
Rear: Upper and lower control arms, sway bar

Wheels and Tires
Aluminium alloy
Tire size: front and rear P255/50VR x 16

Dimensions
Wheelbase: 96.2 in.
Front track: 59.6 in.
Rear track: 60.4 in.
Curb weight: 3192 lb

Appendix 4
Options

The options detailed in this section relate to the heading in Appendix 3 – Technical Specifications. These are not all the options for Corvettes, and should your car be fitted with anything not listed below we suggest that you refer to Michael Antonick's *The Genuine Corvette Black Book* before assuming that it is not an option.

In 1954 and 1955 the options listed in the above book were not really optional but required, and we have therefore started in 1956 when the options really became optional.

1956

Engine
250 horsepower engine with 2 x four-barrel carburettors, 265 cu in.

Transmission
Powerglide automatic transmission
3.27:1 rear axle ratio

Wheels and Tires
6.70 x 15 in. whitewall tires

1957

Engine
245 horsepower engine with 2 x four-barrel carburettors, 283 cu in.
270 horsepower engine with 2 x four-barrel carburettors, 283 cu in.
250 horsepower fuel injection engine, 283 cu in.
283 horsepower fuel injection engine, 283 cu in.

Transmission
Powerglide automatic transmission
Close ratio three-speed manual transmission
Manual four-speed transmission
Positraction rear axle with 3.70:1, 4.11:1 or 4.56:1 ratios

Suspension and Brakes
Heavy duty racing suspension

Wheels and Tires
15 x 5.5 in. steel wheels
6.70 x 15 in. whitewall tires

1958

Engine
245 horsepower engine with 2 x four-barrel carburettors, 283 cu in.
270 horsepower engine with 2 x four-barrel carburettors, 283 cu in.
250 horsepower fuel injection engine, 283 cu in.
290 horsepower fuel injection engine, 283 cu in.

Transmission
Powerglide automatic transmission
Manual four-speed transmission
Positraction rear axle with 3.70:1, 4.11:1 or 4.56:1 ratios

Suspension and Brakes
Heavy duty brakes and suspension

Wheels and Tires
15 x 5.5 in. steel wheels
6.70 x 15 in. whitewall tires

1959

Engine
245 horsepower engine with 2 x four-barrel carburettors, 283 cu in.
270 horsepower engine with 2 x four-barrel carburettors, 283 cu in.
250 horsepower fuel injection engine, 283 cu in.
290 horsepower fuel injection engine, 283 cu in.

Transmission
Powerglide automatic transmission
Manual four-speed transmission
Positraction rear axle

Suspension and Brakes
Heavy duty brakes and suspension

Wheels and Tires
15 x 5.5 in. steel wheels
6.70 x 15 in. whitewall tires
6.70 x 15 in. blackwall tires (nylon)

1960

Engine
245 horsepower engine with 2 x four-barrel carburettors, 283 cu in.
270 horsepower engine with 2 x four-barrel carburettors, 283 cu in.
250 horsepower fuel injection engine, 283 cu in.
290 horsepower fuel injection engine, 283 cu in.

Transmission
Powerglide automatic transmission
Manual four-speed transmission
Positraction rear axle

Suspension and Brakes
Heavy duty brakes and steering

Wheels and Tires
15 x 5.5 in. steel wheels
6.70 x 15 in. whitewall tires

1961

Engine
245 horsepower engine with 2 x four-barrel carburettors, 283 cu in.
270 horsepower engine with 2 x four-barrel carburettors, 283 cu in.
275 horsepower fuel injection engine, 283 cu in.
315 horsepower fuel injection engine, 283 cu in.

Transmission
Powerglide automatic transmission
Manual four-speed transmission
Positraction rear axle

Suspension and Brakes
Heavy futy brakes and steering

Wheels and Tires
15 x 5.5 in. steel wheels
6.70 x 15 in. whitewall tires
6.70 x 15 in. blackwall tires (nylon)

1962

Engine
300 horsepower engine, 327 cu in.
340 horsepower engine, 327 cu in.
360 horsepower fuel injection engine, 327 cu in.

Transmission
Powerglide automatic transmission
Manual four-speed transmission
Positraction rear axle
3.08:1 rear axle ratio

Suspension and Brakes
Heavy duty brakes and steering

Wheels and Tires
15 x 5.5 in. steel wheels
6.70 x 15 in. whitewall tires
6.70 x 15 in. blackwall tires (nylon)

1963

Engine
300 horsepower engine, 327 cu in.
340 horsepower engine, 327 cu in.
360 horsepower fuel injection engine, 327 cu in.

Transmission
Powerglide automatic transmission
Manual four-speed transmission
Positraction rear axle, all ratios
3.01:1 special highway axle ratio

Suspension and Brakes
Power brakes
Sintered metallic brakes

Wheels and Tires
Cast aluminium knock-off wheels
6.70 x 15 in. whitewall tires (rayon cord)
6.70 x 15 in. blackwall tires (nylon cord)

1964

Engine
300 horsepower engine, 327 cu in.
365 horsepower engine, 327 cu in.
375 horsepower fuel injection engine, 327 cu in.

Transmission
Powerglide automatic transmission
Manual four-speed transmission
Positraction rear axle, all ratios
3.08:1 special highway axle ratio

Suspension and Brakes
Special front and rear suspension
Power brakes
Special sintered metallic brake package
Sintered metallic power brakes

Wheels and Tires
Cast aluminium knock-off wheels
6.70 x 15 in. whitewall tires (rayon cord)
6.70 x 15 in. blackwall tires (nylon cord)

1965

300 horsepower engine, 327 cu in.
350 horsepower engine, 327 cu in.
365 horsepower engine, 327 cu in.
425 horsepower engine, 396 cu in.
375 horsepower fuel injection engine, 327 cu in.

Transmission
Powerglide automatic transmission
Manual four-speed transmission
Positraction rear axle, all ratios
3.08:1 special highway axle ratio

Suspension and Brakes
Special front and rear suspension
Power brakes
Drum brakes (substitution credit)

Wheels and Tires
Cast aluminium knock-off wheels
7.75 x 15 in. whitewall tires (rayon cord)
7.75 x 15 in. blackwall tires (nylon cord)
7.75 x 15 in. goldwall tires (nylon cord)

1966

Engine
350 horsepower engine, 327 cu in.
390 horsepower engine, 427 cu in.
450/425 horsepower engine, 427 cu in.

Transmission
Powerglide automatic transmission
Manual four-speed transmission
Close ratio manual four-speed transmission
Close ratio heavy duty manual four-speed transmission
Positraction rear axle, all ratios

Suspension and Brakes
Special front and rear suspension
Power brakes
Special heavy duty brakes

Wheels and Tires
Cast aluminium knock-off wheels
7.75 x 15 in. whitewall tires (rayon cord)
7.75 x 15 in. goldwall tires (nylon cord)

1967

Engine
350 horsepower engine, 327 cu in.
Special front and rear suspension
Power brakes
Special heavy duty brakes

Wheels and Tires
Cast aluminium bolt-on wheels
7.75 x 15 in. whitewall tires
7.75 x 15 in. redline tires

1968

350 horsepower engine, 327 cu in.
390 horsepower engine, 427 cu in.
400 horsepower engine, 427 cu in.
430 horsepower engine, 427 cu in.
435 horsepower engine, 427 cu in.

Transmission
Turbo hydramatic automatic transmission
Manual four-speed transmission
Close ratio manual four-speed transmission
Close ratio heavy duty manual four-speed transmission
Positraction rear axle, all ratios

Suspension and Brakes
Special front and rear suspension
Power brakes
Special heavy duty brakes

Wheels and Tires
F70 x 15 in. white stripe tires (nylon)
F70 x 15 in. red stripe tires (nylon)

1969
Engine
350 horsepower engine, 350 cu in.
390 horsepower engine, 427 cu in.
400 horsepower engine, 427 cu in.
430 horsepower engine, 427 cu in.
435 horsepower engine, 427 cu in.
(aluminium cylinder head available)

Transmission
Turbo Hydramatic automatic transmission
Manual four-speed transmission
Close ratio manual four-speed transmission
Close ratio heavy duty manual four-speed transmission
Positraction rear axle, all ratios

Suspension and Brakes
Special front and rear suspension
Power brakes
Special heavy duty brakes
Heavy duty clutch

Wheels and Tires
F70 x 15 in. white stripe tires (nylon)
F70 x 15 in. red stripe tires (nylon)
F70 x 15 in. white letter tires (nylon)

1970
Engine
350 horsepower engine, 350 cu in.
370 horsepower engine, 350 cu in.
390 horsepower engine, 454 cu in.

Transmission
Turbo Hydramatic automatic transmission
Close ratio manual four-speed transmission
Close ratio heavy duty manual four-speed transmission
Optional rear axle ratio

Suspension and Brakes
Power brakes

Wheels and Tires
F70 x 15 in. white stripe tires (nylon)
F70 x 15 in. white letter tires (nylon)

1971
Engine
330 horsepower engine, 350 cu in.
365 horsepower engine, 454 cu in.
425 horsepower engine, 454 cu in.

Transmission
Turbo Hydramatic automatic transmission
Close ratio manual four-speed transmission
Close ratio heavy duty manual four-speed transmission
Optional rear axle ratio

Suspension and Brakes
Power brakes

Wheels and Tires
F70 x 15 in white stripe tires (nylon)
F70 x 15 in white letter tires (nylon)

1972
Engine
255 horsepower engine, 350 cu in.
270 horsepower engine, 454 cu in.

Transmission
Turbo Hydramatic automatic transmission
Close ratio manual four-speed transmission
Optional rear axle ratio

Suspension and Brakes
Power Brakes

Wheels and Tires
F70 x 15 in. white stripe tires (nylon)
F70 x 15 in. white letter tires (nylon)

1973
Engine
250 horsepower engine, 350 cu in.
275 horsepower engine, 454 cu in.

Transmission
Turbo Hydramatic automatic transmission
Close ratio manual four-speed transmission
Optional rear axle ratios

Suspension and Brakes
Off-road brake and suspension package
Power brakes

Wheels and Tires
Cast aluminium wheels
GR70 x 15 in. white stripe steel belted tires
GR70 x 15 in. white letter steel belted tires

1974
Engine
250 horsepower engine, 350 cu in.
270 horsepower engine, 454 cu in.
Transmission
Turbo Hydramatic automatic transmission
Close ratio manual four-speed transmission
Optional rear axle ratios

Suspension and Brakes
Gymkhana suspension
Off-road brake and suspension package
Power brakes

Wheels and Tires
GR70 x 15 in. white stripe steel belted tires
GR70 x 15 in. white letter steel belted tires

1975
Engine
205 horsepower engine, 350 cu in.

Transmission
Turbo Hydramatic automatic transmission
Close ratio manual four-speed transmission
Optional rear axle ratios

Suspension and Brakes
Gymkhana suspension
Off-road brake and suspension package
Power brakes

Wheels and Tires
GR70 x 15 in. white stripe steel belted tires
GR70 x 15 in. white letter steel belted tires

1976
Engine
210 horsepower engine, 350 cu in.

Transmission
Turbo Hydramatic automatic transmission
Close ratio manual four-speed transmission
Optional rear axle ratios

Suspension and Brakes
Gymkhana suspension
Power brakes

Wheels and Tires
Aluminium wheels
GR70 x 15 in white stripe steel belted tires
GR70 x 15 in white letter steel belted tires

1977
Engine
210 horsepower engine, 350 cu in.

Transmission
Turbo Hydramatic automatic transmission
Close ratio manual four-speed transmission
Optional rear axle ratios

Suspension and Brakes
Gymkhana suspension

Wheels and Tires
Aluminium wheels
GR70 x 15 in. white letter steel belted tires

1978
Engine
220 horsepower engine, 350 cu in.

Transmission
Automatic transmission
Close ratio manual four-speed transmission
Optional rear axle ratios

Suspension and Brakes
Gymkhana suspension

Wheels and Tires
Aluminium wheels
P255/60R x 15 in. white letter SBR tires
P255/70R x 15 in. white letter SBR tires

1979
Engine
225 horsepower engine, 350 cu in.

Transmission
Automatic transmission
Close ratio manual four-speed transmission
Optional rear axle

Suspension and Brakes
Gymkhana suspension
Heavy duty shock absorbers

Wheels and Tires
Aluminium wheels
P255/60R x 15 in. white letter SBR tires
P255/70R x 15 in. white letter SBR tires

1980
Engine
180 horsepower engine, 305 cu in.

Transmission
Automatic transmission
Manual four-speed transmission

Suspension and Brakes
Gymkhana suspension
Heavy duty shock absorbers

Wheels and Tires
Aluminium wheels
P225/60R x 15 in. white letter SBR tires
P255/70R x 15 in. white letter SBR tires

1981
Engine
There were no engine options.

Transmission
Manual four-speed transmission
Performance axle ratio

Suspension and Brakes
Gymkhana suspension
Heavy duty shock absorbers

Wheels and Tires
Aluminium wheels
P255/60R x 15 in. white letter SBR tires
P225/70R x 15 in. white letter SBR tires

1982
Engine and Transmission
There were no engine or transmission options.

Suspension and Brakes
Gymkhana suspension

Wheels and Tires
Aluminium wheels
P255/60R x 15 in. white letter SBR tires
P225/70R x 15 in. white letter SBR tires

1983
No official 1983 model year cars were sold to the public. Instead, cars built from March 1983 were all designated '1984 model year' cars.

1984
Engine
There were no engine options.

Transmission
Manual four-speed transmission
Performance axle ratio

Suspension and Brakes
Delco-Bilstein shock absorbers

Wheels and Tires
16 in. wheels P255/50VR x 16 in. tires

Appendix 5

Color Schemes

Starting in 1963, an exterior color code number appears on a label or plate affixed to the instrument panel support member under the glove box. This runs through 1967 models, and on 1968 through to 1982 models it is affixed to the driver door post. For newer models a label can be found inside the console glove box lid.

Before 1963 no paint identification code was affixed to the body.

If you want to go all the way with originality, you'll use the acrylic lacquer of a similar type to that used by the Bowling Green plant up to 1980. (1981 Corvettes and later have an enamel-type paint with clear coat.) However, most of the early cars up to 1957 used nitro-cellulose lacquer, except for the 1957 model year's Inca Silver with Imperial Ivory coving which was painted in the later style of acrylic lacquer.

From the point of view of the restorer who is not concerned about total originality (where do you find 1957 air for the tires? And 30-year old gas doesn't pull too well!) you are better off having the car repainted by a specialist in modern two-pack lacquer which gives a far longer lasting shine and more durable finish than any other repainting materials currently available.

It is *most important* that you realize that up until 1989 customers were allowed to override the recommended colours, so different exterior/interior paint combinations from those shown in these charts are possible. And there are even some rare examples where variations have taken place from 1989-onward.

IMPORTANT NOTE: Before purchasing paint for your car, establish the correct original color then have your paint supplier check with a paint sample (or 'color chip') to ensure that you are purchasing the color you require. The following is a general guide and not to be used as proof positive of 'correct' color identification.

1953

EXTERIOR BODY color	SOFT-TOP
Polo White	Black

Interior colours available: Red

1954

EXTERIOR BODY color	SOFT-TOP
Black	Beige
Pennant Blue	Beige
Polo White	Beige
Sportsman Red	Beige

Interior colours available: Beige, Red

1955

EXTERIOR BODY color	SOFT-TOP
Corvette Copper	White
Gypsy Red	White/Beige
Harvest Gold	Dark Green
Pennant Blue	Beige
Polo White	White/Beige

Interior colours available: Dark Beige, Light Beige, Red, Yellow

1956

EXTERIOR BODY color	SOFT-TOP
Arctic Blue	Beige/White
Aztec Copper	Beige/White
Cascade Green	Beige/White
Onyx Black	Black/White
Polo White	Black/White
Venetian Red	Beige/White

Interior colours available: Beige, Red

1957

EXTERIOR BODY color	SOFT-TOP
Arctic Blue	Beige/Black/White
Aztec Copper	Beige/White
Cascade Green	Beige/Black/White
Inca Silver	Black/White
Onyx Black	Beige/Black/White
Polo White	Beige/Black/White
Venetian Red	Beige/Black/White

Interior colours available: Beige, Red

1958

EXTERIOR BODY color	SOFT-TOP
Black	Black/White
Charcoal	Black/White
Panama Yellow	Black/White
Regal Turquoise	Black/White
Signet Red	Black/White
Silver	Black/White
Silver Blue	Beige/White
Snowcrest White	Black/Beige/White

Interior colours available: Beige, Charcoal, Red

1959

EXTERIOR BODY color	SOFT-TOP
Classic Cream	Black/White
Crown Sapphire	Black/Turquoise/White
Frost Blue	Black/Blue/White
Inca Silver	Black/White
Roman Red	Black/White
Snowcrest White	Black/Blue/ Turquoise/White
Tuxedo Black	Black/White

Interior colours available: Black, Blue, Red, Turquoise

1960

EXTERIOR BODY color	SOFT-TOP
Cascade Green	Black/Blue/White
Ermine White	Black/Blue/White
Honduras Maroon	Black/White
Horizon Blue	Black/Blue/White
Roman Red	Black/White
Sateen Silver	Black/Blue/White
Tasco Turquoise	Black/Blue/White
Tuxedo Black	Black/Blue/White

Interior colours available: Blue, Black, Red, Turquoise

1961

EXTERIOR BODY color	SOFT-TOP
Ermine White	Black/White
Fawn Beige	Black/White
Honduras Maroon	Black/White
Jewel Blue	Black/White
Roman Red	Black/White
Sateen Silver	Black/White
Tuxedo Black	Black/White

Interior colours available: Blue, Black, Fawn, Red

1962

EXTERIOR BODY color	SOFT-TOP
Almond Beige	Black/White
Ermine White	Black/White
Fawn Beige	Black/White
Honduras Maroon	Black/White
Roman Red	Black/White
Sateen Silver	Black/White
Tuxedo Black	Black/White

Interior colours available: Black, Fawn, Red

1963

EXTERIOR BODY color	SOFT-TOP
Daytona Blue	Beige/Black/White
Ermine White	Beige/Black/White
Riverside Red	Beige/Black/White
Saddle Tan	Beige/Black/White
Sebring Silver	Beige/Black/White
Silver Blue	Beige/Black/White
Tuxedo Black	Beige/Black/White

Interior colours available: Black, Dark Blue, Red, Saddle

1964

EXTERIOR BODY color	SOFT-TOP
Daytona Blue	Beige/Black/White
Ermine White	Beige/Black/White
Riverside Red	Beige/Black/White
Saddle Tan	Beige/Black/White
Satin Silver	Beige/Black/White
Silver Blue	Beige/Black/White
Tuxedo Black	Beige/Black/White

Interior colours available:Black, Blue, Red, Saddle, Silver, White

1965

EXTERIOR BODY color	SOFT-TOP
Ermine White	Beige/Black/White
Glen Green	Beige/Black/White
Goldwood Yellow	Beige/Black/White
Milano Maroon	Beige/Black/White
Nassau Blue	Beige/Black/White
Rally Red	Beige/Black/White
Silver Pearl	Beige/Black/White
Tuxedo Black	Beige/Black/White

Interior colours available: Black, Blue, Green, Maroon, Red, Saddle, Silver, White

1966

EXTERIOR BODY color	SOFT-TOP
Ermine White	Beige/Black/White
Laguna Blue	Beige/Black/White
Milano Maroon	Beige/Black/White
Mosport Green	Beige/Black/White
Nassau Blue	Beige/Black/White
Rally Red	Beige/Black/White
Silver Pearl	Beige/Black/White
Sunfire Yellow	Beige/Black/White
Trophy Blue	Beige/Black/White
Tuxedo Black	Beige/Black/White

Interior colours available: Black, Blue, Bright Blue, Green, Red, Saddle, Silver, White and Blue

1967

EXTERIOR BODY color	SOFT-TOP
Elkhart Blue	Black/Teal-Blue/White
Ermine White	Black/Teal-Blue/White
Goodwood Green	Black/Teal-Blue/White
Lynndale Blue	Black/Teal-Blue/White
Marina Blue	Black/Teal-Blue/White
Marlboro Maroon	Black/Teal-Blue/White
Rally Red	Black/Teal-Blue/White
Silver Pearl	Black/Teal-Blue/White
Sunfire Yellow	Black/Teal-Blue/White
Tuxedo Black	Black/Teal-Blue/White

Interior colours available: Black, Bright Blue, Green, Red, Saddle, Teal-Blue, White

1968

EXTERIOR BODY color	SOFT-TOP
British Green	Beige/Black/White
Cordovan Maroon	Beige/Black/White
Corvette Bronze	Beige/Black/White
International Blue	Beige/Black/White
Le Mans Blue	Beige/Black/White
Polar White	Beige/Black/White
Rally Red	Beige/Black/White
Safari Yellow	Beige/Black/White
Silverstone Silver	Beige/Black/White
Tuxedo Black	Beige/Black/White

Interior colours available: Black, Dark Blue, Dark Orange, Gunmetal, Medium Blue, Red, Tobacco

1969

EXTERIOR BODY color	SOFT-TOP
Burgundy	Beige/Black/White
Can-Am White	Beige/Black/White
Cortex Silver	Beige/Black/White
Daytona Yellow	Beige/Black/White
Fathom Green	Beige/Black/White
Le Mans Blue	Beige/Black/White
Monaco Orange	Beige/Black/White
Monza Red	Beige/Black/White
Riverside Gold	Beige/Black/White
Tuxedo Black	Beige/Black/White

Interior colours available: Black, Bright Blue, Green, Gunmetal, Red, Saddle

1970

EXTERIOR BODY color	SOFT-TOP
Bridgehampton Blue	Beige/Black/White
Classic White	Beige/Black/White
Cortez Silver	Beige/Black/White
Corvette Bronze	Beige/Black/White
Daytona Yellow	Beige/Black/White
Donnybrooke Green	Beige/Black/White
Laguna Grey	Beige/Black/White
Marlboro Maroon	Beige/Black/White
Monza Red	Beige/Black/White
Mulsanne Blue	Beige/Black/White
Ontario Orange	Beige/Black/White

Interior colours available: Black, Blue, Brown, Green, Red, Saddle

1971

EXTERIOR BODY color	SOFT-TOP
Brands Hatch Green	Black/White
Bridgehampton Blue	Black/White
Classic White	Black/White
Mille Miglia Red	Black/White
Mulsanne Blue	Black/White
Nevada Silver	Black/White
Ontario Orange	Black/White
Steel Cities Grey	Black/White
Sunflower Yellow	Black/White
War Bonnet Yellow	Black/White

Interior colours available: Black, Dark Blue, Dark Green, Red, Saddle

1972

EXTERIOR BODY color	SOFT-TOP
Bryar Blue	Black/White
Classic White	Black/White
Elkhart Green	Black/White
Mille Miglia Red	Black/White
Ontario Orange	Black/White
Pewter Silver	Black/White
Steel Cities Grey	Black/White

Sunflower Yellow	Black/White
Targa Blue	Black/White
War Bonnet Yellow	Black/White

Interior colours available: Black, Blue, Red, Saddle

1973

EXTERIOR BODY color	SOFT-TOP
Blue-Green	Black/White
Classic White	Black/White
Dark Blue	Black/White
Elkhart Green	Black/White
Medium Blue	Black/White
Metallic Yellow	Black/White
Mille Miglia Red	Black/White
Orange	Black/White
Silver	Black/White
Yellow	Black/White

Interior colours available: Black, Dark Red, Dark Saddle, Midnight Blue, Medium Saddle

1974

EXTERIOR BODY color	SOFT-TOP
Bright Yellow	Black/White
Classic White	Black/White
Corvette Grey	Black/White
Corvette Medium Blue	Black/White
Corvette Orange	Black/White
Dark Brown	Black/White
Dark Green	Black/White
Medium Red	Black/White
Mille Miglia Red	Black/White
Silver Mist	Black/White

Interior colours available: Black, Dark Blue, Dark Red, Neutral, Saddle, Silver

1975

EXTERIOR BODY color	SOFT-TOP
Bright Blue	Black/White
Bright Green	Black/White
Bright Yellow	Black/White
Classic White	Black/White
Dark Red	Black/White
Medium Saddle	Black/White
Mille Miglia Red	Black/White
Orange Flame	Black/White
Silver	Black/White
Steel Blue	Black/White

Interior colours available: Black, Dark Blue, Dark Red, Medium Saddle, Neutral, Silver

1976

EXTERIOR BODY color
Bright Blue
Bright Yellow
Buckskin
Classic White
Dark Brown
Dark Green
Mahogany
Orange Flame
Red
Silver

Interior colours available: Black, Blue Green, Buckskin, Dark Brown, Firethorn, Smoked Grey, White

1977

EXTERIOR BODY color
Black
Classic White
Corvette Bright Yellow
Corvette Chartreuse
Corvette Dark Blue
Corvette Dark Red
Corvette Light Blue
Corvette Orange
Corvette Tan
Corvette Yellow
Medium Red
Silver

Interior colours available: Black, Blue, Brown, Buckskin, Red, Smoked Grey, White

1978

EXTERIOR BODY color
Black
Classic White
Corvette Dark Blue
Corvette Dark Brown
Corvette Light Beige
Corvette Light Blue
Corvette Red
Corvette Yellow
Silver
Silver Anniversary

Interior colours available: Black, Dark Blue, Dark Brown, Light Beige, Mahogany, Oyster, Red

1979

EXTERIOR BODY color
Black
Classic White
Corvette Dark Brown
Corvette Dark Blue
Corvette Light Beige
Corvette Light Blue
Corvette Dark Green
Corvette Red
Corvette Yellow
Silver

Interior colours available: Black, Dark Blue, Dark Green, Light Beige, Oyster, Red

1980

EXTERIOR BODY color
Black
Dark Blue
Dark Brown
Dark Claret
Dark Green
Frost Beige
Red
Silver
White
Yellow

Interior colours available: Black, Claret, Dark Blue, Doeskin, Oyster, Red

1981

EXTERIOR BODY color
Autumn Red/Dark Claret
Beige
Beige/Dark Bronze
Black
Bright Blue Metallic
Charcoal Metallic
Dark Blue Metallic
Mahogany Metallic
Maroon Metallic
Red
Silver/Charcoal
Silver/Dark Blue
Silver Metallic
White
Yellow

Interior colours available: Camel, Charcoal, Dark Blue, Dark Red, Medium Red, Silver Grey

1982

EXTERIOR BODY color
Black
Bright Blue
Charcoal
Dark Blue
Dark Claret
Gold

Red
Silver
Silver Beige
Silver Blue
Silver Blue/Dark Blue
Silver/Charcoal
Silver/Dark Claret
Silver Green
White
White/Silver
Interior colours available: Camel, Charcoal, Dark Blue, Dark Red, Silver Blue, Silver Green, Silver Grey

1983
No official 1983 model year cars were sold to the public. Instead, cars built from March 1983 were all designated '1984 model year' cars.

1984
EXTERIOR BODY color
Black
Bright Red
Bright Silver Metallic
Dark Bronze Metallic
Light Blue/Medium Blue
Light Blue Metallic
Light Bronze/Dark Bronze
Light Bronze Metallic
Medium Blue Metallic
Medium Grey Metallic
Silver/Medium Grey
White
Interior colours available: Bronze, Carmine, Graphite, Medium Blue, Medium Grey, Saddle

Appendix 6
Check Your Numbers

The following information shows the position of the Vehicle Number (also known as Vehicle Identification No. – VIN), Serial Number or Body-Chassis Number. Each number will be unique to your car.

Location of Vehicle Number
1953–early 1960
Driver side door post

Later 1960–1962
Steering column in engine compartment

1963–1967
Instrument panel supporting block which you can see under the glove box.

Engine Number
From 1957 your Corvette's engine number provides the following information:

The first Character indicates build plant – 'F' for Flint up to 1966, 'V' for Flint after 1966, 'T' for Tonawanda.

The next three or four digits indicate build date, e.g. 04-22 shows that the engine was built on April 22nd.

The last two or three characters (i.e. the suffix number) refer to the engine's intended specification. There are hundreds of these so we advise that you refer to Michael Antonick's *Black Book* for details.

Block Casting Date Code
N.B. This is not to be confused with the block casting number which consists of seven or eight digits. The block date code consists of a character followed by two or three numbers, e.g. F249 indicates that the block was cast on June 24th in either 1959, '69, '79 or '89.

The engine date code is a way of checking that the engine is correct for the car and not a forgery. Obviously this has got to pre-date the year in which the engine and car were built.

The block casting date codes were found in the following positions.

Six-cylinder engines. The date codes are found on the passenger side of the engine, near the starter solenoid.

V8 engines. The code on most V8 engines is found at the rear, passenger-side of the block near to the casting number. There are some exceptions to this: 1968 and early 1969 big blocks have their date codes in front of the starter on the passenger side near to the freeze plug, and 1965 small block Nos. 3858180 has its date code adjacent to the casting number (these cars were cast at Tonawanda but assembled at Flint).

If you have any doubts about the provenance of your Corvette, consult your GM dealer. The vehicle number is also stamped into a Corvette's frame in several places to assist in theft recovery and identification. Also, from 1960, part of the Vehicle Number is stamped into the engine block on a pad just in front of the passenger-side cylinder head.

Appendix 7
Specialists, Clubs and Magazines

There are literally hundreds of Corvette specialists – see your Corvette magazine for a full listing.

SPECIALISTS – USA

Auto Accessories of America
Route 322
Box 427
Boalsburg
PA16827
USA
Tel: 001-814-364-2141
Fax: 001-814-364-9615
USA and Canada Toll Free
1-800-458-3475

Corvette Central
5865 Sawyer Road
Sawyer, MI 49125
800-345-4122

Eckler Industries
PO Box 5637
Titusville, FL 32783-5637
800-327-4868

KLJ Auto Specialists
7505 Connelly Drive
Hanover
Maryland 21076
USA
Tel: 001-410-760-4272

Mid America Designs
Box 1368
Effingham, IL 62401
800-500-8388

Pro Motorcar Products Inc.
22025 US Highway 19 North
Clearwater
FL34625
USA
Tel: 001-813-726-9225
Fax: 001-813-726-6587

Vette Brakes & Products Inc.
7490-30th Avenue North
St Petersburg
FL33710
USA
Tel: 001-813-345-5292
Fax: 001-813-347-4818
USA Toll Free 1 (800) 237-9991

Zip Products
8067 Fast Lane
Mechanicsville, VA 23111-1100
800-962-9632

SPECIALISTS – UK

Avonvale Autobodies
Unit 19, Briar Close Business Park
Briar Close
Evesham
Worcs WR11 4JT
England
Tel: 01386-443939
Fax: 01386-49922

Claremont Corvette
Malling Road
Snodland
Kent ME6 5NA
Tel: 01634-244444
Fax: 01634-244534

CLUBS AND MAGAZINES

Classic Corvette Club (UK)
Contact: Andy Greenfield/Sue Curtis – Membership Secretaries
7 Sudely Gardens
Hockley
Essex SS5 4XQ
England
Tel: 01702-200881

Corvette Fever Magazine
3816 Industry Blvd
Lakeland
FL33811
USA
Tel: 001-813-644-0449

Index

Made in United States
North Haven, CT
04 November 2026

21728197R00124